ISBN 978-0-282-74871-5
PIBN 10862866

This book is a reproduction of an important historical work. Forgotten Books uses
state-of-the-art technology to digitally reconstruct the work, preserving the original format
whilst repairing imperfections present in the aged copy. In rare cases, an imperfection in
the original, such as a blemish or missing page, may be replicated in our edition. We do,
however, repair the vast majority of imperfections successfully; any imperfections that
remain are intentionally left to preserve the state of such historical works.

1 MONTH OF
FREE
READING

at
www.ForgottenBooks.com

By purchasing this book you are eligible for one month membership to ForgottenBooks.com, giving you unlimited access to our entire collection of over 1,000,000 titles via our web site and mobile apps.

To claim your free month visit:

www.forgottenbooks.com/free862866

English
Français
Deutsche
Italiano
Español
Português

www.forgottenbooks.com

Mythology Photography **Fiction**
Fishing Christianity **Art** Cooking
Essays Buddhism Freemasonry
Medicine **Biology** Music **Ancient
Egypt** Evolution Carpentry Physics
Dance Geology **Mathematics** Fitness
Shakespeare **Folklore** Yoga Marketing
Confidence Immortality Biographies
Poetry **Psychology** Witchcraft
Electronics Chemistry History **Law**
Accounting **Philosophy** Anthropology
Alchemy Drama Quantum Mechanics
Atheism Sexual Health **Ancient History**
Entrepreneurship Languages Sport
Paleontology Needlework Islam
Metaphysics Investment Archaeology
Parenting Statistics Criminology
Motivational

ANNUAL REPORT

UPON THE

GEOGRAPHICAL SURVEYS WEST OF THE ONE HUNDREDTH MERIDIAN, IN CALIFORNIA, NEVADA, UTAH, COLORADO, WYOMING, NEW MEXICO, ARIZONA, AND MONTANA,

BY

GEORGE M. WHEELER,

FIRST LIEUTENANT OF ENGINEERS, U. S. A.;

BEING

APPENDIX JJ

OF THE

ANNUAL REPORT OF THE CHIEF OF ENGINEERS FOR 1876.

WASHINGTON:
GOVERNMENT PRINTING OFFICE.
1876.

210625

ERRATA.

Page 2, fifth line from bottom, for "835" read "8035."
Page 6, twenty-third line from top, omit "Aquilae."
Page 10, sixteenth line from top, for "Areitis" read "Arietis."
Page 55, twenty-sixth line from top, for "Cañon City" read "Carson City."
Page 82, first line from bottom, for "10 inch" read "10 second."
Page 83, twenty-second line from top, for "10 inches" read "10 seconds."
Page 116, second line from top, for "10" read "5."
Page 116, thirty-ninth line from top, for "these flows were" read "the ground was."
Page 118, fortieth line from top, omit quotation marks after "1867."
Page 118, forty-fifth line from top, insert quotation marks after "1862."

Page 119, twenty-sixth line from bottom, for "$\frac{a}{P}$" read "$\frac{a}{p}$" for "B" read "b," and for "169" read "1.69."

Page 120, third line from top, omit "foot."
Page 120, eighth line from top, for "V" read "v," for "2508.64," read "2.50864," and for "Q va" read "Q $= va$."
Page 120, thirty-first line from top, for "A" read "a."
Page 120, thirty-fifth line from top, omit "foot."
Page 120, thirty-eighth line from top, for "V" read "v."
Page 120, thirty-ninth line from top, for "$2 = Va$" read "Q $= va$."
Page 135, fifth line from bottom, for "Maclean" read "Nadeau."
Page 137, thirty-eighth line from top, insert "ta" after "come down."
Page 139, twenty-third line from top, for "volcanic" read "isolated."
Page 141, sixth line from bottom, for "Ovagones" read "Aragones."
Page 146, twenty-second line from top, for "Mesa Ricco" read "Mesa Rica."
Page 296, section 3, for "A. cribatus" read "A. cribratus."
Page 299, twenty-third line from bottom, for "Athodius occidentalis" read "Apodius occidentalis."
Page 300, first line from top, for "Dorcus mazana" read "Dorcus mazama."
Page 300, seventh line from top, for "Dichelongcha Backii" read "Dichelonycha Backii."
Page 300, twenty-fourth line from bottom, (right-hand column,) for "Cantharis vividana" read "Cantharis viridana."
Page 300, thirteenth line from bottom, (right-hand column,) for "Pontaria rugicollis," read "Pentaria rugicollis."
Page 300, tenth line from bottom, (right-hand column,) for "Thacolepis inornata" read "Pacolepis inornata."
Page 331, first line from top, for "Khone" read "Kho'ne."
Page 331, eighth line from bottom, for "pekhpetch" read "pekhpétch."
Page 331, fifteenth line from bottom, for "two-two-pau" read "two two pau."
Page 332, fifteenth line from bottom, for "ishg'omo" read "ishgomo."
Page 333, twenty-second line from top, for "tepe" read "tepu."
Page 333, twenty-fourth line from top, for "no'; 'o," read "no'-o."
Page 336, ninth line from top, for "tele'vtchok" read "te'levtchok."
Page 338, ninth line from top, for "tu'be"e" read "tu'b'e."
Page 340, thirty-fifth line from top, for "akhathim" read "akha-thim."
Page 340, thirty-fifth line from top, for "akhathiga" read "akha-thiga."

ERRATA.

[Appendix JJ, Annual Report Chief of Engineers, 1876.]

Page 8, 1st line of table—
For "Fort Cameron" read "New Site Fort Cameron."
For "111° 44' 00".31" read "112° 36' 20".52."
For "38° 16' 53".34" read "38° 16' 40".59."

OFFICE OF THE CHIEF OF ENGINEERS,
Washington, D. C., October 21, 1876.

GEOGRAPHICAL EXPLORATIONS AND SURVEYS OF THE ONE HUNDREDTH MERIDIAN.

Officer in charge, First Lieut. George M. Wheeler, Corps of Engineers, having under his orders First Lieuts. William L. Marshall and Eric Bergland, Corps of Engineers; First Lieuts. William L. Carpenter, Ninth Infantry; Rogers Birnie, jr., Thirteenth Infantry; O. C. Morrison, Sixth Cavalry; C. W. Whipple, Ordnance Corps; and Second Lieut. M. M. Macomb, Fourth Artillery; Acting Assistant Surgeons H. C. Yarrow and J. T. Rothrock, United States Army, who, in addition to their professional duties, were engaged in zoological and botanical labors.

The following scientists have also been attached to the expedition of the past year: Dr. F. Kampf, astronomical observer; Messrs. Jules Marcou, A. R. Conkling, and D. A. Joy, geologists; Dr. O. Loew, mineralogist, and Prof. F. W. Putnam, ethnologist.

At the commencement of the fiscal year the main sections of the survey were engaged in their field of operations, having left the rendezvous at Pueblo, Colo., and Los Angeles, Cal., in the prosecution of their labors of former years in areas as shown by portions of Atlas sheets Nos. 61c, 61 d, 65, 72, 73, 74, 77a, 77 B, 78a and 80a. (See progress map in Appendix JJ.)

The astronomical determinations of the year were confined to those needed to check the several lines of meander within the immediate field of survey.

The several parties were disbanded at Caliente, Cal., the then terminus of the Southern Pacific Railroad, and at West Las Animas, Colo., at the close of the field season in November and December, a number of the assistants repairing to Washington to prepare in the ensuing months the practical results, and at which point a small force of draughtsmen and computers was kept constantly employed in the reduction of field-notes and the production of finished maps.

A report of the special party intrusted to Lieutenant Bergland, Corps of Engineers, to determine as to the feasibility of the diversion of the Colorado River of the West, for purposes of irrigation, at or below the mouth of the Lower Grand Cañon, has been submitted. Examinations were made at such points as it was practicable for the party to reach in the seasons of labor available, and data bearing upon the physical features of the outlying mountainous sections, with their several passes, are contained in the report, which also includes a discussion of practicability of the diversion of this stream.

Of the six quarto volumes authorized to be published by the act of June 3, 1874, as amended by the act approved February 15, 1875, two volumes, namely, III (Geology) and V (Zoology) have been printed, as well as part 1 of volume IV, (Invertebrate Fossils.) ·The manuscript of volume II (Astronomy and Meteorology, including Barometric Altitudes) and the second part of volume IV is ready, while that for volumes VI (Botany) and VII (Ethnology, Philology, and Ruins) is in an advanced stage of preparation, as well as volume I, (Geographical Report.) The manuscript of The Declination of 2,018 Stars, by Prof. T. H. Safford, has also been completed. Seven published sheets have been added to the topographical atlas, and others are completed and in various stages of progress. Six geological-atlas sheets have been published, and three are now ready for, or in the hands of, the engraver.

The results determined by the several scientists connected with the investigation of the subjects of natural history are, in addition to the results primarily intended to exhibit the operations of the survey, namely, the preparation of a detailed map of the sections entered, with a description and delineation, so far as practicable, of their natural resources, and showing the distribution of the arable and arid portions, the former divisible into those sections which are susceptible of cultivation, those in which irrigation can be had, and into mining, timber, and grazing sections, the latter entirely valueless at the present or any near prospective date, for occupation or use in any remunerative capacity from such industry that is likely to venture within its borders.

The topographical maps, that have been prepared upon scales as follows: 1 inch to 8 miles, 1 inch to 4 miles, and 1 inch to 2 miles—forming the principal results of the survey, furnish information of immediate value to the different branches of the military service, and incidentally to the other departments of the Government and to the public.

The regular progress of this important work without interruption, it is believed, will commend itself to the attention of Congress, and it is earnestly recommended.

The amounts required to continue the survey, as estimated by Lieutenant Wheeler, are recommended, viz:

For continuing geographical surveys of the territory of the United States west of the 100th meridian... $95,000 00

For preparing and engraving plates and atlas sheets accompanying reports upon geographical surveys west of the 100th meridian 25,000 00

His annual report, with appendixes and estimates, is appended.

(See Appendix JJ.)

APPENDIX J J.

ANNUAL REPORT OF LIEUTENANT GEORGE M. WHEELER, CORPS OF ENGINEERS, FOR THE FISCAL YEAR ENDING JUNE 30, 1876.

GEOGRAPHICAL SURVEYS WEST OF THE ONE HUNDREDTH MERIDIAN, IN CALIFORNIA, NEVADA, UTAH, COLORADO, WYOMING, NEW MEXICO, ARIZONA, AND MONTANA.

CONTENTS.

REPORT.

APPENDIXES.

ILLUSTRATIONS.

REPORT.

UNITED STATES ENGINEER OFFICE,
GEOGRAPHICAL SURVEYS WEST OF THE 100TH MERIDIAN,
Washington, D. O., June 30, 1876.

SIR: I have the honor to submit the following report upon geographical surveys west of the one hundredth meridian for the fiscal year ending June 30, 1876:

The States and Territories of California, Nevada, Utah, Colorado, Wyoming, New Mexico, Arizona, and Montana, have each been chosen as fields of operation for the different expeditions engaged during the years 1869, 1871, 1872, 1873, 1874, and 1875.

SUMMARY OF FIELD AND OFFICE WORK.

At the close of the last fiscal year, the parties for field-work were actively employed and were commanded as stated in my annual report for that year.

The parties of California section disbanded at Caliente, the present terminus of the Southern Pacific Railroad, in November, 1875, a number of the officers and assistants repairing to Washington, while Lieutenant Whipple traveled southward to Los Angeles in charge of the transportation and supplies. Lieutenant Bergland's party reached Los Angeles on return from the first trip October 4, 1875, and departed again for the field in the southern portion of the valley of the Colorado River on February 13, 1876, and continued the examination of the river southward as far as Pilot Knob. Returning, he crossed the Desert and Coast ranges to Los Angeles, reaching that point May 7, 1876, where, his party having disbanded, the operations of this section of the survey were concluded for the year.

The parties of the Colorado section reached West Las Animas November 25, and proceeded immediately to conclude the operations of the field-season.

The officers on duty with the survey were employed during office-season as follows: First Lieut. W. L. Marshall, Corps of Engineers, in charge of field astronomical and meteorological computations, preparation of meteorological portion of Volume II, survey reports, executive report of field operations for past season, and detailed supervision of a portion of topographical plottings and reductions.

First Lieut. Eric Bergland, Corps of Engineers, in charge of temporary office at Los Angeles, Cal., for plotting and reduction of the topographical and meteorological data derived from the operations of special party for determining feasibility of diverting the Colorado River for purposes of irrigation.

First Lieut. W. L. Carpenter, Ninth Infantry, preparation of reports upon natural-history subjects; examination and disposition of specimens; executive report.

AP. JJ—A

First Lieut. Rogers Birnie, jr., Thirteenth Infantry, assisting Lieutenant Marshall upon field astronomical and geodetic computations, and in immediate charge of topographical room.

First Lieut. C. C. Morrison, Sixth Cavalry, in charge of topographical work, (until relieved by Lieutenant Macomb, April 24, 1876,) special charge of instruments and instrument record, preparing tables of distances, executive report.

First Lieut. C. W. Whipple, Ordnance Corps, in charge of draughting work until relieved.

Second Lieut. M. M. Macomb, Fourth Artillery, in charge of draughting work since April 24, 1876, relieving Lieutenant Morrison.

Acting Assistant Surgeon H. C. Yarrow, United States Army, superintending publication of Volume V, (zoölogy,) examination and classification of collections in natural history, report of operations special field party.

Acting Assistant Surgeon J. T. Rothrock, United States Army, examination and classification of botanical collections, preparation of report special field party, preparation of MSS. for Volume VI, (botany.)

Hospital Steward T. V. Brown, United States Army, meteorological computations.

Dr. F. Kampf, astronomical observer, in computations for latitudes and longitudes. Messrs. George H. Birnie, F. Carpenter, F. A. Clark, W. A. Cowles, Anton Karl, F. O. Maxson, L. Nell, J. C. Spiller, and G. Thompson, (topographers,) reduction and plotting of field-notes. George M. Dunn, F. M. Lee, and William C. Niblack, and Privates Looram and Kirkpatrick, Battalion of Engineers, (barometric recorders,) in computation of field observations. Jules Marcou, A. R. Conkling, and D. A. Joy, (geologists,) preparation of reports. Dr. O. Loew (mineralogist and chemist,) analysis of collections and preparation of reports. H. W. Henshaw, (ornithologist,) identification and classification of collections and reports thereon, and assisting in publication of Volume V. Sereno Watson, (botanist,) and F. W. Putnam, (ethnologist,) examination of collections, and reports thereon. Charles Herman, J. C. Lang, E. Mahlo, and J. E. Weyss, (draughtsmen,) preparation of maps for publication. F. Klett, George M. Lockwood, and J. D. McChesney, correspondence and records, money and property accounts, reports, returns, distribution of reports and maps, &c., &c.

A general summary of the more prominent features of field and office operations is given below.

FIELD.

Sextant latitude-stations 102
Bases measured 6
Triangles about bases measured 50
Main triangulation-stations occupied 111
Secondary triangulation-stations occupied 273
Three-point stations occupied 436
Camps made 825
Miles meandered 9, 463. 3
Miles traversed, not meandered 4, 799. 9
Stations on meanders 835
Magnetic variations observed 222
Monuments built 237
Cistern-barometer stations occupied 707
Aneroid stations occupied 5,553

Mining camps visited 22
Mineral and thermal springs noted 21
Geological and mineralogical specimens collected 380
Paleoutological specimens collected............... 107
Botanical specimens (species) collected 350
Mammals, specimens collected 90
Birds, specimens collected 710
Other ornithological specimens collected 57
Reptiles, lots collected 67
Fishes, lots collected........... 29
Insects, lots collected......................... 325
Shells, lots collected.. 12
Crustacea, lots collected................................. 11
Radiates, lots collected.......... 5
Ethnological specimens collected........................... 363

OFFICE.

Astronomical positions computed. 102
Stations adjusted by method of least squares 158
Triangles computed.... 644
Distances computed.......... .. .. 1, 288
Latitudes and longitudes computed............................ 293
Azimuths computed... 586
Sheets plotted 1 inch to 2 miles 9
Special sheets plotted, (various scales) 13
Cistern-barometer altitudes computed........................... 930
Aneroid-barometer altitudes computed......................... 5, 013
Atlas maps, (1 inch to 2 miles,) published 1
Atlas maps, (1 inch to 4 miles,) published................... 3
Atlas maps, (1 inch to 8 miles) 3
Alas maps, (1 inch to 4 miles,) nearly ready for publication..... 5
Atlas maps, (1 inch to 4 miles,) partially completed 4
Reports published...... 3
Reports distributed 1, 951
Reports in course of publication 1
Maps distributed... .... 3, 240

Personnel 1875–'76.

	Time employed.
OFFICERS, ACTING ASSISTANT SURGEONS, ETC.	
First Lieut. George M. Wheeler, Corps of Engineers, in charge.	The whole year.
First Lieut. William L. Marshall, Corps of Engineers..........	Do.
First Lieut. Eric Bergland, Corps of Engineers.................	May 21, 1875, to end of year.
First Lieut. W. L. Carpenter, Ninth Infantry.................	May 22, 1875, to May 1, 1876.
First Lieut. Rogers Birnie, jr., Thirteenth Infantry............	The whole year.
First Lieut. Charles C. Morrison, Sixth Cavalry................	Do.
First Lieut. C. W. Whipple, Ordnance Corps....	July 1, 1875, to January 15, 1876.
Second Lieut. M. M. Macomb, Fourth Artillery................	April 24, 1876, to end of year.
Acting Assistant Surgeon H. C. Yarrow, (zoölogist)	July 1, 1875, to January 1, 1876.
Acting Assistant Surgeon J. T. Rothrock, (botanist)	July 1, 1875, to June 1, 1876.
Hospital Steward T. V. Brown, (recorder and computer).......	July 1, 1875, to November 22, 1875.
ENLISTED MEN.	
First-class privates, John F. Kirkpatrick and William Looram, Company C, Battalion of Engineers, (barometric recorders.) Sergeant Eugene Farnum, Company G, Twelfth Infantry, and seven enlisted men, Company G, Twelfth Infantry.	
ASSISTANT ENGINEERS.	
Dr. F. Kampf, astronomer..................................	The whole year.
George H. Birnie, topographer	Do.
Frank Carpenter, topographer..............................	Do.
F. A. Clark, topographer...................................	Do.
W. A Cowles, topographer..................................	Do.
F. O. Maxson, topographer.................................	Do.
Louis Nell, topographer....................................	Do.
J. C. Spiller, topographer..................................	Do.
Gilbert Thompson, topographer	Do.
HYPSOMETRIC RECORDERS.	
George M. Dunn	The whole year.
F. M. Lee	Do.
William C. Niblack	July 1, 1875, to November 30, 1875.
GEOLOGISTS.	
A. R. Conkling..	The whole year.
Jules Marcou	July 1 to 31, October 1 to Dec. 31, 1875
Douglas A. Joy...	July 1, 1875, to November 30, 1875.
MINERALOGIST AND CHEMIST.	
Dr. Oscar Loew	July 1, 1875, to June 1, 1876.
COLLECTOR IN ZOÖLOGY.	
H. W. Henshaw, (ornithologist)...........……...........	The whole year.
ETHNOLOGIST.	
F. W. Putnam...	December 20, 1875, to close of year.
DRAUGHTSMEN.	
Charles Herman...........................…	year.
J. C. Lang	
Emil Mahlo	to close of year.
John E. Weyss..	he year.
CLERICAL.	
Francis Klett	whole year.
George M. Lockwood......................................	Do.
J. D. McChesney	Do.

The gentlemen named in my last annual report have continued to extend cheerful assistance in the completion of results to the point of final publication in the natural-history branch, and I take this occasion to tender to them a merited recognition of their kindness.

The officers of the supply departments of the Army and of the military departments, districts, and posts touched by the different parties of the survey, have extended valuable assistance.

ASTRONOMICAL.

In this branch of the survey no work was accomplished during the year at main stations, since a large number of these had been occupied in former years, and measured and developed bases in the immediate vicinity of those bordering on the fields of survey had been completed. The usual observations for latitude at selected points along the routes of travel were made by the officers in charge of parties, and the results therefrom are given in a tabulated statement, as well as those of prior years. It was found impracticable to complete the dome and middle room of the observatory at Ogden for want of time and means. The telegraph lines southward from Santa Fé having reached Mesilla, New Mexico, and that west from San Antonio, Texas, Fort Stockton, it is improbable that during the fiscal year El Paso will have been connected both north and east, and further work accomplished in the direction of making a circuit from that point in the vicinity of the thirty-second parallel to San Diego. No further steps have been taken in the direction of securing sites for more or less permanent field observatories, but my recommendations of the past year are renewed.

Herewith is given the latitude of points determined by parties of the expeditions during several years and not elsewhere published.

AP. J J—1

GEOGRAPHICAL POSITIONS.

Sextant and transit for time; sextant for latitude; telegraphic time-signals for difference of longitude.

Year.	Station.	Atlas sheet No.	Object observed.	Longitude west from Greenwich.	Latitude.	Altitude.	Variation of the needle, E.	Observer.	Computer.	Remarks.
				° ′ ″	° ′ ″	*Feet.*	° ′ ″			
1869	Camp Hallock, Nev.	40	Sun / Polaris / β Scorpii / α Scorpii	115 19 34.05	40 48 34.05	5789.7	16 21 24	Lieut. Wheeler / Maj. Roberts	Lieut. Wheeler / Maj. Roberts	
1869	Camp Ruby, Nev.	49	Sun / Aquilæ / Polaris / Ophiuchi	115 31 06.75	40 03 38.63	6152.6	17 09 04	Lieut. Wheeler / Lieut. Lockwood / Maj. Roberts	Lieut. Wheeler / Maj. Roberts / Dr. F. Kampf	
1872	Deep Creek, Utah	49	Sun / Polaris / α Aquilæ / α Boötis	113 57 16.05	40 06 01.71	5336.6		Lieut. Hoxie / E. P. Austin	Lieut. Marshall	
1869	Elko, Nev.	40	Sun / Polaris / β Scorpii / α Aquilæ	115 45 37.20	40 49 38.44	5148.4	17 35 03	Lieut. Wheeler / Maj. Roberts	Lieut. Wheeler / Maj. Roberts / Dr. F. Kampf	
1872	Fillmore, Utah	59	Sun / Polaris / ε Pegasi / α Boötis	112 16 54.93	38 57 14.94		16 15 00	Lieut. Hoxie / Lieut. Wheeler / E. P. Austin	Lieut. Marshall / E. P. Austin	Camp three miles east of, on Chalk Creek.
1869	Hamilton, Nev.	49	Sun / Polaris / α Aquilæ	115 25 06.24	39 15 48.87	7601.3	16 43 29	Lieut. Wheeler / Maj. Roberts	Lieut. Wheeler / Maj. Roberts	
1872	Kanab, Utah	67	Sun / Polaris / α Cygni / α Tauri	112 31 39.00	37 09 25.43	4909.0	14 23 00	Lieut. Marshall / E. P. Austin	Lieut. Marshall	Monument in square; front of bishop's residence.
1869	Monte Christo Mill, Nev.	49	Sun	115 34 49.20	39 13 16.83	7596.0	17 05 06	Lieut. Wheeler / Maj. Roberts	Lieut. Wheeler / Maj. Roberts / Dr. F. Kampf	
1869	Peko, Nev.	40	Sun / Polaris	115 30 14.60	40 55 46.35	5180.0		Lieut. Wheeler / Maj. Roberts	Lieut. Wheeler / Maj. Roberts	
1872	Pipe Springs, Ariz.	67	Sun / Polaris / α Lyræ / α Cygni	112 49 57.00	36 51 36.34	5397.2		Lieut. Marshall / E. P. Austin	Dr. F. Kampf / Lieut. Marshall	Camp near.

Year	Station		Objects observed	Longitude	Latitude	Elevation	Observers		Remarks
1872-73	Provo, Utah	50	Sun, Polaris, α Cygni, α Lyræ, α Bootis, α Virginis, α Jupiter	112 40 27.00	40 13 47.84	4567.3	Lieut. Hoxie...... J. H. Clark......	Lieut. Hoxie...... J. H. Clark......	Old Camp Rawlins; longitude determined by Lieut. Hoxie, 1873; latitude by Lieut. Marshall, 1872.
1873	Richfield, Utah	50	α Herculis, Sun, Polaris, α Serpentis, α Lyræ, α Bootis, α Aquilæ, α Ophiuchi	112 08 27.00	38 46 11.40	5982.5	Lieut. Hoxie...... J. H. Clark......	Lieut. Hoxie...... J. H. Clark......	
1872	Toquerville, Utah	67	Sun, Polaris	113 16 20.00	37 15 19.88		Lieut. Wheeler...... Lieut. Marshall...... E. P. Austin......	Lieut. Marshall......	Barometric record lost.

Geographical positions by measurement from or by trigonometrical connection with main astronomical points.

Year.	Station.	Atlas-sheet No.	Connected with astronomical station at—	Longitude.	Latitude.	Altitude above sea-level.	Variation of needle, E.	Connection made by—	Remarks.
				° ′ ″	° ′ ″	Feet.			
1879	Fort Cameron, Utah	59	Beaver, Utah	111 44 00.31	38 16 53.34	6057.7	......	Louis Nell	Difference in longitude and latitude determined by triangulation in 1872. At this time no flag-staff had been erected, and the object fixed was a building, the only one then in course of construction.
1873	Camp Douglas, Utah, (sun-dial)	41	Salt Lake City, Utah	111 50 13.92	40 45 47.47	5094.3		G. Thompson	By chaining from monument in Mormon Observatory, Temple square, Salt Lake City.
1873	Camp Douglas, (old flag-staff)	41	......do......	111 50 14.07	40 45 47.58			G. Thompson	By chaining; old flag-staff since removed, but stump left standing as bench-mark.
1873	Fort Ellis, Montana	94	Bozeman, Montana	110 59 49.97	45 40 08.00		-	......	Approximate difference between astronomical observing pier and fort, taken from Land-Office plate. The position of the monument as located by J. H. Clark and the center of fort as given by Land-Office assumed as position of flag-staff.
1873	Fort Fred Steele, Wyo., (flag-staff)	43	Fort Fred Steele, Wyo	106 56 54.27	41 46 50.63			J. E. Weyss	Difference between monument and flag-staff determined from plats and measurements by J. E. Weyss, and are given as approximate.
1873	Fort Sanders, Wyo	43	Laramie, Wyo	105 36 07.66	41 17 96.89			J. E. Weyss	

9

GEOGRAPHICAL POSITIONS.—*From sextant-observations in the field, for latitude and time.*

Year.	Station.	Atlas-sheet No.	Objects observed.	Latitude.	Altitude above sea-level.	Variation of needle.	Observer.	Computer.	Remarks.
				° ′ ″	Feet.	° ′ ″			
1873	La Veta Creek, Colorado	63	Polaris, α Lyræ, α Arietis	37 31 42.40		14 07 17	Lieut. Marshall	Lieut. Marshall	Two miles above junction with Cucharas River.
1874	Abiquiu, N. Mex.	69 D	Sun, Polaris, ε Pegasi, α Andromedæ	36 12 29.40	5930.1	13 54 01	Lieut. Birnie	Lieut. Birnie	River bank below town.
1872	Adamsville, Utah	59	α Aquilæ, Polaris, α Coronæ, ε Pegasi	38 15 00.27	5600.0		Lieut. Hoxie	Lieut. Hoxie	
1874	Alamosa Creek, Colorado	61 D	α Aquilæ, ε Pegasi	37 20 51.70	11156.4		Lieut. Wheeler	Lieut. Whipple	Near head South Fork.
1873	Alma City, Colorado	59 D	Polaris, α Lyræ, α Boötis	39 18 23.76	10254.0	14 35 00	Lieut. Marshall	Lieut. Marshall	Camp southeast of town one-quarter mile.
1869	Antelope Springs, Nev.	49	Sun, α Aquilæ	39 25 42.19	7201.0	17 00 27	Lieut. Wheeler, Lieut. Lockwood, Lieut. Marshall	Lieut. Wheeler, Lieut. Lockwood, Lieut. Marshall	
1872	Antelope Springs, Utah	59	Ophiuchi, α Andromedæ, Polaris	37 46 26.50	5850.0	16 20 00			On road from Cedar City, Utah, to Pioche, Nev.
1873	Animas City, Colorado	61 C	Polaris, α Cygni, α Lyræ	37 24 22.70	6662.3		Lieut. Marshall	Lieut. Marshall	
1873	Arkansas River, Colorado	61 B	Polaris, α Cygni, α Boötis	38 28 30.00	7006.5	14 41 00	Lieut. Marshall	Lieut. Marshall	Near Mouth of Badger Creek, Colorado.
1871	Camp Apache, Ariz.	83	Sun, α Coronæ, α Cygni, β Persei, ε Pegasi, α Aquilæ, α Andromedæ	33 47 18.51	5000.9	14 10 42	Lie t. Wheeler, Lieut. Lockwood	Lieut. Lockwood	
1873	Camp Apache, Ariz	83	Polaris	33 47 18.70			Lt. Tillman	Lieut. Tillman	Camp separated by a short distance and near crossing of White Mountain Creek.

Geographical positions from sextant-observations, &c.—Continued.

Year.	Atlas-sheet No.	Station.	Objects observed.	Latitude.	Altitude above sea-level.	Variation of needle. E.	Observer.	Computer.	Remarks.
				° ′ ″	Feet.	° ′ ″			
1872	59	Asay's Ranch, Utah	Polaris, α Aquilæ	37 33 55.98		16 51 00	Lieut. Hoxie	Lieut. Hoxie	
1871	66	Ash Meadows, Nev	Sun, Polaris	36 31 06.90		14 06 00	Lieut. Wheeler	Lieut. Lockwood	
1871	66	At-too-bah, or Cañon Spring, Ariz.	Corona, Pegasi	35 14 43.98			Lieut. Lockwood	Lieut. Lockwood	
1873	59 D	Argentine Pass, Colorado	Polaris	39 38 40.90	11018.4	16 09 15	Lieut. Marshall	Lieut. Marshall	Camp near.
1872	59	Basin rim, Utah	Lyræ, Polaris, α Aquilæ, Persei	37 41 05.60	7467.0		Lieut. Marshall	Lieut. Marshall	West of Last Bluff, at spring in break of Wahsatch plateau.
1871	75	Beaver Creek, Ariz.	Polaris, β Andromedæ	34 44 02.59	3871.4		Lieut. Lockwood	Lieut. Lockwood	Camp on.
1874	61 D	Beaver Creek, Colorado	Lyræ, Polaris, Arcturus, Pegasi	37 36 09.60	8415.5		Lieut. Marshall	Lieut. Marshall	Mouth of tributary of South Fork Rio Grande.
1871	66	Beaver Dam, Nev	Sun, Polaris	36 53 57.61			Lieut. Lockwood	Lieut. Lockwood	
1872	59	Beaver Lake district, Utah	Polaris, α Coronæ Bor., Pegasi	38 31 25.93			Lieut. Hoxie	Lieut. Marshall	
1869	58	Benson's Creek, Nev	Polaris	38 40 41.33	6064.6	16 24 00	Lieut. Wheeler	Lieut. Wheeler	
1873	63	Big Hills, Ariz.	Sun, β Andromedæ, α Andromedæ, α Coronæ Bor	33 23 07.70	3702.5	13 06 08	Lieut. Lockwood, Lieut. Tillman	Lieut. Lockwood, Lieut. Tillman	
1872	59	Black Rock Spring, Utah	Polaris, α Boötis, Pegasi	38 49 33.59		16 09 00	Lieut. Hoxie	Lieut. Marshall	
1873	52	Blue River, Colorado	Polaris, α Lyræ, α Boötis	39 49 54.97	8925.7	14 37 10	Lieut. Marshall	Lieut. Marshall	Eight miles from junction Ten-Mile Creek.

				34 33 08.54	5400.1	14 51 93	Lieut. Lockwood	Lieut. Lockwood	Lieut. Lockwood	
1871	Honohta Fork, Aria	76	γ Andromeda α Cygni Polaris α Persei							
1872	Box Cañon Spring, Utah	59	α Lyrae α Polaris	37 30 39.39	6509.0		Lieut. Marshall	Lieut. Marshall	Lieut. Marshall	On trip from Potato Valley to Paris settlement.
1873	Buffalo Slough, (divide at head of.)	61 B	α Polaris α Cygni	38 47 28.80	9783.9	14 24 49	Lieut. Marshall	Lieut. Marshall	Lieut. Marshall	
1873	Camp Bowie, Aria	89	Sun Maria γ Pegasi α Corona Bor α Andromedæ	39 10 16.2	4871.6		Lieut. Tillman	Lieut. Tillman	Lieut. Tillman	
1872	Olá Camp Floyd, Utah	50	β Andromedæ Sun Polaris Boötis α Aquilæ	40 15 54.63	4866.5	16 59 30	Lieut. Hoxie	Lieut. Hoxie	Lieut. Marshall	Near Fairfield in Cedar Valley.
1871 1871	Camp Pinal, Aria Camp Verde, Ariz	83 75	β Polaris Maria	33 21 01.45 34 34 20.19	3150.7		Lieut. Lockwood Lieut. Wheeler Lieut. Lockwood	Lieut. Lockwood Lieut. Lockwood		
1872	Camp L, Utah	59	β Lyrae α Polaris	37 30 03.70			Lieut. Hoxie	Lieut. Hoxie		
1872	Camp S, on Virgin River, Utah,	67	α Lyrae α Tauri Earls α Herculis	37 08 06.60		15 29 00	Lieut. Hoxie	Lieut. Hoxie		
1872	Camp 17, Utah	50	α Aquilæ α Polaris	39 17 59.50			List. Hoxie	List. Hoxie		
1873	Cañada Alamosa, N. Mex	84	α Lyrae α Polaris	33 39 15.00	6540.0		Lieut. Tillman	Lieut. Tillman		
1869	Cave Valley, Nev	58	α Lyrae Sun α Aquilæ	38 39 00.69	6463.8	16 16 13	Lieut. Wheeler Lieut. Lockwood	Lieut. Wheeler Lieut. Lockwood		Camp near entrance to cave.
1871	Cedar Creek, Ariz	76	Orl ole α Pegasi	34 04 03.44			Lieut. Lockwood	Lieut. Lockwood		Altitude at mouth, 4037.2
1872	Cedar Springs, Utah	50	α Polaris	39 54 70	5100.0	17 09 00	Lieut. Marshall	Lieut. Marshall		Camp 1¼ miles north of east of town.
1874 1874	Cement Creek, Colo Cañon de Chaco, N. Mex	61 C 68 D	Polaris Sun Maria	37 54 04.30 36 06 31.90	11483.6 5838.7	14 58 00 13 29 42	Lieut. Marshall Lieut. Birnie	Lieut. Marshall Lieut. Birnie		Head of creek.
1874	North of Cañon de Chaco, N. Mex.	68 D	Pegasi α Andromedæ α Polaris ε Andromedæ	36 03 18.80	6396.2	13 46 57	Lieut. Birnie	Lieut. Birnie		Mess north of cañon.

Geographical positions, from sextant-observations, &c.—Continued.

Year	Station	Atlas-sheet No.	Objects observed	Latitude	Altitude above sea-level	Variation of needle. E.	Observer	Computer	Remarks
1873	Camp Chase, N. Mex	84	Polaris / α Andromedæ / α Lyræ	33 16 37.70	5374.2		Lieut. Tillman	Lieut. Tillman	Headwaters Cuchilla Negra.
1874	Chico Rito, N. Mex	70 B	Polaris / ε Pegasi / α Lyræ	36 40 17.30			Lieut. Blunt	Lieut. Blunt	East of Clifton.
1873	Cienega San Simeon, N. Mex	89	α Andromedæ / Polaris / α Corone	33 04 08.50	3854.8	19 56 53	Lieut. Tillman	Lieut. Tillman	
1874	Cimarron, N. Mex	70 A	α Andromedæ / β Corone / Polaris / α Pegasi	36 01 40.60	6418.9		Lieut. Birnie	Lieut. Birnie	Camp in cañon near.
1872	Circleville, Utah	59	α Corone Bor / Polaris / α Andromedæ	38 09 51.90	5694.0	21 30 00	Lieut. Marshall	Lieut. Marshall	
1869	Clear Creek, Nev	58	Polaris	38 50 07.73	6092.5	16 26 44	Lieut. Wheeler	Lieut. Wheeler	Spring Valley, eastern side.
1869	Clover Valley, Nev	58	α Pegasi / Polaris	37 30 27.00	4902.0	14 25 19	Lieut. Wheeler / Lieut. Lockwood	Lieut. Wheeler / Lieut. Lockwood	
1869	Cold Spring, Nev	49	α Aquilæ / Polaris	40 04 01.88	6173.1	17 12 27	Lieut. Wheeler	Lieut. Wheeler	
1872	Cottonwood Cañon, Utah, (head of.)	59	α Lyræ / Polaris / α Arietis	37 25 50.80	5973.0		Lieut. Marshall	Lieut. Marshall	
1871	Cottonwood Springs, Nev	66	Sun / Polaris / α Pegasi / α Corone Bor	36 03 10.53	3449.5		Lieut. Wheeler / Lieut. Lockwood / Lieut. Lyle	Lieut. Lockwood	
1873	Cottonwood Creek, Utah, N.	50	Polaris / α Lyræ / Boötis / γ Ursæ Majoris	39 13 38.50	5790.8	16 50 00	Lieut. Hoxie	Lieut. Tillman	Camp on creek, Joe's Valley.
1873	Cottonwood Creek, Utah, S	50	Jupiter / Polaris / Boötis	39 04 46.50	6055.1	16 16 00	Lieut. Hoxie	Lieut. Marshall	Camp on creek.
1873	Cow Spring, N. Mex	84	α Lyræ / Polaris / α Andromedæ / α Lyræ	32 22 02.00	4954.1	11 54 45	Lieut. Tillman	Lieut. Tillman	

13

Year	Station	No.	Stars observed	Latitude/Longitude	Elevation	Time	Observer	Computer	Remarks
1874	Coyote Creek, N. Mex	70 C	Polaris, α Coronæ Bor., β Pegasi, α Aquilæ, Polaris, α Andromedæ, α Lyræ	30 08 27.40	7054.7	14 15 00	Lieut. Blunt	Lieut. Blunt	7 miles west and 5 miles north of Guadalupita.
1873	Fort Craig, N. Mex	84	Sun, Polaris, α Andromedæ	33 38 27.00	4619.0	12 59 09	Lieut. Tillman	Lieut. Tillman	
1873	Cucharas River, Colorado	68 C	Polaris, α Persei	37 30 15.60		14 00 38	Lieut. Marshall	Lieut. Marshall	
1873	Cummings, Fort, N. Mex	84	α Cygni, Polaris, α Andromedæ	39 26 54.00	4777.7	12 29 47	Lieut. Tillman	Lieut. Tillman	
1873	Currant Creek, Colorado	61 B	α Lyræ, Polaris	38 39 29.30	8036.8	14 24 07	Lieut. Marshall	Lieut. Marshall	Two miles below Mound Springs.
1873	Deer Spring, N. Mex	76	α Cygni, Polaris, α Aquilæ, α Coronæ Bor.	34 50 32.65	5981.9	13 53 32	Lieut. Hoxie	Lieut. Tillman	
1872	Deseret City, Utah	50	β Pegasi, Polaris	39 14 04.70	4642.0	16 14 00	Lieut. Hoxie	Lieut. Hoxie	
1871	Desert Springs, Cal	73 A	α Cygni, α Aquilæ, α Bootis	35 18 26.10	1989.0	15 31 00	Lieut. Lockwood	Lieut. Lockwood	
1872	Desert Spring, Utah	52	Polaris, α Ophiuchi, α Andromedæ	37 49 24.90	5867.0	16 20 00	Lieut. Marshall	Lieut. Marshall	
1871	Desert Station, Ariz	89	α Persei, Polaris, α Lyræ, Polaris	39 30 08.80	2135.2		Lieut. Lockwood	Lieut. Lockwood	
1871	Diamond Creek, Ariz	67	Sun, Polaris	35 45 19.11	1,350.4		Lieut. Wheeler, Lieut. Lockwood, Lieut. Marshall	Lieut. Lockwood	Mouth of.
1874	Diana Creek, Colorado	61 C	Sun, Polaris	37 49 12.00	9473.0	14 39 00	Lieut. Marshall	Lieut. Marshall	Under Engineer Peak.
1873	Dirty Devil Cañon, Utah	59	Polaris, α Serpentis	38 17 24.70	5517.8	16 18 00	Lieut. Hoxie	Lieut. Tillman	Camp in.
1873	Dirty Devil River, Utah	59	α Lyræ, Polaris, α Serpentis, α Lyræ	38 16 21.70	5056.9	16 20 00	Lieut. Hoxie	Lieut. Tillman	Camp in cañon of.
	Dirty Devil River, Utah	59	α Bootis, Polaris, α Bootis, α Aquilæ	38 16 43.00	5480.5		Lieut. Hoxie	Dr. F. Kampf	Near Camp 93.
1871 1874	Disaster Rapids, Ariz. / Dolores River, Colorado	66 / 61 C	Sun, β Pegasi, Sun	35 55 59.10 / 37 46 53.60	9653.2	14 09 34	Lieut. Wheeler / Lieut. Whipple	Lieut. Lockwood / Lieut. Whipple	At foot of, on Colorado River. Near headwaters.

Geographical positions, from sextant-observations, &c.—Continued.

Year.	Atlas-sheet No.	Station.	Objects observed.	Latitude.	Altitude above sea-level.	Variation of needle E	Observer.	Computer.	Remarks.
				° ′ ″	Feet.	° ′ ″			
1874	61 C	Dolores River, Colorado	Polaris, α Andromedæ, α Lyræ, α Pegasi	37 30 37.40	7074.7	14 00 00	Lieut. Whipple	Lieut. Whipple	
1874	61 C	Dolores River, Colorado	Polaris, α Andromedæ, α Lyræ, α Pegasi	37 25 27.40	6534.1	....	Lieut. Whipple	Lieut. Whipple	
1873	59	Dome Spring, Utah	Polaris, α Serpentis, α Bootis	37 30 41.90	3774.7	15 34 00	Lieut. Hoxie	Dr. Kampf	
1873	84	Doubtful Apache, N. Mex	α Lyræ, Polaris, α Andromedæ	39 53 54.50	5546.7	....	Lieut. Tillman	Lieut. Tillman	
1873	68	Dry Camp, on trail from Paria, Utah, to Oraybe, Ariz.	α Pegasi, Polaris, f Pegasi, α Corono, α Aquilæ	35 58 07.90	4955.2	14 49 47	Lieut. Hoxie	Lieut. Marshall	Camp 56.
1873	83	Dry Camp, N. Mex	Polaris, f Pegasi, α Lyræ, α Andromedæ	39 20 57.00	4363.2	....	Lieut. Hoxie	Lieut. Marshall	Near Ralston.
1873	61 A	East River, 1 mile above junction with Gunnison.	Polaris, α Corono Bor	38 40 52.90	7991.6	....	Lieut. Marshall	Lieut. Marshall	
1873	61 A	East River trail	Polaris, α Corono Bor	38 49 07.60	9462.8	....	Lieut. Marshall	Lieut. Marshall	On Spring Creek.
1869	66	Eldorado Cañon, Nev	Polaris, α Aquilæ	35 43 55.36	898.0	....	Lieut. Wheeler, Lieut. Lockwood	Lieut. Wheeler, Lieut. Lockwood	Camp near bank of Colorado River.
1874 1873	69 D 83	Embuda, N. Mex, Escudilla Peak, near.	Polaris, α Corono Bor, Polaris, α Aquilæ, α Pegasi	36 11 13.90 33 59 07.40	5691.0 7368.4	13 15 07 13 39 58	Lieut. Birnie, Lieut. Hoxie	Lieut. Birnie, Lieut. Tillman	Base of, near Arizona and New Mexico boundary.
1872	50	Eureka City, Utah	α Bootis, Polaris	39 58 08.90	6400.0	17 09 00	Lieut. Marshall	Lieut. Marshall	One and one-half mile east of, between Eureka and Homansville.

Year	Station	No.	Star	Latitude	Elevation	Time	Observer	Observer	Remarks
1872	Fairfield, Utah	50	Sun / Polaris / a Bootis / a Aquilæ / Polaris	40 15 54.53	4866.5	16 59 30	Lieut. Hoxie	Lieut. Marshall	Camp near, and old Camp Floyd.
1873	Fair Play, Colorado	59 D	Polaris / a Lyræ / a Bootis / Polaris	39 11 35.30	10054.0	14 35 00	Lieut. Marshall	Lieut. Marshall	Camp west of, under Bald Peak.
1872	Faust's ... Utah	50	Polaris / a Pegasi	40 11 48.33	5986.0	16 51 48	Lieut. Hoxie	Lieut. Marshall	
1872	Fiddlebridge Mountain, Utah	59	Polaris / a Tauri	38 59 19.86	5339.0		Lieut. Hoxie	Lieut. Marshall	
1873	...sh	59	Polaris / a Aquilæ / a Coronæ Bor	38 39 09.70	8763.2		Lieut. Hoxie	Dr. ...	
1872	Fish Spring, Utah	50	Sun / Polaris / a Aquilæ	39 59 11.98	48.0	17 04 48	Lieut. Hoxie	Lieut. ...	
1871	Flax, Aris	68	a Perseī / a Cygni / Polaris	33 09 32.53			Lieut. Lockwood		
1873	Florida River, Colorado	69 A	Polaris / a Lyræ / a Polaris	37 16 16.00	6918.0		Lieut. Marshall	Lieut. Marshall	On Macomb's trail.
1873	Gila Spring, N. Mex.	76	a Coronæ Bor / a Pegasi / a Aquilæ	34 01 35.60	7924.8		Lieut. Hoxie	Lieut. Tillman	
1873	Garland, Fort	69 C	Sun / Polaris / a Perseī / a Lyræ / Polaris	37 25 98.11	7863.7	14 07 08	Lieut. Marshall	Lieut. Marshall	See also Lieutenant Wheeler's observations at Smith's Island, a little north of west.
1873	Geyser Spring, N. Mex.	83	a Aquilæ / a Pegasi / Polaris	33 07 58.10	89.7	13 59 54	Lieut. Hoxie	Lieut. Tillman	
1872	Gila River, N. Mex	83	a Andromedæ / Polaris	33 33 92.00	3477.3		Lieut. Hoxie	Lieut. Marshall	
1873	do	83	Pegasi / Polaris / a Aquilæ	38 41 00.00	3731.5		Lieut. Hoxie	Lieut. Marshall	
1874	Godwin Creek, Colorado	61	a Pegasi / a Coronæ Bor	37 59 38.60	9430.1	15 00 00	Lieut. Marshall	Lieut. Marshall	Five miles below head.
1872	Gould's Ranch, Utah	67	a Cygni / Perseī	37 09 14.70	4059.5		Lieut. Marshall	Lieut. Marshall	Old Fort Defiance.
1873	Green Springs, Aris	67	a Coronæ / Polaris / a Pegasi	36 11 13.00	4931.2	15 98 24	Lieut. Hoxie	Lieut. Tillman	
1872	Grass Valley, Utah	59	a Aquilæ / a Ophiuchi / Sun	38 33 98.36	6857.0	17 45 00	Lieut. Marshall	Lieut. Marshall	Camp near trail to Fish Lake.

Geographical positions, from sextant-observations, &c.—Continued.

Year.	Atlas-sheet No.	Station.	Objects observed.	Latitude.	Altitude above sea-level.	Variation of needle. E	Observer.	Computer.	Remarks.
				° ′ ″	Feet.	° ′ ″			
1872	59	Grass Valley, Utah	Polaris, α Andromedæ, α Lyræ	38 03 58.56	6507.0	17 45 13	Lieut. Marshall	Lieut. Marshall	
1872	50	Gunnison, Utah, rendezvous camp at.	Sun, α Corone Bor, β Pegasi, α Andromedæ, α Aquilæ, α Polaris	39 08 58.70	5144.0		Lieut. Wheeler, Lieut. Marshall	Lieut. Marshall	For position of astronomical monument, see list of main astronomical positions.
1873	59	Gunnison's Trail, Utah	Polaris, α Serpentis, α Lyræ	38 47 37.90	7635.4	16 00 00	Lieut. Hoxie	Lieut. Tillman	
1872	59	Gunnison Valley, Utah	α Bootis, α Coronœ, Polaris, β Pegasi	38 51 06.18	6732.3		Lieut. Marshall	Lieut. Marshall	
1869	59	Hawawah Spring, Utah	Sun, Polaris	38 22 47.46	5550.0	16 39 56	Lieut. Wheeler, Lieut. Lockwood	Lieut. Wheeler, Lieut. Lockwood	
1874	61 B.	Hayden Creek, Colorado	Polaris, Sun	38 20 24.60	7197.3	14 05 30	Lieut. Marshall	Lieut. Marshall	
1872	59	Hay Spring, Utah	α Ophiuchi, α Bootis, Polaris, α Coronœ Bor, α Aquilæ	38 19 18.96	5092.0	16 15 36	Lieut. Hoxie	Lieut. Marshall	
1873	62 A	High Creek, Colorado	β Pegasi, Polaris, Sun	38 41 10.00		15 00 43	Lieut. Marshall	Lieut. Marshall	
1869	56	Homer, Cedar Valley, Nev	α Lyræ, α Cygni, β Pegasi, α Aquilæ	38 03 23.40	5881.0	17 40 27	Lieut. Wheeler	Lieut. Wheeler	
1869	49	Ice Creek, Nev	Polaris, α Aquilæ	39 09 26.23	7084.9	16 35 06	Lieut. Wheeler, Lieut. Lockwood	Lieut. Wheeler, Lieut. Lockwood	
1869	66	Indian Spring, Nev	Sun	36 34 01.04		15 41 29	Lieut. Wheeler, Lieut. Lockwood	Lieut. Wheeler, Lieut. Lockwood	
1872	50	Indian Spring, Utah	Polaris, Polaris, ε Pegasi	39 59 26.72	5771.0		Lieut. Hoxie	Lieut. Marshall	

Year	Station	No.	Stars	Latitude	Elevation	Time	Observer	Observer	Remarks
1872	Iron City, Utah	59	a Andromeda a Ophiuchi	37 33 11.18	6099.0	18 30 00	Lieut. Marshall	Lieut. Marshall	Camp two miles south of, on Pinto and Cedar City road.
1871	Ivanpah, Nev	74	Polaris a Corone Bor.	35 33 14.54	4483.0		Lieut. Lockwood	Lieut. Lockwood	
1874	La Jara Valley, N. Mex.	69	a Pegasi Polaris	36 05 02.70	6988.4	14 05 54	Lieut. Birnie	Lieut. Birnie	Camp in.
1873	Jefferson, South Park, Colo	53 C	a Andromeda Polaris	39 24 36.40	9962.5		Lieut. Marshall	Lieut. Marshall	
1873	Joe's Valley, East side, Utah	50	a Lyrae a Bootis Polaris a Bootis γ Ursæ Majoris	39 24 33.00	8420.3	17 00 00	Lieut. Hoxie	Lieut. Tillman	
1874	Laguna del Ojo Hediondo, N. Mex.	69 A	a Lyrae Polaris a Aquilæ a Corone Bor.	36 37 42.50	7181.2	14 15 49	Lieut. Birnie	Lieut. Birnie	On east side.
1869	Lake Creek, Source of, Nev.	58	Sun Sun	38 40 38.45	5464.0	15 57 42	Lieut. Wheeler Lieut. Lockwood Lieut. Marshall	Lieut. Wheeler Lieut. Lockwood Lieut. Marshall	
1873	Lake Creek, Colo., Cabins at forks of.	52	Polaris a Corone Bor.	39 02 33.70	10795.0	14 53 00	Lieut. Marshall	Lieut. Marshall	
1873	La Loma, Colo	61 D	Sun Polaris	37 40 59.30	7742.5	14 10 23	Lieut. Birnie	Lieut. Birnie	On Rio Grande, below town.
1874	Largo Cañon, N. Mex.	69 A	Polaris a Aquilæ Sun	26 39 33.40	5650.8		Lieut. Birnie	Lieut. Birnie	Junction cañons Largo Blanco and Cerexal, (Camp 23.)
1869	Las Vegas Ranch, Nev	66	a Andromeda Sun Polaris	38 11 15.15	2074.0	15 08 11	Lieut. Wheeler Lieut. Lockwood Lieut. Lockwood	Lieut. Wheeler Lieut. Lockwood Lieut. Lockwood	
1871	Leach's Point, Cal	73	Sun a Aquilæ a Bootis	35 33 32.46	3408.6	15 06 00	Lieut. Tillman	Lieut. Tillman	
1873	Leidendorf's Wells, N. Mex	89	a Andromeda Polaris	32 14 04.00	4601.4	12 17 49	Lieut. Tillman	Lieut. Tillman	
1873	Limestone Water Pocket, Ariz.	67	a Corone Bor Polaris	36 32 18.40	5405.4	15 15 45	Lieut. Hoxie	Lieut. Marshall	
1872	Lincoln District, Utah	59	a Aquilæ a Pegasi a Corone a Aquilæ	38 15 22.00			Lieut. Hoxie	Lieut. Hoxie	

Geographical positions, from sextant-observations, &c.—Continued.

Year.	Atlas-sheet No.	Station.	Objects observed.	Latitude.	Altitude above sea-level.	Variation of needle. E.	Observer.	Computer.	Remarks.
1873	70 A	Line Colorado and New Mexico.	Polaris, α Persei, α Lyra	36 59 55.80	8757.4	14 01 16	Lieut. Marshall	Lieut. Marshall	On creek south of Francisco Pass.
1871	76	Little Colorado Crossing, (near mouth of Puerco.)		34 53 16.80	5083.3	....	Lieut. Lockwood	Lieut. Lockwood	
1874	61 C	Los Pinos River, Colo.	α Andromedæ, Polaris, α Lyra, ε Pegasi	37 29 10.85	9774.6	14 35 37	Lieut. Whipple	Lieut. Whipple	Tributary San Juan.
1874	61 B	Los Pinos Indian agency, Colo.	Sun, Polaris, β Pegasi	38 11 36.50	9064.8	14 50 00	Lieut. Marshall	Lieut. Marshall	Six-tenths mile southeast.
1873	58	Mammoth Mill, Utah	Polaris	38 04 43.90	8947.0	15 59 00	Lieut. Hoxie	Lieut. Hoxie	
1872	59	Meadow Creek, Utah	α Cygni, α Corona Bor, ε Pegasi	38 51 20.12	5992.5	16 11 00	Lieut. Hoxie	Lieut. Marshall	
1872	59	Mill Spring Station, Utah	Polaris, α Cygni	38 17 26.80	6504.0	17 20 00	Lieut. Hoxie	Lieut. Hoxie	
1872	59	Minersville, Utah	α Herculis, α Andromedæ	38 12 54.59		16 30 00	Lieut. Marshall	Lieut. Marshall	
1873	67	Moen-copie Cañon, Ariz	Polaris, ε Pegasi, α Aquilæ, α Corona Bor	36 06 03.00	4984.1	14 23 43	Lieut. Hoxie	Lieut. Marshall	Mormon wagon-road.
1873	84	Monica Spring, N. Mex	Polaris, α Lyræ, β Andromedæ	33 55 09.90	7735.3	....	Lieut. Tillman	Lieut. Tillman	
1869	58	Monument Cañon, Nev	Polaris, α Aquilæ	38 38 06.00	6114.0	16 31 54	Lieut. Wheeler, Lieut. Lockwood	Lieut. Wheeler, Lieut. Lockwood	Indian Springs.
1874	70	Mora Cañon, N. Mex	Polaris, ε Pegasi, α Lyræ	35 48 49.60	6597.6		Lieut. Blunt	Lieut. Blunt	8.5 miles south, 25 miles east of Fort Union.
1874	70 C	Mora River, N. Mex	α Andromedæ, Sun	35 59 03.50	7042.7	14 40 00	Lieut. Blunt	Lieut. Blunt	0.5 mile south, 1.75 miles ,east of Lower Mora settlement.
1869	66	Mormon Cañon, Meadow Creek, Nev.	Polaris, Sun	37 16 23.00	3092.9	16 34 30	Lieut. Wheeler, Lieut. Lockwood	Lieut. Wheeler, Lieut. Lockwood	

Year	Station	No.	Star	Latitude	Elevation	Time	Observer	Observer	Remarks
1873	Montezuma, Colo	59 D	Polaris / Lyræ / Sun	39 30 23.70	9051.8	15 49 15	Lieut. Marshall	Lieut. Marshall	On Snake River, north of town.
1871	Mosquito Spring, Cal	73	a Aquilæ / a Bootis / a Polaris	35 23 03.99	2009.8	15 30 00	Lieut. Lockwood	Lieut. Lockwood	
1872	Mountain Meadows, Utah	59	Sun / a Polaris / a Aquilæ	37 33 18.10	5741.8		Lieut. Hoxie	Lieut. Hoxie	
1873	Mount Pleasant, Utah	50	a Andromedæ / Sun / a Lyræ / a Polaris / γ Ursæ Majoris	39 33 17.40	6089.6	17 10 00	Lieut. Hoxie	Lieut. Marshall	
1873	Mule Spring, N. Mex	84	a Bootis / a Polaris / a Andromedæ / a Lyræ	32 37 36.50	5981.8	13 08 49	Lieut. Tillman	Lieut. Tillman	
1869	Mud Spring, Nev	66	a Polaris	37 11 06.88	4900.0	16 09 45	Lieut. Wheeler	Lieut. Wheeler	
1873	Muddy Creek, Utah	59	a Bootis / a Serpentis / a Lyræ	38 58 58.90	6360.0	16 00 00	Lieut. Lockwood / Lieut. Hoxie	Lieut. Lookwood / Lieut. Tillman	
1873 / 1869	Muddy Creek, Colo / Murray's Creek, Nev	62 / 49	Sun / a Polaris	37 56 22.10 / 39 15 15.90		14 07 27 / 16 35 18	Lieut. Marshall / Lieut. Wheeler	Lieut. Marshall / Lieut. Wheeler	
1873	Navajo, Rio	69 B	a Polaris / a Persei	37 01 51.40	6411.0		Lieut. Lockwood / Lieut. Marshall	Lieut. Lookwood / Lieut. Marshall	Near Pagosa Spring and Tierra Amarilla trail
1873	Navajo Spring, Ariz	67	a Cygni / a Polaris / a Pegasi / a Corone Bor	36 46 19.10	4410.9		Lieut. Hoxie	Lieut. Tillman	
1871	New Creek, (Ives's,) Ariz	75	a Pegasi / a Corone Bor / a Polaris	35 36 51.00			Lieut. Lockwood	Lieut. Lockwood	Pahroach Spring.
1871	Nelson's Tanks, Ariz	75	a Aquilæ / β Andromedæ / a Lyræ	34 46 20.42	6916.5		Lieut. Lockwood	Lieut. Lockwood	Mogollon mesa.
1873	...Fork, Utah. Pit	50	a Polaris / a Virginis / a Lyræ	39 51 11.40	5046.2		Lieut. Hoxie	Lieut. Tillman	Price River.
1873	Nutria Spring, N. Mex	76	a Polaris / a Ophiuchi / a Pegasi	35 18 13.65	6334.0	14 16 09	Lieut. Hoxie	Lieut. Tillman	
1872	Oak Creek Cañon, Utah	50	a Aquilæ / a Bootis / a Polaris	39 21 10.70	5158.0		Lieut. Marshall	Lieut. Marshall	Two miles east of Oak Creek City, Utah.

Geographical positions, from sextant-observations, &c.—Continued.

Year.	Station.	Atlas-sheet No.	Objects observed.	Latitude.	Altitude above sea-level.	Variation of needle E.	Observer.	Computer.	Remarks.
				° ′ ″	Feet.	° ′ ″			
187?	Ojo Spring, N. Mex	76	Polaris / α Coronæ Bor / α Aquilæ	34 03 18.90	7946.4	12 34 42	Lieut. Hoxie	Lieut. Tillman	
1874	Ocate River, N. Mex	70 C	α Coronæ Bor / α Pegasi / Polaris	36 10 90.50	7076.6	14 15 06	Lieut. Blunt	Lieut. Blunt	Near Las Gallinas.
1874	Ojo Caliente Creek, N. Mex. (Camp on, No. 19.)	69 D	α Aquilæ / Sun / Polaris	36 17 07.40	6943.7	13 15 00	Lieut. Birnie	Lieut. Birnie	One mile south of springs.
1873	Oraybe, Ariz	68	ε Pegasi / α Andromedæ / Polaris	35 59 57.60	4756.8		Lieut. Hoxie	Lieut. Marshall	Tanks near.
1872	Otter Creek, Utah, head of	59	ε Pegasi / α Aquilæ / α Coronæ Bor	38 49 35.90	7447.7		Lieut. Marshall	Lieut. Marshall	
1874	Pagosa Springs, Colo	69 A	β Pegasi / Polaris / Markab / Sun / Altair / α Arietis	37 15 41.40 / 15 44.10	7108.6	14 39 30	Lieut. Wheeler / Lieut. Marshall / Lieut. Whipple	Lieut. Marshall / Lieut. Whipple	
1871	Pahghun Pahghun Springs, Ariz	66	Sun / α Coronæ Bor / β Pegasi / β Capricorni / Polaris	36 24 51.83	2281.9		Lieut. Lyle / Mr. Marvine	Lieut. Lockwood	
1871	Pah-reach Spring, Ariz		α Coronæ Bor / α Pegasi / Polaris	35 36 51.00		14 15 00	Lieut. Lockwood	Lieut. Lockwood	New Creek of Ives.
1872	Pauquitch, Utah	59	Sun / Polaris	37 49 34.00	6973.3		Lieut. Marshall	Lieut. Marshall	Camp 3 miles north of, on Sevier.
1872	Paragoonah, Utah	59	α Coronæ Bor / Polaris / α Andromedæ	37 54 44.97	6232.0	19 30 00	Lieut. Marshall	Lieut. Marshall	One and one-half mile north of.

Year	Locality	No.	Star	Latitude	Elevation		Observer	Observer	Remarks
1872	Paria, Utah	67	Mira / Polaris / a Perseï	37 10 49.07	4582.1	14 13 00	Lieut. Marshall	Lieut. Marshall	River-bank at mouth.
1872	Paria River, Utah, mouth of	67	a Cygni / Polaris / a Perseï	36 51 13.90	3017.6		Lieut. Marshall	Lieut. Marshall	Where Paria and Kanab road leaves river.
1872	Paria River Cañon, at great bend	67	Sun / Polaris / a Perseï	36 58 34.94	3873.5		Lieut. Marshall	Lieut. Marshall	
1872	Paria River, camp near head, Utah	59	Polaris / a Aquilæ	37 32 11.57			Lieut. Hoxie	Lieut. Hoxie	
1872	Paria River Cañon	67	Aquilæ / Polaris	37 14 11.94		14 13 00	Lieut. Hoxie	Lieut. Hoxie	
1873	Pasainyo Camp, N. Mex	84	Polaris / Lyræ	38 43 46.00	4444.8		Lieut. Tillman	Lieut. Tillman	
1873	Peach Orchard Spring, Ariz.	68	Polaris / Pegasi / a Aquilæ / a Corona Bor	35 46 42.40	6297.5		Lieut. Hoxie	Lieut. Hoxie	
1869	Pearl Creek, Nev	49	a Aquilæ / Polaris	40 17 10.74	5965.1	16 18 51	Lieut. Wheeler / Lieut. Lockwood	Lieut. Wheeler / Lieut. Lockwood	
1874 / 1873	Pecos Cañon, N. Mex. / San Pedro River, Ariz.	70 / 83	Sun / Polaris / a Aquilæ / a Ophiuchi	35 46 25.00 / 32 42 39.70	7884.2 / 5874.5	12 49 20	Lieut. Blunt / Lieut. Hoxie	Lieut. Blunt / Lieut. Hoxie	Confluence of main branches.
1871	Picacho Station, Ariz	82	Pegasi / a Perseï / Polaris	39 44 90.67	1750.2		Lieut. Lockwood	Lieut. Lockwood	
1871	Picket Post, Ariz.	83	a Cygni / Polaris	33 17 01.97	9669.6		Lieut. Lockwood	Lieut. Lockwood	
1872	Piermont, Nev	49	Sun / Orionis	39 38 42.15	5581.2	16 47 00	Lieut. Hoxie	Lieut. Marshall	Camp near O'Neil's Ranch.
1871	Pinal Creek	83	Pegasi / a Perseï	33 39 26.46	3112.2		Lieut. Lockwood	Lieut. Lockwood	Wheatfields.
1871	Pinal Mountains, Ariz	83	Cygni / Antares / Herculis	33 23 10.94	3925.5		Lieut. Lockwood	Lieut. Lockwood	
1872	Pine Valley, southwest of, Utah.	59	Polaris	37 23 33.80		16 00 00	Lieut. Marshall	Lieut. Marshall	Near Earle's ranch.
1873	Palomas Creek, N. Mex	84	Polaris / a Lyræ	33 09 18.00	5415.5		Lieut. Tillman	Lieut. Tillman	

AP. JJ—2

Geographical positions, from sextant-observations, &c.—Continued.

Year.	Station.	Atlas-sheet No.	Objects observed.	Latitude.	Altitude above sea-level.	Variation of needle. E	Observer.	Computer.	Remarks.
				° ′ ″	Feet.	° ′ ″			
1871	Portage Rapids, Ariz........	66	Polaris........ a Aquilæ ...... a Pegasi ...... a Lyræ.........	35 48 35. 90	....	....	Lieut. Wheeler ...	Lieut. Wheeler ...	Above.
1872	Potato Valley, Utah, (head of.)	59	Polaris .... a Persei	37 39 22. 02	7331. 9	....	Lieut. Marshall ...	Lieut. Marshall ...	
1872	Potato Valley, Springs in Mountains, south of head, Utah.	59	a Cygni .... Polaris .... a Persei	37 37 37. 90	8669. 6	....	Lieut. Marshall ...	Lieut. Marshall ...	
1873	Prieto Crossing, Ariz	83	Polaris .... a Andromedæ...	33 33 47. 30	5332. 8	12 36 19	Lieut. Tillman......	Lieut. Tillman......	
1873	Pueblo Colorado, Ariz	68	Coronæ .... Sun........... Polaris .... ε Pegasi ...... a Coronæ Bor.	35 42 10. 40	6400. 9	14 31 16. 3	Lieut. Hoxie.......	Lieut. Hoxie.......	
1873	Pueblo Viejo, Ariz..........	83	a Aquilæ ...... Sun .... β Andromedæ Polaris .... a Aquilæ ...... a Pegasi ...... a Coronæ Bor.	32 49 00. 10	2711. 6	14 10 47	Lieut. Tillman	Lieut. Tillman	Safford post-office.
1871	Puerco River, mouth, Ariz...	76	Altair .... γ Andromedæ Polaris	34 53 16. 80	5083. 3	....	Lieut. Lockwood..	Lieut. Lockwood.	
1873	Purgatoire River, Colo.,(head north fork.)	70 A	Polaris .... a Lyræ......... Arietis	37 18 24. 60	8433. 2	14 22 46	Lieut. Marshall...	Lieut. Marshall...	West of Spanish peaks, on road from Cucharas River, Colo., to Elizabethtown, New Mex.
1869	Quinn Cañon, Nev	58	Sun .... Polaris .... a Serpentis ... a Boötis	37 58 18. 29	6326. 0	16 20 15	Lieut. Wheeler ... Lieut. Lockwood.	Lieut. Wheeler ... Lieut. Lockwood.	
1873	Rabbit Valley, Utah, (foot of.)	59	Polaris .... a Lyræ.........	38 19 00. 10	6520. 8	16 20 00	Lieut. Hoxie......	Lieut. Marshall ...	
1873	Ralston, New Mex..........	89	Sun .... Coronæ ...... Polaris .... a Pegasi ...... a Andromedæ a Lyræ.........	32 16 10. 50	4487. 7	....	Lieut. Tillman...... Lieut. Hoxie......	Lieut. Tillman...... Lieut. Marshall...	

			Star					Lieut. Lockwood	Lieut. Lockwood	Lieut. Lockwood	
1871	Rattlesnake Cañon, Aris	75	β Andromeda .. α Ophiuchi ...... Polaris	34 35 49.93	....		14 49 00				Spring Valley.
1869	Rattlesnake Spring, Nev ...	58	Sun...	38 57 91.17	6038.9	16 17 54		Lieut Wheeler Lieut. Lewood	Lieut Wheeler ... Lieut. Lockwood ...		
1871	Beller Springs, Aris	75	Polaris .... α Ophiuchi ... α Pegasi	35 06 34.92	3596.6	14 98 00		Lieut. Lockwood	Lieut Lockwood ...		Lockwood Springs.
1874	Reynolds, Old Fort, Colo	62 A	Sun ... Polaris ...	38 12 03.90	4300.0	....		Lieut. Bht	Lieut. Blunt		
1874	Rio Grande, Colo............	61 C	α Pegasi .... Polaris ... α Corona Bor ... α Pegasi ...	37 45 10.10	0816.0	14 50 00		Lieut Marshall ...	Lieut Marshall ...		Camp two miles below mouth of Pole Creek.
1874	Rio Grande, South Fork, at mouth of Beaver Creek.	61 D	α Aquilæ ... α Pegasi ... Arietis ...	37 36 09.60	8415.5	....		Lieut Marshall ...	Lieut Marshall ...		
1874	Rio Grande, Colo	61 D	α Andromeda ... Polaris ...	37 44 93.50	9787.4	....		Lieut Whipple ...	Lieut Whipple ...		Jennison's ranch.
1873	Rio Mimbres, N. Mex.......	84	α Lyræ ... α Pegasi ... β Polaris ...	39 30 46.00	4935.7	19 58 00		Lieut. Till sm.....	Lieut. Tillman		
1874	Rio Piedra, Colo............	61 C	α Lyræ ... Polaris ... bd omp.	37 98 15.90	7947.9	....		St. Marshall ...	Lieut. Mhall ...		Camp on West Fork on trail from Rio Grande Cañon.
1874	Rio San Juan, Colo, under Station 92.	61 C	α Aquilæ ... α Pegasi ... Sun...	37 95 99.40	7776.7	....		Lint Marshall ...	Lieut Marshall ...		At point where Pagosa and Del Norte trail leaves river.
1869	Rio Virgen, mouth of, Nev.	66	α Pegasi ... Polaris ... α Aquilæ ...	36 08 45.54	1900.0	15 47 11		Lint Wheeler ... Lint Lewood	Lieut Wheeler ... Lieut. Lockwood ...		
1869	Rose Valley, Nev............	56	Sun ... Polaris ... α Pegasi ...	37 54 51.90	64.0	17 50 09		Lit Wheeler Lieut. Lockwood ...	Lit Wheeler ... Lieut. Lockwood ...		
1869	Santo District, Nev ...	49	α Aquilæ ... Polaris ...	39 09 46 06	6574.0	16 97 99		Lieut Wheeler ...	Lieut Wheeler ...		
1874	Ege Plains, Glo............	60	α Andromeda ... α Lyræ ... α Pegasi ... Polaris ...	37 95 27.40	5684.8	14 34 38		Lieut Whipple	Lieut Whipple ...		
1873	Saguache, Colo., (Craig's Ranch)	61 D	Sun...	38 09 10.60	7463.8	14 34 43		Lieut Marshall ...	Lieut Marshall ...		
1872	Salina, Utah	59	α Corona Bor ... Polaris ... α Pegasi ...	38 57 10.56	....	....		Lieut Marshall ...	Lieut Marshall ...		

Geographical positions, from sextant-observations, &c.—Continued.

Year.	Atlas-sheet No.	Station.	Objects observed.	Latitude.	Altitude above sea-level.	Variation of needle. E.	Observer.	Computer.	Remarks.
1873	69 B	San Antonio Creek, Colo..	Polaris / " Persei / " Cygni	37 01 08.10	7680		Lieut. Marshall	Lieut. Marshall	Near Los Pinos Plaza south of Conejos.
1873	83	San Francisco River, N. Mex	Polaris / " Aquilæ / " Ophiuchi / " Pegasi	33 11 44.90	3798.3	13 31 23	Lieut. Hoxie	Lieut. Tillman	
1873	83	......do	Polaris / " Aquilæ / " Ophiuchi / " Pegasi	33 02 27.80	3484.7	12 22 22	Lieut. Hoxie	Lieut. Tillman	
1872	50	San Francisco Spring, Utah.	Polaris / " Cygni	38 26 47.05	6597.1	16 58 00	Lieut. Hoxie	Lieut. Hoxie	
1873	83	San Francisco River, N. Mex.	Polaris / " Aquilæ / " Pegasi	33 26 31.50	5177.9	13 49 90	Lieut. Hoxie	Lieut. Tillman	
1873	83	......do	Polaris / " Ophiuchi	33 15 15.90	4706.6	12 51 43	Lieut. Hoxie	Lieut. Tillman	
1874	78 A	San Geronimo, N. Mex	Polaris / " Pegasi / " Andromedæ	35 39 90.30	6794.5		Lieut. Blunt	Lieut. Blunt	On Tocalote Creek.
1874	69 A	San Juan River, N. Mex	Polaris / " Coronæ Bor. / " Aquilæ	36 49 14.90	5474.1	14 06 05	Lieut. Birnie	Lieut. Birnie	One mile South Cañon Largo.
1874	61 D	San Juan River, Colo	Sun	37 25 98.40	7776.7	15 01 00	Lieut. Marshall	Lieut. Marshall	At point trail to Del Norte leaves main fork.
1874	68 B	San Juan River, N. Mex	Polaris / " Coronæ / " Aquilæ	36 44 68.00			Lieut. Birnie	Lieut. Birnie	Ten miles west of eastern boundary of Navajo reservation.
1874	61 D	San Juan River, Colo., Head of East Fork.	Pegasi / " Pegasi / " Polaris	37 23 05.40	9011.9	14 58 94	Lieut. Wheeler / Lieut. Whipple	Lieut. Whipple	
1874	69 A	San Juan River, N. Mex	Aquilæ / " Pegasi / " Polaris / " Andromedæ / " Lyræ	36 59 06.90	4808.4	14 10 31	Lieut. Whipple	Lieut. Whipple	

Year	Station	No.	Stars	Latitude	Elevation		Observed by	Computed by	Remarks
1874	San Juan River, N. Mex., (mouth of La Plata)	60 A	Pegasi / Polaris / Andromedæ	36 45 59.40	3836.3	14 34 36	Lieut. Whipple	Lieut. Whipple	
1874	San Juan River, N. Mex., (mouth of Mancos)	60 A	Lyræ / Pegasi / Andromedæ / Polaris	36 56 48.80	4781.2	14 .. 31	Lieut. Whipple	Lieut. Whipple	
1873	San Juan mines, Colo	61 C	Sun / Polaris / Arietis	37 50 19.90	9397.0	14 38 37	Lieut. Marshall	Lieut. Marshall	Howardsville.
1872	Santaquin, Utah	50	Lyræ / Aquilæ / Sun / Polaris	39 58 36.06	......	17 96 00	Lieut. Marshall	Lieut. Marshall	One mile north of.
1871	Saratoga Springs, Cal	66	Bootis / Polaris / Coronæ Bor / Pegasi	35 41 04.53	963.6	15 05 00	Lieut. Lockwood	Lieut. Lockwood	
1872 / 1872	Scipio, Camp near, Utah / Sevier River, head of East Fork, Utah.	50 / 59	Sun / Polaris	39 10 50.93 / 37 30 05.33	5113.0 / 8814.7	......	Lieut. Marshall / Lieut. Marshall	Lieut. Marshall / Lieut. Marshall	Four miles south of.
1872	Sevier River Cañon, East Fork, Utah.	59	Cygni / Coronæ / Polaris	38 10 54.70	6081.2	......	Lieut. Marshall	Lieut. Marshall	
1872	Sevier River, East Fork, Utah.	59	Andromedæ / Aquilæ / Polaris	37 39 51.73	7300.0	......	Lieut. Hoxie	Lieut. Hoxie	
1872	Sevier Lake, Camp near, Utah.	59	Polaris / Lyræ	38 49 48.60	4949.0	17 28 00	Lieut. Hoxie	Lieut. Hoxie	On bench, southwest side.
1872	Sevier Pass, Utah	50	Aquilæ / Coronæ / Polaris	39 32 33.90	4767.0	17 00 00	Lieut. Marshall	Lieut. Marshall	At west side, (mouth of cañon.)
1869 / 1872	Schafer Spring, Nev / Short Creek, Ariz	58 / 67	Sun / Polaris	37 33 12.88 / 36 59 27.16	6186.0 / 5082.7	16 10 45	Lieut. Lockwood / Lieut. Marshall	Lieut. Lockwood / Lieut. Marshall	
1869	Sheep Ranch, Cedar Valley, Nev.	58	Polaris / Aquilæ	38 13 48.00	7072.7	16 46 26	Lieut. Wheeler	Lieut. Wheeler	
1874	Simpson's Peak, Camp under, Colo.	61 C	Polaris / Coronæ Bor. / Pegasi	37 40 37.70	10800.7	14 30 00	Lieut. Marshall	Lieut. Marshall	Peak 14055.9 feet.
1873	Skull Camp, N. Mex	84	Aquilæ / Polaris / Lyræ	31 00 55.00	5371.1	12 00 00	Lieut. Tillman	Lieut. Tillman	River Animas, N. Mex.
1872	Skumpah, Utah	67	Andromedæ / Polaris / Capella / Cygni	37 16 14.79	6142.1	......	Lieut. Marshall	Lieut. Marshall	One and one-half miles north of west of settlement.

Geographical positions, from sextant-observations, &c.—Continued.

Year.	Atlas-sheet No.	Station.	Objects observed.	Latitude.	Altitude above sea-level.	Variation of needle E.	Observer.	Computer.	Remarks.
1869	49	Slough, Long Valley, Nev	Polaris / Sun	39 49 27.28	6915.6	16 59 55	Lieut. Wheeler / Lieut. Lockwood	Lieut. Wheeler / Lieut. Lockwood	
1874	69 A	Smith's Island, Colo	α Aquilæ / Polaris / α Coronæ / α Aquilæ / α Pegasi / α Polaris	37 25 42.40	7868.0		Lieut. Wheeler	Lieut. Whipple	Near Fort Garland, Colo.
1869	49	Snake Creek, Nev	Polaris	39 00 05.28	5369.0	16 37 50	Lieut. Wheeler	Lieut. Wheeler	
1872	49	Snake Valley, Nev	Polaris / α Bootis / α Pegasi	39 00 42.66		16 38 00	Lieut. Lockwood / Lieut. Hoxie	Lieut. Lockwood / Lieut. Marshall	
1873	50	Soldier's Fork, Utah	α Aquilæ / Jupiter / α Bootis / Polaris	39 59 15.10	5397.5		Lieut. Hoxie	Lieut. Tillman	
1873	50	So...er's Fork, Head of, Utah	α Lyræ / Jupiter / α Bootis / α Lyræ / Polaris	39 55 06.50	6397.9		Lieut. Hoxie	Lieut. Tillman	
1873	59 D	South Fork S. Platte, Colo., near Weston's Pass	Polaris	39 04 27.70	10117.3	14 17 00	Lieut. Marshall	Lieut. Marshall	
1872	50	Spanish Forks, Utah, (camp near.)	α Bootis / Polaris	40 05 10.80			Lieut. Marshall	Lieut. Marshall	
1869	58	Spring below Panacca, Nev	Sun / Polaris	37 45 27.07	4718.1	16 58 51	Lieut. Wheeler	Lieut. Wheeler	
1872	49	Spring Valley, Nev	α Aquilæ / α Bootis / Polaris	39 45 24.14	6102.4		Lieut. Hoxie	Lieut. Marshall	
1869	65	Saint Thomas, near	Sun	36 26 33.43	1600.0	15 47 99	Lieuts. Wheeler and Lockwood / Lieut. Marshall	Lieuts. Wheeler and Lockwood / Lieut. Marshall	
1872	59	Sulphur Springs, Utah	Sun / α Ophiuchi / α Andromedæ	38 01 04.30	5400.0				
1873	84	Sunset Camp, Ariz	Polaris / α Lyræ / β Andromedæ	33 13 24.00	5976.2		Lieut. Tillman	Lieut. Tillman	

Year	Station	No.		Stars observed	Latitude	Elevation	Time	Observer	Observer	Remarks
1871	Sunset Crossing, Ariz., (of Colorado Chiquito.)	74		Polaris; γ Andromedæ; a Aquilæ	34 59 41.70	4691.5		Lieut. Lockwood	Lieut. Lockwood	
1871	Sunset Tanks, Ariz	76		Polaris; β Andromedæ	34 54 34.40	5797.2		Lieut. Lockwood	Lieut. Lockwood	
1871	Tanks at the Pines, Ariz	83		a Lyræ; a Persei; Polaris	33 49 31.30	5648.7		Lieut. Lockwood	Lieut. Lockwood	
1873	Tarryall Creek, near mouth of Rock Creek, Colo.	52	D	a Lyræ; Polaris; a Bootis	39 18 08.65	9006.3	14 49 00	Lieut. Marshall	Lieut. Marshall	
1873	Taylor River, Colo., at head of North Fork.	61	B	a Coronæ Bor	38 59 39.60	10209.0		Lieut. Marshall	Lieut. Marshall	
1873	Taylor River, Colo., on South Fork.	61	B	Polaris; a Coronæ Bor	38 47 52.80	9666.3		Lieut. Marshall	Lieut. Marshall	Main tributary to Gunnison River.
1873	Thousand Lake Mountain, east base of, Utah.	59		Jupiter; Polaris; a Lyræ; a Bootis	38 28 40.90	9248.4		Lieut. Hoxie	Lieut. Tillman	
1873	Three Water Pockets, Utah	59		a Serpentis; Polaris	37 44 49.70	4673.1		Lieut. Hoxie	Dr. Kampf	
1873	Tierra Amarilla, N. Mex	69	B	a Lyræ; a Bootis; Polaris; Markab	36 41 29.90	7466.1	13 42 96	Lieut. Marshall	Lieut. Marshall	
1874	Nutritas Plass	69	B	a Persei; Polaris	36 41 29.40			Lieut. Wheeler	Lieut. Whipple	
1873	Timber Cañon, Utah	50		a Cygni; a Lyræ; Sun; a Leonis; a Bootis	39 44 43.15	7978.7		Lieut. Hoxie	Lieut. Tillman	
1873	Topographical Station and Camp, N. Mex.	76		Polaris; a Aquilæ; a Pegasi; a Ophiuchi	34 56 38.00	6291.9	13 58 12	Lieut. Hoxie	Lieut. Tillman	Near Ojo Caliente.
1871	Truxton Springs, Ariz	75		Sun; Polaris	35 24 52.51	3885.5		Lieut. Lockwood; Lieut. Lyle	Lieut. Lockwood	
1873	Tule Springs, N. Mex	76		a Aquilæ; a Coronæ; a Pegasi	34 31 30.30	5924.7	13 36 23	Lieut. Hoxie	Lieut. Tillman	
1873	Fuleroes, Fort, N. Mex	83		Polaris; a Aquilæ; a Ophiuchi	33 52 47.90	6740.4	13 17 42	Lieut. Hoxie	Lieut. Tillman	Old fort.
1874	Tunicha Creek	68	D	a Pegasi; Polaris; a Andromedæ	36 13 41.60	5510.0	14 00 41	Lieut. Birnie	Lieut. Birnie	On mesa, between Tunicha and Vaca Creeks.

Geographical positions, from sextant-observations, &c.—Continued.

Year	Atlas-sheet No.	Station	Objects observed	Latitude	Altitude above sea-level.	Variation of needle. E.	Observer.	Computer.	Remarks.
				o ′ ″	Feet.	o ′ ″			
1869	66	Vegas Wash, Nev., (mouth of.)	Polaris	36 06 34.85		16 01 05	Lieut. Wheeler	Lieut. Wheeler	
			a Aquilæ				Lieut. Lockwood	Lieut. Lockwood	
1874	70 A	Vermejo Creek, N. Mex.	Sun	36 41 30.60	6249.6	14 30 00	Lieut. Lockwood	Lieut. Blunt	Crossing Santa Fé and Trinidad roads.
1871	66	Virgin Hill, Nev.	Polaris	36 37 45.29			Lieut. Wheeler	Lieut. Wheeler	
			a Coronæ Bor				Lieut. Lockwood	Lieut. Lockwood	
1872	59	Virgin River, Utah, North Fork.	Pegasi	37 21 47.37			Lieut. Hoxie	Lieut. Hoxie	
			Polaris						
			a Pegasi						
			a Aquilæ						
			a Herculis						
1872	67	Virgin River, Utah, Camp 8.	Polaris	37 08 06.60		15 29 00	Lieut. Hoxie	Lieut. Hoxie	
			a Lyræ						
			a Andromedæ						
1869	40	Walker's Ranch, Nev.	Polaris	40 43 50.67	5145.9		Lieut. Wheeler	Lieut. Wheeler	On trail to El Vado de los Padres.
1872	67	Warm Creek, Utah.	Perseï	37 03 23.80	4063.9		Lieut. Marshall	Lieut. Marshall	
1873	83	Water-Hole, N. Mex.	Lyræ	33 12 49.30	5642.9	13 30 00	Lieut. Hoxie	Lieut. Tillman	
			Polaris						
			a Aquilæ						
			a Ophinchi						
			a Pegasi						
1873	59	Water-Pocket, Utah.	Polaris	37 39 09.80	3559.6	15 48 00	Lieut. Hoxie	Dr. Kampf	East of Escalante River.
			a S-rpentis						
			a Lyræ						
			a Bootis						
1873	60	Water-Pocket, Utah, (on ridge near.)	Polaris	37 27 28.70	4781.2	15 29 00	Lieut. Hoxie	Dr. Kampf	East of Escalante River.
			a Bootis						
1873	59	Welcome Creek, Utah	Sun	37 34 13.30	3640.9	15 07 00	Lieut. Hoxie	Dr. Kampf	Camp.
			Polaris						
			a Serpentis						
			a Lyræ						
1869	66	West Point, Nev.	Sun	36 40 33.56	1754.0	15 18 59	Lieut. Wheeler	Lieut. Wheeler	Camp.
			Polaris				Lieut. Lockwood	Lieut. Lockwood	
1873	69 A	Wet Mountain Valley, Colo	Sun	38 09 22.60	8000.0	14 20 00	Lieut. Marshall	Lieut. Marshall	Near Garnier's ranch.
			Polaris						
1873	84	Wet Camp, N. Mex.	Polaris	33 27 43.00	6016.5		Lieut. Tillman	Lieut. Tillman	Camp.
			β Andromedæ						
			a Lyræ						

Year	Station	No.	Object	Latitude	Elev.		Observer	Observer	Remarks
1860 1873	Union Peak .... Price River, Utah, (branch of North Fork.)	¹⁄₈ 50	Sun .... Jupiter ...... β Bootis Polaris.	38 58 93.01 39 55 98.40	13030.0 7911.0	....	Lieut. Wheeler ... Lieut. Hoxie.......	Lieut. Wheeler ... Lieut. Tillman.....	Summit of Peak.
1873	Price River, (bluff west of junction of North and South Forks,) Utah.	50	α Lyræ .... α Bootis Polaris ..... α Lyræ .... β Ursæ Majoris γ Ursæ Majoris	39 48 11.30	7088.0	....	Lieut. Hoxie.......	Lieut. Tillman.....	Camp.
1869	Wild Hop Creek, (Pioneer Cañon, Nov.)	58	Polaris ...... α Aquilæ	38 23 16.90	6998.0	15 59 99	Lieut. Wheeler....	Lieut. Wheeler ...	
1869	Willow Creek, Nev.........	49	α Aquilæ ...... Polaris	40 31 13.91	5518.9	17 97 07	Lieut. Wheeler.... Lieut. Lookwood.	Lieut. Wheeler ... Lieut. Lookwood.	
1873	Wingate, Fort, N. Mex......	77 A	α Ophiuchi ... α Pegasi ..... α Aquilæ	35 28 49.47	6982.1	14 51 27	Lieut. Hoxie....... L. Nell...........	Lieut. Tillman..... L. Nell...........	Flagstaff.
1871	Young's Spring, Ariz........	75	α Corone Bor... α Pegasi ..... α Ophiuchi..	35 39 04.98			Lieut. Lookwood.	Lieut. Lookwood.	
1873	Yucca Camp, N. Mex	89	Polaris....... α Andromedæ	39 18 48.00	4374.0	12 37 49	Lieut. Tillman....	Lieut. Tillman.....	
1875	Station A, Cal	65 D	α Lyræ .... Me α Cygni ...... α Corone ..... β Pegasi ..... α Cephi ..... Polaris	36 4 44.92	1232.8	15 10.36	Lieut. Birnie......	Lieut. Birnie......	Base-station, Panamint Valley, Cal.
1875	Alamo Gordo, N. Mex	78 A	Ursæ Me α Arietis α Pegasi ..... Polaris	34 48 45.70	4963.4	....	Lieut. Morrison...	Lieut. Birnie......	Drain of.
1875	Animas Valley, Colo	61 C	α Aquilæ ...... Polaris	37 21 19.10	6561.3	....	Lieut. Marshall...	Lieut. Birnie......	
1875	Cañon Pajarito, N. Mex	78 A	α Aquilæ ...... α Arietis α Pegasi ..... Polaris	34 59 44.90	5099.0	....	Lieut. Morrison...	Lieut. Birnie......	
1875	Bear Creek, Cal.............	65 C	α Lyræ .... ε Pegasi ..... Polaris	35 56 59.30	6145.3	....	Dr. Kampf........	Lieut. Birnie......	Head of.
1875	Bennett's Well, Cal.........	65 D	α Corone ..... α Pegasi ..... α Aquilæ ...... Polaris	36 09 49.19	−5.8	....	Lieut. Birnie......	Lieut. Birnie......	Death Valley, Cal., below sea-level.
1875	Burrough's Park, Colorado...	61 C	α Aquilæ ...... Polaris ...... Sun	37 56 97.90	10878.9	....	Lieut. Marshall...	Lieut. Birnie......	

Geographical positions, from sextant-observations, &c.—Continued.

Year.	Station.	Atlas-sheet No.	Objects observed.	Latitude.	Altitude above sea-level.	Variation of needle. E	Observer.	Computer.	Remarks.
				° ′ ″	Feet.	° ′ ″			
1875	Cartago, Cal	65 C	β Ceti Polaris	36 18 35.30	3569.0		Lieut. Birnie	Lieut. Birnie	Owen's Lake. Time by equal altitudes of meridian-star.
1875	Cerro Cuervo, N. Mex	78 A	α Aquilæ α Arietis Pegasi Polaris	35 07 38.50	4882.5		Lieut. Morrison	Lieut. Birnie	
1875	Cerro Cueva, N. Mex	78 A	α Aquilæ α Arietis Pegasi Polaris	35 10 35.60	5338.4		Lieut. Morrison	Lieut. Birnie	Foot of.
1875	Cerro Gordo Landing, Cal.	65 D	α Coronæ Pegasi α Aquilæ Polaris	36 27 36.55	3656.1	15 18 42	Lieut. Birnie	Lieut. Birnie	
1875	Cherry Creek, Colo	64 A	α Coronæ Pegasi α Aquilæ Polaris	37 18 31.90	7449.4		Lieut. Marshall	Lieut. Birnie	
1875	Cimarron Seco, N. Mex	70 B	α Aquilæ α Arietis Pegasi Polaris	36 50 06.70			Lieut. Morrison	Lieut. Birnie	
1875	Laguna, Colo	77 B	α Aquilæ α Arietis Pegasi Polaris	35 13 01.90	6999.8		Lieut. Morrison	Lieut. Birnie	
1875	Cottonwood Cañon, Cal	65 B	α Coronæ Pegasi α Aquilæ Polaris	36 31 46.88	2992.7		Lieut. Birnie	Lieut. Birnie	
1875	Cottonwoods, Cal	73 D	α Lyræ Boötis Serpentis Polaris	34 48 01.78	2487.8		Lieut. Birnie	Lieut. Birnie	Mojave River.
1875	Egan's Falls, Cal	65 D	Boötis Cygni β Ophiuchi Polaris	36 19 42.75			Lieut. Birnie	Lieut. Birnie	Darwin Cañon.

	Station		Stars	Latitude		Time			Remarks
1875	Estoros, N. Mex	78 A	α Aquilæ / α Arietis / α Pegasi / Polaris	35 09 30. 30	5310. H	......	Lieut. Morrison	Lieut. Birnie	Laguna de los.
1875	Forks Los Angeles and Caliente roads, Cal.	73 A	β Pegasi / β Ceti / β Polaris	35 07 49. 00	2752. 0	14 47. 43	Lieut. Birnie	Lieut. Birnie	Near Tehachipi Pass.
1875	Old Fort Tejon, Cal	73 A	α Bootis / α Cygni / α Ophiuchi / Polaris	34 59 03. 40	3945. 7	14 54. 30	Lieut. Wheeler	Lieut. Birnie	
1875	Furnace Creek, Cal	65 D	α Corona / α Pegasi / α Aquilæ / Polaris	36 28 21. 50	465. 1	15 41. 30	Lieut. Birnie	Lieut. Birnie	One and a quarter mile from mouth.
1875	Glendale, Cal	65 C	α Corona / α Pegasi / α Aquilæ / Polaris	36 29 50. 50	9923. 5	15 41. 18	Dr. Kampf	Lieut. Birnie	
1875	Glenville, Cal	65 C	α Andromedæ / α Polaris	35 43 18. 10	3094. 3	......	Dr. Kampf	Lieut. Birnie	
1875	Granite Wells, Cal	73 B	α Pegasi / α Polaris / α Lyræ / α Bootis / α Serpentis / Polaris	35 29 31. 90	4015. 9	15 06. 30	Lieut. Birnie	Lieut. Birnie	Near Pilot Knob.
1875	Huntington's, Cal	73 D	α Lyræ / α Bootis / α Serpentis / Polaris	34 32 16. 08	2998. 6	15 44. 49	Lieut. Birnie	Lieut. Birnie	Mojave River.
1875	Hunter's Ranch, Cal	65 B	α Corona / α Pegasi / α Aquilæ / α Arietis / Polaris	36 33 40. 85	6974. 7	15 44. 49	Lieut. Birnie	Lieut. Birnie	Jackass Spring.
1875	Juan de Dios, N. Mex	78 A	α Aquilæ / α Arietis / α Pegasi / Polaris	34 48 45. 70	4819. 5	......	Lieut. Morrison	Lieut. Birnie	Head of Arrojo.
1875	Kern River, Cal	65 C	α Corona / α Pegasi / α Aquilæ / Polaris	36 23 50. 90	9328. 8	15 37 48	Dr. Kampf	Lieut. Birnie	Branch of.
1875	Kern River, Cal	65 C	α Corona / α Pegasi / α Aquilæ / Polaris	36 21 38. 90	8917. 1	15 51 48	Dr. Kampf	Lieut. Birnie	Branch of.
1875	Kern River, Cal	65 C	α Andromedæ / α Polaris / α Lyræ / α Pegasi / Polaris	36 20 37. 90	6442. 1	16 45 48	Dr. Kampf	Lieut. Birnie	Branch of.

Geographical positions, from sextant-observations, &c.—Continued.

Year.	Station.	Atlas-sheet No.	Objects observed.	Latitude.	Altitude above sea-level.	Variation of needle. E.	Observer.	Computed.	Remarks.
				° ′ ″	Feet.	° ′ ″			
1875	Kern River, Cal	65 C	α Lyræ, α Andromedæ, ε Pegasi, Polaris	36 10 29.10	7578.8	....	Dr. Kampf	Lieut. Birnie	West Fork.
1875	Kernville, Cal	65 C	α Andromedæ, α Lyræ, ε Pegasi, Polaris	35 45 3.30	2707.7	16 43 00	Dr. Kampf	Lieut. Birnie	Near post-office.
1875	La Motte's, Cal	65 C	α Coronæ, α Aquilæ, Polaris	35 53 15.90	6460.8	15 51 00	Dr. Kampf	Lieut. Birnie	
1875	Lime Creek, Colo	61 C	Sun	37 49 28.90	9581.2	....	Lieut. Marshall	Lieut. Birnie	
1875	Little Lake, Cal	65 C	α Andromedæ, α Lyræ, ε Pegasi, Polaris	35 55 56.96	3086.4	14 45 18	Lieut. Birnie	Lieut. Birnie	
1875	Lone Pine, Cal	65 A	α Lyræ, ε Pegasi, Polaris	36 35 58.68	3810.1	15 19 42	Lieut. Birnie	Lieut. Birnie	North Plaza.
1875	Los Angeles, Cal	73 C	α Lyræ, α Boötis, Ursæ Majoris, Polaris	34 03 34.41	325.6	14 30 30	Lieut. Wheeler and Lieut. Birnie,	Lieut. Birnie	Rendesvous Camp.
1875	Menatchey Valley, Cal	65 C	α Coronæ, ε Pegasi, α Aquilæ, Polaris	36 12 37.90	9503.2	15 16.24	Dr. Kampf	Lieut. Birnie	
1875	Nadeau's Station, Cal	65 C	ε Pegasi, α Ceti, β Ceti	35 29 10.11	2294.4	....	Lieut. Birnie	Lieut. Birnie	Red Rock Cañon.
1875	Oasis Valley, Nev	65 B	α Coronæ, α Pegasi, α Aquilæ, Polaris	36 53 02.15	3085.2	15 20.24	Lieut. Birnie	Lieut. Birnie	Springs, south end of.
1875	Olancha, Cal	65 C	α Boötis, α Cygni, β Ophiuchi, Polaris	36 16 51.79	....	14 54.12	Lieut. Birnie	Lieut. Birnie	Post-office.

33

1875			Stars	Latitude			Observer	Computer	Remarks
Olancha Peak, Cal	65	C	α Coronæ, α Pegasi, α Aquilæ, α Polaris	36 14 44.90			Dr. Kampf	Lieut. Birnie	Foot of.
Olancha Peak, Cal	65	C	α Coronæ, ε Pegasi, α Aquilæ, α Polaris	36 16 12.70	8743.9	16 07.18	Dr. Kampf	Lieut. Birnie	Northwest side of.
Oso Meadows, Cal	65	C	α Andromedæ, ε Lyræ, ε Pegasi, α Polaris	36 11 48.60	5981.7	15 45.00	Dr. Kampf	Lieut. Birnie	West mesa.
Pino, N. Mex	78	A	α Aquilæ, α Arietis, ε Pegasi, α Polaris	35 16 33.40	5504.4		Lieut. Morrison	Lieut. Birnie	Where the Ute trail comes down.
Rio Grande, Colo	61	C	α Andromedæ, Sun, α Polaris	37 48 37.00	11660.3		Lieut. Marshall	Lieut. Birnie	Camp 110.
Rule Creek, Colo	62	D	α Aquilæ, α Arietis, ε Pegasi, α Polaris	37 54 17.00			Lieut. Morrison	Lieut. Birnie	Camp 109.
Rule Creek, Colo	62	D	α Aquilæ, α Arietis, ε Pegasi, α Polaris	37 35 09.00			Lieut. Morrison	Lieut. Birnie	
Saguache, Colo	61	D	α Cygni, α Boötis, α Polaris	38 04 58.90	7723.1		Lieut. Marshall	Lieut. Birnie	Death Valley.
Salt Wells, Cal	65	D	α Coronæ, ε Pegasi, α Aquilæ, α Polaris	38 44 16.15	117.1		Lieut. Birnie	Lieut. Birnie	Las Valles de.
San Augustin, N. Mex	78	A	ε Lyræ, ε Pegasi, α Polaris	35 30 35.30	6002.0		Lieut. Morrison	Lieut. Birnie	Springs.
Say-qui-ta, Cal	65	B	Equalaltitude ☉'s upper limb, ☉'s upper limb.	36 55 01.50	5553.9		Lieut. Birnie	Lieut. Birnie	
Silver Springs, Cal	65	D	α Coronæ, ε Pegasi, α Aquilæ, α Polaris	36 14 02.50	4110.2		Lieut. Birnie	Lieut. Birnie	
Stone's Ranch, N. Mex	78	A	α Arietis, ε Pegasi, α Polaris	35 21 43.30			Lieut. Morrison	Lieut. Birnie	

Geographical positions, from sextant-observations, &c.—Continued.

Year	Station	Atlas-sheet No.	Objects observed	Latitude	Altitude above sea-level.	Variation of needle. E.	Observer.	Computer.	Remarks.
				° ′ ″	Feet.	° ′ ″			
1875	Tehachipai Valley, Cal	73 A	α Andromedæ, α Lyre, α Pegasi, Polaris	35 06 49.80	3830.6	14 12.00	Lieut. Whipple	Lieut. Birnie, Lieut. Macomb	
1875	Trout Meadows, Cal	65 C	α Coronæ, α Pegasi, α Aquilæ, Polaris	36 01 58.90	5998.3	16 28.48	Dr. Kampf	Lieut. Birnie	
1875	Twenty Mile Station, Cal	65 D	Bois, α Cygni, β Oph nddi, Polaris	36 01 10.40			Lieut. Birnie	Lieut. Birnie	Shepherd's Cañon.
1875	Uhl's Ranch, Cal	65 C	α Andromedæ, α Lyre, α Pegasi, Polaris	35 52 09.90	9661.5		Dr. Kampf	Lieut. Birnie	
1875	Unaweep, Colo	61 C	α Aquilæ, Polaris	38 07 48.50	7475.8		Lieut. Marshall	Lieut. Birnie	
1875	Ute Creek, N. Mex	70 C	α Aquilæ, sia, α Pegasi, Polaris	36 16 31.50			Lieut. Morrison	Lieut. Birnie	Upper.
1875	Walker's Creek, Cal	65 C	α Pegasi, α Aquilæ, Polaris	35 45 17.90	3937.5		Dr. Kampf	Lieut. Birnie	
1875	Weldon, Cal	65 C	α Bootis, α Andromedæ, α Aquilæ, α Pegasi, Polaris	35 40 03.70	2716.9	16 13.18	Dr. Kampf	Lieut. Birnie	Nichols's Ranch.
1875	Whitney, Cal	65 C	α Coronæ, α Pegasi, α Aquilæ, Polaris	36 29 30.30	11051.2	15 34.48	Dr. Kampf	Lieut. Birnie	Foot of mount.
1875	Whitney Meadows, Cal	65 C	α Andromedæ	36 25 44.66	9371.4		Lieut. Birnie	Lieut. Birnie	

1875	Wild Rose Spring, Cal	65 D	α Coronæ ....; α Pegasi ......; α Aquilæ ......; α Polaris	30 15 54. 37	4663. 4	15 19. 12	Lieut. Birnie	Lieut. Birnie .....
1875	Willow Spring, Cal	73 A	α Andromedæ ...; α Lyræ ....; ε Pegasi ....; Polaris	34 59 38. 70	9530. 8	....	Lieut Whipple	Lieut. Birnie ......; Lieut. Macomb ...

GEODETIC AND TOPOGRAPHICAL.

Measured and developed bases were laid out at Los Angeles, Weldon, and Panamint Valley, in the California section. None were found necessary or requisite in the Colorado section.

The sketch herewith shows the progress made in the development and number of the secondary triangulation belts in the Colorado section, and will at an early day be supplemented by a map showing the development of triangles from the Los Angeles base to the eastward. Triangulation observations have been made by each one of the moving field-parties over large areas in California, Nevada, Colorado, and New Mexico.

A list of geographical positions, other than those given in the present report, will appear in the list of astronomical positions in Volume II; other positions in the appendix of part second, Volume II, giving altitudes of prominent positions, and the remainder in Volume I of the survey publications.

The areas in Colorado, New Mexico, California, Nevada, and Arizona were occupied in accordance with the project submitted to and approved by the Chief of Engineers and Secretary of War.

Many connections have been made with the stakes and other marks of the public-land surveys. The degree of progress in topographical accuracy of delineation is one of the notable features of the year. The outgrowth from the surveys of the parties under Lieutenant Marshall admitted of the construction of a special sheet of the San Juan mining region, on a scale of 1 inch to 2 miles, being an independent map of one-fourth the area of the southwest quarter of atlas-sheet 61, which area has also been reproduced and published, on a scale of 1 inch to 4 miles, during the year. The mountainous portions of the country having intricate drainage areas are now all delineated and published, upon a scale of 1 inch to 4 miles, while plateau and semi-desert areas are drawn upon a scale of 1 inch to 8 miles. The larger scales are also susceptible of reduction to the uniform scale proposed for the atlas of the entire section west of the one-hundredth meridian. Many monuments have been built, and the positions occupied are susceptible of identification and further use in connection with a extended scheme of triangulation, or for the purpose of fixing accurately the positions of minor points that may spring into importance in succeeding years.

ROUTES OF COMMUNICATION.

Tables of distances and road-profiles have been computed over the principal traveled routes and important trails in those portions of Colorado and New Mexico, Southern Califorinia, Utah, and Arizona over which the surveys have extended. These are based upon odometer and barometer measurements of the survey. In addition to the simple distances, the facilities for camping are noted.

Such tables, together with material gathered in the field-season of 1876, will furnish the basis of an extended set of tables, and supply to a great extent a need long felt for authoritative distances over a country where the distribution of water is such that the traveler needs to know where he may find wood, water, and grass or grain to guide him in his selection of routes and camps. These tables, with profiles of the more important routes, it is proposed to publish in separate form at an early day, as stated in report of last year.

PROGRESS-MAP.

The progress-map herewith shows by colors the different stages of prosecution of the map results. Appropriations being small and the season limited, only a small expedition comparatively can be put in the field.

ITINERARY.

The following is a brief *résumé* of the result of such observations bearing upon the characteristics and natural resources of the section visited as I was able to make while in charge of main party No. 1 of the California section during the field-season of 1875.

The routes pursued were along the regularly-traveled road from Caliente to Los Angeles, through Tehachipi Pass, across the northwestern arm of the great outlier of the Mohave Desert, via Elizabeth Lake, &c., from Los Angeles to Santa Monica and Wilmington on the coast; thence via Cahuenga Pass, through San Fernando Valley and the valley of Santa Clara, to old Fort Tejon and its neighborhood, and thence by the southern end of Tulare Valley to Caliente. Besides the routes above mentioned, that from Los Angeles, via Cajon Pass to Camp Mohave, had been traversed in the latter part of the year 1868, from which a somewhat clear idea of the characteristics of that portion of Southern California adjacent to these lines could be formed.

The results of the operations of the other parties of the California section of 1875 are set forth in the various reports submitted by the executive officers. (See reports of Lieutenants Bergland, Birnie, and Whipple, also that of Dr. Oscar Loew.)

Los Angeles is at present the principal point for commercial exchange in the county of that name, and is favorably located in the center of a part of the somewhat level arable territory south of the passes through the Southern Sierras, west of the coast range, forming a basin of considerable dimensions most favored by nature of any of the so far developed portions of Southern California.

Its outgoing connections are by the Southern Pacific Railroad north to the San Fernando Mountains, through which a tunnel (being a function of said road) is being constructed, that, upon completion, will connect it with the southern terminus of that branch of the same road now traversing Tulare Valley, and through Tehachipi Pass to the arm of the desert, thence via Soledad Pass to the northern entrance of the San Fernando Tunnel, thus making a connection that, once accomplished, will stand as a landmark of the skill and energy of these railway constructions.

The southern section of this same road is being built via San Gorgonio Pass toward the desert, in direction of Fort Yuma, with great vigor. Other branches of the same corporation lead direct to Wilmington, with a branch to Anaheim, being part of the route bordering upon the coast already reconnoitered to the southward as far as San Diego. The Los Angeles and Camp Independence Railroad, under the auspices of a corporation formed in pursuance of the laws of the State of California, is built from Santa Monica a point a little south of west of Los Angeles to this city, with the expectation of being pushed eastward through the Cajon Pass in the direction of the mining-districts in vicinity of Death Valley and Camp Independence.

It will be seen that the outgoing connections from Los Angeles are plentiful, and that it is likely to become a prominent point upon the through line of transit so soon as the Southern Pacific shall have connected its line of road from San Francisco to the Colorado River. The distance from Los Angeles to Santa Monica by rail is 12½ miles, and at the latter point a little town has sprung up, consequent upon the advantages of commercial relations, transit of supplies, &c., from the north, the evidences of a good beach, convenient as a resort for the people of this section and such others as may be desirous of taking advantage of the more genial climate of this part of the California coast. Wilmington, until lately the point from which military supplies have been forwarded into Arizona, is at the head of a little lagoon or estuary not navigable for large crafts, at the lower part of which estuary, by means of a breakwater lately built by and under the direction of officers of the Engineer Corps, after the plans of Lieut. Col. B. S. Alexander, Corps of Engineers, and in charge of Maj. G. H. Mendell, with Lieut. Clinton B. Sears as assistant, has been partially opened for navigation, so that several landings may be effected near the old town of San Pedro, to which it is understood the line of railroad will be carried at an early day.

The route outward from Los Angeles was via the Cahuenga Pass leading into the San Fernando Valley, or plains, as they have been termed, a large grazing section, on the stage-road leading hence to Santa Barbara, an oasis in its way facing the broad valley. The buildings lie nestled in groves of magnificent oaks, the soil itself being a heavy, rich, dark loam, evidently capable of producing most luxuriant crops, could sufficient water for irrigating purposes be had. It is not unlikely that, in some of the ravines leading back into the foot-hills, the search for artesian water would prove successful, and that it would be found sufficient for the successful irrigation of a large part of the valley.

Across the valley brings us to the old Mission of San Fernando, which was described by the earlier travelers to this region, near which has been built a little valley town, by the name of San Fernando, at the present terminus of the line of rail leading from Los Angeles to this point. The remains of the old conduits for irrigation and other hydrographic purposes, made by the Jesuit fathers with the aid of Indian labor, are still visible, and attest the degree to which agriculture was carried in those days by this rude kind of labor, guided by the padres, and it is not difficult, after a closer inspection of the remains, to believe in the tales of wonderful fertility of the fields well watered and covered with foliage, and of the large herds of cattle and sheep reputed to have been the property of these early missionaries.

The whole valley of San Fernando is one level plain of large extent, of a strong, naturally productive soil, as is evinced by the luxuriant growth of herbage of varying size, that needs only the advent of a proper amount of water to bring it into the condition of a garden. Much water falls in the neighboring mountains, but it mostly passes out of sight, except in the seasons of freshet, and under the beds of alluvium to the drainage of Los Angeles River.

The Big and Little Tujunga and Pacoima Creeks are all streams of considerable size, varying in different seasons, but soon sink upon reaching the plain. The storage of these waters in the vicinity of the mouths of the cañons through which they debouch would serve to act as reservoirs sufficient in size and extent to hold the necessary supply to make fertile all the outlying lands, of which there are at least 150,000 acres cultivatable by irrigation. In this valley alone were seen only two ranches, where but a few acres of land are cultivated. Fields of wheat barley, and oats were noted, lying between the two, where, without the aid of irrigation, crops have been produced in favorable years. The route from San Fernando north takes one across a range of this name into Santa Clara Valley, the axis of which, bearing north in its eastward course, connects the coast with the desert and the interior section leading toward the Colorado River. The Soledad Pass at its head, the summit of which is 3,332 feet above the level of the sea, is the lowest connection between this portion of the southern coast of California and the interior, being the lowest summit of the main coast range, and, indeed, the lowest summit found between the drainage reaching the Pacific from the Columbia and the basins north of the fortieth parallel to the Mexican border upon the Pacific slope. It is unfortunate that no good harbor exists near this point from the Pacific coast. Were such the case, we might easily expect to see another large city like San Francisco spring up at an early day.

Agricultural facilities are being utilized ; new mines in the interior are discovered from time to time ; grazing has long been notable in this section of the State, and as its climatic status is becoming better understood, it, with other points of interior California, are likely rapidly to advance and become from year to year more thickly settled.

The route to the northward crosses the divide near the vicinity of Elizabeth Lake ; thence, continuing northward, follows the edge of the northwestern arm of Mohave Desert, reaching, via Lievre ranch, the southern end of the Canada de las Uvas, or one of the passes between the Colorado drainage and that of the Sacramento via the great Tulare Valley. The profile of this pass is much gentler than that of Tejon or Tehachipi Pass, and could be made use of for a railroad leading to the Pacific coast near the thirty-fifth parallel, the same to be carried northward, with a terminus at San Francisco. It is believed that no practicable profile so far has ever been found for a railroad north of the Santa Clara Valley, because of the impracticable ridges of the coast range immediately encountered in going north. This fact may prove of consequence in connection with some later railroad undertaking, for a pass once entered and held determines for the parties in possession that no near competition can be effected, as no parallel road could naturally, with any reasonable expense, be constructed and maintained.

The springs and little creeks running part or all the year on the southern side of this cañon have, of late years, as is reported by settlers in this region, grown less and less, until many of them have entirely disappeared.

Lievre ranch itself, an old Spanish grant, reputed to have been one of the finest in this section of the country, has along its southeastern borders become entirely covered with beds of drifting sand, covering up and killing out the annual grasses of this region, and adding to the area of desolation commanded by the outlying desert.

Old Fort Tejon, once the scene of military operations against the Indians of this quarter, was reached during our trip, where we found the old adobe buildings still standing in fair repair. It is on a level spot, but geographically not well situated, being in the jaws of a side cañon entering the main cañon of the creek upon which the post was situated, from which the lives of the garrison and the safety of the animals could at any time have been easily jeopardized. The elevation, being 3,245.7 feet above the level of the sea, afforded an agreeable climate as compared with that in the great Tulare Valley, that lies to the northward, the elevation of which (Bakersfield) is 465 feet. The reservation has now passed into the hands of private parties, and is used for stock-raising.

The outlying mountains show mineral wealth, gold, silver, lead, and antimony, although little prospecting and less developing has been done. It may be said that very few of the mountain ranges in the interior of California have been fully and fairly prospected. Prospecting-parties usually traverse well-known routes or trails, and have in view certain objective points, to which their attention is entirely directed. Failing in good results there, they usually return to their old stamping-ground and wait for an occasion to call them in another direction. It need not astonish any one to hear of most remarkable mineral discoveries in the coast range and Sierra Nevada for many years to come, or if further developments upon mining-claims taken up heretofore and abandoned as worthless shall prove the contrary. The creek, the main source of which is the one that flows by old Fort Tejon, thence down the cañon toward the great plains, and through which until the railroad had reached Caliente, on Tehachipi Creek, the stage-line to the south had usually passed. The road from the mouth of this creek to Caliente passes a little north of Tejon ranch, and likewise of the old Indian reservation, which has now passed into private hands, and traverses the passes of a series of foot-hills of remarkably pleasant contour, well grassed, and covered irregularly with copses of oak and pine. Although brown and seared at the date of our visit, still the landscape presented was one not easily to be forgotten.

Buena Vista Lake, lying in the great Tulare plain in front of us, acts as the reservoir of Kern River, evidently like Lake Tulare, which during seasons of great rain-fall and the spring melting of snows overflows its banks. Tulare Lake, however, as a reservoir will reach approximately to 100 square miles in size, and loses by evaporation throughout the months of the year characterized by non-precipitation or by a minimum rain-fall all that is obtained from the regular river source of supply that reaches it. I am informed that it was found by Maj. R. M. Brereton, late of the royal engineers, that the average amount of annual evaporation from its surface was approximately 6 feet. The question of irrigation has of late years been agitated in the State of California by the gentleman to whom allusion has been made and others, and his aid is likely to promote the same by his engineering skill and experience in works of this class in India. The lines of proposed canals and opportunities for utilizing water-supply gathered in the various basins, especially those upon the western flanks of Sierra Nevadas, can be found in the report of the commissioners upon irrigation of the great valleys of of California. (See Executive Document No. 290, 43d Cong., 1st sess.)

NATURAL RESOURCES.

As time has permitted, a number of the assistants have been employed in laying down upon the preliminary maps and in tabulating the areas whose natural resources permit of use for agricultural, mining grazing, and timber purposes as in contradistinction to those absolutely worthless, being arid and barren. The table, with remarks, is submitted herewith, and it is proposed to gather complete statistics upon these important points in all the areas traversed hereafter, and as time shall permit to make comparison of larger areas after material shall have been worked up in connection with the surveys of prior years. A number of colored maps, graphically illustrating the same, can then be prepared.

In addition to the determination of the natural resources of the sec tion surveyed, as denoted by its superficial area, it is proposed, as far as practicable with the force at the disposal of the officer in charge, to have noted the prominent streams, lakes, and springs, the area of basin-drainage in which they are situate, with an accurate delineation of the perimeters of the same, and, when practicable, the determination of the rain-fall for a period, or annually, when it shall have been observed, and further, what bearing this practical information shall have upon the subject of irrigation by canals, reservoirs, or artesian wells within any of the basins visited.

Atlas-sheet number.	Total area.	Agricultural, irrigable, and arable.	Timber.	Grazing.	Barren.	Land available for agriculture, timber, or grazing.	Remarks.
59 (D.) Central Colorado.	4,584.8 square miles. 2,703,872 acres.	Zero.	57 per cent. 2,406 square miles. 1,541,307 acres.	38 per cent. 989 square miles. 594,559 acres.	31 per cent. 887 square miles. 567,813 acres.	79 per cent. 3,337.8 square miles. 2,136,059 acres.	
61 (B.) Central Colorado.	4,974.8 square miles. 3,735,872 acres.	3 per cent. 138 square miles. 89,076 acres.	61 per cent. 2,608 square miles. 1,668,881 acres.	17 per cent. 797 square miles. 465,099 acres.	19 per cent. 819 square miles. 519,817 acres.	81 per cent. 3,469 square miles. 2,216,055 acres.	
61 (D.) Central Colorado.	4,394 square miles. 2,767,360 acres.	37 per cent. 1,600 square miles. 102,392 acres.	27 per cent. 1,167 square miles. 747,187 acres.	31 per cent. 908 square miles. 581,145 acres.	15 per cent. 648 square miles. 415,104 acres.	65 per cent. 3,676 square miles. 2,352,356 acres.	
69 (C.) Central Colorado.	4,394 square miles. 2,767,360 acres.	4 per cent. 173 square miles. 110,694 acres.	43 per cent. 1,859 square miles. 1,189,964 acres.	52 per cent. 2,249 square miles. 1,439,007 acres.	1 per cent. 43 square miles. 27,673 acres.	99 per cent. 4,351 square miles. 2,739,687 acres.	
61 (C 1.) Central Colorado.	1,098.3 square miles. 702,912 acres.	½ per cent. 5.49 square miles. 3,515 acres.	47 per cent. 516.2 square miles. 330,369 acres.	9½ per cent. 104.34 square miles. 66,779 acres.	43 per cent. 472.26 square miles. 302,249 acres.	57 per cent. 625 square miles. 1,406,053 acres.	
65. Eastern portion Eastern California and Southern Nevada.	Partial area 10,283 square miles. 6,581,120 acres.	2 per cent. 205.66 square miles. 131,622 acres.	6 per cent. 617 square miles. 394,967 acres.	88 per cent. 9,049 square miles. 3,931,360 acres.	4 per cent. 411 square miles. 263,344 acres.	96 per cent. 9,872 square miles. 317,876 acres.	
69 (A.) Southwestern Colorado and Northern New Mexico.	Total area 4,372 square miles; surveyed, 2,920 square miles. 2,798,208 acres.	15 per cent. 438 square miles. 280,390 acres.	44 per cent. 1,295 square miles. 729,372 acres.	41 per cent. 1,197 square miles. 766,108 acres.	Zero.	100 per cent.	
69 (B.) Southern Colorado and Northern New Mexico.	4,372 square miles. 2,798,208 acres.	5 per cent. 219 square miles. 130,904 acres.	60 per cent. 2,623 square miles. 1,678,846 acres.	99 per cent. 4,328 square miles. 2,770,397 acres.	1 per cent. 44 square miles. 279,810 acres.	99 per cent. 4,328 square miles. 2,770,397 acres.	Timber land is also grazing land.

68 (C.) Northwestern New Mexico.	4,419 square miles. 2,828,160 acres.	½ of 1 per cent. 11 square miles. 7,040 acres.	25 per cent. 1,105 square miles. 707,040 acres.	75 per cent. 3,314 square miles. 2,121,120 acres.	Zero.	100 per cent.	Timber land is also grazing land.
69 (D.) Central New Mexico.	4,419 square miles. 2,828,160 acres.	9 per cent. 398 square miles. 254,720 acres.	60 per cent. 2,651 square miles. 1,696,640 acres.	99 per cent. 4,375 square miles. 2,800,000 acres.	1 per cent. 44 square miles. 28,381 acres.	99 per cent. 4,375 square miles. 2,709,879 acres.	Timber land is also grazing land.
70 (A.) Northern New Mexico and Southern Colorado.	4,372 square miles. 2,798,906 acres.	9 per cent. 394 square miles. 251,827 acres.	62 per cent. 2,710 square miles. 1,734,309 acres.	99 per cent. 4,328 square miles. 2,770,327 acres.	1 per cent. 44 square miles. 27,361 acres.	99 per cent. 4,328 square miles. 2,770,327 acres.	
70 (C.) Central New Mexico.	4,419 square miles. 2,828,160 acres.	4 per cent. 177 square miles. 113,125 acres.	30 per cent. 1,326 square miles. 848,640 acres.	66 per cent. 2,916 square miles. 1,866,565 acres.	Zero.	100 per cent.	
75 Central and Western Arizona.	17,954.6 square miles. 11,490,944 acres.	25 per cent. 4,488 square miles. 2,875,238 acres.	10 per cent. 1,795 square miles. 1,149,094 acres.	30 per cent. 5,386 square miles. 3,438,289 acres.	35 per cent. 6,284 square miles. 4,021,550 acres.	65 per cent. 11,670.6 square miles. 7,469,114 acres.	
Totals.	100 per cent. 71,404.5 square miles. 45,698,880 acres.	11.5 per cent. 8,236.5 square miles. 5,271,360 acres.	31.7 per cent. 22,670 square miles. 14,506,800 acres.	55.7 per cent. 39,810 square miles. 25,478,400 acres.	13.5 per cent. 9,690 square miles. 6,201,600 acres.	86.5 per cent. 61,714 square miles. 39,497,280 acres.	Timber land in sheets 69b, 69d, and 70a is also grazing land.

52 D.—Almost the entire section is mountainous, the elevation being above 8,000 feet; it is too high for agricultural land.

The *principal valleys* are South Park, altitude from 9,500 to 10,000; Upper Arkansas Valley, 9,000 to 10,600; Blue River basin, 8,700 to 10,600; Ten-mile Creek and Blue River Valley, 7,800 to 11,000 feet.

The *timber* found up to timber-lines on all the mountains is spruce and pine; it is good for lumber. That along the streams is mainly cottonwood.

The *grazing* in the valleys is very good. It is usually necessary, however, to seek the lower lands for wintering cattle.

The land is mainly valuable for its *mines*. In the vicinity of Georgetown, Bakersville, and Gray's Peak, in northeastern part of sheet, silver-mines are worked. Silver-mines of Mount Lincoln and gold placer-mines near Granite Post-Office, and the Fairplay and Alma placers, are in the central part of sheet. In the western part of sheet are gold-mines on tributaries of Tennessee Creek. The water-supply, mainly of mountain-streams, is very good.

61 B.—Constituting a portion of Central Colorado. The greater portion of the sheet is mountainous, very little being under 7,000, in places running up to over 14,000 feet. The *principal valleys*, those in which the agricultural land is situated, are South Park, on the Platte, 8,500 to 8,800 feet; Upper Arkansas, 7,300 to 8,800 feet; Pleasant Valley, 7,000 feet; San Luis Valley, 7,700 to 8,500 feet; and Tumichi Valley, 7,500 to 8,000 feet. Crops are more or less uncertain, owing to the altitude.

In the northeastern portion of the sheet, and on the lower slopes of the mountains, throughout its whole extent, the *grazing* is good.

The *timber* in Saguache, Elk, and Sangre de Cristo ranges is pine and spruce, suitable for lumber. The scattered timber on foot-hills is mainly piñon, suitable only for fuel.

The *mines* at head of Chalk Creek, head of San Luis Valley, are silver and lead. Gold is found in place and placer-mines about Upper Arkansas, also in Union Park at head of Gunnison River; salt-wells at head of Salt Creek.

The *water*-supply is very good, from the South Platte, Arkansas, Gunnison, and San Luis Creek, with their many branches.

The *barren land* is above timber-line and along sides of cañons, where the bare rocks are developed.

61 D.—Half-mountainous. *Timber* on western rim of San Luis Valley. Piñons near valley on western foot-hills. Elsewhere spruce and pines. Large cottonwoods along Rio Grande, Alamosa, and La Jara Creeks. Sangre de Cristo range within limits of sheet, well wooded to timber-lines.

Agricultural lands on Rio Grande below mouth of South Fork, and in San Luis Valley along all running streams, also on Upper San Juan. Elsewhere land too high for cultivation.

Grazing lands.—San Luis Valley; hills of western boundary of San Luis Valley; on headwaters of the Cochetopa; on Upper San Juan. Large part of San Luis Valley covered with sage, but susceptible of irrigation.

Mines.—Summit and Decatur districts, on headwaters of Alamosa Creek, near divide between Rio Grande and San Juan waters. Gold and silver, principally the former. Gold found in Sangre de Cristo range; (not worked extensively.)

Barren land lies above timber-lines.

62 C.—The sheet lies in Central Colorado. In the western part the Sangre de Cristo, Greenhorn, and Culebra ranges limit the great plain extending eastward from their base. The *cultivatable land* is found in the valleys of the southern tributaries of the Arkansas, the Saint Charles, Greenhorn, Apache, Huerfano, Cucharas, Santa Clara, and Purgatoire; also along the Ute, Sangre de Cristo, and Trinchera Creeks. There is heavy pine and spruce *timber* on the mountains.

The *grazing land* in the mountain-valleys is very good, on the plains but indifferent, but not so poor as to be called barren. The little barren land of the sheet is above timber-line.

The *Rosita silver*-mining district is rapidly developing. Silver has also been found in Spanish Peak. Gold is found in the Sangre de Cristo range, iron on Grape Creek, and good coal in northwestern part of sheet.

The *water*-supply in the mountains is good, and on the plains sufficient for grazing lands, with enough for irrigation along the streams.

61 C, sub.—Central Colorado. Mountainous, except small tracts, perhaps 3 per cent. of level ground.

Timber.—Area well timbered with spruce upon the mountain-sides, and below 8,500 feet with large yellow pines and cottonwood. Scrub-oak, reaching 20 feet in height and 10 inches in diameter, found on Animas and Uncompahgre.

Agricultural lands are situated in the Uncompahgre Valley in north section of area, and on Lake Fork of the Gunnison. Good crops of wheat, corn, and vegetables may be produced in these small valleys.

Mines.—Extensive deposits and veins of gold, silver, lead, copper, manganese, and iron exist. Gold, silver, lead, and copper found in Upper Animas.

On headwaters of Uncompahgre, silver, lead, copper, and iron.

On headwaters of Lake Fork of the Gunnison, silver, lead, copper, no iron.

On headwaters of Dolores, galena-ores.

On headwaters of La Plata, gold veins, and below the mountains an extensive placer. Limestone for flux is abundant; also in the Uncompahgre a fine flux in the shape of fluorspar is found.

Grazing lands.—Uncompahgre Park, Unaweep Valley in the north. The valley of the San Miguel to the west. The valleys and the rolling hills along the Animas on the south, and in the valley of the Rio Grande. Of these the San Miguel and Rio Grande Valleys are too high for winter grazing, reaching above 8,000 feet.

The *barren land* lies above timber-line, and along sides of cañons in inaccessible places.

65.—The sheet takes in a portion of Eastern California and Southern Nevada. That portion of the atlas sheet considered, lies east of the main range or Sierra Nevada. With their axis parallel to that of the main range, run the Grapevine and Funeral mountains: the Panamint and Telescope ranges; the Inyo, Cerro Gordo, Darwin, and Coso ranges. On all of these, excepting the northern portion of Panamint and Funeral mountains, and on the lower hills, piñon suitable for fuel only, is found. In the vicinity of Mazurka Cañon there is pine timber.

The *agricultural* districts lie mainly in Owens River Valley, with occasional minor tracts about springs and in Oasis Valley.

The *grazing* land extends off from the bases of the mountains limited by the lower valley land, such as Death Valley, Salinas and Butte Valleys.

Mines have been worked. Silver in the Panamint, Rose Spring, Lee, Cerro Gordo, Inyo mountain districts, and gold and silver in the Sherman, Lookout, and the district near Camp Independence. Silver and copper in the Ubahebe district, and gold in the old Coso and Owens Valley districts.

The *barren land* lies in the low valleys.

69 A.—This sheet covers a portion of Southwestern Colorado and Northwestern New Mexico. A little over one-half of the sheet has been surveyed.

The lands capable of *irrigation* lie along the San Juan River, and its branches coming from the north flowing from the San Juan Mountains, viz, the La Plata, Los Animas, Florida, Los Pinos, Piedra Blanca, and Navajo. The grazing land extending back from the valleys into the foot-hills; much of the grazing, particularly in the southern portion of the sheet, is very poor.

Timber found on the mesas and low hills is mainly piñon and cedar, of little value for lumber, but fine fuel.

Coal is found near Laguna Hedionda.

69 B.—This sheet covers a portion of Southern Colorado and Northern New Mexico, with the Rio Grande running through the eastern portion, *irrigable* land is found in the northeastern portion along the valley of this river; also the valleys of its tributaries, the Conejos, La Jara, the Culebra, Costilla, Colorado, Cristobal, and Montes, up the San Antonio and Los Pinos Creeks; in the western part on the Navajo, Blanca-Chama and its branches, the Brazos, Nutritas, and Cangilon. Scant grazing found in the San Luis Valley, and very fine *grazing* in the spurs of the San Juan range and in the Rio Chama drainage-basin.

Fine pine timber and cottonwood found in the spurs of the San Juan.

Indications of silver in same; coal is found in southwestern part, near Laguna Hedionda.

69 C.—This sheet constitutes a portion of Northwestern New Mexico. It is very dry, almost entirely destitute of permanent water. In the eastern part the headwaters of the Puerco of the East give enough water for a little irrigable land, and at Ojo Nuestra Senora, in the center of the sheet, there is about a square mile of good land.

The country is not entirely barren, but the grazing is by no means good.

The timber is small, being piñon. Along the Chaco, water can be had by digging wells. Lignite is found along the Chaco and at the head of Cañon Blanco.

69 D.—In North Central New Mexico. The sheet is cut through by the Rio Grande, giving much irrigable land in its valley and those of its tributaries, Taos Creek, Embuda Creek, Cañada, Pojuaque, and Santa Fé on the east, Jemez and Chama, with its branches, Ojo Caliente, El Rito, Cañones, Cangilon, and Coyote on the west. Up from these, extending into the timber of the mountains, there is fine grazing.

On the Valles Mountains and Santa Fé and Vegas ranges fine spruce and pine timber is found. On the foot-hills there is piñon.

Indications of silver and gold are found in the Santa Fé and Las Vegas ranges, although there are no worked mines. Lignite is found on Santa Fé Creek. Copper is found in the Gallinas Mountains, also north of Abiquiu.

There is little barren land excepting that above timber-line.

70 A.—Sheet covers a portion of Northern New Mexico, and Southern Colorado. It is well *watered* by the Purgatoire, the Canadian, with its branches, Willow, Vermejo, Van Brummer, Ceroso, Ponil, Cimarron, Moreus, and Cineguilla Creeks on the east,

and the Culebra and Costilla on the west. In the valleys of all these, good irrigable land is found.

Grazing land extends over nearly the whole section, the grass being very fine.

Timber is pine and spruce in the mountains, and piñon and cedar on the low hills.

In Moreno and Ute Valleys place and placer gold-*mines* are found. Silver found on Colorado Creek in southwestern part of sheet. Gold and silver on Baldy Peak in Taos Mountains. Coal and iron found near Trinidad, and lignite croppings from Elizabethown to Trinidad in all cañons.

The little *barren land* lies above timber-line in the mountains.

70 C.—Situated in North Central New Mexico. The western third is mountainous. It is well *watered* by the Canadian and its branches, Cimarroncito, Rayado, Ocate, Sweetwater, and Mora, with its branches, the Sapello, Manuelitos, Cebolla, and Coyote, in the valleys of all of which is irrigable land capable of growing crops for three times the present population.

The *grazing* is wonderfully good, winter and summer, throughout the whole area excepting a little above timber-line.

Pine and spruce *timber* is found in main range and Turkey Mountains.

There are indications of *silver, gold,* and copper throughout the range, but no worked *mines.*

Sheet 75.—Parts of Central and Western Arizona. Much broken by low mountains and mesas with no well-defined ranges. There are no long watersheds to make large streams. The *water-supply* is not great. The Colorado Chiquito in the northeast runs through a cañon.

The *agricultural* land is found mainly about small springs in vicinity of San Francisco Mountains, in Chino and Williamson's Valley of the Rio Verde drainage, in vicinage of Camp Verde, about the heads of Bill Williams Fork, and on Hassayampa Creek and old Camp Date Creek and Skull Valley.

Pine and piñon *timber* is found on the mountains and piñon on the lower hills.

Grazing found in nearly all the valleys excepting in southwestern and northwestern part, where, as well as on the mountain-tops, it is arid.

Place and placer-*mines* of gold and deposits of silver are found in the Bradshaw Mountains.

BAROMETRIC HYPSOMETRY.

Lieutenant Marshall submits a report on this subject showing the operations carried on during the year, which have been principally with a view to the determination of altitudes alone, a large number of which have been fixed by both cistern and aneroid barometric measurements. A new set of instructions, enlarged, revised, and improved by Lieutenant Marshall during the year, have been published, and are sufficient for a complete manual that will answer for a long time to come. (See Appendix A.)

Dr. O. Loew presents a report on the meteorological condition of the Mohave Desert, and compares the results of his observations with those made in other countries.

The widely different climatic conditions existing in a comparatively limited area in Middle and Southern California make the observations of Dr. Loew of considerable value and interest. (See Appendix B.)

NATURAL HISTORY.

Geology.—Prof. Jules Marcou submits a report on the geology of a portion of Southern California; Dr. O. Loew, on the geological and mineralogical character of Southern California and adjacent regions; and Mr. A. R. Conkling, on portions of New Mexico and Colorado.

The field was not an entirely new one to Professor Marcou, he having visited the region in connection with the surveys for a Pacific railroad along the thirty-fifth parallel; yet his report herewith will be found to contain new and valuable information. (See Appendix H 1.)

Dr. Loew was for most of the season attached to the party of Lieutenant Bergland, operating in connection with examinations to determine the feasibility of diverting the waters of the Colorado, and his notes on the geology and mineralogy of the region traversed, it is believed, will furnish information regarding the same heretofore but meagerly known. Attached to his report is a list of the rock and mineral specimens collected by himself and Mr. Conkling. He also submits a report on the chemical composition of the alkaline lakes, thermal and mineral springs of Southern California met with during the season. (See Appendixes H 2 and H 3.)

Mr. Conkling submits a report on the geology of the mountain ranges from La Veta Pass to the head of the Pecos River. (See Appendix H 4.)

Paleontology.—Collections in this branch were made by Prof. Jules Marcou and Messrs. A. R. Conkling and D. A. Joy, which have received a preliminary examination only. Whenever the necessary facilities will allow an analytical examination and report upon them will be made.

MINING IMFORMATION.

As in the preceding years, examinations into the general character, condition, &c., of mining districts and mines located within the areas entered by parties of the expedition have been made, and the following is a condensation from the results of the investigations for 1875-'76 of mines located in California, Nevada, Colorado, New Mexico, and Arizona.

BLIND SPRING DISTRICT, CALIFORNIA.

Examined by Dr. O. Loew, October, 1875.

The mines of this district are in the Blind Spring Mountains, an isolated range about eight miles long and three and a half wide, which lies between the Sierra Blanca and Sierra Nevada. The trend of its longer axis is north and south, following the direction of the mountains. Its croppings cover a large area, approximately 7,000 acres   Benton is the post-office. The nearest railway town is Carson, Nev, one hundred and sixty-five miles distant, with which it is connected by a freight and stage line. The roads in the neighborhood are through deep sand. This district was discovered in 1865, organized in the same year, and has been worked continuously since.

In 1871, a geological investigation was made by Dr. Hoffman of this survey. The direction of the lodes is north and south. The wall-rock is of granite. In the vicinity the azoic formation prevails. The main ores are argentiferous galena and argentiferous copper glance; but these are accompanied by other valuable minerals in large and small quantities, among which are partzite—a very rare silver ore, stetefeldtite, cerussite, malachite, cerargyrite, and chrysocolla. These metallic combinations are imbedded in masses in the ledge-matter, which consists of ferruginous clay and quartzites, and varies in width from 3 to 8 feet. Among the base ores, iron pyrites, cuprite, and zincblende in large and well-formed crystals occur. Indications of gold, varying from a trace up to $5 per ton, are found.

Some of the ores are treated by direct milling; others are previously roasted. Salt is added to the ores in roasting in order to convert the sulphates into chlorides; the escaping gases and fumes have a strong odor of sulphurous chloride.

The water-level has not yet been reached. The Comanche is the principal lode. It is a ledge of clay and quartzite from 3 to 8 feet wide, through which the ore is distributed. Its direction is north and south. The dip is 73°. It is between smooth granite walls, apparently slickensided. It has a shaft 475 feet deep, with four levels. The vein of the Diana Mine dips 60° to the east. Its direction is north and south. It averages 5 feet in width, and contains free milling ore. The shaft is 590 feet deep, with five levels. The Eureka, Cornucopia, and Sierra Blanca lodes were worked extensively from 1865 to 1867, but since that time only at intervals.

There is a mill connected with the Diana Mine, and at the Comanche lode there are a mill and a furnace with two improved pans for amalgamation. All lead and copper found in this connection are lost, owing to the expense of transporting the apparatus and chemicals necessary to save them. The salt and soda used are obtained from Columbus, Nev., and Black Lake Flat, Cal.

A custom-mill will be erected soon. The cost of a 10 stamp mill will be about $25,000; with the stetefeldt furnace, $40,000. Freight from the railroad is 5 cents per pound. Other expenses will average as follows: Cost per ton for mining the ore, $15; for reducing, $20; mining labor per diem, $3.50; milling labor, $4; running a tunnel on main veins, per foot, $4 to $6; sinking a shaft, $6 to $18; running a drift, $4. One man can stope from one to three tons of ore per day.

The facilities for raising farm-produce are poor, as the soil is of coarse sand, and the water is insufficient for irrigation. Hay is worth $40 per ton; in the mountains there is some piñon and cedar timber. About 100 yards from the town of Benton there is a large hot spring, furnishing good palatable water and forming the source of a small creek. There are a few cattle in the vicinity, and some deer, rabbits, and quails. The inhabitants number about 600, besides two or three dozen Pah-Ute Indians.

HARDSCRABBLE DISTRICT, COLORADO.

Examined by Dr. O. Loew, August, 1874, and A. R. Conkling, November, 1875.

The Hardscrabble or Rosita district lies on the western slope of the Wet Mountains, near the edge of the park known as Wet Mountain Valley, and in the neighborhood of the town of Rosita, which is its post-office. Its nearest railroad-town is Cañon City, 29 miles distant, with which it is connected by freight and stage lines. Fare to Cañon City, $3; cost of freight, 1¼ cents per pound. The country roads are good. The district was discovered and organized in 1872, since which time it has been worked at intervals.

A previous examination was made in 1873 by Prof. J. J. Stevenson of this survey. The district is very large, extending north to Grape Creek, and from the Wet Mountain Valley on the west to the plains on the east. It is in a region of rolling mountains of moderate elevation, whose general trend is north and south. The veins run northwest and southeast. Dip of main vein, 23° to the southwest. In a few cases they are natural fissure veins; generally, they are deposits which have been disturbed by volcanic forces. They are usually richer on the side of the hanging wall. The pay-streak is quite regular in form, and is seldom more than 6 or 8 inches thick. In the well-defined veins, the wall-rock is a porphyry; in other cases, quartzite. Barite in white tabular crystals occurs with the ore. In age the country-rock is azoic and paleozoic.

There are two classes of ore, for smelting and for milling, but principally of the latter class. Average yield about $80 per ton. A trace of gold has been frequently found in assaying. In the Virginia Mine, the yield of gold has been as high as 1 ounce per ton. The ores are argentiferous galena, chloride of silver in clay and gypsum, copper glance, malachite, stephanite, stromeyerite, ruby silver, native silver in leaf-form, gray copper, chalcopyrite, and pyrite in cubes and pentagonal dodecahedrons. The base metals are lead and copper.

Water-level is reached at a depth of 30 feet, and marks a richer grade of ore. The examination of 1874 included 10 mines.

West Virginia, siliceous and clayey deposit, containing carbonates of copper, galena, chloride of silver, and manganese.

Humboldt, same character of ore; dip of strata, 75° south-southwest.

Pocahontas, same character of ore, with gypsum-spar, iron, and copper pyrites, galena, carbonate of lead, and chloride of silver.

Del Norte, crevice, 2 feet in width.

Leviathan, crevice, 11 feet.

Chieftain, crevice, 3½ feet.

Pioneer, shaft, 70 feet.

Minnesota, gypsum-spar and clay, with chloride of silver, pyrites of iron, and galena; assay, $60.

Pelican, crevice, 8 feet; assay, $60; dip, 72°.

Senator, strike, east and west; dip, 44°; vein, 3 to 8 feet wide; wall-rock, porphyritic trachyte; ore, principally galena in quartzite; assay, from $25 to $1,200 per ton; resembles a true fissure-vein.

In 1875, the assay of the Pocahontas was reported to be 220 ounces; of the Virginia, 228 ounces. Between March and November, 1875, $50,000 were expended on the Pocahontas, and $120,000 extracted. The reduction-works began operations in July, 1875. There is one mill with a Dodge crusher and four or five pans; capacity, four tons per day. The amalgam is strained cold.

The cost of a 10-stamp mill would be $30,000. Other expenses will average as follows: Mining the ore per ton, $10; reducing, $40; mining labor per diem, $3; milling labor, $3; running a tunnel on main vein per foot, $8; sinking a shaft, $7.50; running a drift, $3.50.

There are poor facilities for farming, but in the Wet Mountain Valley, seven miles distant, there is good grazing country, supporting 25,000 cattle. Hay is worth $20 per ton; grain, 4 cents per pound. There is an abundance of wood and water, and coal within twenty miles. The inhabitants number about 1,000. Plenty of game is found, including the deer, bear, mountain-sheep, grouse, turkey, prairie-chicken, duck, fox, beaver, mink, wolf, and badger.

PANAMINT DISTRICT, CALIFORNIA.

Examined by Dr. O. Loew, October, 1875.

This district is situated in the Panamint range of mountains, between Death Valley and Panamint Valley. Its croppings cover an area of about 6 square miles. The ledges traverse the main range, occurring chiefly on the western slope, upon which there are but few spurs and foot hills. The post-office of the district is Panamint. It is connected with the railway at Caliente by a stage-road via Shepperd's Cañon and Indian Wells. The distance is 160 miles; fare, about $40. The cost of freight from San Francisco, via Caliente, from which point wagons run, is 5 cents per pound. The completion of the Independence Railroad will reduce the cost of transportation. The roads to the mines, upon the mountain-slopes, are very good, but the routes to Caliente and Darwin lead through deep sand.

The lodes are confined to the higher portions of the range. Their direction is north-northeast and south-southwest. The ore occurs in masses of different sizes, imbedded in the quartz ledge, whose position is nearly vertical between walls of crystalline limestone and calcareous conglomerate. The veins are well defined and large, and the walls are generally slickensided. The mass of the mountain range is granite. In many places in this there are dikes of intrusive rock, such as diorite, porphyry, and serpentine. In the vicinity of the mines, especially, the granite is accompanied by talcose schist, primitive limestone, primitive clay slate, hydraulic limestone, and calcareous conglomerate. No fossil shells are found in the higher portions of the range. In the lower portions occur more recent formations, such as the quaternary, which covers the Panamint Valley. In this valley there are also hot springs, and saline deposits from which salt is obtained for the roasting purposes of the mines.

The products of these mines are partly free milling-ores and partly roasting-ores; the latter are milled after the completion of the roasting process. Their average yield per ton is from $90 to $120. No gold is found. The silver ores are principally chloride of silver, stromeyerite, stetefeldtite, and argentiferous copper glance, with native silver, and, rarely, argentiferous galena. Copper is the chief base metal. A little antimony accompanies the ore, and, exceptionally, some iron pyrites and zinc-blende.

The principal mine of this district is the Hemlock, whose ledge has been traced for 1¼ miles, appearing on the other side of the hill as the Alabama lode. The ledge dips 80° to the southeast, and varies in width from 1 to 6 feet. Though in some places barren, it is again very rich. It has been pierced by two tunnels, 75 feet and 229 feet in length, from which drifts and winzes have been pushed. Next in importance is the Wyoming lode. This ledge, running through limestone, varies in width from 2¼ to 10 feet, with a branch vein from 2 to 6 feet wide, diverging from the main lode at an angle of 15°. Two drifts, 240 feet and 500 feet in length, and two tunnels, 160 feet and 180 feet long, have been run. The Surprise lode is large and well defined, 6 feet in width, and is apparently a continuation of the Wyoming. The World's Wonder and Steyart's Wonder are worthy of mention, and, in addition to these, there are a dozen others, equally as good, which are not worked at present. No positive data can be given concerning the expenditures and receipts of this district, but the product of the furnace for one month was 40,000 ounces of silver.

The Sunrise Company runs a 5-stamp mill, but, having no roasting furnace, works only free milling ore. The Surprise Valley Company has a stetefeldt furnace and a 20-stamp mill. Each stamp weighs 900 pounds, and makes 85 drops per minute. There are two crushers, working 25 tons of ore per day. There are ten amalgamating pans, each holding a ton of roasted ore. The amalgam is strained cold. It is generally composed of 1 part silver to 5 part mercury, with a little lead and copper. The furnace is constructed for working 40 tons of ore per day. It is discharged every 1¼ to 2 hours. From 3 per cent. to 10 per cent. of salt is added to the crushed ore in roasting.

The cost of machinery is unusually high, owing to expensive transportation. The stetefeldt furnace cost $18,000. Other items of expense will average as follows: Mining labor per diem, $4; milling labor, $5.50; running a tunnel on main vein, $10 to $15. The present sources of grain supply are Los Angeles and Caliente. Hay is brought from the northern portion of Inyo County. With an influx of settlers these commodities may become cheaper, as there is some good farming-land in that region. The timber is chiefly of piñon and juniper; its quantity is small, but sufficient. The water is very good and pure, and there is enough in the vicinity to run another 30-stamp mill. The Surprise Valley Company owns a large spring near the mines, and the Sunrise Company has a well 40 feet in depth. There are about 500 inhabitants in the district, besides some Pah-Ute Indians.

CHARLOTTE DISTRICT, CALIFORNIA.

Examined by Lieut. C. W. Whipple, July 20, 1875.

This district is at the head of the Tejongo Creek, near the summit of the San Gabriel Mountains. It is about 40 miles from Los Angeles, its principal source of supply, and 32 miles from San Fernando, which is the most convenient post-office and railway-station. It is now reached by a difficult trail up the Tejongo, over which supplies are transported on burros. An easier trail, by way of the Arroyo Seco, is believed to be practicable and will be attempted.

It was discovered in 1870 by Mexican herders. Since that time it has been worked at intervals by these people and by the proprietors of the Harding Mine. No geological examination has been made, but the main vein has been followed a distance of 4 miles in a southeast direction. The rock of the walls and in the vicinity is almost entirely quartz, but farther down the stream it is a coarse, friable granite. No silver has yet been found. The vein is of gold-bearing quartz, in which some iron pyrites occur. In places this is hard and compact, and again it is soft and earthy.

The mountain range has here a general east and west trend, with spurs projecting to the south, which are traversed by the several veins. The Charlotte and Las Animas Mines are upon the same vein. The former,* upon which work is now in progress, has been sunk to a depth of 11 feet. The main vein is about 3 feet wide, and is crossed diagonally by another, 18 inches in width, with an inclination of 30°. The Poyorena, the Union, and Mejicana are on a second vein. Work upon the Poyorena is about to be resumed. Its shaft has reached a depth of 30 feet. The vein is 3 feet thick, extending without fault from surface to bottom. The Union is at a depth of 40 feet. The Pacific, on a third vein, was not visited. The Harding Mine has reached a depth of 40 feet.

The quality of ore seems to improve with increase of depth. In places it is very rich, the managers of the Charlotte Mining Company claiming that these culled pieces will assay $300 per ton. This company has completed two arrastras, with which they expect to work two tons of ore per day. It is anticipated that it will cost $30 per ton to extract and work the ore.

The district is peopled by about fifty miners and prospectors. The supply of water in summer is scanty, proceeding from the springs at the heads of the valleys. On the north exposure of the mountains, near their summit, there is good pine timber. Bear and deer are plenty.

The principal impediment to the development of this district is its inconvenience of access. With a new road, which would allow the introduction of improved machinery, it might be very profitably worked.

SAN EMIDIO DISTRICT, CALIFORNIA.

Examined by Douglas A. Joy, November 9, 1875.

This district is in the San Emidio Mountains, at the head of Plato Creek, on the divide between this stream and the San Emigdio Cañon.

---

* This mine has since been abandoned.—ED.

It was discovered in 1871, since which time it has been continuously prospected. It is connected by trail with the San Emigdio Cañon, and thence by wagon-road with Bakersfield, 45 miles distant, which is the most convenient post-office and railway station.

The outcroppings follow the top of the ridge, which has a general east and west trend. They have been traced more than 3 miles, but the exact area of the district is uncertain. The strike of the vein is due north, and its dip, according to the best estimate, is 90°, vertical. The country-rock is massive granite. The dependence of the quality of the vein upon the local character of the country-rock has not been investigated.

The wall-rock is granite. Its dip is 90°. Tertiary sandstones overlie this rock about 4 miles down the ridge. The character of the vein has not yet been determined. None of the ore extracted up to date has been worked. It is proposed to smelt it in a reverberatory furnace to crude antimony, (pan sulphuret.) The ore extracted assays 75 per cent. of sulphuret.

The ore is nearly pure sulphuret of antimony, with some silver and iron. Some quartz is irregularly intermixed, and also a little feldspar. The greater part of the ore is colored red on the outside by oxide of iron, and, until broken, does not show the characteristic black metallic appearance. It is hard, and but little decomposed.

The district has been worked in but one place, and not extensively there. The water-level is not yet reached. Several tunnels, from 15 to 45 feet in length, have been run, but nothing more has been done. This mine is supposed to have been worked in former times by the Spaniards, as there are indications of previous labors here, and in the adjacent valley of Plato Creek an old furnace has been recently discovered.

The vein-matter should be followed in running shafts and drifts. On an average, eight tons of ore can be stoped per diem by one man. Water is abundant in the neighboring cañons. The country roads are good. On the surrounding mountain slopes there is good pasturage and excellent pine timber. Deer, grouse, and mountain-sheep abound in the mountains, and antelope on the plains.

GREEN MOUNTAIN DISTRICT, CALIFORNIA.

Examined by Douglas A. Joy, October 6, 1875.

The Bright Star Mine is the principal one in this district. It is situated at an elevation of about 8,000 feet, within a mile of the summit of the Pah-Ute Mountains, which lie in a clump between the Kern River and Kelso Valley, about 15 miles south of Kernville. The district was discovered in April, 1866, by a Mr. Rains. It was organized in the same year, and has been worked continuously since that time. Its most convenient post-office is Havilah, with which it is connected by trail. The nearest railway station is Caliente, which, by wagon-road, is 85 miles distant. The country roads are good, though steep. Cost of freight from Caliente is $3\frac{1}{2}$ cents per pound.

The exact area of this district is not known, as no geological surveys have been made. The superintendent of the Bright Star Mine claims to have traced the vein-wall a distance of $1\frac{1}{2}$ miles, but admits that the substance of the vein is not entirely quartz. Its general trend is northeast and southwest. There is no other mining ledge in the immediate vicinity.

The walls are slickensided, and there are other indications of a true fissure-vein. It dips at an angle of 82°. Its strike is northeast, directly perpendicular to the bedding of the country-rock. The rock is primitive, granite and syenite. The mineral constituents are very irregularly scattered. Sometimes they are almost purely quartz and feldspar, and again are chiefly hornblende, or mica schist.

The native metals alone are collected. This is done by amalgamation in stamps. The average yield per ton is about $22. The bullion obtained is an alloy of gold and silver, one part silver to two parts gold, whose average worth is $14 per troy ounce. Water leaks in at various levels, especially at the lowest, which is 280 feet below the surface. No change has been observed in the character of the ore, which is a compact quartz, containing native gold and silver, with which are mixed sulphurets of arsenic and antimony, and pyrites in small quantities. As it comes from the mine, the ore is mixed with clay and country-rock from the horses and walls. The vein-rock is but little decomposed. The native metal can occasionally be distinguished with the naked eye. The sulphurets occur both in specks and in masses of considerable size.

About $150,000 has been expended on the Bright Star Mine, from which 1,200 pounds troy of bullion, valued at $14 per ounce, have been taken. Its ores are worked by one mill, with three batteries of five stamps each, which make 85 drops per minute. The entire mill is run by one small engine of 35-horse power. There is a great scarcity of water, the only source of supply, aside from two or three small springs, being the mine itself, which furnishes enough for the engine, the stamps, and culinary purposes. The sulphurets have not been treated yet, but are allowed to accumulate and await the completion of a furnace in which they may be roasted.

The cost of a 10-stamp mill on the grounds is about $6,000. Other expenses will average as follows : Cost per ton for mining ore, $2 ; for reducing the same, $2 ; mining labor, per diem, $3.37 ; milling labor, $4 ; running a tunnel on main vein, per foot, $6 ; sinking a shaft, $12 ; running a drift, $6. One man will stope two tons of ore per day.

There is but little farming-land in the immediate vicinity, and but few cattle or other stock. The present sources of supply are Kernville, Havilah, Kelso Valley, and small valleys in the mountains. Grain is worth 5 cents per pound ; hay, $25 per ton. The mine is surrounded by very heavy pine timber. Deer and quail abound.

CLARKE DISTRICT, CALIFORNIA.

Examined by Dr. O. Loew, July, 1875.

The Clarke Mountain, near Ivanpah, is a part of the Opal Mountains. On both sides of this, but especially on the eastern slope, are the mines of the Clarke district, whose croppings cover an area of great extent. This district was discovered in 1869, was organized in 1870, and has been constantly worked since. Its post-office is Ivanpah. The nearest railway town is San Bernardino, which is 200 miles distant, via Soda Lake and the Mohave River. Cost of freight from this point is 5 cents per pound.

The mines were previously examined by Lieutenant Wheeler in 1871. The direction of the lodes is east and west, transverse to the general trend of the district. Quartzite and paleozoic limestone form the wall-rock ; the vein-matter is quartz, and, in some instances, calcspar. In

AP. JJ—4

age the country rock is paleozoric and azoic. Syenite, granite, porphyry, mica schist, slate, sandstone, and limestone are present.

Some of the ores are free milling; others require roasting. In the quartzite, the principal silver ores are argentiferous galena, stromeyerite, stetefeldtite, and pyrargyrite; in the limestone, they are massicot, minium, cerussite, malachite, and cerargyrite, or chloride of silver. Gold is present in the argentiferous quartz.

The water-level has not been reached. The most important mines are the Beatus, with a shaft 300 feet, and 3 levels 100 feet each; the Snow Storm, rich in carbonate of lead, with a shaft 80 feet deep; the Lisa Bullock, whose ore is chiefly stromeyerite, assaying $600 per ton, whose shaft is 100 feet deep, with two levels, and whose vein is 4 feet wide, with walls slickensided; the Lucky Boy, with shaft 61 feet deep and vein 5 or 6 feet wide, whose principal yield is galenite, assaying from $20 to $50 per ton; the Stonewall, with shaft 50 feet deep; and the Savage, whose shaft is 40 feet deep, and whose vein, which has been traced for 1,000 feet, is 5 feet in width. The ore of the latter is chiefly carbonate of lead and minium, assaying from $30 to $40 per ton. The Green-Eyed Monster and Copper World are also good mines. About 12 miles south of Ivanpah, but still in Clarke district, is the Bullion Mine, containing malachite, minium, and chloride of silver. There are one small mill of ordinary construction and one smelting-furnace attached to these mines. The yield has been about $300,000.

The cost of a 10 stamp mill, including transportation, would be $25,000. Other expenses will average as follows: cost of mining the ore per ton, $20; reducing, $25; mining labor, per diem, $4; milling labor, $3; running a tunnel on main vein, $5 to $6; sinking a shaft, $10. One can extract 500 pounds of ore per diem.

The supply of water is limited, and there are but few facilities for farming. Some piñon and juniper grow on the mountains, and pine timber is found 15 miles north of Ivanpah. Mountain-sheep are the only game. There are a few hundred domestic animals in the vicinity. The country roads are tolerable. The inhabitants number about 100, besides 40 or 50 Pah Ute Indians.

WASHOE DISTRICT, NEVADA.

Examined by Dr. O. Loew, October, 1875.

The Washoe district, celebrated for its Comstock lode, is on the eastern slope of the Virginia range, and in the immediate vicinity of Virginia City, which is its post-office. It is connected by rail with Reno on the Central Pacific Railroad.

Thorough geological examinations have been previously made by different parties, and especially by James D. Hague, of Clarence King's survey, in whose able reports on which this district is treated at length. The ore of the Comstock lode occurs in pockets, and as impregnation in quartzitic ledges, which traverse the country-rock in a northeasterly direction. Frequently the quartzitic vein matter is accompanied by clay. The country-rock is principally azoic; syenite is it, and also that feldspathic porphyry is also frequent. In some instances it shows an advanced degree of decomposition, that is, converted into clay, even while retaining the original

The ores are chiefly free milling, although undoubtedly some are benefited by previous roasting. The average assay yield is from $50 to $60. Water-level has been reached in all shafts. In the Ingersoll Mine hot water has been encountered.

The principal ores are chloride of silver, or cerargyrite, brittle silver, or stephanite, pyrargyrite, polybasite, free gold, native silver, argentiferous galena, copper glance, zinc blende, and iron and copper pyrites. Antimony, copper, and lead are the chief base ores. Gold occurs in all of the silver ores, and forms an important share of their values.

Only two mines, the Belcher and the Imperial Empire, were visited. The shaft of the Belcher is 850 feet deep vertical, then 750 feet inclined. Work is being prosecuted on the 1,600-foot level. Four hundred tons per day are extracted. This amount is the working capacity of the mills at the time of high water in the Carson River. At other seasons not more than 150 or 200 tons can be reduced, and less ore is then extracted. Since 1860 166,000 tons have been taken from this mine. The ledge varies in width from 25 to 50 feet. Its strike is northeast and southwest. The ore is found in pay-streaks, and also occurs as impregnation.

In the Imperial Empire the vein is 25 feet wide; the strike northeast and southwest; the shaft 2,000 feet deep; the average yield, 40 tons per day of good milling ore.

Since work began on the Belcher, in 1860, silver and gold bullion to the value of $28,000,000 has been extracted, and dividends have been paid to the amount of $15,000,000. The yield for 1874 was $9,150,000; in 1875 it was $200,000 per month. In 1870 the total product of all mines in the vicinity did not reach that amount.

California is the chief source of supply for produce, &c. The nearest timber of importance is in the mountains near Cañon City. Some water is found in the mines, but the whole supply for Virginia City is not great.

HUALAPAIS DISTRICT, ARIZONA.

Examined by Dr. O. Loew, August, 1875.

The mines of this district are in the Cerbat range of Arizona, principally on the western slope. The trend of the mountains is northwest and southeast; of the district and of the veins, nearly the same. The croppings cover a belt 10 miles long and 2 miles wide, lying between Sacramento Valley and Hualapais Valley. Mineral Park is the district post-office. The stage-line from Prescott to Hardyville passes here. Cost of freight overland from California is 8 cents per pound; by vessel via Yuma to Hardyville, and thence by wagon 30 miles to Mineral Park, it is 5 cents per pound. The roads are very good. This district, formerly called Sacramento district, was discovered and organized in 1863, but, in consequence of Indian troubles, has been worked but desultorily.

These mines were examined by Lieutenant Wheeler in 1871. The veins are well defined. Their angle of dip is from 70° to 80°. They appear to be richer where the country-rock is more micaceous. Some are contact veins, between syenite foot-wall and porphyry hanging-wall. The vein matter is chiefly quartz. The rock is azoic. Its principal constituents are granulite, granite, apatite, talcose schist, syenite, and quartzite, tinged with iron and manganese. Dikes of intrusive porphyry rock are occasionally met.

There is but little free-milling ore; it is chiefly roasting. Assays show a yield of $60 to $1,400 in silver per ton. Some gold occurs with the silver, and in some mines the gold predominates in value. The ores are galenite, pyrargyrite, or ruby silver, cerargyrite, or chloride of silver, chrysocolla, pyrite, cerussite, chalcocite, and sphalerite, or zinc blende. Base metals are copper, lead, antimony, iron, and zinc.

The Keystone lode is from 2 to 5 feet wide, with seams of ruby silver and separate seams of base sulphurets, which in this ore never mix with the high sulphurets. The ledge is of great extent, with seven located claims. Its shaft is 185 feet deep, with 2 levels, 40 feet long. The Lone Star has a shaft 200 feet deep, with 2 levels. Gold abounds in its ore. The main quartz ledge of the Metallic Accident is from 12 to 20 feet wide; shaft 60 feet deep. The main vein is composed of 2 parallel ledges, which are cut by a third at right angles nearly. Its ore is chiefly chloride of silver, assaying on an average $800 per ton. The Sixty-Three lode has 3 tunnels, averaging 250 feet in length, with shafts 200 feet deep. The Montezuma contains chrysocolla and chloride of silver. The Cerbat lode assays $150 per ton. Of this two-thirds is gold. Much ore has also been taken from the Cupel and Little Tiger mines.

About $60,000 worth of bullion has been taken from the Sixty-Three lode; expenditures nearly the same. Further estimates are not reliable. There is one 5-stamp mill, and a second in course of construction. Each stamp weighs 850 pounds, and makes 90 drops per minute. There are 2 furnaces and 2 5-foot combinationa pans. If much lead is present, the amalgam is strained hot; otherwise, cold.

A 10-stamp mill, with 2 furnaces, will cost, including transportation, $45,000. Other expenses will average as follows: Cost of reducing the ore, per ton, $50; running labor, per diem, $4; milling labor, $4.50; mining the ore, per ton, $12; running a tunnel on main vein, $15; sinking a shaft, $25; running a drift, $12.

No farming is done. There is some good grazing country, but not much stock. Grain is brought from California. Its cost is 8 to 10 cents per pound; cost of hay, $20 per ton. Timber is scarce. The water at the mines contains sulphate of lime and magnesia; but pure water is obtained from the mountains, 4 miles away. The mountain sheep, antelope, deer, quail, and rabbit are found. The Hualapais Indians live in the vicinity. About 100 whites people the district.

NEW COSO DISTRICT, CALIFORNIA.

Examined by Lieut. Rogers Birnie, jr., and Dr. O. Loew, 1875.

This district is in the main ridge of the Coso range, on both slopes of the same, in the vicinity of the town of Darwin, which is post-office to the district. It is of great area, lying between Panamint Valley on the east and Owen's Lake on the west. It is connected by stage with Caliente, 120 miles distant; fare, $30. Cost of freight from the same point, by the Cerro Gordo Freighting Company, is 5 cents per pound. The district was discovered and organized in 1874.

The dip of the veins is 60°. The wall-rock is principally limestone, in addition to which granite and basalt are found. A ledge of limestone, varying from 200 feet to an indefinite width, runs northwest and southeast, which is also the general trend of the mountains and of the veins. The country-rock is chiefly azoic. Dikes of basalt and hornblende porphyry occur in the limestone. No fossil remains are found.

All ores are treated by the smelting process, which, however, is not entirely satisfactory, as much silver of the chloride is volatilized and lost. Assays yield from $30 to $600 per ton. Gold is found in considerable quantities, from $5 to $140 per ton of ore. The yield of silver is from 40 to 50 ounces per ton. Iron forms 4 per cent. of the ore, and lead from 40 to 65 per cent. of the same. Saltpeter is found in the southern portion of the district.

Silver is present as sulphide in the galenite, and as chloride in the cerussite and massicot; these ores are intermixed. Mingled with these in small quantities are blue and green carbonates of copper, (azurite and malachite,) silicate of copper, (crysocolla,) black and red oxides of copper, (melaconite and cuprite,) oxides of iron and manganese, and iron pyrites. The fissures in the limestone are filled with solid chunks of galena, imbedded in a pulverulent mass composed of yellow and red oxides of lead and carbonate of lead. Here and there calc spar and seams of reddish argillaceous limestone occur in the vein matter, and also occasional bowlders of limestone. The walls are frequently slicken-sided, and in some instances show "breaks," or hollow fissures, probably caused by earthquakes.

The Defiance lode is situated one-half mile east of Darwin, in a belt of limestone about 500 feet wide, which is bordered on the east by granite and a dike of hornblende porphyry, and on the west by slate. The lode consists of four broad, parallel veins, of 45, 27, 47, and 28 feet in width. Dip, 40° to southwest; course, northwest and southeast. The average assay yield is $50. Expenditures, $4,000; receipts, $5,000. One tunnel is in course of construction.

The vein of the Bella Union is from 5 to 8 feet wide. It has a feeder uniting with it at a very acute angle. The walls are slickensided, especially the foot-walls, and a series of lines like scratches, with a dip of 25°, are seen on them. The ore assays from $60 to $75. Over 160 tons have been extracted from the feeder, and 400 from the main vein. Expenditures, $4,000. No bullion has yet been obtained. The Lucky Jim and Christmas Gift are two other large mines, worked at present.

There is one smelting-furnace in operation and another in process of construction. The furnace of the Defiance Company will cost $20,000. It will be an iron furnace, of 50 tons capacity. The boiler will be of 42-inch diameter and 16 feet in length. Other expenses will average as follows: Cost of mining the ore, per ton, $2; smelting, $10; mining labor, per diem, $4; milling labor, $4; for cross-cutting a tunnel, per foot, $20; cross-cutting a drift, $20; sinking a shaft, $12. One man can stope or extract 5 tons of ore per day.

There are no facilities for farming, as there is not enough water for irrigation. Hay and grain come from Owen's Valley. Hay costs $60 per ton; grain, 6 cents per pound. Some piñon timber grows 8 miles from Darwin. Water is very scarce and is sold for 2½ cents per gallon in Darwin, whither it is brought in pipes from the mountains several miles distant. There are about 1,000 cattle and 200 horses in the district, which affords poor grazing. Rabbits, quails, and deer are the kinds of game. About 500 people live here.

BANNER DISTRICT, CALIFORNIA.

Examined by Geo. H. Birnie, April 19, 1876.

This district is in the central part of San Diego County, in a cañon of the Quiamaca Mountains, and on the foot-hills on both the east and west slopes of the same. It covers about 36 square miles. The trend of its longer axis is north 8° west. The course of the mountains is north 10° west. The Banner district was discovered in 1870 by Louis B. Redman. It was organized in the same year, and has been worked continuously since.

Its post-office is Banner. The nearest railway is at Seven Palms, in the Coahuila Valley, but there is as yet no practicable route thither. The railway station most convenient of access is Anaheim, distant 116 miles. The district is connected by stage-line with San Diego; fare, $5. The cost of freight from San Diego is 1¾ cents per pound. The local roads are tolerable.

A geological investigation has previously been made by Mr. Buoy, of San Francisco. The lode dips to the east at an angle of 65°. Its direction is north 10° west. The wall-rock is of slate and granite; the vein is richer in the slate. No fossils have been discovered in the vicinity. A little galena has been found. The ore is a laminated quartz, containing some arsenic and antimony. It is worked by the wet process, yielding from $20 to $50 per ton, principally gold, but with an alloy of 4 per cent. of silver. The water-level has been reached, but it marks no change in the nature of the ore.

The principal mines now worked are the Ready Relief, Hubbard, and Kentuck. The Golden Chariot, which has reached the greatest depth, is now suspended, with the intention of resuming operations soon. They are all on the mountain-side, with shafts on an incline of about 65°. Up to date the Golden Chariot has expended, approximately, $200,000 and extracted $300,000, and the Ready Relief has expended $100,000 and extracted $150,000. The mill of the Ready Relief has 10 stamps, with settling-pans. The Golden Chariot mill is similar. The Whitney mill has 10 stamps. There are no crushers. Each stamp, weighing 750 pounds, makes 80 drops per minute. Each mill has three settlers. Cold amalgam is used.

The cost of a 10-stamp mill without roasting-furnace, including transportation and erection, is about $12,000; other expenses will average about as follows: Cost per ton for mining the ore, $10; for reducing the same, $2; mining labor, per diem, $3; milling labor, $3; running a tunnel on main vein, $16; sinking a shaft, $25. One man can stope or extract one-half of a ton of ore per day.

In the vicinity there are some stock, chiefly cows, sheep, and horses. Barley is valued at 3 cents per pound; hay, at $40 per ton. San Diego is the principal source of supply. There is a plenty of water and of oak and spruce-pine timber. Black-tailed deer, quails, rabbits, hares, and pigeons abound. The district numbers about 75 inhabitants, besides a body of 100 Diegueños Indians.

NEW ELDORADO DISTRICT, CALIFORNIA.

Examined by Douglas A. Joy, October 8, 1875.

The ledges of this district lie in a range of hills between Kelso Valley and the Mohave Desert. Its principal opening, the St. John's Mine, is on the divide between Kelso Valley and Kelso Cañon, east of Pah-Ute Mountain. It was discovered by Mr. St. John in 1866, organized the same year, and worked continuously since that time. Caliente, the nearest railway station, is distant 75 miles by wagon-road.

No regular geological examination has been made, but the district has been traced by croppings 2½ miles, and may extend farther. The general direction of the vein is northwest and southeast. The general dip in the main shaft is 37°. Both foot and hanging walls are of massive granite, with no decided bedding. The direction of their slope is very irregular, but, in general, is 37° northeast and southwest. Though slickensided and possessing other features of a true fissure vein, it is undoubtedly not such.

The material of the country-rock is granite. The mineral constituents are very irregularly distributed, both mica and feldspar often disappearing, leaving quartz and mica or quartz and feldspar only. The mica is often replaced by hornblende, forming a syenite, and again the rock seems to be composed entirely of hornblende.

At the water-level, 300 feet down, the ore is greatly decomposed, requiring only a pick and shovel to extract it. But at the lowest level, 700 feet, the quartz is very hard and at its richest quality. The ores are worked by the wet milling process. The supposed average yield is $30 per ton. Some ores, found in small quantities, have assayed as high as $60 per ton.

The ore contains about half as much silver as gold. The gold generally occurs in native condition. At the lowest level it is found in leaf state, and is readily seen with the naked eye. Some yellow sulphurets of iron and black sulphurets of arsenic and antimony are found, but these are not sufficient in quantity to be profitably utilized.

The St. John Mine has been worked nearly ten years, in which time, by estimate, $80,000 has been expended and bullion to the amount of $200,000 has been extracted. It possesses a 12-stamp mill, each stamp weighing 750 pounds, and making 82 drops per minute. The amalgam is strained cold, through canvas bags, no pan amalgamation being carried on.

The cost of a 10-stamp mill, including transportation from San Francisco and erection on the grounds, is about $4,500. Other expenses average as follows: Cost per ton of mining the ore, $4; reducing the same, $4; mining labor per diem, $2 and board; milling labor, $3 and board; running a tunnel on main vein, per foot, $8; sinking a shaft, $16; running a drift, $8. One man will stope or extract two tons of ore per diem.

The neighboring country roads are good, but steep. The water used is obtained entirely from the mine, there being no large stream nearer than Kern River. Kelso Valley is well adapted for farming and grazing purposes, and upon this plain and in its vicinity there are, perhaps, 10,000 cattle. On the summit of Pah-Ute Mountain, about five miles distant, there is an unlimited quantity of heavy pine timber. The principal kinds of game are deer, grouse, and bear.

JULIAN DISTRICT, CALIFORNIA.

Examined by George H. Birnie, April 19, 1876.

The Julian district is near the center of San Diego County, on the headwaters of San Dieguito River. It is 122 miles from Anaheim, by the old Government road from Wilmington to Fort Yuma. The Southern Pacific Railroad at Seven Palms, in the Coahuila Valley, is nearer, but there is no practicable road to it. There is a mail stage tri-weekly to San Diego; distance, 65 miles; fare, $5. The post-office is Julian. The mountain-roads in the vicinity are very good.

The first discovery in this district was made in 1870, by Messrs. Julian and Bailey. It was organized in the same year, since which time it has been worked continuously. The district is 6 miles long and 2 miles wide, following the summit-line of the Quiamaca Mountains, which have a north-northeast trend. The direction of the lodes is north-northeast. The dip is 45; the wall-rock is of granite and slate; the veins are found to be richer in slate. Water has been reached at a depth of 260 feet, at which level the ledges are better defined and of a superior quality. At a depth of 400 feet the veins continue good.

The average yield per ton is $20; of this amount one per cent. is silver, and the remainder is gold. The ore is worked by the wet process. It is a laminated quartz, nearly pure. Sulphurets of iron and arsenic are found in small quantities, but nothing more.

The principal mines are the Helvetia, Tom Scott, Pride of the West, and Washington; all on the summit of the mountain-range. Of these, by approximation, the Helvetia has expended $40,000 and extracted $55,000; the Tom Scott has expended $10,000 and extracted $20,000; the Washington has expended $10,000 and extracted $20,000; and the Pride of the West has expended $5,000 and extracted $10,000. There are two mills running. The Helvetia has 10 stamps, weighing 650 pounds each, and making 90 drops per minute. It has no crushers and no settlers; cold amalgam is used. The engine is of 15 horse-power. The Reynold's (custom) mill has but 5 stamps; in other respects it is similar to the Helvetia.

A 10-stamp mill will cost, on the grounds, about $12,000. Other expenses will average as follows: Cost of mining the ore per ton, $6; reducing the same, $2; mining-labor per diem, $3; milling-labor per diem, $3; running a tunnel on main vein per foot, $10; sinking a shaft, $25; running a drift, $9.

The surrounding country is excellently adapted for farming and grazing, and great numbers of stock are raised. All necessary produce is grown in the district. Barley is worth $2\frac{1}{4}$ cents per pound; hay, $15 per ton. The timber, evergreen-oak and spruce-pine, is abundant. There is plenty of water in wells and springs, and in Vulcan Creek.

The district has 300 inhabitants. A body of Mission Indians, 150 in number, live here. There are many kinds of game, including the deer, bear, rabbit, hare, quail, California lion, pigeon, gray squirrel, duck, goose, and wild hog.

CASTLE DOME DISTRICT, ARIZONA.

Information obtained by George H. Birnie, March, 1876.

Castle Dome district lies in the foot-hills and on the western slope of a range of mountains in Arizona, 18 miles east of the post-office of Castle Dome Landing, on the Colorado River. The eastern side of these mountains has not yet been prospected. The general trend of the range is north 25° west, with other ridges running at right angles to this direction. The district, as already traced, is 2 miles in width and 7 in length, following the trend of the mountains.

· It was discovered and organized in 1863 by Messrs. Snively and Conner. Except a portion of 1872, it has been worked constantly since that time. The most practicable route thither is from Seven Palms railway station to Ehrenberg, 110 miles, and thence by river to Castle Dome Landing. Supplies may be sent to this point by the Colorado Steam Navigation Company. Mule-teams, for the conveyance of freight, run from the river to the mines. Cost of such transportation, $6 per ton.

A geological investigation has been made by Professor Blake, of Connecticut. The lodes are found to run north 25° west. Some follow and others intersect the stratification. The rich veins are found in fluor spar or talc. The wall-rocks are of mixed slate and porphyry. They are perpendicular, with an east and west slope. Fissure-veins are found at a depth of 350 feet. The country rock is basaltic. No fossils occur.

The ores are worked by smelting and by the process of iron precipitation. The ores are galena, carbonate of lead, and anglesite. An average yield of $32 per ton in silver is obtained. Lead forms 67 per cent. of the ore. Iron is the other base metal. Traces of gold are found. Water has not yet been reached.

The principal mines are the Flora Temple, Castle Dome, William Penn, Caledonia, Don Santiago, Little Willie, and Norma. The general character of the ore is the same throughout all of these, except the Castle Dome and Caledonia, in which anglesite and carbonates are found. Their claims vary in extent from 200 feet by 1,000 feet to 600 feet by 1,500 feet, and lie in the foot-hills and the mountains above. From the main lodes about 6,000 tons of ore have been extracted and taken away for reduction, at an expenditure of about $200,000. On the Colorado River, 18 miles from the mines, the Castle Dome Smelting Company has a blast-furnace, with an engine of 20 horse-power. Its capacity is 20 tons per diem.

The cost of a furnace constructed at the mines would be $12,000. Other expenses average as follows: Cost per ton for mining the ore, $8; for reducing the same, $9; mining labor per diem, $2; smelting labor, $2; running a tunnel on main vein, $3.50; sinking a shaft, $4; running a drift, $3. One man can stope from one to five tons of ore per diem, according to the size of the vein, or can extract six tons per diem. Expense will be reduced by the completion of the Southern Pacific Railroad to the Colorado River.

There are a few horses, mules, and burros in the vicinity. There are no facilities for raising produce. Barley is worth $4 per hundred-weight. Alfalfa, wheat, corn, oats, sugar-cane, vegetables, cotton, fruits, and wild hemp can be procured at Yuma. The varieties of timber are cotton-wood, suwarrow, iron-wood, willow, and mesquite. Water is scarce, the nearest supply being a good well 10 miles from the mine. The deer, mountain-sheep, antelope, quail, rabbit, and hare abound. There are 200 inhabitants in the district, besides about the same number of Date Creek Apache Indians.

COLORADO DISTRICT, NEVADA.

Examined by Dr. O. Loew, July 31, 1875.

These mines are near the Colorado River, along the foot-hills of the Black Mountains, in some cases approaching the summit of the same. In area this district is 3 miles wide by 7 miles long. A small island in the Colorado River forms its approximate southeast corner. Saint Thomas is the post-office. San Bernardino, Cal., 300 miles distant, is the nearest railway station. Freight from that point is 8 cents per pound. By steamer from San Francisco it is $75 per ton. This would be reduced by the extension of the Fort Yuma and Fort Mohave steam-boat route to this point, which is 80 miles above Mohave. The district was discovered in 1861 by N. S. Louis. It was organized in the same year, and has been worked at intervals since.

Previous geological investigations have been made by an assistant of Professor Silliman, in 1865; by the State geologist of Nevada, in 1869; and by a party of this survey in 1871. The direction of the mountains is north and south; of the lodes, northeast and southwest. The veins are very distinct from the wall-rock, and conform, in dip and strike, with the slates in which they are situated. The gangue consists of quartz and calc spar, with talcose slate occasionally intermixed. The adjacent rock is azoic and volcanic. No shells are found. The neighboring mountains are largely made up of trachyte and rhyolite. Assays average $50 per ton. The ores are reduced by roasting and milling. In the roasting process, sulphur and salt are added. The salt is obtained from the extensive deposits in the valley of the Virgen River. Sulphur was recently discovered on the Muddy River.

The silver ores consist of sternbergite, with some argentiferous galena and copper glance. Chloride of silver has been found, but only at the surface of the lodes. Iron and copper pyrites occur with the silver ores. The principal base metal is iron. Copper and lead, with some antimony, are also present. Assays show the presence of gold in the silver-ores.

The principal mine is the Tekehetukup, whose vein is from 6 to 8 feet wide. Its shaft is 200 feet deep, with two levels, 200 feet and 264 feet in length. The ore is found in pockets of various sizes. The Queen City Lode contains galena in talc. Its shaft is 95 feet deep; its vein is 2 feet in width. There is one mill of ordinary structure, with roasting-furnace attached. The stamps make eighty drops per minute. There are four pans and two settlers in the works. The amalgam is strained cold. Up to date, about $100,000 has been extracted from one mine.

A 4-stamp mill, at the mines, will cost $1,500. Other expenses will average as follows: Mining the ore, per ton, $5; reducing the same, $25; mining labor per diem, $4; milling labor, $2.50; running a tunnel on main vein, $20; sinking a shaft, $25; running a drift, $15. One-half of a ton of ore can be stoped by one man in one day.

Grain is worth 7 cents per pound; hay, $70 per ton. Las Vegas, Nev., is the source of supply. Timber is very scarce. The Colorado River furnishes the requisite water. Mountain-sheep and rabbits are the principal kinds of game. There are about a dozen men in the district, besides a few Pah-Ute Indians.

CERRO GORDO DISTRICT, CALIFORNIA.

Examined by Dr. O. Loew, October, 1875.

The mines of the Cerro Gordo district are northeast of Owen's Lake, on the east and west slopes of the Inyo Range, which is the southern extension of the White Mountains. The district covers about 4 square miles. Its croppings are scattered over 800 acres or more. The trend of the mountains is northwest and southeast; of the district, the same. Its post-office is Cerro Gordo. It is 150 miles from Caliente, the present terminus of the railroad, with which it is connected by a freight-line; cost of freight, 5 cents per pound. The country roads are rough. This district was discovered in 1866, by Pablo Flores and companions. It was organized in 1867, and has been worked continuously since.

Previous geological examinations have been made by Clarence King, by the California Geological Survey, and by a former party of this survey. The lodes run northwest and southeast. The lead-veins, on the west side of the range, correspond with the stratification of the country rock, which is of Silurian limestone; the copper ores occur in syenite, on the east side of the mountains. The veins are true fissure-veins, in some cases contact-veins between limestone and slate. A thin seam of clay frequently covers the foot-wall. At times calcite and clay are found in the vein-matter, and these are intermixed with galena and the carbonates. The richest silver ores generally lie near the center of such mixed veins. The mountains are largely made up of Silurian limestone. Encrinites are found in this in great numbers. Some spirifers also occur, but other fossil-shells are rare. Dikes of intrusive rock, such as diorite and syenite, abound, and the strata are much displaced, in some cases standing on edge. On the east side of the main peaks there are large foot-hills entirely made up of syenite, evidently of an eruptive character. Quartz-ledges, with stetefeldtite and chloride of silver, traverse this rock. This eruptive syenite consists chiefly of oligoclase, orthoclase, and hornblende, with quartz in small quantities. Signs of former glacial action are evident.

The ores are worked by the smelting process. The average yield is 400 pounds of lead and 60 ounces of silver per ton of galena, while the argentiferous copper ores give 75 ounces of silver per ton. On the west side of the main peak the ores are galena, massicot, cerussite, anglesite, and mimetite, accompanied by sulphide of arsenic. Cerargyrite and argentite occur, and nuggets of gold, in value from $60 to $100, have been found.

On the east side of the peak the ores are gray copper, stromeyerite, and copper glance, with some native silver and argentite. This ore, especially of the Buena Suerte lode, contains gold to the amount of $20 to $30 per ton. Iron, arsenic, antimony, zinc, lead, and copper are the base metals.

The principal mines west of Cerro Gordo Peak are the Union, whose lode is 12 feet in width, with a tunnel 600 feet long and a shaft 800 feet in depth; the Santa Maria, 6 feet in width, with which is connected the Omega Tunnel, 900 feet long, and a shaft 300 feet deep; the San Felipe, with a vein from 1 to 6 feet in width; the Ignacio, a wide quartz ledge, containing argentiferous copper glance and galena; and the Jefferson, with the Buena Vista Tunnel, 800 feet long. The ore of the latter lode is 35 per cent. lead and yields 80 ounces of silver per ton. It contains a number of "chimneys of soft ore;" carbonate of lead and massicot.

East of the peak are the Belmont, Wittekind, and Buena Suerte lodes. The ledge of the latter is from 2 to 5 feet thick, and contains bodies of ore which assay from $150 to $400 per ton. Its walls are slickensided. It has four tunnels, averaging 150 feet in length. The Potosi Tunnel, 4,400 feet long, to be completed in two years, at a cost of $97,000, is intended to connect with the leading mines at Cerro Gordo.

Altogether, about $5,000,000 of silver have been exported from this district since the introduction of machinery, some seven years ago. There are two smelting-furnaces; of simple construction, but no mills. The reduction-works have a working capacity of 8 tons. The lead and silver of the bullion are not separated here, but are sent elsewhere for cupellation. Great loss is sustained in smelting the ores, as hardly two-thirds of the assay yield is obtained in silver.

Mining expenses will average as follows: Extracting the ore, per ton, $50 to $100; labor, per diem, $4; running a tunnel on main vein, $10; sinking a shaft, $20; running a drift, $5. Owen's Valley and the San Joaquin Valley are the sources of supply for hay ($60 per ton) and grain, (5 cents per pound.) Some piñon and juniper timber grows on the Inyo range. There is no pine nearer than the Sierra Nevada, 60 miles distant. The water is supplied from a source 4 miles away, from which three powerful engines produce 10,000 gallons a day, lifting it 2,200 feet. The inhabitants of the district number 500, in addition to 100 Pah-Ute Indians. There is no game in the vicinity.

NEW ALMADEN MERCURY-MINES, CALIFORNIA.

Examined by Dr. O. Loew, May, 1875.

These mines are situated in the Santa Cruz Mountains of Santa Clara County. There is no regularly organized district, but the area covered by croppings is 3 miles long and 500 feet wide. They have been constantly worked since 1840, when they were discovered by the whites. Previous to this date the cinnabar obtained here was in use among the Indians as a paint. The post-office is called New Almaden. The mines are 12 miles from the railroad at San José, to which town a stage-line runs. Cost of freight from that point is $2 per ton. The roads are very good.

A geological investigation has already been made by Professor Whitney. The trend of the ore-bearing rock, like that of the mountains, is southeast and northwest. There are no regular veins. The cinnabar occurs in pockets and chunks in the ledge, which is partly serpentine and partly quartz, and as "impregnation" of serpentine, quartzite, and sandstone. The main rock in the vicinity is ferruginous sandstone. Tertiary strata extend along the foot-hills, but the mines and the adjacent rock are of greater age, probably azoic.

The ores are reduced by roasting, and the mercury distills over into chambers, where it condenses. Water is found at a depth of 300 feet. Cinnabar is the only ore met with. The principal mines are the New Almaden, with a shaft 1,200 feet deep and a tunnel 800 feet long; the Enriqueta, with a tunnel 700 feet long, and the Guadalupe, with a shaft 300 feet in depth. In 1874 the New Almaden Mine alone yielded 11,000 bottles of mercury, weighing 76½ pounds per bottle. There are 6 Postamente furnaces of old construction, with 18 condensing-chambers, and one Idria furnace of the latest style, with water-condensers. The condensing-chambers are of brick, communicating with each other by channels. They are about 30 feet high and from 10 to 15 feet in length and width.

There is but little timber in this region. The Alamita Creek furnishes the necessary water. Deer, quail, and rabbits abound. Domestic animals are abundant, and produce of all kinds is cheap. There are 3,000 people in the immediate vicinity.

U-BE-HE-BE DISTRICT, CALIFORNIA.

Examined by Lieut. Rogers Birnie, jr., October, 1875.

The deposits of this district lie about 72 miles a little north of east from Cerro Gordo, on the northern slopes of a spur of the Panamint range, which trends to the west nearly at right angles to the main range, and separates Panamint and Salinas Valleys.

Ore was first discovered here by William Hunter, of Cerro Gordo, on July 2, 1875. The district was organized July 8, 1875. There are now eight locations recorded.

The ores are principally carbonates of copper, containing more or less silver. The ledges run with the country-rock, northeast and southwest. The main ledge lies between limestone and granite, the outcrop showing for more than 2 miles. Mineral deposits are found through an extent of 10 miles.

Wood and water are scarce, and are brought a distance of 10 miles. Wood is obtained from the higher portions of the spur alluded to above. Water comes from one of the tributary valleys of the Salinas.

Locations of mineral deposits, containing silver in quantity, have also been made in or near Grapevine Cañon, and named Armstrong district. None of these ledges have been developed, and very little work has been done.

The outcrop at the U-be-he-be district is encouragingly spoken of, but at present the region is not very accessible.

SUMNER DISTRICT, CALIFORNIA.

Examined by Douglas A. Joy, October, 1875.

This district lies about 1 mile north of Kernville, on the west side of the North Fork of Kern River, on the eastern slopes of the foot-hills leading down into the cañon of the same. Its post-office is Kernville. It is 40 miles to the railroad at Caliente, with which town it is connected by freight and stage lines. Cost of freight from that point, 1¼ cents per pound. The roads are very good.

The veins are true fissure-veins, with slickensided walls. The wall-rock is granite. The adjacent country is of azoic formation. Ores are worked by milling and amalgamation. The average assay in the Sumner lode is $16 per ton. The bullion consists of 66 per cent. gold, 31 per cent. silver, and 3 per cent. of base metals. A newly-discovered lode, the Mineral King, is said to assay from $100 to $400 per ton in silver and $50 in gold.

In the Sumner lode the rock is principally a gold-ore. The quartz contains free gold, ruby silver, and pyrites. Antimony, arsenic, and iron are the base elements.

Water-level is reached at a depth of 118 feet. In the Sumner Mine the shaft is 290 feet deep, with 3 levels, each 300 feet in length. The vein is from 30 to 40 feet in width. A plunger-pump is connected with this mine. The vein of the Mineral King is 22 feet in width, and contains ruby silver and free gold. It was struck at a depth of 109 feet. Its tunnel is 50 feet in length.

At these mines there is one mill with 80 stamps and one with 16. Each stamp weighs 900 pounds, and makes 80 drops per minute. Each stamp will crush a ton of ore per day. There are six pans and settlers. Amalgam is strained cold. The mill is run by the water-power of Kern River.

Cost of milling and mining labor at Kernville is $3 per diem. Grain is 4 cents per pound. Good pine timber grows on the surrounding mountains. The water of Kern River is abundant and good. There are 1,200 inhabitants in the district, including Kernville. Domestic animals are numerous, and game is not scarce; the deer, bear, quail, and rabbit are found.

ROSE SPRING DISTRICT, CALIFORNIA.

Examined by Lieut. Rogers Birnie, jr., 1875.

This district is in the Panamint Range, between Panamint Valley and Death Valley. It is bounded on the north by Cottonwood Cañon and on the south by Rose Spring Cañon. The ledges lie in the main range and in the foot-hills and spurs which slope into Death Valley on the east. The district is of irregular shape. Its outcroppings and developments are extensive. It was discovered by Joseph Nossano in 1873, and has since been worked continuously, but not vigorously. As early as 1860 antimony was mined in this region. Its post-office is Panamint. It is connected by wagon-road with Tate's station on the Panamint and Caliente stage and freight line. Its total distance from Caliente is 154 miles. A number of geological investigations have been hitherto made in this district.

The country-rock is of granite, slate, porphyry, and limestone. No fossils are found. The veins run with the country-rock, and are generally richest in the slate and porphyry, which are the predominant constituents. The trend of the mountains is north and south ; of the veins, east and west.

The principal ores are chlorides, bromides, and sulphurets of silver; some fine specimens of horn-silver were found in the dump of the North Star mine. Traces of gold are noticed. But little smelting ore has been discovered. The free and wet processes are both used, but the wet milling ore is most abundant. Some of the ore is roasted. An average of the assays of eight mines in the district gave a yield of $919 per ton, silver.

There have been 154 locations made in the district. Some of the most important are as follows :

North Star Mine.—Vein about 2½ feet thick ; trend, a little north of east ; dip, 60° ; free milling-ore ; a contact vein, with walls of porphyry and slate ; chloride is the most prominent ore.

Garibaldi Mine.—Vein 50 feet thick in places ; trend, east and west ; dip, 64° ; milling ore, with trace of antimony and copper, and a small percentage of lead ; hanging-wall, slate ; foot-wall, probably metamorphic rock.

Annie Mine.—Incline 75 feet long ; trend, northeast and southwest ; ore, stromeyerite.

Polar Star.—Trend, north and south ; shaft, 14 feet ; cut, 20 feet ; yield, chlorides and stromeyerite.

Maria.—Shaft, 20 feet ; yield, chloride, native silver, and stromeyerite.

Star of the West.—Trend, north and south ; dip, east ; yield, stromeyerite ; wall-rock, limestone and porphyry.

Nellie Grant.—Shaft, 35 feet ; ore, chloride ; trend, east and west ; dip, north ; wall-rock, slate or limestone.

Mary Ann.—Shaft, 5 feet ; trend, north and south ; dip, west ; ore, stromeyerite ; wall-rock, limestone and slate.

A 10 stamp mill, erected in the district, would cost $25,000. Cost for mining the ore, $4 per ton ; mining labor, per diem, $4 ; running a tunnel, per foot, $15 ; sinking a shaft, $18 ; running a drift, $15. Expenses would be decreased by a railroad. Panamint is now the general source of supply. There are but few facilities for agriculture. There is but little stock in the vicinity, although there is grazing enough for a moderate number. In the southeastern part of the district timber is plenty. Water is scarce, being obtained from springs. Many of the mines are 4 or 5 miles from water. There are 25 inhabitants in the district, besides 30 Pah-Ute Indians. Mountain-sheep, rabbits, and deer abound.

AZTEC DISTRICT, NEW MEXICO.

Examined by Alfred R. Conkling, July ,1875.

The Aztec District, also known as the Creek District, is situated between Baldy Peak on the west and the Cimarron plains on the east.

One square mile will include the area marked by mineral croppings; half an acre will cover all the surface yet broken for mines. North and south is the general trend of the mountains from the main range, of which minor ridges and spurs project to the east. The district was organized in 1868, but placer-mining had been in progress prior to that time. It has been worked continuously. Its post-office is Ute Creek. The nearest railway town is Las Animas, 200 miles distant by wagon-road. Freight from that point is 3¼ cents per pound.

The Aztec lode is in a wall-rock of slate and granite. Its direction is from northeast to southwest. The veins are quite irregular, and dip into the mountains. No fossils are found in the immediate vicinity of the ore-deposits, but just east of them cretaceous limestone occurs. The ores are worked by the wet process, without roasting. The average yield is $60 per ton, principally in gold. The ores are auriferous quartz, galena, chalcopyrite, malachite, pyrite, and calcite in rhombohedrons and in the form of geodes.

The water-level has not been reached. As the depth increases, the ore grows poorer. In the Aztec mine, which is the principal one, a depth of 180 feet has been attained. This was formerly worked by a company with mills, but it is now occupied by individuals, who reduce the ore in arrastras. The veins are 12 in number; branches join the main vein from the east and south. The width of the main vein is from 4 inches to 6 feet. Ore is found in pockets. The gulch-gold is worth $18.50 per ounce, mine-gold $17.

According to reports, which are not very reliable, a total amount of $1,000,000 has been extracted. In 1874, a sum of $15,000 was taken out, and about an equal amount in 1875. It is said that the placer-mines yielded $86,000 in 1869. When the mill was running regularly, 15 tons of ore were worked in 24 hours. It was a 15-stamp mill, each stamp weighing 400 pounds, and making 24 drops per minute. There were three pans, but no settlers. The amalgam was strained cold.

The cost of a 15-stamp mill, delivered at the mines, is $30,000. A smelting-furnace with one stack can be made for $500. Other expenses will average as follows: reducing the ore, $6 per ton by arrastra, or $3.50 by mill; mining labor per diem, $2.50; milling labor, $2; running a tunnel on main vein, $25; sinking a shaft, $18. Grain costs from 3 to 4 cents per pound; hay from $12 to $15 per ton. Facilities are favorable for all produce except grain.

Timber is plenty, and there is an abundance of good water. The country roads are good. The inhabitants of the district number about 60, besides wandering bands of Ute and Apache Indians. There are several varieties of game, including the deer, turkey, bear, and grouse.

On the west side of the mountains placer-mining is extensively practiced. The water is brought from the Spanish range, through a gigantic flume, 47 miles long, which was built at a cost of $250,000, and has since been sold for $20,000.

LOOKOUT DISTRICT, CALIFORNIA.

Examined by Lieut. Rogers Birnie, jr., August, 1875.

The mineral cropping of this district is found throughout the eastern slopes of the Argus range, from Darwin Cañon on the north to Shepherd's Cañon on the south. The trend of the range is north and south; of the foot-hills, in which the deposits mostly lie, east and west. The district was discovered in May of the present year, by Jerome Childs. It was organized in July, and has been worked since that time. It may be reached from Caliente by following the road of the Cerro Gordo Freighting Company to Tate's Station, and thence to the district, distance 156 miles. Or another route is by way of the Olancha, Independence, and Caliente road to Darwin, 152 miles, and thence by pack-trail to the district, 12 miles.

As far as discovered, the lodes and deposits have no definite direction. They are found in a belt of limestone, 3 miles wide, which runs from north to south. The veins are richest in white limestone. When they approach the granite, the ore changes from galena to copper—galena in schist is found in the granite. The country-rock along the summit of the range, from Shepherd's Cañon to a few miles south of Darwin, is granite; thence to Darwin Wash it is slate. At an altitude of 5,000 feet limestone appears, and at irregular intervals on the eastern slope granite and limestone occur.

Some ores are reduced by smelting, others by the milling process. Galena is the principal product. The base metals are lead, iron, and a little copper. Assays from the mines on Lookout Hill average $150 per ton. The assay yield, in general, varies from $25 to $3,000 per ton. Iron is a constituent of some ores to the extent of 30 or 40 per cent. Gold is present, sometimes to an important degree, and 12 or 14 locations of gold ledges have been made. Some assays have shown $9,000. The iron sulphides are reduced by the roasting process.

About 60 locations have been made, most important of which are the Lookout, Minnietta, Modoc, Confidence, Minute Gun, Lone Star, Antelope, Bismuth, Capital, Royal Arch, and De Soto. The Lookout and Confidence, on the north side of Lookout Hill, run nearly parallel and a little north of east. The Modoc, on the south side, runs southeast. The ledges apparently bear toward the center of the hill. The Lookout and Minnietta contain galena that is nearly pure. Gray carbonate and chloride are also found in the Minnietta. Quartz shows most plainly in the Confidence. In general, a great amount of ore is in sight along the ledges, but the mines are not yet sufficiently developed to decide concerning the permanency of the vein. A granite vein thrusts out of the limestone at one-half mile below Silver Springs.

A ten-stamp mill, erected in the district, will cost $50,000. This would be greatly decreased by the completion of the Los Angeles and Independence Railroad. Cost of mining or milling labor per diem is $4. Cost of sinking a shaft is $10 or $12 per foot. Supplies are brought from the railroad at Caliente, and from Owen's River Valley, by way of Darwin. Timber is 8 or 9 miles distant. Water is obtained chiefly from springs. The Minute Gun Spring, above the principal mines, flows two inches of water, miner's measurement; Silver Spring, on a level with the mines, is of equal size, and the Lillian Spring yields one inch of water. Other supplies could be brought in pipes from Egan's Falls on the north, or Snow's Cañon on the south. There is but little game in the district. It is peopled by 40 persons. The roads are poor. No freight-lines are yet organized.

TEMESCAL TIN MINES, CALIFORNIA.

Examined by Lieut. Eric Bergland, May 4, 1876.

These mines, not worked at present, are in the southern part of San Bernardino County, on the southwestern slope of the Temescal Mountains, about 7 miles from the village of Temescal. They may be reached most easily from the railway at Anaheim, or from Colton, a station on the Southern Pacific Railroad. A good road leads from the village up to the mines.

As the mines were abandoned, it was difficult to obtain information concerning them. Three shafts were found. As they were partially filled with water, their depths could not be ascertained. In one of these the water was 90 feet below the surface, and in the other two, 40 feet. The latter were closely boarded; the walls of the former were unprotected, but no tin-ore was visible near the top of the ground. The ore had been raised by windlasses worked by hand or by horse-power. One shaft is vertical, the others are inclined at angles of 65° and 75° to the horizontal.

The vein extends a considerable distance, and is said to appear in the range on the south side of the valley. Last winter one of the shafts was cleared of water, and five tons of ore were extracted and shipped to England. The ore is reported to contain 18 per cent. of tin.

Two houses and a blacksmith-shop had been built at the mines. In an adjacent valley water and a small patch of grass were found.

Economic Botany and Agriculture.—Dr. J. T. Rothcock, acting assistant surgeon, United States Army, submits a report on this subject, including portion of California, and the islands of Santa Barbara and Santa Cruz, which will be found interesting and instructive, especially to those who have not an intimate knowledge of the varied resources and productions of that region. (See Appendix H, 5.)

The report of Dr. Loew on the physical and and agricultural features of Southern California, the coast counties, island of Santa Cruz, and the Mohave Desert, valleys, oases, and soils, will add not a little to the interest attaching to the Pacific Coast. (See Appendix H, 6.) To the botanist, especially, will his notes on the geographical distribution of vegetation in the great Mohave Desert be of value. (See Appendix H, 7.)

Zoölogy.—Mr. H. W. Henshaw, who has been connected with the survey for several years, presents a report upon the results of his observations and collections. (See Appendix H, 8.) Mr. S. H. Scudder, of Cambridge, Mass., kindly consented to examine the specimens of orthoptera collected during the season, and his report thereon is herewith. (See Appendix H, 9.) Dr. John L. Le Conte cordially consented to examine the collection of coleoptera, and his classified report is appended. (See Appendix H, 10.) Lieut. W. L. Carpenter submits a report on the Alpine insect fauna of the Rocky Mountains. (See Appendix H, 11.)

Mr. Henshaw presents some notes upon the mammals taken and observed in California during the season. (See Appendix H, 12.)

Ethnology.—Dr. H. C. Yarrow, acting assistant surgeon, United States Army, was placed in immediate charge of a special party for the purpose of making ethnological researches in the vicinity of Santa Barbara, Cal., and his report, prefaced by an historical account, as given by Cabrillo, a Portuguese, who visited this coast in 1542, is appended. (See Appendix H, 13.) The collection obtained by this party by excavating in mounds and graves is of varied interest, and the examination of and report upon the same by Prof. F. W. Putnam, curator of the Peabody Museum of Archæology, will, when it appears in Volume VII of the survey reports, it is believed, prove a considerable addition to the early history of this section.

Dr. Loew also submits a report of an ethnological character, regarding the Indian tribes visited, their customs and relations, &c. (See Appendix H, 14.)

Dr. O. Loew presents a report upon the physiological effects of a very hot climate, which, while somewhat foreign to the general subjects of investigation by the expedition, may yet, as stated by him, be one " deserving attention in connection with the exploration or occupation of the Colorado Valley, and one not heretofore treated upon to any great extent." (See Appendix H, 15.)

Philology.—Mr. Alb. S. Gatchet has examined several vocabularies collected by members of the expedition, and presents an analytical report upon idioms spoken in Southern California, Nevada, and on the Lower Colorado River. (See Appendix H, 16.)

PUBLICATIONS.

During the year the following maps have been published :
Progress Map.
Crayon Atlas Sheets 49 and 67.
Geological Atlas Sheets 50, part of 58, and 66, 59 and 67, 75 and 83.
Map showing restored outline of an ancient fresh-water lake, (" Lake Bonneville.") Geological Atlas title-sheet.
Topographical Atlas Sheets 61 B, 61 C, 61 C, (sub.,) 69 D, 75, 76, and 83.
And the following reports :
Preliminary Report, 1869. (Revised edition.)
Volume III, (Geology.)
Volume IV, (Paleontology.) Part I. Invertebrates.
Volume V, (Zoölogy.) (Not ready for distribution.) Progress has been made upon Volumes I, II, VI, and VII, and it is expected to have the MS. of three of them in the hands of the printer by January 1, 1877. The manuscript for Catalogue of Declination of 2,018 stars has gone forward.

DIVERSION OF THE COLORADO RIVER.

By Department letter of May 6, 1875, in addition to the regularly organized work of the season, a special preliminary examination as to the feasibility of diverting the waters of the Colorado River of the West for purposes of irrigation was authorized, and, as stated in my last annual report, Lieutenant Bergland was intrusted with the charge of a separate party for this purpose.

Both a summer and a winter trip were made. In submitting his report, Lieutenant Bergland, after describing the organization of the party, the routes followed, and discussing the question of the diversion of the river, decides, as the result of his first trip, that it cannot be successfully done at any point between the foot of the lower Grand Cañon and the head of the Colorado Valley, an approximate distance of 326 miles, because of the outlying ridges with high passes, through some of which the river has cut its way ; and as a result from his second trip, he concludes that no such diversion can be successfully made at any point along the present channel of the river within the territory of the United States.

As shown in my report to the Chief of Engineers of April 27, 1875, in pursuance of a communication of Mr. E. F. Beale to the President, under date of March 20, 1875, no diversion of the river on a large scale could practically be made between the junction of the Green and Grand rivers and the point of its emergence from the Grand Cañon, near which

Lieutenant Bergland began his survey. He has examined the general course of the river for a distance of 326 miles, and caused sections to be made near the mouth of the Virgin and Camp Mohave. Observations were made at Liverpool Landing and about Fort Yuma, with profiles from Fort Yuma westward to the depressed area near Indian Wells, and from Algodones to the westward, in vicinity of the bed of New River, so called. The volume of water measured at Stone's Ferry, near the mouth of the Virgin, was found to be, in August, 1875, approximately 18,410 cubic feet per second, or sufficient in amount to irrigate 3,682,000 acres, assuming one cubic foot per second for each 200 acres.

At Camp Mohave in September the flowage was 8,680 cubic feet per second, or sufficient at the assumed rate for the irrigation of an area of 1,736,000 acres. At Fort Yuma in March, 1876, the volume of water was found to be 7,658 cubic feet per second, sufficient for the irrigation of 1,531,600 acres. The estimate of the volume of water necessary for the irrigation of a single acre for one crop, (one two-hundredth part of a cubic foot per second, used 200 days,) is taken from the report of the commissioners, Messrs. Alexander, Mendell, and Davidson, upon the irrigation of the San Joaquin and Tulare Valleys of California. By assuming the increased area of the cross-sections of the river at high water as shown by the section made at Stone's Ferry, the velocity of discharge remaining the same as that noted as the mean velocity at that point in August, the increase of the volume of discharge would be 31,439 cubic feet per second, making the total volume of discharge nearly 49,849 cubic feet per second, or a little less than three times the volume observed in the middle part of the heated season. Could this amount be utilized upon ordinarily compact soil, which, however, is partially impracticable, approximately 10,000,-000 acres of land could, by its instrumentality, be brought under cultivation. The increase of discharge upon the assumption of a mean velocity at high water, the same as that observed in September at Camp Mohave, would give 34,274 cubic feet at that point; because of the large area of bottom-lands overflowed at high stages, the velocity would be increased but slightly.

At Fort Yuma the increase, on account of estimating with increase of cross-sections of the river at these points, as shown by cross-section, assuming an unchanged mean velocity, would be 14,244 cubic feet, giving nearly a double volume of flow, while at Camp Mohave it is tripled, and at Stone's Ferry nearly doubled.

Lieutenant Michler, while engaged upon the Mexican Boundary-Survey in the winter of 1854-'55, estimated the volume of discharge at the

NOTE FROM MEMORANDA OF LIEUTENANT BERGLAND.—At Stone's Ferry and Fort Yuma it is seen that the increase in discharge corresponding to the observed high-water marks is about double that of the measured discharge, while the increase at Camp Mohave is three times that of the measured discharge.

As the highest observed water-marks given at the three places do not belong to the same year, no direct comparisons can be made as to the increase in discharge, but it is highly probable that the large approximate increase at Camp Mohave is nearer the true increase than that at the other places, since at Camp Mohave the velocity would not be greatly augmented, while at Stone's Ferry there must be a considerable increase in the mean velocity due to the rise. This increase of velocity would be less at Fort Yuma, since a portion flows outside the sections, but it would be much greater than at Camp Mohave.

Fort Yuma, Cal., March 20, 1876.—Area 2,726.5 square feet. Width 461 feet. Hydraulic radius or mean depth 5.85 feet. Mean velocity 2.809 feet per second. Discharge 7,658.74 cubic feet per second. High-water mark of 1862 above surface of river 10.19 feet. Increase of area of section at high water, 5,059 square feet. It is to be noted that when the water reaches high-water mark of 1862 the bottom-lands are more or less flooded, and all the water does not pass through the section. The high-water marks given are the highest observed at each place. Increase in discharges

junction of the Colorado and Gila Rivers 6,249 cubic feet per seco
with a velocity of 3 feet per second. The estimated evaporation
Camp Mohave as given by Lieutenant Bergland, deduced from obser
tions extending over a small interval, is a little less than 8 feet ; t,
would be increased at Yuma, while at Stone's Ferry it would be sim
in amount.

A lake required to contain the influx of all the waters of the rivel
the stage mentioned near Fort Yuma, the only point near which it
be diverted, would be somewhat less than one thousand square m
The area of depressed region (approximately) to which a channel ca
cut from the river is 1,600 square miles. Hence, it appears that du
the low-water stage this lake would act as a reservoir without ou
while at a high stage the overflow at its maximum would cause a r
lar channel to be cut, making its own outlet to the gulf and forn
along its route a series of lagoons during seasons of no flood, or
greater part of each year; but should it be possible to introduce wa
from the gulf to the lake thus to be formed in the depressed area,
same would be transferred into a tidal lake with a current settin
toward the gulf by way of the conduit thus formed, receiving i
waters from the river and affected by the changes of the high and
waters from the sea. No change of climate can be expected to ensue
the formation of a lake or number of lakes, aggregating the above a
The amount of evaporation from the surface would be insufficien
change noticeably the relative humidity of the surrounding atmosph
while the prevalent winds, both from local and far-distant sources, we
diffuse and disseminate the amount of moisture thus received with g
rapidity. The soils along portions of the route, as shown by Dr. Lo
report, (see Appendix H 6,) are, many of them, of an arable nature, n
ing moisture only to be made productive.

In the lower parts of the valley of the Colorado, cotton, coffee, su
tea, and flax could undoubtedly be grown with success, while a s
change for the better in the relative humidity at local points could
likely be brought about by planting the eucalyptus, and trees of
strength of foliage, along the narrow valleys bordering the river.
mulberry could also doubtless be made to flourish. The unfortu
climate to which this portion of the Southwest is at present treate
the hand of nature is likely to retard its rapid settlement, even if w
was plentifully available; still at some future time in the settleme

through section = 14,244 cubic feet per second. Here the velocity throughou
section would be increased at time of high water, and a large quantity would
outside of the section.

Stone's Ferry, Nevada, August 12, 1875.—Area of section = 5,723 square feet. W
= 480 feet. Hydraulic radius, or mean depth = 11.89 feet. Mean velocity = 3.21
per second. Discharge = 18,410.38 cubic feet. High-water mark of 1871 is 17.0
above surface of water at time of observations. Increase of area at high water 9,
The whole discharge at high water takes place through the section. Supposin
mean velocity to remain the same as August 12, 1875, the increase in discharge w
be 31,439.7 cubic feet per second; but as in reality there would also be an increa
the velocity, the increase in discharge would be somewhat greater than this, but
much greater cannot be determined without direct experiments.

Camp Mohave, A. T., September 2, 1875.—Area of section = 4,628 square feet. W
1,116 feet. Mean depth (hydraulic radius) = 4.144 feet. Mean velocity = 2.50
per second. Discharge = 11,610.93 cubic feet per second. High-water mark of
above surface of river = 8 feet. Increase of area of section at high water 13,6
feet, (excluding overflow on flats.) Increase in discharge through the section wou
34,274 cubic feet, but as a considerable quantity of the bottom beyond the sect
then covered with water, this will not represent the total increase. Here there w
be but slight increase in the velocity, as the water has a chance to spread ove
bottom-lands, (see plot of section.)

the West, each cubic foot of the waters of the Colorado is likely to become valuable in agricultural, mining, or other pursuits.

The topographical data obtained by Lieutenant Bergland's party, as partially shown by the sketches herewith, (see Appendix B,) although apparently meager, is yet sufficient to illustrate the route of the party and to locate the principal points referred to in the report. It will all be incorporated in the regular atlas maps covering the regions traversed, to which it will prove a valuable contribution. A sketch of the region south of latitude 37° 20′ is here introduced, showing the diversion between the direct coast-drainage and that forming part of the great valley of the Colorado and of the interior basins, those lying in California and Nevada. The total number of square miles shown is 104,300, of which 88,300 square miles approximately belong to California, and 16,000 square miles to Nevada; 42,600 square miles of this area is a part of the great valley and coast drainage, or 48 2-10 per cent. Of the remaining area about 45,700 square miles, of which less than 2 per cent., is now arable, or could be made arable with all the water available within its limits. Of the amount constituting the coast and valley drainage, allowing that 60 per cent. can be made valuable for agricultural purposes, (an amount presumedly largely in excess of what would be found to be true after a rigid examination and survey of these areas,) it appears that of this portion of California south of latitude 37° 20′, only 29.4 per cent. can ever, with the best facilities, be made useful for agricultural purposes. The area in California of a desert character, the 45,700 square miles above mentioned, is much larger than the area of the same character in Arizona, the latter having been charged as the acme of the desert of the United States, whereas the Great Colorado plateau and plateau-ridges about the upper waters of the Salt and Gila Rivers, and the sources of certain minor streams, as the San Pedro and Santa Cruz, are sections that are quite the reverse of deserts. While the southwestern portion of Arizona must always remain sparsely inhabited because of its want of water and stretches of sandy waste, yet the amount thus withdrawn from all hope of settlement, except at little water-stations, is comparatively less than the desert portion of California lying south of latitude 37° 20′ N., as shown by the sketch. This statement will in no wise invalidate the claims of the remainder of this great State to the rank it has so well taken among the greater grain-producing States of the Union, no more than it accuses nature in being wanton in the disposition of its fields for the uses of mankind.

That part lying east of the summit of the Sierras, where the rainfall immediately changes from one as great as 40 to 50 inches to 6, 8, and 10 inches annual fall, is not included, and is in addition to the area of 45,700 square miles mentioned. The part of Nevada shown is that portion least inhabitable from an agricultural point of view. While in all these desert sections mines have been and are still being found, oftentimes in sufficient proximity to water to warrant their being worked, still little underground exploration, even, has been undertaken, except in a few localities. Roads across this section of the country have been few, and the danger of exploring away from the main ones is the principal hinderance to the hardy prospector in his labors. Dr. Loew speaks of the wonderful change noticed by every traveler in crossing the summits of the Coast range south of Tehachipi Pass, found to exist between the desert and coast flora. This report will doubtless prove interesting. (See Appendix H 7.)

The basin drained by the Colorado River of the West comprises approximately 241,965 square miles, or approximately 154,857,600 acres,

composed of valley, plateau, and mountain section in wonderful
riety, portions of which have been laid out into the following politi
divisions: California, Nevada, Utah, Wyoming, Colorado, New Me
co, and Arizona; and prior to its exit into the Gulf of California
washes the eastern shore of Lower California and the western sh
of part of Northwestern Mexico. The length of the river from t
junction of the Grand and Green is approximately 875 miles. I
elevation at Hanlon's Ferry, near Fort Yuma is 120 feet; at t
grand bend to the south, near head of Black Cañon, 900 feet:
the junction of the Green and Grand rivers, 3,860 feet, from wher
the name Colorado begins. It is essentially a cañon river until
leaves the territory of the United States, when its character in t
regard materially changes, and with it the peculiarities of eros
and alluvial depositions in vicinity of its shifting bed, while opp
tunities for diverting the same are more likely to be found.
climate along its banks varies partly with the elevation, but mo
largely with the amounts of rainfall, which, until reaching the
Grand Cañon, may be said to vary from $\frac{1}{2}$ to 10 inches annually.
In portions near the sources of Grand River, in the high, mountainous regions of Colorado, the rainfall increases somewhat in proportion to the altitude, and without any specific data on the
subject it is safe to say that it reaches 40, if not a larger number of
inches annually; but the areas showing the larger amounts of precipitation are comparatively small, and confined to the narrow valleys of
the main stream and their side branches within the mountainous portions proper. No such amount of rainfall is known in any part of the
Green River Basin, even at its source. Very little of this valley is
available for agricultural purposes, and it would be difficult to improve
large tracts of land along the main stream or any of its immediate tributaries. The districts, then, into which the entire valley-drainage of the
Colorado area may be divided are as follows:

First, the more desert parts, bounded on the east by the western mesa
wall of the Lower Grand Cañon, which is limited in extent on the north by
the rim of the great interior basin near the Nevada and Utah line, and extending southward by the heads of the Salt and Gila rivers to the continental divide. Within this area there are strips of considerable size
not desert, but diversified between mountain and desert, the northern
portions of which, especially in the vicinity of the Salt and Gila rivers,
are susceptible of cultivation, forming some of the finest grazing-fields
in the world. with large patches of pine and other timber, and admitting
of considerable settlement. Ridges traverse these portions, those running north and south being usually mineral-bearing, that have been
prospected, but can scarcely be said to have been worked for the precious minerals.

The next is the plateau and cañon district, which has been delineated
along its western line in the earlier reports and maps of the survey, the
eastern limit running along the continental divide north as far as latitude 37°; from that point to the mouth of the Green and Grand rivers,
thence in a nearly due west line to the great mesa-wall, passing northward of the Lower Grand Gañon, with an arm of the great interior basin
in the vicinity of latitude 38° and longitude 113°, approximately.

The third portion is the province of the mountains, with their out
lying foot-hills, being the basins, respectively, of the Green and Grand
rivers, whose peculiarities have been noted by earlier explorers, and
are being examined from time to time by Government parties.

The first or desert province is approximately 72,889 square miles; the plateau province approximately 83,986 square miles; the mountain province approximately 85,190 square miles.

The majority of the land within the drainage of this river and its tributaries is still owned by the Government. The uses to which it may be applied must be confined largely to grazing and mining purposes, while local spots will admit of cultivation by irrigation process, in connection with the gradual development of the country, yet the husbanding of water becomes a matter of great import to all those who may at some future time occupy this portion of our interior domain. For a distance of 435 miles from the junction of the Green and Grand rivers it traverses territorial domain; and all that part of the Grand River still traversing public lands, as well as the basin of the Green River, is now owned by the Government, with few exceptions, and the disposition of its waters is a subject over which the General Government should assume entire control.

Legislation may be had defining more clearly and with greater certainty the rights of persons settling along the banks of main and tributary streams. It will be seen that the amount of precipitation, except in the mountainous province, is very small, and that little which is collected and, for which the river is the main channel of discharge, should be utilized in the most efficient manner. That can only be done by using its waters, so far as practicable, upon the land along its banks and within the immediate valleys adjacent thereto, and by artificial reservoirs. The measurement of the waters of this stream at proper points within the three provinces is a matter of importance, and the same should be ordered at an early day, so that the volume of water that could be made available for irrigation at stated intervals of the year, and especially during the seasons best adapted to crops, might become known, to the end that general legislation could direct the *pro rata* uses of the same for purposes of irrigation, when needed by actual settlers who have acquired title from the Government, or who are about to do so in pursuance of the homestead-laws. There being no law upon this subject to protect the rights of settlers upon streams as against prior occupants, nothing now prevents the diversion of the water of any stream, not alone of this great river above the head of navigation, but of all those flowing through Government lands in this and all other drainage-basins of the Western States and Territories, by interested proprietors, whose *locus* may have been selected at the point most likely to control the maximum of the waters of the immediate water-shed. It has been decided that water-rights guaranteed to settlers under the law of July 26, 1866, (Revised Statutes, p. 432, sec. 2339,) by a late opinion of the Supreme Court, rendered in the case of Basey *vs.* Gallagher, may inure under certain conditions. (See the American Law Times and Reports, March and April, 1875, for decision of the court.) The conditions of the case in question are best shown by the following extracts from the brief of the case, viz:

"4. In the Pacific States and Territories a right to running waters on the public lands of the United States for purposes of irrigation may be acquired by prior appropriation as against parties not having the title of the Government. The right exercised within reasonable limits, having reference to the condition of the country and the necessities of the community, is entitled to protection. This rule obtains in the Territory of Montana, and is sanctioned by its legislation.

"5. By the act of Congress of July 26, 1866, which provides 'that whenever, by priority of possession, rights to the use of water for mining, agricultural, manufacturing, or other purposes have vested and accrued, and the same are recognized and acknowledged by the local customs, laws, and decisions of courts, the possessors and owners of such vested rights shall be maintained and protected in the same,' the customary law with respect to the use of water, which had grown up among occupants of the public land under the peculiar necessities of their condition, is recognized as valid. That law may be shown by evidence of the local customs, or by the legislation of the State or Territory, or the decisions of the courts. The union of the three conditions in any particular case is not essential to the perfection of the right by priority; and in case of conflict between a local custom and a statutory regulation, the latter, as of superior authority, will control."

And portions of the opinion as delivered by Justice Field:

"The question on the merits in this case is whether a right to running waters on the public lands of the United States for purposes of irrigation can be acquired by prior appropriation as against parties not having the title of the Government. Neither party has any title from the United States; no question as to the rights of riparian proprietors can therefore arise. It will be time enough to consider those rights when either of the parties has obtained the patent of the Government. At present both parties stand upon the same footing; neither can allege that the other is a trespasser against the Government without at the same time invalidating his own claim."

* * * * * * *

In the case of Tartar *vs.* The Spring Creek Water and Mining Company, decided in 1855, the supreme court of California said: "The current of decisions of this court go to establish that the policy of this State, as derived from her legislation, is to permit settlers in all capacities to occupy the public lands, and by such occupation to acquire the right of undisturbed enjoyment against all the world but the true owner. In evidence of this, acts have been passed to protect the possession of agricultural lands acquired by mere occupancy; to license miners; to provide for the recovery of mining claims; recognizing canals and ditches which were known to divert the water of streams from their natural channels for mining purposes, and others, of like character. This policy has been extended equally to all pursuits, and no partiality for one over another has been evinced; except in the single case where the rights of the agriculturist are made to yield to those of the miner where gold is discovered in his land. The policy of the exception is obvious. Without it the entire gold region might have been inclosed in large tracts, under the pretense of agriculture and grazing, and eventually what would have sufficed as a rich bounty to many thousands would be reduced to the proprietorship of a few. Aside from this, the legislation and decisions have been uniform in awarding the right of peaceable enjoyment to the first occupant, either of the land or of anything incident to the land." (Per Heydenfeldt, J., 5 Cal., 397.) The case here laid confirms the occupant by possession as against the parties subsequently appropriating the waters under like conditions, and declares the doctrine which must determine, in other cases of occupancy after perfection of title, a more thorough and complete recognition of the rights to a specified amount of water, because of prior appropriation, which amount can be limited by nothing short of the entire volume of discharge. That this right has been claimed and maintained, to the detriment of incoming settlers, in the State of California, is well known. The difficulty arises because of the actual severing of the water from the land, its natural bed and mate, by legislation looking to the protection of settlers and other owners in their vested rights, real or supposed.

NOTE.—I am indebted to Hon. Montgomery Blair for reference to the decision of the Supreme Court regarding water-rights under the law of 1866.—G. M. W.

Unless this can be remedied, it is plain that the area west of the one hundredth meridian, and, indeed, still other areas lying between this line and the Mississippi River, cannot be settled except at local points, which will be selected principally because of the opportunities to use the entire water-supply with the greatest certainty, and the several tracts will thus fall into the hands of speculators and other large holders, to the detriment of that class of settlers most likely to enter and occupy these outlying lands, and establish the nuclei of a continuous cordon of settlement from east to west through the entire interior; small settlements though they might be, still they would be susceptible of a healthy increase by immigration, provided that the rights to each settler of an equable distribution of the water should be guaranteed by law. This cannot, however, be completely accomplished, until the gauging of the streams and measurement of the outlying districts are made by meandering the water-sheds, estimating the geographical conditions of each, bringing all into districts arranged according to the physical configuration of the country—the division being usually made by drawing lines normal to the general course of the main stream at each point selected for observation—and by determining the arable and arid portions within these districts, with the general profile of the country, a law could finally be framed that would provide for the entire equable distribution of the waters within each basin or water-shed. To this end it is suggested that, at an early day, the gauging of the rivers and principal creeks be commenced, not alone in the valley of the Colorado, but in all the basins of drainage of the entire western country. This can only be done by establishing a certain number of stations that can be made meteorological stations, furnishing a part of their data for the use of the Weather Bureau, when they can be reached by telegraph, and at which the gauging and changes in the streams can be ascertained, as well as the areas of the water-shed and the source of the main stream, with profiles of lines of irrigation-canals, and marked depressions fitted for storage-reservoirs, with estimates of evaporation, while the amounts of arable and timber and grazing sections can be determined by the moving field-parties engaged in the prosecution of geographical surveys, so that prior to actual occupation an exact knowledge can be had of the amounts of water needed to irrigate certain classes of soils in the several localities. That this can be done in connection with the prosecution of geographical surveys of the interior, with comparatively little increase of expense, is patent to me, and a plan looking to the initiation of such additional work can be submitted at any time; should appropriations be sufficient, and authority granted, the same may be taken up during subsequent seasons.

SUMMARIZED HISTORY OF THE SURVEY.

The following is a brief summarized history of the various points bearing upon the development of this work:

The officer at present in charge, while serving upon the staff of Brig. Gen. E. O. C. Ord (commanding Department of California) in 1869, conducted a reconnaissance looking to the development of routes of communication through Southern and Southeastern Nevada. While completing the results of that work, in compliance with instructions, a project was presented for the prosecution of similar work in other portions of the Military Department of California. In 1870 this work was suspended, owing to lack of funds. In the early spring of 1871 it was re-organized, and funds from appropriations for "surveys for military defenses" and other sources were placed at disposal for its prosecution in Nevada, California, and Arizona. The expedition of

this season was disbanded at Tucson, Ariz., and certain of its members ordered to Washington to prepare the necessary results. In
1872, fresh means having been appropriated by Congress, a more
complete organization was had. The operations of this season were
confined to a lesser area, principally in Utah, partially in Arizona and
Southeastern Nevada. In 1873, the expedition was again sent into the
field, and the character of the work was advanced measurably, in pursuance of the policy of perfecting the methods needed in a fully-
developed geodetic survey of the entire western mountain region.
During the season of 1873, in order to complete and connect the work
extending from Nevada, Eastern California, and Northern Arizona to
the east base of the Rocky Mountains, it was found essential to occupy
points easily accessible by railroad-communication from the east. In
1874, connection between the eastern and western portions of the work
was made more complete. The scheme of triangulation, based upon
the positions of the several astronomical points determined along the
east base of the Rocky Mountains, and the bases developed and measured therefrom, was extended to the west. In 1875, the survey for field-
operations was organized in two divisions, designated respectively as
the California and the Colorado sections. The Colorado section operated to extend the work in Colorado and New Mexico, begun in 1873
and further prosecuted in 1874, while the operations of the California
section were laid in portions of the Southern Sierras and that branch
of the coast range or ranges lying south of latitude 37°, and extending
toward the Colorado desert southward of San Gorgonio Pass. One
party was charged with an examination of the Colorado River, looking
to a determination of the possibility of its diversion from its present
channel for purposes of irrigation.

The area occupied by the survey, and lying mostly south of the fortieth parallel, is as follows:

	Square miles.
The part lying in Nevada	58,940
The part lying in California	40,625
The part lying in Arizona	60,120
The part lying in Utah	44,015
The part lying in Colorado	37,550
The part lying in New Mexico	53,236

Points have been determined by astronomical means in Nebraska and
Montana.

The maps issued up to the present date represent an area of square
miles as follows:

	Square miles.
On a scale of 1 inch to 8 miles	192,217
On a scale of 1 inch to 4 miles	13,028
On a scale of 1 inch to 2 miles	1,091

CONCLUSION.

In submitting the report for the year it is deemed proper to invite
attention to the area covered by the survey in relation to the total area
west of the one hundredth meridian, for which a detailed atlas is proposed as the final result.

It may not be improper to speak of the position that the work has reached,
as regards its methods, *personnel*, class of instruments, &c., resulting in an
organization prepared for service in any portion of the entire territory
west of the one hundredth meridian. Officers of the Corps of Engineers
and line of the Army, being subject to changes of assignment whenever
the exigencies of the service may require, cannot be considered as permanent assistants for a term of years; while most of the civilians whose

services prove satisfactory can only be retained provided continuous appropriations for the field and office work are kept up.

The supply of instruments of that character calculated to be used for the class of observations needed for work in the mountain regions increases year by year, and the resulting accuracy of the work due to the improvements in those employed, and the experience of the observers, renders each year's results more and more satisfactory. Unfortunately, during the long session of Congress, inasmuch as the work of the survey must be conducted during the summer months, (except in cases where it should be advisable to send an expedition to Southern New Mexico or Arizona,) no preliminary arrangements other than of a very meager character can be made for the expedition of a season until action by Congress. Hence each season's field-work is liable to be limited.

Should estimates be asked for looking to the completion of the entire survey after a certain fixed and definite standard, and with an organization selected for this express work, its systematic and vigorous prosecution would be heightened, should the same receive the approval of Congress and obtain a certain reasonable annual appropriation for its continuation.

As the season of 1877 may be made a long one, it is submitted that in order to complete it the amounts estimated should be appropriated for in full.

ESTIMATES.

For continuing geographical survey of the territory of the United States west of the one hundredth meridian, for the fiscal year ending June 30, 1878, being for field and office work $95,000 00

Distributed as follows:

For parties in the field	40,000 00
For office force	13,920 00
For transportation and purchase of animals	10,000 00
For material and outfits	6,500 00
For subsistence in the field	6,000 00
For forage, winter herding, fuel, storage, &c	9,500 00
For repairs of instruments	1,500 00
For contingencies, (erection of observatories and monuments at astronomical and geodetic stations, &c.)	7,580 00
Total	95,000 00
For preparation, engraving, and printing the maps, charts, plates, cuts, photographic, plate and other illustrations for reports, and for additional office-room in Washington, for fiscal year ending June 30, 1878	25,000 00

FINANCIAL STATEMENT.

Amount remaining unexpended from the appropriation for continuing the geographical survey of the territory of the United States west of the one hundredth meridian, for fiscal year ending June 30, 1876	$2,214 75
Amount remaining unexpended from the appropriation for engraving and printing the plates and atlas sheets accompanying the report of the geographical surveys west of the one hundredth meridian, for fiscal year ending June 30, 1876	10,217 86
Amount appropriated for continuing the geographical survey of the territory of the United States west of the one hundredth meridian, for the fiscal year ending June 30, 1877	20,000 00
Amount appropriated for preparing, engraving, and printing maps, cuts, plates, charts, and other illustrations for reports upon the geographical survey of the territory of the United States west of the one hundredth meridian, for the year ending June 30, 1877	10,000 00

All of which is respectfully submitted.

GEO. M. WHEELER,
First Lieut. Corps of Engineers, in charge.

Brig. Gen. A. A. HUMPHREYS,
Chief of Engineers, U. S. A.

EXECUTIVE AND DESCRIPTIVE REPORT OF LIEUTENANT WILLIAM L. MARSHALL, CORPS OF ENGINEERS, ON THE OPERATIONS OF PARTY NO. 1, COLORADO SECTION, FIELD-SEASON OF 1875.

UNITED STATES ENGINEER OFFICE,
GEOGRAPHICAL SURVEYS WEST OF THE 100TH MERIDIAN,
Washington, D. C., April 15, 1876.

SIR : I have the honor to submit the following executive report of the operations of party No. 1, Colorado section of the expedition for geographical surveys west of the one hundredth meridian, during the past field-season, together with a brief description of the topography and resources of the region surveyed and profiles of present and prospective routes of communications within this area.

PART I.—EXECUTIVE REPORT.

The Colorado section of the expedition consisting of the parties under the command of Lieut. W. L. Carpenter, Ninth United States Infantry, C. C. Morrison, Sixth Cavalry, and myself, was organized at South Pueblo, Colo., in the early part of June, 1875, and the area proposed by you for survey subdivided among the three parties as follows :

To Lieutenant Carpenter was assigned the completion of the triangulation partially measured the previous year along the Spanish and Raton ranges, south of the latitude of Fort Garland and north of Santa Fé, with certain portions of atlas-sheets 69*d* and 62*o* which had not been sufficiently examined the previous season, with directions to rigidly correct by triangulation the bases measured under your direction in 1874, at Trinidad, Fort Union, Las Vegas, and Santa Fé, and to make careful barometric profiles of all present or prospective routes across the southern extension of the Sangre de Cristo range between Fort Garland and Santa Fé. In addition to the executive charge of the party, Lieutenant Carpenter was assigned as naturalist to this party, and directed also by your instructions to make such paleontological collections in the field discovered by Prof. E. D. Cope in 1874, in the San Juan basin, as opportunity and time would admit.

Lieut. C. C. Morrison, in accordance with your recommendations, was instructed to complete the survey of such portions of the San Juan range south of the headwaters of the Conejos as were left incomplete the preceding year ; to seek for a wagon-route from the valley of the Rio Chama near Tierra Amarilla, via the Washington Pass and the headwaters of the Bonito, to the west, and to complete the survey of the atlas-sheets 69*c*, 76*b*, 77*a*, 77*b*, and 78*a*, already partly surveyed. This programme necessitated long and rapid marches over areas already surveyed, to such isolated and in many cases almost inaccessible tracts which had been passed by in former years by parties of the survey on account of their rough character and the lack of water and grass for their animals. I respectfully refer to the reports of those officers, who do not report through me, for information as to the detailed manner in which those general instructions were carried out. To each of the above parties were assigned one field-astronomer, one topographer, one barometric observer and recorder, one aneroid and odometer recorder, and the necessary number of packers and cooks.

Each were provided with one sextant and artificial horizon, one triangulation-instrument reading to 10″ of arc by Vernier, one topographer's transit reading by Vernier to 1′ of arc, one odometer-vehicle and three odometers to be used in connection with the topographer's transit and the aneroid in meandering and profiling the roads, hand-compasses for meandering unimportant drainage-lines, two sets psychrometers, two cistern-barometers, two aneroids and pocket-thermometers for the use of such persons as require them, and with the necessary ruled and headed blank-books and forms for properly recording their observations; printed instructions as to the use of books and instruments and as to the methods of survey to be followed, accompanying them.

This organization of parties having been effected and the necessary arrangements made for breaking up the depot which had been established at this point the preceding season, the parties took the field in the following order: Lieutenant Carpenter's party on June 9, 1875 ; Lieutenant Morrison's, June 12, and my own on June 15, 1875.

The organization of the party under my immediate charge was as follows : First Lieut. W. L. Marshall, Corps of Engineers, executive officer and field-astronomer ; Assistant Engineer J. C. Spiller, topographer ; Mr. George M. Dunn, meteorological observer ; First-class Private Wm. Looram, Company D, Battalion of Engineers, aneroid and odometer recorder ; D. Y. Mears, chief packer ; A. R. Mitchell and Harry Gregg, packers ; Thomas Norman and Allen Smith, cooks, or in all nine men.

From Pueblo the party proceeded up the Arkansas River to Oil Creek, thence by way of the Twin Creek Pass to Tarryall Creek, where a small area left unsurveyed in 1873 by my party was filled out and the meander of Tarryall Creek continued until near the point it was left by Mr. Young's meander in 1873. Having completed this, the party proceeded south of the Platte and Arkansas divide and completed the survey of about 250 square miles of the broken and rolling plateau lying in atlas-sheet 61b, between Cottonwood and Badger Creeks and south of Poncho Park, bordering the Arkansas Cañon. The road via the Currant Creek Pass was meandered until the old (1861) wagon-road from Cañon City to the Puncha Pass was reached, and this road then followed to the Arkansas River at the mouth of the South Arkansas, connecting with our meanders of 1874–75 at this point.

From Puncha Pass the pack-train was sent to Saguache, and the topographer and myself proceeded to make stations upon Antoro Peak, a high mass south of the headwaters of Poncho Creek, and upon such other peaks along the continental backbone as were necessary to secure the drainage and detailed topography lying between the headwaters of the Tumichi and the northern tributaries of Saguache Creek.

We rejoined the main party at Saguache on July 10. On the 15th of July, having received the necessary supplies, we followed the southern fork of the Saguache, the tributaries of which were meandered to their heads, and, in addition, eleven triangulations and topographical stations were made by Mr. Spiller upon the high plateau surrounding the headwaters of Saguache and Lagarita Creeks, and dividing their waters from those of the Rio Grande and those sinking in the San Luis Valley north of Del Norte, after which the party crossed from the valley of the Saguache to the Cochetopa, striking this stream where the new Saguache and Lake City wagon-road crosses it. After meandering this road to the summit of the range, the main portion of the party was sent to Lake City, Mr. Dunn meandering the road from the agency to Lake City, while the topographer and myself, with a small party, turned to the southwest, and, attaining the continental water-shed, made two stations on high peaks, standing upon the volcanic plateau which forms the northern rim of the Rio Grande Loup; then proceeded westward, following the Cannibal Plateau, which forms the dividing ridge between the Lake Fork of the Gunnison and the Cebolla, to its northern terminus, and then, crossing the deep cañon of the Lake Fork of the Gunnison, attained the summit of the ridge culminating in the lofty Uncompahgre Peak on the west side of the Lake Fork, which, after making two stations, we left and went for supplies to the camp at Lake City August 3.

On the 5th day of August I left Lake City, and with a portion of the pack-train went to Antelope Park for rations which had been forwarded to that point, leaving the topographer to continue topographical work as directed by me. He made a triangulation-station upon the Uncompahgre Peak, re-occupying that station to perfect angles which had not been read a sufficient number of times the preceding season. Upon another peak, five miles farther east, meandered and profiled the road from Lake City to the head of the Lake Fork; made stations upon Red Cloud Peak, the highest of the Lake Fork group; upon Hanclie's Peak, a lofty mass at the head of the Lake Fork, and then followed the divide between the waters of the Lake Fork of the Gunnison and those of the Rio Grande del Norte, making frequent stations en route as far as to the point where the Lake City and Antelope Park trail crosses the continental divide, and returned to camp at Tellurium Post-Office, at the head of the Lake Fork, on the 15th of August, having made in twelve days three stations over 14,000 feet and six others approaching or exceeding 13,000 feet in altitude, besides the necessary meanders, profiles, and minor stations in an exceedingly rugged and difficult region.

On the 16th of August we proceeded via the Forks of the Animas and the incipient town of La Plata, here located, to the head of Hensen Creek, a tributary of the Lake Fork of the Gunnison, in which vicinity the topographer made several stations, to complete the topography of this region; after which we followed an old Indian trail to the valleys of the Uncompahgre and Unaweep, over a very high and rough country, the mountains breaking down very suddenly and the streams all flowing in excessively deep and rocky gorges. The great gorge of the Uncompahgre and that of Ibex Creek will equal or surpass in ruggedness of scenery and in depth any similar gorge in the United States. The descent into the valley of the Uncompahgre is very abrupt, rendering it improbable that this trail will be of any value, other than as furnishing horsemen a possible way of avoiding the Uncompahgre Cañon. From the head of Ibex Creek to the rim of the Uncompahgre Park, the trail is almost entirely above timberline, and in the next three miles descends nearly five thousand feet to the park below.

After meandering the Uncompahgre to the northern boundary of atlas-sheet 61c, (38° 10′ latitude,) we entered the valley of the Unaweep, a tributary of the Uncompahgre. Two topographical stations were made on low hills in the Unaweep Valley, and a triangulation-station upon the most westerly of the range of rugged peaks which divides the headwaters of the Unaweep from those of the San Miguel, to extend our triangulation to the southwest; and then on August 26 crossed to the drainage-area of the San Miguel River, following the trail around the western extremities of the Unaweep group of peaks.

From our camp near the junction of the two forks of the San Miguel, where this river plunges into its deep cañon, Mr. Spiller with a small party crossed the cañon and made stations upon several of the nearly isolated points near the western border of atlas-sheet 61c, while I meandered the Gold Fork of the San Miguel to its head and made a station upon the divide between this stream and the Uncompahgre. Returning to camp, a station was made by me at the forks of the San Miguel, and then I meandered the trail to the Trout Lake, near the head of the South Fork, where I was rejoined by Mr. Spiller, September 1. After making two stations upon high peaks in this vicinity, the party proceeded over the divide to the drainage-area of the Dolores, but on the 5th September it began to rain, and continued raining and snowing with but short intermissions until September 21. Advantage was taken of every intermission, and during this interval we succeeded in successfully making two stations in the divide between the Dolores River and Hermosa Creek, one of the tributaries of the Animas River, and in meandering the trails over the heads of the Dolores Cascade and Lime Creeks; but the delay thus caused, coming upon us while in the midst of the area we desired to complete, caused us serious embarrassment and loss of time when it was too late in the season to regain it. Having completed the drainage of the upper tributaries of the Animas, the trail to Hermosa was meandered, and on September 22 the party started over to the Rio La Plata, which stream was afterward meandered to its head, and two topographical stations were made by Mr. Spiller and myself upon prominent peaks of the Sierra La Plata. The highest of the La Plata peaks, however, which had been selected for a principal station in the scheme of triangulation, was not accessible from the east, so that the party accordingly proceeded to the Rio Mancos, the western fork of which we meandered, and made a very successful station upon the La Plata Peak at its head.

Retracing our course, a topographical station was made upon the northern edge of the Mesa Verde, and upon minor points between the La Plata and Animas Rivers. On October 1 and 2, we crossed the low ridge between the Animas and Florida, and, after following the latter several miles to the northeast, crossed another low divide and entered the drainage-area of the Rio Los Pinos. The party was then divided, and the main party sent through the Los Pinos Cañon, which Mr. Spiller meandered; while I followed the trail about the heads of the western tributaries of the Piedra, along the divide between the Piedra and Los Pinos River, making the necessary topographical stations en route, and rejoined the party at the head of the Los Pinos Cañon. From this point the topographer and small party proceeded to make stations and secure topographical details about the headwaters of the western fork of the Rio Los Pinos, whence my party were driven by snow in 1874, while I went for supplies to Antelope Park, meandering en route the Rio Grande from the Ute Pass to San Juan City, and, in returning, profiled the wagon-road over the head of Crooked Creek as far as to the entrance to the pass; a topographical station was also made by me on a low peak on the western edge of Antelope Park.

Upon returning to camp at the head of the Los Pinos, October 9, I left directions for Mr. Spiller to complete the meanders of the tributaries of the Piedra, to make a triangulation-station upon the double-capped peak north of Pagosa Springs, and upon lower points in the Piedra basin, and then to cross over into the Rio Grande basin and proceed with its survey until rejoined by me. This programme was satisfactorily carried out by Mr. Spiller.

On October 10 I took one packer and a pack-mule, and, attaining the summit of the continental backbone at the head of the eastern fork of the Rio Los Pinos, followed the divide around the headwaters of the Piedra and the southern tributaries of the Rio Grande as far as to the headwaters of the Rio San Juan, making topographical stations upon ten of the highest peaks, including one five miles north of the Pagosa Peak, in the dividing range between the San Juan and Piedra, and Macomb's peak, named in honor of Col. J. N. Macomb, Corps of Engineers, who first explored the San Juan basin.

Under this peak heads the main or west fork of the San Juan, the east fork of the Piedra and Hot Spring and Thunder Creeks, tributary to the Rio Grande.

Although not of very great height, barely exceeding 13,000 feet, this is the culminating point of this portion of the Atlantic and Pacific divide, and a very marked and noteworthy feature of the landscape; a long ridge of brown trachyte, surmounted by a cap of the same material, which is vertical on the south and west sides for many feet; its summit is attained by climbing over the loose masses of trachyte which lie under very steep slopes on the eastern sides. This point was occupied in 1874, but snow and cold prevented any successful notes. A station was made the next day by Mr. Spiller, whom I met near Macomb's Peak, on a high point some miles northeast of Macomb, between the two forks of Hot Spring Creek.

From Antelope Park, beginning October 23, the topographer meandered and profiled the new wagon-road from Alden's Junction to Lake City, meandered Clear Creek and its tributaries, the road from Antelope Park to Del Norte, and the tributaries of the Rio Grande, which had not been completed last year, making en route the necessary three-point and topographical stations, to accurately locate the points adjacent to his lines and to check up his meander, while I made a station upon Bristol Head, taking repeated sights with 10-inch theodolites to our triangulation-stations, from which the position of this point has been accurately determined; and then proceeded to the head of the west branch of the south fork of the Rio Grande and made a topographical station upon a high and sharp peak seven miles east of Mount Macomb. Snow had meanwhile fallen to the depth of from 10 inches to 1 foot upon the mountains, rendering mountain-work very disagreeable, and oftentimes dangerous, from slipping. On November 2, we were compelled thereby to bring our work in the high mountains to a close, after four and a half months of continuous and almost excessive labor in the high masses forming the Atlantic and Pacific divide and its outlying spurs.

From Del Norte the road via the Mosca Pass was meandered, the Huerfano traced to its mouth, and the Arkansas meandered thence to West Las Animas, where I disbanded the three parties of the Colorado section about the 25th day of November, 1875.

During the field-season the system prescribed by the printed instructions issued from the office of the survey in 1874 was carried out. Nineteen triangulation-stations were occupied by Mr. Spiller and myself, the principal of which were selected for their well-defined and sharp peaks as well as on account of their position in the scheme of triangles. The principal angles were repeated from six to twelve times. As soon as a peak was occupied, a substantial stone monument was built to a height of from 6 to 14 feet, and these monuments taken as targets thereafter. Prior, however, to the occupation of peaks, the natural object itself was taken as the target. In most cases, however, these points were very sharp and well-defined, so that monuments were scarcely necessary. The angles were read from a 6-inch Würdemann theodolite, graduated to read to 10 inches, the stations being from 15 to 60 miles apart. Where practicable the sides were near 30 miles in length, but in carrying the triangles across the wide stretches of mesa-country in the San Juan basin, where but few well-defined points were visible, long sights were necessary. Probably the largest triangle which has ever been measured falls in Colorado and New Mexico, viz:

La Plata Creek; Banded Peak; Mount Taylor.

	Miles.
From La Plata to Banded it is	83.905
From Banded to Mount Taylor	139.876
From La Plata Peak to Mount Taylor	154.499

From Gray's Peak, near Denver, to Mount Taylor, the azimuth is carried by four lines of sight:

	Miles.
1. Gray's Peak to Hunt's Peak	86.250
2. Hunt's to Uncompahgre	71.515
3. Uncompahgre to La Plata Peak	55.200
4. La Plata to Mount Taylor	154.499
Total	367.464

Also,

	Miles.
1. Gray's Peak to Pike's Peak	65.614
2. Pike's to Sierra Blanca	90.338
3. Sierra Blanca to Banded	70.791
4. Banded to Mount Taylor	139.876
Total	366.619

The first line of sights was carried by my parties, Mr. Nell and Mr. Spiller measuring the angles, from Gray's Peak to Mount Taylor; the second, as far as to the Sierra Blanca. Banded Peak was occupied by Lieutenant Morrison's party in 1875 and Mount Taylor by Lieutenant Price's party in 1874, and re-occupied the past season by Lieutenant Morrison's party, the angles being read by Messrs. Thompson and Clarke. Sierra Blanca was re-occupied 1875 by Mr. Maxson, of Lieutenant Carpenter's party.

In addition to the triangulation-stations, which have all been computed, seventy-seven high peaks were occupied as topographical and secondary triangulation-stations. Upon these the angles were read to the nearest minute, repeated to primary stations to give sufficient reliable data for the computation of the position of the more important of these points. Seventy-four lower points were also occupied, which were mathematically fixed, and numerous check-stations introduced upon the meander-lines. Upon each triangulation and topographical station a panoramic view of the horizon was made in perspective and sights taken to all points not too distant, such as peaks, ends of marked spurs, junction-courses, and bends of streams, towns, ranches, and other artificial features; and these points accurately located in relative horizontal position by intersections, and vertically by angles of elevation or depression. In the immediate neighborhood of the stations horizontal plans showing the local topographical details were made. Since a large tract of country was covered by the party, the topographical stations were made on an average of one every seven miles. Points located by intersections from these stations or computed from the measured angles in connection with the bases measured in 1874 are taken as the frame-work or skeleton for the map now being constructed by Mr. Spiller; and the minor details, roads, streams, &c., filled in from meander-lines and sketches and bearings made along the trails; the distances being measured by odometer, checked by frequent three-point stations, and the bearings from a topographer's transit. Each of the meander-stations, which numbered about seventeen hundred, was also a barometric station, so that continuous profiles are secured over the entire route of the party.

The method of covering large areas by sketches alone from high stations, these stations made on an average of from 8 to 12 miles apart, is defective—

1st. Because it can only be applied in a mountainous country, where well-defined points exist which may serve as vertices for the triangles and for commanding positions from which the necessary sketches may be made; at best, then, it is only applicable to limited areas.

2d. Except in very exceptional districts not more than one-half the points sighted from one station can be recognized or seen from another; particularly is this true of the lower points sighted.

3d. The details are good only in the immediate neighborhood of the points occupied, i. e., in the most inaccessible and practically the least important portions of the territory to be mapped. Any one at all cognizant with mathematical drawing can at once understand the difficulty and the absurdity of attempting to represent the details included in from 49 to 144 square miles of territory from sketches and bearings made from a single station; of representing properly the sinuosities of roads and streams from occasional sights at prominent changes of direction, &c., and of representing by accurate contours the slopes of the country from such data as can be obtained from so few primary barometric bases, upon such a large scale as required for detailed topographical maps.

In a partially mountainous region, watered by numerous streams, traversed by roads, and quite well settled, as parts of New Mexico and Arizona, where peaks above timber-line are rare, and well-marked points are not very numerous, the method would prove a failure as far as detailed maps are concerned, for, in addition to the objections to the system where commanding points can be obtained, the foundation of the system itself would be insufficient for its requirements, where such stations do not exist. In level districts or on the rolling plains it is worthless.

The mountain-stations are, as they have always been used on this survey, essential to a good map; they are necessary for securing the mathematical accuracy needed in the location of points; for gaining a knowledge of the headwaters and upper drainage of minor streams which cannot be meandered, and for impressing upon the topographer's mind a good *general* idea of the topography and the relations of great topographical features to each other; but to rely upon the knowledge gained from them for an accurate *detailed* map of the artificial as well as the natural features of the country is folly, reasoning as topographical features and stations *are* and *must be*, and not theoretically as they *should* be, found.

As previously stated, the high stations were made during the past season as numerous as have ever been practiced by the advocates of this method, and its insufficiency showing itself in the lack of practical details in the lower and inhabited and traveled regions, close odometric meanders of all the roads and principal streams and careful profiles were made for furnishing the information absolutely necessary for the work, and which could not possibly, without immense physical labor and expenditure of time and money, be secured from topographical stations.

The odometer, if checked by the three-point problem every 8 or 10 miles, will give results which will compare very favorably with actual chaining, at a great gain of time and expense, and the bases measured by it serve well for the location of all points within a few miles of the trails followed, and which may not be better located from triangulation or topographical stations. Some form of this instrument must necessarily always form part of the outfit of a party engaged upon geographical work in unsettled regions, its importance increasing with the number of marked topographical stations decreased. In nearly every portion of the West, however, which still remains to be mapped, there is always to be found a sufficient number of marked natural objects to serve as vertices for a system of triangles, which may be located easily and at not too great expense, to be used as checks upon odometer-meanders to prevent the accumulation of errors by false measurements.

Where the camps could not be located trigonometrically, latitude-checks were introduced by me; but these were seldom necessary in the mountainous region traversed by the party the past season.

BAROMETRIC WORK.

Cistern-barometers and thermometers were used at the hours prescribed by the printed instructions, in camps and on the march, at culminating points of the trails.

Aneroid-barometers and thermometers were observed at all meander-stations; the aneroid being compared with the standard every morning and evening, and, when a considerable elevation was crossed, at its summit also. When stations were made upon peaks, simultaneous observations were taken at base and summit on cistern-barometers and psychrometers. General meteorological observations were also taken.

Profiles of all routes by which the San Juan mines can be reached were taken, and will be found appended to my descriptive report of routes of communication.

I wish here to tender my acknowledgments to Mr. J. C. Spiller and Mr. G. M. Dunn, for the care and pains taken to perfect their notes and secure abundant and good material in their several branches of work. The former gentleman brought from the field very full and elaborate sketches and topographical details, and the records of the latter were kept throughout the season in good shape. Mr. Dunn also succeeded in transporting his barometer throughout the season and bringing it back to the office without injury, which is quite a success, considering the rough character of the country surveyed and the means of preserving the instrument from almost inevitable falls and consequent injury.

PART II.—GENERAL DESCRIPTION OF THE AREA SURVEYED.

The Platte and Arkansas divide, or that portion of it surveyed this year between the southern edge of South Park and the Arkansas River, is a high rolling area, covered along its northern surface with basaltic and trachytic overflows, with but few high peaks and but scantily timbered. Ridges and low cones of lava appear here and there above the general level to the north, heavily grassed with the nutritious bunch-grasses of the mountains, and quite well watered by tortuous streams. A portion of the divide reaching from near the head of Currant Creek to Basalt Peak—a black mass sloping like an inclined plane to the west, but abrupt on other faces—rises to over 11,000 feet above the sea. Between Basalt Peak and Trout Creek Pass, however, the elevation is slightly above the general level of South Park. Over this portion of this high rolling area are scattered many little parks and grassy swales, the most extensive of which is Puncha Park, south of Basalt Peak, and the basin at the head of Badger Creek. Several cattle-ranches are located in these beautiful little parks, the cattle running at liberty over the hills during the summer-months, but are driven to lower altitudes for winter-herding. The streams bordering on the south side of the divide or those flowing into the Arkansas cut quite deep cañons in the plateau, attaining the level of the Arkansas from 2,000 to 3,000 feet below the summit of the limiting rim of the cañou of that stream. The general level of that portion of the plateau south of Basalt Peak exceeds that of the northern portion. South of Puncha Park is a short range of peaks rising nearly to timber-line, or to about 11,500 feet, which furnish water to Tallahassee, Gorell, and Badger Creeks. The basin of Badger Creek, save toward the north, is inclosed in quite well-defined ridges, and its cañon for miles above its mouth is a very formidable one. Stratified rocks appear, in quite extensive development in this southern portion of the Arkansas plateau, and predominate over the trachytic overflows. There is no agricultural land, except narrow strips along the cañons of Current, Tallahassee, and Gorell Creeks, the general level exceeding 9,000 feet above the sea.

The scarcity of timber, save upon the high ridge between Basalt Peak and 39-Mile Mountain and the ridge bordering Badger Creek, is quite a marked feature of the Arkansas plateau. The southern slopes of the hills are well grassed, but there are very few trees; these occur, where at all, on the northern slopes. This is probably due to the pervious nature of the trachytic soil and the greater dryness of the southern slopes. Spruce is the principal timber, but this gives place, near the Arkansas, to juniper and piñon. The old (1861) Puncha Pass and Cañon City toll-road leaves Currant Creek near the Soda Spring, and passes through Puncha Park, and via the headwaters of Badger Creek, to the Arkansas River, near the mouth of the South Arkansas. It is now abandoned, the more southerly route, via the Arkansas Cañon and Pleasant Valley, being now used as the mail-route. It is easy, however, to travel in nearly any direction over the divide, save near the Arkansas River, where it is too much cut by cañons. The country bordering the Arkansas River has been already well described by Professor Stevenson in his report; it is barely sufficient to say that at the mouth of the South Arkansas the valley of the Upper Arkansas, which lies between the massive Saguache range and the edge of the Arkansas plateau, and extends for some twenty miles above the point named, is closed by the foot-hills of the Sangre de Cristo range, through which the Arkansas has cut a short but narrow cañon. This valley is not of great importance; the agricultural lands are limited by the edges of terraces, which, breaking down in succession from the slopes of the high peaks to the west, close in within a comparatively short distance of the river, and by their height, and by the fact that they are cut transversly by the cañons of the streams emerging from the mountains far below the level of their upper surfaces, irrigation of any considerable proportion of this valley, at reasonable expense, is precluded.

Several ranches, however, exist, and fair crops are raised. On the South Arkan

AP. JJ—6

sas itself are many little farms, and the climate is quite mild, as is evidenced by the number of cacti and other southern plants which abound over the surface of the valley. Grass is very scant, and as a grazing region it is insignificant. The scenery of this portion of atlas-sheet 61b is unsurpassed. The wide valley of the Arkansas, with the Saguache range rising abruptly from its western terraces to over 14,000 feet altitude, massive and snow-crowned; the Sangre de Cristo range, as a spur shooting off from the Saguache range at Hunt's Peak, (a lofty, sharp point south of the South Arkansas River,) trending around to the southeast and closing the valley; the Arkansas plateau to the east, with its almost uniform surface, and to the north the rounded humps of Buffalo Peak, offer to the view the greatest variety of landscape and the gentlest as well as the most imposing of natural features. Some 6 miles above the mouth of the South Arkansas this stream is joined by Puncha Creek, which is notable as offering the only practicable pass into the San Luis Valley from the east, between its head and the sand-hills, 60 miles distant. The approaches to the pass from the Arkansas Valley are narrow and quite tortuous, the road crossing and recrossing the stream at frequent intervals, but the grades as established are quite uniform and a very good wagon-road exists. The ascent averages 250 feet to the mile; hence it is too steep for railroad purposes. Over the head of Puncha Creek, between Hunt's Peak and Mount Antoro, the last high peaks of the Saguache range, it is practicable to build a road to the west for wagons, and, now that the valley of the Gunnison is being rapidly filled with settlers, it will be an obvious advantage to build one at this point. The ascending gradient is about the same as in the Puncha Pass from the South Arkansas, but on the west side there is a steep pitch from the summit, which, however, may be avoided by carrying the road around the hillsides, lengthening the road, and thereby diminishing the gradient. Upon the western side in a few miles from the summit the country opens out into the Tumichi Valley with very gentle slopes. It may also be practicable to build a road over some of the headwaters of the South Arkansas. The range runs no lower down, but it is much wider in that vicinity, and there would be consequently a larger stretch of mountain-road even if practicable gradients can be secured, which is not yet ascertained. Capt. J. W. Gunnison, in vol.— of the Pacific Railroad Surveys, remarked upon the apparent existence of passes at the heads of the Tumichi and Carnero Creeks, which led me more closely to examine this vicinity than perhaps I should otherwise have done.

The summit of the Puncha Pass, 8,945 feet, being attained, one finds himself at once in the San Luis Valley proper, the Homan's Park of Gunnison. It is very narrow in this vicinity, being inclosed by the northern extremity of the Sangre de Cristo range and the broken volcanic overflow which covers the region about the headwaters of Kerber Creek. Gradually widening out toward the south, it attains at its maximum a width of some forty-five miles at Del Norte and ends about the southern border of Colorado, where overflows of volcanic matter break the general level. On the eastern side it is bounded by the Sangre de Cristo range, a very decided and well-marked sierra gradually rising toward the south from 9,000 feet at the Puncha Pass to 14,300 feet at the Sierra Blanca. Many of its sharp peaks attain 13,000 feet, and several, notably the ragged mass called the Three Tetons, exceed 14,000. The continuity of the range is unbroken and, though narrow, (barely twelve miles from the San Luis Valley to the Wet Mountain Valley on the eastern side,) is, on account of the steepness of its slopes and the ragged nature of its crest, impassable, save at the depressions called the Hayden, Music, Sand-Hill, and Mosca Passes. Of these the Mosca Pass is probably the only one which will be used in ordinary travel and traffic. The Hayden and Music are too steep, and on the eastern side the approaches to the Sand-Hill Pass are bad and on the west it is choked with sand. On the San Luis Valley side the Sangre de Cristo range is very abrupt, shooting up at once from the plain; from the east, however, its summit is very much more easily attained. Long slopes extend from near timber-line far into the Wet Mountain Valley, which is itself considerably higher than the San Luis. Wet Mountain Valley indeed resembles more a slightly-hollowed-out glacial bench upon the flanks of the Sangre de Cristo, than a true valley of depression between the Sangre de Cristo and Wet Mountain ranges.

At the Sierra Blanca, the most massive and imposing group in Colorado Territory as far as I have observed, which rises nearly 7,000 feet above its base, the Sangre de Cristo range ends; more properly or orographically speaking, it ends at the Mosca Pass; and the Sierra Blanca should be perhaps named as a separate division of the Rocky Mountain system, beginning at the Sangre de Cristo Pass and ending at the Mosca.

From the Sierra Blanca the bounding ridge of the San Luis Valley changes its direction to the eastward, and, after encircling the heads of the Sangre de Cristo Creek, continues its course to the west of south. South of the Sangre de Cristo Pass the range changes its character from the sharp sierra of the Sangre de Cristo proper, and is now a mountain-range, with more massive and rounded peaks and longer foot-hills and slopes. Its high and impassable character is preserved, however, and no passes worthy of the name exist south of the Abeyta until the Taos Pass in New Mexico is reached. These passes fall within the area surveyed by other parties of the expedition, and reference is made to their reports for description. It is, however, to be remarked that, since the southern continuation of the Sangre de Cristo range holds snow throughout the year, and its summits and slopes are less steep and consequently averaging more high land, and therefore less affected by the heated air of the San Luis Valley and the plains to the eastward, and less rapidly drained also, the streams which flow down their slopes are more constant, and, instead of soon sinking in the sand and volcanic *débris* of the valley, reach the Rio Grande on the surface.

The character of the b ᵁunding ridge west of the San Luis Valley is altogether different from that just described. Although a component part of the Atlantic and Pacific divide, the part facing and bounding the San Luis Valley is not imposing from its great height, and is remarkable only for the lack of marked points.

Due west of the Puncha Pass, the massive and peaked Saguache range ends at Mount Antoro, a rounded trachytic mass about 13,450 feet in height. The divide from this point trends to the westward, with rounded, wooded hills, for some 15 miles, and then turns to the southwest around the heads of the Saguache for 40 miles; then, making the remarkable tongue-like loup about the headwaters of the Rio Grande, again continues its course to the southwest, on the western border of San Luis Valley, below Del Norte.

That portion of this divide seen from the San Luis Valley is an extensive trachytic overflow, of a plateau rather than a mountain character, cut by streams which flow in cañons of greater or less width, sometimes inclosing within their walls beautiful little valleys and parks, heavily grassed and well wooded, at other points narrow and gorge-like impassable box cañons, its summit covered here with piñons, there by groves of pines, spruce, and quaking-aspen, and again by long, sweeping, rolling stretches, bounded perhaps by vertical ledges of trap, but beautifully and luxuriantly grassed, affording now shelter and pasturage for numerous deer and few elk, and in future destined to be, like the Arkansas plateau, the summer range of more domestic cattle.

The few mountains deserving the name in this portion of the San Luis Valley rim are rounded, wooded, dome-like masses, seldom rising above timber-line, save near the heads of La Garita and Carnero, where they attain considerable height. That portion of this overflow between the headwaters of the Saguache and the San Luis Valley is cut up by the drainage-lines and cañons of the streams into numerous little hills and limited benches, which, though presenting from the valley the gently-sloping and sweeping outlines characteristic of such formations, offer serious difficulties to travel by frequent almost vertical bluffs and many ascents and descents along any given line.

The streams rising in this broken plateau north of the Rio Grande and flowing into the San Luis Valley sink in the sand or swell the volume of the San Luis swamp. South of Del Norte, the Conejos and Alamosa, which rise far back in the Sierra San Juan, invisible from the valley, reach the Rio Grande on the surface; others, as the Piedra Pintada and La Jara and Rito San Francisco, sink in the gravel.

From the head of the San Luis Valley, looking south, the view is similar to what is seen in the dryer valleys of the great basin; the Sangre de Cristo range upon the eastern border; the broad expanse of plain, with the rounded summits of the Cerro de las Utas and the Cerro San Antonio just appearing above the horizon or lifted by the mirage, together with the flat-topped and limited basaltic mesas between them, far into the air; the cerritos, or little hills of lava, about the gate-way of the Rio Grande at Del Norte quivering through the vibrating air; the glaring white sand-hills, with their wavy crests and outlines piled in front of the Mosca Pass, all suggest the Great Basin rather than the region of the Rocky Mountains.

The surface of the San Luis Valley is very interesting and quite varied in its characteristics. In the northern end, San Luis Creek, and Kerber Creek, its tributary, and Saguache Creek, a large mountain-stream, flow out into the center, and, here uniting and spreading out over the flat surface, give rise to the San Luis swamp. It is a swamp, properly speaking, only for 8 or 10 miles; but there is a succession of pools and small lakes of slightly alkaline water and covered with myriads of water-fowl near the eastern edge of the valley, from about 5 miles below Saguache, and in certain seasons from Saguache as far as to the mouth of the Mosca Pass. In the northern portion of the area covered by this swamp, where quite extensive tracts are kept constantly wet, there is a most luxuriant growth of swamp-grass and sedges. White-tail deer abound, and large herds of cattle find sustenance in these sedges and grasses. These are cut quite extensively for hay, and are said to be valuable for that purpose. Certainly some use other than is now made of this abundant material will be found. The cattle seem hardly to make an impression upon it; it grows to the height of a man's waist, is very thick and heavy upon the ground, and covers a large area. From the commencement of the swamp far up into the cañon of the Saguache, or for about 20 miles of its course, it is bordered by the richest of soils, producing most abundant crops of wheat, oats, and vegetables, and is by far the most inviting agricultural region of the valley. In the cañon of Saguache, the bottom-lands, as far as the bend of this stream, easily irrigated at slight expense, will average one-fourth of a mile in width, capable of cultivation for 12 or 15 miles above Saguache, and above that point furnishing, in connection with the bunch-grasses of the rolling and plateau like region bordering it, abundant grass for stock nearly to its head. The streams, other than those named above, tributary to the San Luis swamp, but which reach the center of the valley under the surface, are bordered by but narrow strips of agricultural lands and furnish but little water for irrigation. The intervals between the streams in this portion of the valley are almost destitute of vegetation other than sage and chico, of which there is quite an abundant growth. About midway of the length of the valley, the Rio Grande emerges from the mountains to the west, and, after following its course to the south of east for 30 miles, turns to the southward, and soon enters a cañon with low walls of volcanic material. It is joined by the Trinchera, Culebra, and Costilla Creeks from the east, and by the Conejos and Alamosa from the western side. These are all quite large streams, and along their banks are quite extensive strips of land, not difficult of irrigation, and of good soil. This is particularly true of the Conejos, which is quite thickly settled along its banks by Mexicans; and the Upper Culebra, where a quite extensive and fertile valley, watered by numerous little streams tributary to it, is cultivated. Beyond the strips of agricultural land bordering the streams, the southern portion of the San Luis Valley is now a desert. The soil is decomposed trachytes, resembling in texture gravel and coarse sand. There are extensive areas, covered originally not very thickly with short grass, where large herds of sheep are herded by the Mexican inhabitants and keep the face of the earth bare wherever they are driven.

Undeniably the greater part of this soil is capable of producing good crops, its fertility being apparent in the luxuriant growth of sage and greasewood, which in great part covers it.

Indeed, the San Luis Valley, if its resources of water were utilized to the full requirements of the land susceptible of irrigation, is capable of sustaining any population the extensive mines of Colorado in its vicinity are likely to attract. In the southern portion good crops of corn and vegetables are produced and in the northern the usual small grains. The marshy and arid portions of the valley are capable of furnishing the necessary flesh-food. Agriculture, however, at present is not profitable, and the people who have settled it are mainly stock and wool growers. The present state of agriculture cannot, therefore, be taken as at all representative of the capabilities of this valley.

The entire valley is treeless, save along a few of the larger streams, where groves of cottonwood exist. On the northern and western rims, near the valley, piñon, a fine fuel, is abundant, and farther back spruces and pines for timber. The Sangre de Cristo range from the edge of the valley to the timber-line, wherever the slopes are sufficiently gentle to retain soil, are well wooded with spruce and cottonwood; the latter, however, of small size, fit only for poles and fencing.

This large area, comprising over 2,000 square miles, is but thinly settled, compared with its resources; but the richness of the mines in its vicinity and the consequent demand for its products is causing quite an influx of population.

Saguache and Del Norte are quite prosperous towns, and the agricultural lands, especially about the former, are nearly all taken up by settlers.

With the exception of the Americans at the two towns mentioned, the United States troops at Fort Garland, and a few rancheros in the valley, the population is Mexican, and is mainly distributed about numerous small plazas on the Conejos, Culebra, and Costilla Creeks, where they live in poverty, ignorance, and idleness; dwelling in wooden stockade-like pens plastered with mud, or in more or less pretentious huts of adobe, cultivating the ground even yet in primitive fashion with wooden plows, drawn by oxen, and raising barely sufficient crops of corn, oats, onions, and chili or red pepper for their support, their poor sheep and the jack-rabbits of the sage-fields furnishing their meat.

From the time we left the San Luis Valley, in July, until the first part of November, the party were almost constantly in the mountains surrounding the headwaters of the Rio Grande. A minute description of our work and daily travels, and of the local peculiarities of the small areas examined daily by us, would perhaps give a very vivid idea of the personal toils and trials of the party in mountain-travel, but would perplex the mind with unimportant details, and prevent its comprehending the general topographical structure of this interesting region. For this reason I will continue a general description, referring to particular features only where they are remarkable and specially noteworthy.

A little to the north of Saguache the Cochetopa Pass (first examined and reported upon by Capt. J. W. Gunnison, United States Topographical Engineers, in 1853) offers the most practicable passage-way from the Atlantic to the Pacific slope along the Rocky Mountain barrier from Cheyenne to New Mexico. South of this pass the divide sweeps round to the west, and then, at about 107° 35', returns upon itself, inclosing in a wide and deep tongue-shaped cul-de-sac the headwaters of the Rio Grande del Norte. The portion of the divide forming this *cul-de-sac*, with its slopes, may be subdivided for the sake of description into three divisions:

1st. The northern, extending from the Cochetopa Pass to the depression at the head of the western tributary of the Cebolla Creek.

2d. The bottom of the Loup, extending from this depression to the pass at the head of the Rio los Pinos; and

3d. From the headwaters of the Los Pinos to the Sierra San Juan and the Summit mining-district southwest of Del Norte.

About the bottom of the Loup are numerous short ranges, determined by the gorges and cañons of the streams flowing from them, which, for convenience and because they are connected intimately with this portion of the divide, will be described in full with it.

The first and third of these divisions are similar, inasmuch as they exhibit more clearly the plateau character of the volcanic overflow reft by the cañon of the Upper Rio Grande, and cut out by water and ice into cañons and gorges, until now portions of the plateau exhibit mountain-peaks and short ranges. This first division, however, more distinctly retains its plateau-like character. Beginning at the Cochetopa Pass, the summit of the divide gradually increases its height from 10,000 feet at the pass to approximately 13,600 feet at the head of the Saguache, exhibiting the character already described in speaking of the western rim of the San Luis Valley. The east branch of the Saguache heads in tremendous cup-like cavities and box-canons under mesa-like portions of this plateau, which here may be followed for miles around the headwaters of the Saguache, La Garita, Carnero, and Embargo Creeks, upon nearly a flat surface, strewn with trachytic rocks, and covered, where decomposition of these rocks has sufficiently taken place to furnish a soil, with a thick mat of Alpine grasses and flowers.

Upon the Rio Grande side the descent from the head of the Saguache to the river, a few miles distant, is very abrupt; here vertical ledges and bluffs close in upon the river and form the narrow and short cañon called Wagon-wheel Gap, 30 miles above Del Norte. From the heads of La Garita and Embargo Creeks, where the flat mesa-like surface ends, eastward to the San Luis Valley and southward to the Rio Grande, the country breaks down in detached hills and bluffs, and near the valleys sharp points or cerritos of basalt.

West of the deep crateriform cavity, in which heads the eastern of the two main branches of the Saguache, the plateau is crowned by a short ridge of volcanic material, curving sharply down to the surface of the grassy plateau in steep slopes of loose fragments to the south, and inclosing in perpendicular walls, in semicircular sweep to the north, the headwaters of the main or western branch of the Saguache. These peaks surmount the plateau for some 12 miles farther west, several of them approaching 14,000 feet in altitude, and end in conical masses which slope gradually down to the upper surface of Bristol Head, a plateau-mountain V-shaped in plan, the opening of the V turned to the north and the vertex near the Rio Grande. This mass is of, perhaps, 30 square miles' area, bounded by a vertical bluff from 2,500 to 3,000 feet on the southwestern side of the V, and by slopes more or less gradual or abrupt on the eastern. It is nearly 13,000 feet in altitude, and a marked feature of the landscape. It is composed mainly of a brownish trachyte.

The country to the north of these peaks slopes down to the Gunnison, the slopes being cut by the cañons of the Cochetopa and Cebolla and their tributaries into subordinate ridges of not great height, and even crest.

West of the Bristol Plateau and the peaks standing upon it as a pedestal, there is a depression extending some eight or ten miles about the heads of western forks of the Cebolla, a great part of which is below timber-line. Being rather imperfectly drained, it is covered by numerous marshy spots. From the lowest part of this depression, over which the Indian trail from the old Los Pinos agency crosses into the Rio Grande Basin, the country—a rolling series of hills well grassed—slopes gradually upward to the rim of the Clear Creek Basin, which also forms the western wall of the cañon of the Lake Fork of the Gunnison. A break in this latter rim at the headwaters of the western branch of the Cebolla offers a passage-way for the new wagon-road, which leaving the valley of the Rio Grande under Bristol Head, first crosses the Atlantic and Pacific divide to this branch of the Cebolla, which it follows to its head and over the break mentioned to the cañon of the Lake Fork. North from this breaks a round-topped, rolling stretch of elevated country, ranging in altitude above sea-level from 12,000 to 12,800 feet, and named by us the Cannibal Plateau—in memory of the horrible butchery and acts of cannibalism which were practiced under its western edge in 1874—extending for about twelve miles, through the eastern slopes of which the Cebolla, which has first collected the numerous branches flowing down from the mountains north of Bristol Head, cuts a formidable cañon.

From a few miles south of the Cochetopa Pass, around this northern rim of the Rio Grande Loup to the depression northeast of Bristol Head, or for the entire portion of this rim in the first of the divisions into which this loup is divided, the range is impassable, transversely, for other means of transportation than lightly-laden pack-mules, and for those in but very few places.

In the valleys of the Cochetopa and the Upper Saguache, above the walls of its cañon, and on the rolling hills southwest of the Cochetopa Pass, is much fine grazing-land, which, however, is too high for winter herding. The same may be said of the valley of Cebolla, above its cañon. All of the streams flowing from this portion of the loup to the northward, cañon up a short distance from their heads, so that in the upper portions of their courses there is but little bottom-land. This is at too great an altitude for cultivation.

Below the cañon of the Cebolla, however, there are several ranches where potatoes and the hardier cereals were cultivated, but whether they would mature cannot be said, since it was an experiment being tried when we passed. Timber and fire-wood are abundant. No mines have as yet been discovered in this division, and beyond a few ranches along the Lake City and Saguache road, there are no inhabitants on the northern slopes of the range.

2.—THE BOTTOM OF THE RIO GRANDE LOUP AND THE NEIGHBORING RANGES AND MASSES.

This division of the continental backbone, with its spurs and attendant ranges and mountain masses, is the most interesting of the mountainous portion of Colorado, both on account of the high and rugged character of its mountains, and the great area covered by their peaks, and the number and character and prospective value of the mines here discovered and now being rapidly developed. Of the mines but little will be said, since the character of the work precluded any investigation of the subject from want of time. The examination of a single mine, including the time necessary to reach it, would occupy an entire day, and since but little over a month could be given to the survey of the entire complicated mass of mountains in which they are situated, this time could not be spared for each of the many mines which would have to be visited; besides, they are only partially developed, and not much can be said as to the persistency and value of the lodes discovered, whatever may be their prospects. The topography of the country and the engineering questions relating to the routes of communication, and those relating to the capacity of the regions in the neighborhood of the mines to subsist a large population, were the subjects which more particularly engaged the attention of the party, and it is of these that we will principally speak.

West from Bristol Head the country slopes up from the valley of Clear Creek to the summit of the rounded peaks which border the cañon of the Lake Fork of the Gunnison by not abrupt gradations. Clear Creek heads not far from the Rio Grande, and runs a little north of east from its head under a rounded mass near the head of Crooked Creek, for some 8 miles of its course bordered by a grassy though narrow valley. Numerous tributaries enter it through cañons from the rounded mountains to the northwest.

Between Clear Creek and the Rio Grande is an area of some 50 square miles, with nearly level upper surface, cut by the valleys, or rather cañons, of the tributaries of Crooked Creek. This bench is about 9,600 to 10,200 feet in altitude, and ends at the base of the rounded mountain just referred to. Along the southern edge of this bench the Rio Grande runs in a cañon bounded by walls 1,200 to 2,000 feet in height on the southern side, but only 800 to 1,000 on the northern. The rounded summit under which Clear Creek heads, divides it from Lost Trail Creek, which with Pole Creek runs down from the edge of the Lake Fork Cañon, which here becomes more mountainous and peaked in its general appearance; the divide increases in height and in ruggedness, from the heads of Clear Creek, inclosing, besides Pole and Lost Trail Creeks tributary to the Rio Grande, numerous short tributaries of the Lake Fork of the Gunnison in its northern ravines. Toward the south, however, it retains its high altitude for several miles, or until near the narrow valley of the Rio Grande, to which level it breaks down abruptly by bluffs. On the Lake Fork side the fall from the divide to the level of the stream is very marked and abrupt. The rim is strongly marked and linear, parallel to the stream, with ravines running down its steep slopes.

Near the head of Pole Creek a branch of the Lake Fork, named by Mr. Prout, Snare Creek, takes its rise. Between this stream and the headwaters of the Lake Fork and the Animas is a group of four peaks, exceeding 14,000 feet altitude, of which Handies Peak is the culminating mass. South of this group there are no high peaks in the divide, but there is quite a broad plateau bordering the eastern side of the Animas and separating it from the heads of Pole Creek, which extends to Mount Canby, an irregular red mass of volcanic material near the headwaters of the Rio Grande.

About the immediate head of the Rio Grande itself, the country is an extensive area above timber-line, exhibiting no marked peaks of height above the general surface on the western side; but to the south, and in fact in every direction, save to the east, towering peaks, high above timber, and with bold sky-lines and profiles, can be seen, but these are in subordinate groups and not in the main divide itself. On the southern side of this semicircular division, about the heads of Hines Fork and Ute Creek, are peaks reaching 13,500 feet.

Between Hines' Fork and the Los Pinos Pass are very well-marked and steep spurs, with well-defined peaks, and slopes broken by bluffs of trap. On the western side of Ute Creek the spur between this stream and the Los Pinos Pass is of nearly uniform height, save that near its center, rising from its upper surface, as from a suitable pedestal, is one of the handsomest and most symmetrical cones in Colorado. It is about 5 miles from the Rio Grande River, and is a marked feature, noticeable and prominent from every point of view, the more so, perhaps, because there are so few peaks in its immediate vicinity, the other mountains in the same spur having rounded summits, or apices, with very gently-sloping upper surfaces; the lower slopes are bluff-like. This cone was ascended by my party in 1874, and a primary triangulation-station made upon it. It was named by me Simpson's Pyramid.

On its eastern side the Rio los Pinos takes its rise and flows south through a tremendous cañon toward the San Juan. East of Simpson's Pyramid the Ute trail crosses from the valley of the Rio Grande to the head of the Rio los Pinos; its summit is flat and marshy for some miles, or until the fall of the Los Pinos is sufficient to give good drainage. The descent to the Rio Grande side is exceedingly steep for about half a mile, the fall in this distance being 1,000 feet. The pass will never be valuable as a wagon-route unless the discovery of very rich mines about the headwaters of the Los Pinos makes it a necessity.

The minor streams in this division all have rapid fall, and, except Hines' Fork, near their mouth flow through narrow cañons. Their headwaters, however, as a rule, are in Alpine meadows with not excessive slopes; covered, however, with a thick growth of mountain-willow, which much impedes travel. In places, also, the surface of these more gradual but limited slopes is broken by short ledges of trap, from 4 to 50 feet in height, which readily puts an end to mule transportation or causes a very winding course to be followed.

As seen from the Rio Grande Cañon, however, the characteristic of this division is abruptness, and indeed all the streams save Clear Creek are bordered by lofty escarpments and steep slopes.

Although on the Rio Grande side of this division as far as to the summit of the divide, there are, with the exception of Simpson's Pyramid, no very remarkable topographical features, upon the Pacific slope are numerous sharp ranges and spurs of gigantic mountains, peaked, cut by deep cañons and frightful gorges, rising far above the limit of arborescent vegetation, covered the greater part of the year with snow, the locus now of one of the most extensive and rapidly developing mining regions in the United States.

These ranges and spurs are:
1st. The Lake Fork group of peaks.
2d. The Uncompahgre Peak group.
3d. The Unaweep range.
4th. The Sierra San Miguel.
5th. The "Needles." Floridas Comb, or "Sierra Los Pinos" of Newberry, about the heads of the Rio los Pinos, and the masses of mountains within which heads the Animas River.

In these groups of high and rugged mountains, which cover approximatively with their peaks the area between longitudes 107° 20′ and 108°, latitude 37° 35′ and 38° 08′, or about 1,400 square miles, the Animas, Rio los Pinos, and Rita Florida, tributary to the Rio San Juan, the Dolores, San Miguel, Uncompahgre, and Lake Fork of the Gunnison, and other minor streams tributary to the Gunnison, and the Rio Grande del Norte with its headwaters intertwined with the above, take their rise. Perhaps within the limits of the United States there cannot be found another area of the same superficies so intricate in topographical detail, so rugged and steep and difficult of access, and requiring so much patience, energy, and hard physical labor in its survey.

In the center of this mountain region the Animas River takes its rise and drains with its tributaries Cement and Mineral Creeks, an approximatively elliptical area with the longer axis northeast and southwest, or from the head of the Animas to the head of the south fork of Mineral Creek. The Animas River itself runs parallel to the eastern portion of the periphery of this ellipse, bending around from its headwaters, first to the south, then southwest, closely bordering the bounding elliptical divide, and finally cuts through this rim in a tremendous cañon some 15 or 16 miles in length, and from 2,500 to 4,000 feet in depth. The southwestern vertex is drained by the south fork of Mineral Creek, while the north fork of this stream drains the western portion of the periphery. Cement Creek, lying between the Animas and the north fork of Mineral Creek, drains but a very small portion of the periphery, and is a comparatively insignificant stream. The larger axis of this ellipse is 20 miles, the shorter about 14 miles in length.

Following the divide around from the head of the Animas, we have the Lake Fork of the Gunnison heading near the northeastern vertex. It flows first in a south-easterly direction, then, making a great sweeping bend, flows north to the Gunnison River. Its tributary, Henson Creek, heads just north of the headwaters of the Animas, and, flowing east, joins it near where it begins its northward course, flowing through a very deep and gorge-like cañon.

Turning westward, the Uncompahgre headwaters are found against the head of the Animas, and the tributaries of this stream, all flowing in deep and abrupt cañons, drain the northern portion of the divide, in which Cement and the North Fork of Mineral Creek take their rise. The divide between the Animas and Uncompahgre is not difficult, nor that between the headwaters of the North Fork of Mineral Creek and the Red Fork of Uncompahgre; but the tributaries of this stream, and the stream itself, can be followed in any case but a few miles from their heads. To all intents and purposes, the headwaters of the Uncompahgre are shut off from the outer world save by way of the headwaters of the Animas, the Uncompahgre gorge which begins at the mouth of Red Creek being utterly impassable. This gorge is from 4 to 6 miles in length, through which the stream flows in its course to the north-northwest.

The western side of the divide is drained by the San Miguel, which by its two main branches collects the waters, and then uniting and plunging into a deep cañon, with sandstone walls, flows northwest to the Gunnison.

The southern side is drained by Cascade and Lime Creeks, tributary to the Animas, below its cañon; and the eastern by the Rio Grande and the southern tributaries of the Lake Fork flowing eastward.

From the outer periphery of the elliptical divide, then, we see the waters flowing in radial directions to all points of the compass as from a dome. Within the periphery, spurs from it fill up with mountain-forms the entire area, save the narrow cañons and valleys of the streams. In these valleys there is no surface of any extent which is approximately level save Baker's Park, a small area of perhaps 2½ square miles; just above the point the Animas cuts through the southern rim of its upper basin. Upon Mineral Creek also are several small flats of inconsiderable extent, mostly marsh land, or else covered with bowlders.

Baker's Park has been described by Prof. J. J. Stevenson in Vol. III of the survey reports, and needs no extended description here, even if minute description were within the proposed scope of this paper. Since his visit, however, Baker's Park, which was then almost uninhabited, has become the center of quite a large mining population. Where my camp was established in 1873, at the mouth of the Cunningham Gulch, where then there were no houses, has sprung up a town of log cabins called Howardsville, built upon both sides of the stream; and at intervals of 4 or 5 miles along the Animas, above and below Howardsville, are incipient mining towns; of these Silverton, near the mouth of Crescent Creek, is quite well built, the houses being in many cases quite handsome frame structures. It is well provided with stores, blacksmith-shops, and bar-rooms. Eureka, 4 miles above Howardsville, and La Plata City, at the Animas Forks, are being quite rapidly built up. Baker's Park will probably within the next five or ten years be well covered with houses and mills.

The most noteworthy feature of the drainage-basin of the Animas is the excessive steepness of its mountains, or at least of those directly bordering the main streams. The height of Baker's Park above sea-level averages 9,400 feet, and the peaks within a mile or so of its surface, exceed 13,000 feet altitude. Slopes of 45° are common, and nearly vertical bluffs for hundreds of feet are not infrequent, as seen along the Animas River above Eureka and on the South Fork of Mineral Creek.

The peaks in this elliptical divide, the dome of the San Juan country, will average over 13,000 feet in height. Those at the vertices of the ellipse, near the headwaters of Lake Fork of the Gunnison and Mineral Creek, approach or exceed 14,000 feet.

About the headwaters of Cement and Mineral Creeks the Animas and Uncompahgre divide, the mountains are lower, and, with the exception of one or two peaks between Cement and the headwaters of the Western Fork of the Upper Animas, which are steep knife-edges of volcanic *débris*—loose fragments of rhyolite or trachyte under as steep gradients as the material will stand—are of rounded slopes, which, though still retaining the characteristic steepness common to the mountains of this entire region, are, as a rule, covered with Alpine grasses, and the soil, though often boggy from melting snow, offers foot-hold for animals in their attempts at climbing or crossing them. The most marked feature in this portion of the divide are the two brilliant scarlet-red peaks between the headwaters of Cement Creek and the Red Fork of the Uncompahgre. They are not of very great height, but the decomposition of the pyrites in the trachytes composing them leaves the entire surface of those beautiful cones a brilliant red, which contrasts strongly with the green bald pates or the sombre brown of the trachytic masses of the neighboring peaks. They attract the eye instantly from any point of view by their brilliancy, and are a well-known landmark. At their bases on the Red Fork of the Uncompahgre is a small area, probably 200 acres, of nearly level ground, which is the locus of a new mining town called, I believe, Park City. The valley is named Red Mountain Valley, and is attained only from the Animas side, except by a rough and steep trail from the headwaters of the Uncompahgre; but this question of communication will be treated in full hereafter in this paper.

From the head of the North Fork of Mineral Creek to the southwestern vertex of the ellipse there is a very bold range of mountains, unbroken, save in one place, by depressions, but throughout its course marked by exceedingly sharp, and in many places pinnacled, ridges and peaks of brown or bright-red color, and varying in height from 13,400 to 14,000 feet. The single break mentioned occurs at the head of a short branch of the North Fork of Mineral Creek, which flows down from the west about 1½ miles above the junction of the two main branches of this stream. Here is a low place, reaching not far above timber-line, over which the range may be crossed to the head of a tributary of the Lake Fork of the San Miguel. Elsewhere the range is impassable, save at the southwest vertex of the ellipse, where a steep trail crosses to the trout lakes. At this southwest vertex is quite a group of lofty mountains, in which the Dolores also, in addition to the San Miguel, Mineral, and Cascade Creeks, takes its rise. Between this group and the eastern rim of the Animas Basin runs a very high ridge, broken only by two depressions, one at the head of Cascade Creek, the other west of Sultan Mountain, the double-capped peak which closes Baker's Park to the southward. Neither of these points are below timber-line, and are not, practically speaking, *passes*, but wriggling trails pass over each. This portion of the rim is noticeable as presenting on its southern side the only stratified rocks in the entire inclosing-walls of the Upper Animas Basin, the gray and red limestones and sandstones of the Carboniferous series, the formation elsewhere being mainly rhyolite and trachyte, seamed with quartz and metalliferous veins.

The eastern portion of the divide presents a clump of peaks about the head of Cunningham Gulch, including Kendall, Blair, Hazleton, and Galena Peaks, approaching or exceeding 13,000 feet in altitude. To the north of these, until we reach the lofty group already mentioned south of Handie's Peak, the divide is rolling, of nearly even crest, cut through by the cañon of the Upper Animas about Eureka Gulch, where bluff-like walls are presented. Within the periphery of this inclosing-rim the only peak worth mentioning is King Solomon's Peak, or Tower Mountain, a rude but symmetrical cone, rising from a broad base north of west of Howardsville to a height of 13,550 feet above sea level. Its nearly perpendicular slopes facing Howardsville are seamed with quartz veins, several of which are silver-bearing.

Spurs and subordinate ranges.

Between the streams flowing outward in every direction from the oval-shaped rim of the Animas Basin, extend short spurs and ranges of mountain-peaks, in some cases exceeding in beauty of outline, in height, and in boldness, any portion of the dome itself. Beginning at the northeast vertex of the ellipse, as before, we have, first, between the two main forks of Lake Fork of the Gunnison the Lake Fork range, trending north 85° east; next, the "Uncompahgre group," the drainage-axis trending north 45° east, which, with minor ridges, such as the Ibex or Mountain Sheep spur, fill up the space between the Lake Fork and the Uncompahgre. West of the Uncompahgre, springing directly from the dome at the northwestern extremity of its shorter axis, and trending north 75° west, is the Unaweep Range. From near the southwestern vertex the Sierra San Miguel, separated from the dome by a low depression, trends to the west between the Dolores and San Miguel. From the southwestern vertex, trending 20° west, the rib forming the divide between the Animas and Dolores, and ending in the high group of the Sierra La Plata, extends for twenty-five or thirty miles. From the southeastern edge of the rim, and trending S.° east, the Needles, or Sierra Los Pinos of Newbury, extends about the headwaters of the Los Pinos and Florida, cut by the cañon of the former stream. On the eastern side the two prongs of the continental divide complete the circuit.

Analyzing the topography of this division, then, without reference to geological axes or their origin, but regarding simply the topographical features as they exist, the seemingly confused mass of mountains about the headwaters of the Animas River takes the definite form of an oval or oblong nucleus, with radial spurs as the frame or skeleton; secondary spurs, and long sweeping slopes of the original plateau cut by water and ice fill up the area with mountain-forms which at first confuse the mind and give rise to a feeling of bewilderment. Proceeding, however, in the survey of this area with a knowledge of the analysis above, it is comparatively easy to unravel the mazes of this labyrinth and to depict them.

The central rim is first defined, and afterward, by winding in and out around the extremities of the radial-divides, the intermediate valleys and their limits, i. e., the spurs, themselves, may be defined one by one.

The headwaters of the Lake Fork of the Gunnison and of Hensen Creek.—The Lake Fork Range.

The Lake Fork heads in quite a wide and long cup-shaped valley on the west side of Handie's Peak, above timber-line, where, after collecting the numerous small rills which flow down the steep slopes surrounding the American Basin, it flows out first in a northerly direction, then sweeping around to the southeast, and finally to the east and north. toward the Gunnison. describes. in its course from its head to the mouth of Henson Creek. a large and symmetrical reversed S. Just below its first bend. and at its junction with a small stream from the north. is a narrow valley. probably 200 yards wide. called Burroughs Park. which extends. with few hummocks over its surface. for 2 miles down the stream. which then begins a rapid fall in a secondary cañon. from which it emerges near the point the stream changes its course from southeast to east. Here begins another narrow park-like area. covered with grass and willows at the point Lake Creek is joined by Cottonwood Creek. which drains the eastern slopes of the Handies Peak mass. and flows down in a gorge-like cañon from the south of west. This park-like area extends with but one break nearly to the upper end of San Cristobal Lake. a beautiful sheet of water which entirely fills up the bottom of the Lake Fork Cañon for 2 miles of its course at the western extremity of the Lake Fork Range.

From the lake to the mouth of Henson Creek the stream is in narrow secondary cañons. and the surface of the bottom of the great cañon is filled with rolling hillocks of debris from the mountain-slopes. From this point. where there is a small tract—probably 300 acres—of level ground. to the point the wagon-road leaves the valley of the Gunnison. the stream is alternately bordered by small parks insignificant in extent and by abrupt walls of volcanic rocks of small height. the lower extremities of the slopes from the rims of its cañon cut through by the stream. Throughout this entire portion of its course the Lake Fork flows in a deep cañon. tolerably wide at top. varying in depth from 5.000 to 3.000 feet. bordered on the south and east by the well-defined rim already described of the plateau north of Clear Creek. and by the Cannibal Plateau. and on the north and west by the Lake Fork group of peaks and the summit of the Uncompahgre plateau. The width of the cañon at top will average about 4 miles.

The stream below the lake will average from 50 to 100 feet in width where its current is not unusually rapid. and from 2 to 4 feet in depth. Between the lake and the mouth of Henson Creek it plunges in succession over two picturesque falls. the lower 50. the upper over 100 feet in vertical height.

Henson Creek is about one-half the volume of the Lake Fork. rises in a wide extent of rolling highlands north of the headwaters of the Animas. and plunging rapidly downward. flows east. cutting the tremendous cañon in some places 5.000 feet in depth. bounded by abrupt walls. and very narrow at bottom. which separates the Lake Fork group from the plateau upon which stands the Uncompahgre Peak : at its mouth. on the small flat just mentioned, Lake City, the most promising town of the San Juan country, is located.

The mountains and plateaus bordering on the cañons of the Lake Fork and Henson Creek are from 12,000 to 14,000 feet in height, and the Lake Fork range is situated between the two gorges.

Lake Fork range.—Of the groups of peaks and drainage axes radiating from the Animas divide, the Lake Fork range is one of the few which conveys to the mind of the spectator the impression of veritable mountains. Here the mountain-forms are massive, with clearly-cut and distinct peaks, sharp and decided ridges, and slopes with beautiful and graceful horizontal contours. None of those pinnacles, those thin vertical walls with horizontal tops, those bluffs surmounted with rounded upper surfaces, those flat-topped benches and long-sweeping rolling areas limited by bluffs and cañons. which are so often encountered in this region and convey the idea of eroded plateaus rather than of mountain-chains, are seen here.

Though short, barely 12 miles from the head of the Lake Fork to its bend, there is no portion of the Sierra Madre of Colorado, even, which will compare with it in boldness or beauty. Brilliantly-painted peaks, red, orange, and greenish, with short but full and rapid mountain-torrents fed by the perennial snows of its summit. dashing. falling. and foaming down its steep ravines, with its flanks bordered and limited by the cañons and deeply-cut gorges of the Lake Fork and Henson Creek, with no foot-hills softly shaded with verdure sloping down to gentle valleys, but everywhere sharp. decided, and bold.

From any point of view not near the rims of the cañons of the Lake Fork and Henson Creek, the summits only of this ridge are seen. and seem to be cones rising from sweeping plateaus. On coming nearer, a strange, beautiful, and awe-inspiring sight bursts at once upon the view. A mountain-range, perfect in its details, magnificent in contour, sublime in height, beautiful and gorgeous in color, nearly covered in bass-relief, its base thousands of feet below the general level of the country, sunk out of sight in narrow and seemingly bottomless cañons!

The culminating point of this range forms an appropriate central figure for this masterpiece of artistic nature, this magnificent basso-relievo, and in its symmetry, in its coloring, in its freedom from anything not massive and appropriate, in its silvery setting of mighty snow-banks and rushing torrents, is unapproachable.

The peak itself is pointed and well defined. From the summit a long ridge, sharp and graceful in outline, runs to the east of south for a mile or more, when the mountain falls away in rounded, sweeping curves, in steep slopes, to the bottom of the cañon of the Lake Fork, nearly 4,000 feet below.

On the north side the slopes to the cañon of Hensen Creek are more abrupt, but of the same general character. The peak occupies the vertex of an angle where the divide sharply bends round the head of a tributary of Hensen Creek. It is of a brilliant red color, which continues nearly to its base, but few trees finding foot-hold upon its flanks, save on the southeastern side. This color is due probably to the red oxides of iron in the pyrites of the decomposed trachytes, covering with ashen material or gravelly slides its steep slopes. After crossing the pass at the head of Cinnamon Gulch, a tributary of the Animas, to the head of the Lake Fork, this peak is in full view throughout its entire southern slopes, and, the Lake Fork here taking a bend to the southeast before its final bend to the north, seems to fill up with its slopes the cañon of that stream.

A triangulation-station was made by Mr. Spiller upon this beautiful peak, and he gave it the name of Red Cloud. Its altitude, from cistern-barometer observations referred to camp at its base, is 14,092 feet.

The average height of the peaks of this range, which in all number twenty-seven, is 13,500 feet, many of them exceeding this height.

Uncompahgre Peak and Group.

At the head of Hensen Creek is an extensive area, above the limit of arborescent vegetation, covered with mountain grasses and flowers, separated from the Animas and Uncompahgre Rivers by a quite steep ridge, rising some 500 feet above the general level of this area. It is easy to ride around the dividing line, however, between the northern tributaries of Hensen Creek and the streams flowing north to the Uncompahgre and Gunnison, save at three points where the otherwise nearly even crest of this treeless area is broken by abruptly-rising masses of volcanic material, which will, of course, cause detours to be made to avoid them; but, in spite of the few rough places, it is entirely practicable to ride from the headwaters of Hensen Creek along or near the Uncompahgre drainage-axis as far to the northeast as latitude 38° 10', or about 17 miles, where the portion above timber-line ends. To attain to the summit of this axis, however, is at best a difficult matter, except by way of the headwaters of the Animas River. The streams flowing down from it to Hensen Creek and the Lake Fork, as well as those running in a northerly direction, all are in deep, rocky, and steep cañons, but few of which are passable, and none of which can be followed without great labor and fatigue in climbing over bogs, rock-slides, and fallen timber.

Between the tributaries flowing northward to the Gunnison or Uncompahgre, sharp, almost rectilinear spurs, exhibiting ridge-like peaks, truncated pyramids, or house-like walls of bare rock, with deep and rocky-sided cañons, extend for from 8 to 12 miles north from the divide, and then break down to even-crested ridges covered with timber, and continuing to the cañon of the Gunnison. This general evenness of crest and the truncated aspects of the mountains appearing above it give that plateau-like impression to this portion of the elevated area about the headwaters of the streams in Southern Colorado, in common with the whole northern rim of the Rio Grande basin.

The three masses mentioned above as rising above the general crest of the Uncompahgre spur are, first, the *Wild Horse*, at the head of the western fork of Ibex Creek, the *Wetterhorn*, and further east, the highest mass in Colorado Territory, the *Uncompahgre* Peak. These three peaks are similar in general outline in so far as that each slopes more gradually—if slopes is a proper word to use for such declivities—to the south and are terminated toward the north by nearly vertical bluffs, the Wild Horse and Wetterhorn seeming to imitate their more massive neighbor, and the latter of the two certainly overdoing the character. The Uncompahgre Peak presents a truncated, terraced aspect, seemingly stratified from successive flows of lava, rising about 1,800 feet above the general level of the divide. On the northern side there is a vertical bluff of nearly 2,000 feet; lesser bluffs extend around the east and west sides, leaving a broken, narrow ridge extending to the south, sloping like the curved edge of a snow-drift to the east, but on the western bounded by vertical ledges of trap and steep slopes of rocks and *débris*. The peak was ascended along this ridge with but little difficulty, except at one point near the summit, where a vertical wall of small height affords a good excuse for an awkward climber to roll down the *débris* slopes to the westward.

In 1874 a large cinnamon bear and her cub were found sportively tumbling and rolling from the summit of Uncompahgre at this point, and came near occasioning the loss of our theodolite and of one of the packers who was carrying it. Just as he raised his head above the ledge the bear happened to be about to look down over the same place and both animals, each rather disconcerted at the proximity of the other, tumbled off the cliff together. Both bear and packer, however, happily escaped further injury than a good fright and a few bruises. This peak, though not a sharp point, makes an excellent triangulation-station, since all of the higher points within a radius of 80 miles from the Sierra La Plata around the horizon can be seen from it, and the peculiar shape of its summit makes it readily recognized from any of the points in return. Its altitude is 14,409 feet above sea-level, as determined by cistern-barometer and psychometer observations referred to camp at Lake City. The observations of 1874 make it 14.447 feet, but, since barometric observations taken in August generally give results too high, the lower altitude has been adopted.

A description of the horizon as seen from this peak is not given, since this report is already exceeding its proper limits, and the ranges have their physical peculiarities described elsewhere herein.

The Wetterhorn, to the south of west a few miles from Uncompahgre Peak, is a shark's nose in form, and its ascent being unnecessary for topographical purposes was not attempted. It exceeds 14,000 feet in altitude and appears inaccessible.

The Uncompahgre Peak was reached by us by way of the lower cañon of Hensen Creek, and by following up the tributary of that stream heading under it. This is a rough and difficult trail, and it is recommended to any one desirous of viewing the mountain masses in Southwestern Colorado from the summit of their highest peak to reach it either by way of the heads of the Animas and Hensen Creek, or else, after following the Lake Fork of the Gunnison 9 miles below Lake City, to turn to the west, following an Indian trail (which here comes down) as far as to the summit of the divide, and then turn to the left and follow the ridge to Uncompahgre Peak. This is the longer route from Lake City than that up the cañon of Hensen Creek and its tributary. but will occasion less wear and tear upon man and mule and try the temper of either less.

The Uncompahgre spur and group, as defined herein, form the divide between the Lake Fork of the Gunnison and the streams flowing nearly parallel to it to the north. Of these, between the Lake Fork and the Uncompahgre, there are four heading in this divide, which were traced for only about 7 miles north of the Uncompahgre Peak. For this distance they flow in deep-cut gorges and cañons without bottom-lands, but it is not known into what tributaries of the Gunnison they empty before reaching that stream, nor by what names they are known to either whites or Indians. The most western of the four, however, which flows probably into the Uncompahgre, was named Ibex Creek, for here large herds of mountain-sheep were seen grazing, or else bounding over the rocky slopes bordering this stream. I have heard the name "Cebatta" given to the eastern and "Cimarron" to the middle of the three remaining, but do not know whether they are so known.

HEADWATERS OF THE UNCOMPAHGRE.—UNCOMPAHGRE PARK.—UNAWEEP VALLEY AND RANGE.

The Uncompahgre—the stream with the "valley of fountains"—drains the northwestern side of the Upper Animas River. Crossing the divide at the head of the Animas to the northwest, or at the head of Cement Creek east fork, or at the head of the north fork of Mineral Creek, one finds himself upon tributaries of the Uncompahgre. Near the headwaters of these streams (with the exception of that heading against Cement Creek) are small areas of a few acres sufficiently level for explorers, miners, and donkeys to sleep upon, but even these are often boggy and covered with marsh-grass. Mineral City, near the headwaters of that tributary heading against the Animas, is situated on one of these flats near timber-line, and of course is in a bog. In Red Mountain Valley, already mentioned, quite an extensive park (for this region) of several hundreds of acres affords a good town-site, but is effectually shut off from wagon-communication with the rest of the drainage-area of the Uncompahgre by high and ragged mountains, and can consequently be of little service to miners other than those owning property on this fork.

The Poughkeepsie Fork, heading against Currant Creek, east fork, is quite open near its lower portion, but at its head is impracticable even for mules until a trail is cut in the steep mountain-sides.

The entire drainage-area of the Upper Uncompahgre, including Ibex Creek, already mentioned, is a mass of rugged peaks, bluffs, cañons, and gorges. For about 6 miles from its head the main stem of the stream, which heads opposite the Animas, may be followed, but it then enters that tremendous gorge, impassable even for men on foot, which shuts off communication with the beautiful little Uncompahgre Park below. Vertical walls for hundreds of feet inclose the river, which has cut, through the sandstones of Carboniferous age underlying the trachytes and volcanic breccias and scoriæ, a still narrower secondary cañon, or a cañon within a cañon, the upper surface of which being cut to an equal depth by the cañons of streams joining the main stream, prevents passage above the river along its banks, while the bed itself is filled with bowlders and rocks and gorged with fallen trees and logs.

To climb further than the top of the secondary cañon is an impossibility, for here are sheer precipices of thousands of feet in total height. In fact, for its length, the gorge of the Uncompahgre will excel in the rugged grandeur of its scenery, and in its terrible appearance of ruin and desolation, the grand cañon of the Colorado at its deepest point. From the summit of the frustum, a truncated mass upon the western border of the gorge, to the level of the stream, it is nearly 6,000 feet; to the summit of Blaine's Peak, at the head of one of the tributaries of the Uncompahgre, it is about 400 feet more.

The mountains, or rather ruins, bordering the gorge, especially on the western side and giving it its depth and grandeur, are simply indescribable. If the god of desolation ever exercised his wildest freaks on earth he chose this spot, and cut these lofty masses into those strange forms and weird shapes; those yawning chasms with their red jaws; those beetling precipices with plutonic brows horridly frowning, capping all with slender columns and spires under different angles of inclination to the horizon, which, projected against the sky, seem to be black figures of supernatural origin dancing in glee over the ruin below.

Of the impassable nature of this gorge we speak from sad experience. Bent upon "exploring" it, we succeeded only in tumbling our mules into the cañon, and, after leaving one therein, chose a little worse place to better our trail and succeeded in causing a other roll on the part of our pack-mules of some 1,500 feet, which left us without adequate transportation; and had it not been that we found a prospector here with stock, which we purchased, we might have been stuck in that gorge until this day.

The Uncompahgre gorge about 4 miles from its head opens out into first a wide cañon with flat bottom, covered with stones and sand, and finally into a small park-like area, probably a mile and a half in width and from 8 to 10 miles in length, bordered on the west side by a straight line of cliffs of sandstone, capped with volcanic material, gradually decreasing in height toward the north, and on the east by slopes more or less steep from the Uncompahgre group and its spurs. This park, from the Animas River or the headwaters of the Uncompahgre, is reached by a trail over the highest part of the ridg eeast of the Uncompahgre gorge, and about 2 miles from it, which has already been described in my executive report.

The upper end of the park is quite well wooded along the stream with cottonwood. Scrub-oak attaining 6 or 8 inches in diameter and yellow pines are also found. It is generally well grassed with tall mountain bunch-grass, with wild oats and blue-joint. About midway of the length of the park are hot sulphur-springs impregnated with iron and salts of lime and the alkalies. They have formed a red deposit over 6 or 8 acres of ground which is incrusted with saline efflorescences and bare of vegetation. I observed no outlet to the springs. Their temperature exceeds 120° Fahrenheit; this was the maximum register of our thermometers. Near the head of the park are said to be quite a large group of these springs from which the river and its fork derived their Indian name, "The Valley of Fountains." The entire surface of the park is below 7,500 feet and the greater part of it is susceptible of irrigation.

The Uncompahgre River which flows along the western border of the park is here about 60 feet in width (in August.) and 3 feet in maximum depth.

The lower end of the park is nearly bare of vegetation, save sage, mountain-tea, and cactus of the flat-leaved or prickly-pear variety.

Near the northern limit of atlas.sheet 61c, (38° 10',) the Uncompahgre is joined from the southwest by the Unaweep and the park is closed by ledges which close in upon the stream. Indian trails, here broad and well worn, follow on down the Uncompahgre, with branches running through the Unaweep Valley southwest to the San Miguel, and beyond to the Navajo and Moquis country, crossing the old Spanish trail from Los Angeles to Santa Fé near the bend of the Dolores.

The Unaweep Creek, at its junction with the Uncompaghre, is about two-thirds the volume of that stream, of pure, bright, sparkling water, flowing in a valley, which just abov this junction widens out into quite an extensive area covered with sage and scattered bunch-grasses.

I..e stream about 4 miles above its mouth flares out in fan-shape, its numerous branches draining the northern slopes of the Unaweep range. The entire drainage-area of the Unaweep, with the exception of the flat valley at its mouth, is a series of rounded rolling hills, heavily grassed and without timber other than a few scattered clumps of pines, and along the streams cottonwood, until the steeper foot-hills at the base of the Unaweep spur are reached, which are quite heavily timbered with spruce. The entire valley is a splendid range for cattle, the lower extremity not too high for wintering them.

Several thousands of acres, near the mouths of the streams in the valleys mentioned, are susceptible of irrigation, and would undoubtedly produce good crops.

Beyond the limits of the valleys of the Uncompahgre and Unaweep, the country is a portion of the Colorado Plateau system, the portion of it falling in atlas-sheet 61c averaging 9,000 feet altitude, but few rounded peaks north of the Unaweep, and a few miles from it, breaking the general level. These rise only to a very small height above the surface of the country, and are seemingly trachytic or basaltic cones.

The Unaweep spur shoots off from the rim of the Animas Basin near the head of Mineral Creek, and trending north 75° west, forms the dividing ridge between the headwaters of the San Miguel south fork and those of the Unaweep. Its northeastern slopes form the grotesque western walls of the Uncompahgre gorge.

This spur is short, and a well-marked serrated range of peaks. On the Unaweep side its slopes below timber-line are long and gradual, but very sharp and steep from the 10,000-feet curve to the summits. On the side of the San Miguel, however, the slopes are more abrupt as far as to the narrow valley of that stream. The eastern mountains of this spur, especially the group in the northeast, drained entirely by the Uncompahgre, are very much chopped, weathered, and eroded, and exhibit very peculiar features. In some instances the mountain-summits are thin walls of even crests, quite long, and vertical for from 500 to 1,200 feet below their summits. In others these walls are so weathered that prisms and spires only are left surmounting the thin slabs. Several of them are truncated at from 13,700 to 13,900 feet above sea-level, presenting the appearance of pyramidal frustums. In the midst of these strange forms Blaine's Peak, a sharp, decided cone on the extreme northern edge of the range, is situated. It is 14,249 feet in height, exceeding by about 400 feet the masses to the east and south. As just stated, it is in the extreme northern part of the Unaweep range, and is drained altogether by the Unaweep and Uncompahgre; the main stem of the former heading in an extensive cup-like amphitheater southwest of Blaine's Peak, of which this peak occupies the northeastern part of the rim. The southern part of the rim of this crateriform cavity is one of the wall like ridges before referred to; the eastern, a thin and sharp divide between the western fork of the Uncompahgre, Crescent Creek, and the Unaweep. The northern rim is broken down for nearly a mile of its crest in a V-shaped notch, through the narrow bottom of which the headwaters of the Unaweep escape. Blaine's Peak was not ascended, since it was not necessary as a station, the topography having been secured from more easily-attained points. The northern side of this peak is inaccessible on account of the nearly vertical ledges and buttresses of this face. From the south it is necessary to cross the high rim of the amphitheater before beginning the ascent, which must be made along the sharp divide on the eastern side of the cavity. The easiest way to attain the summit would be to cross over the head of the most northerly of the three extreme eastern tributaries of the San Miguel to the southern rim of Crescent Creek, and then to follow this stream to its junction with the main stream and up its seemingly open cañon to the divide mentioned.

This course can be taken upon mules, at least as far as timber-line on Crescent Creek, whence the last 1,800 feet can be made upon foot. Crescent Creek, within 2 miles of the Uncompahgre, cañons, and its course cannot be followed from that stream, even if its mouth could be reached. It empties into the Uncompahgre in the most formidable portion of the great gorge, and is unattainable from that stream save near its head, which can be reached by crossing from Red Mountain Valley west, to the head of the tributary mentioned above. Blaine's Peak may be also reached in this way, but all trails in this neighborhood are excessively steep, and in many places rocky from slides and choked with fallen trees.

SAN MIGUEL RIVER.

The San Miguel River receives the greater part of its waters from its two tributaries, the northern of which drains the southern slopes of the Unaweep group of peaks and portions of the rim of the Animas Basin opposite the headwaters of the north fork of Mineral Creek. The southern, heading against the headwaters of the south fork of Mineral Creek, drains the western portion of the outer periphery of this oval-shaped rim, and, flowing north, unites with the northern, or Gold Run, from 12 to 15 miles from the head of either branch. The northern of these two streams flows for 8 miles of its course in a flat valley 400 yards wide, running with sluggish flow in serpentine folds, and bends from side to side of this valley.

The edges of the valley are bluffs of Cretaceous sandstones, capped on the north by brown trachytes. At the head of the valley is a semicircular cul-de-sac, with walls on the eastern and southern sides from 450 to 1,200 feet in vertical height, over which dash in spray the headwaters of this fork. The northern of the three forks into which Gold Run is here divided, flows from its head, 2½ miles distant, in a succession of cataracts, rapids, and falls, the fall being nearly 4,000 feet in 2 miles. Just above the junction of the two forks, Gold Run plunges into a cañon, and, in a series of cataracts, attains in 1 mile a level 1,000 to 1,200 feet lower than the valley above. In the cañon, cut through sandstone, it is joined by the south fork, and flows off in a west of north direction to the Gunnison, as far as can be seen in a deep cañon with shelving sides of red sandstone. The south fork, from the lake 4 miles below its head, flows in steep and deep ravines, and 2 miles below the lake, just above its junction with Turkey Creek, rushes in a series of cataracts into quite a deep cañon, with narrow, grassy bottom, along which runs a trail to the lately-discovered gold-placers on the Gold Fork, at its mouth.

Trout Lake is a beautiful sheet of water, one-third of a mile long, nearly inclosed in brilliant red mountains, skirted with green. Its vicinity furnishes one of the most picturesque and warmly-tinted pictures of mountain scenery in Colorado. The lake abounds with trout, which, however, are not of as firm flesh as the trout of the mountain streams, nor do they attain to as great size. This may be due to the effects of the vegetable mold of the bottom of the lake and the bubbles of gas which are continually rising therefrom through the water. It cannot be seen how the trout reached the lake, since lesser difficulties than the falls and cataracts below effectually prevent their ascending other streams in this area. Southwest of Trout Lake is a wide timbered pass leading to the Dolores River, which, with the Mineral Creek and Cascade Creek, drains the high group of peaks at the head of the south fork of the San Miguel. The Dolores, a few miles below its head, is inclosed in a cañon with steep sandstone walls, from which it does not emerge until the great bend is reached. A difficult and nearly-obliterated trail, traversed last year by Lieut. C. W. Whipple's party, leads down this stream.

San Miguel Mountains.

Between the Dolores and the San Miguel is a detached group of high peaks, three of which exceed 14,000 feet in height, hitherto called the Sierra San Miguel. The drainage-axis is nearly east and west, cut through by the San Miguel River near its cascade. This group reaches its greatest height in a massive peak, with extensive fields of snow on its northeastern flank. This peak was ascended in 1874 by Mr. Spiller, and in our notes was called Glacier Point. Mr. Wilson, of the Geological Survey of the Territories, also ascended it, and called it Mount Wilson. It is the most massive and imposing mountain in Southern Colorado, with the single exception of the Sierra Blanca, near Fort Garland, and is 14,243 feet in height. Two miles east of this great peak is a strange freak of nature. This is a thin column of trap-rock, about 230 feet in height, shooting vertically from a rude cone, forming a slender and well-proportioned column, with rounded summit, cleft in twain to a depth of perhaps 10 feet. Its summit is about 13,150 feet above sea-level, and offers a fine point for triangulation, with the single objection that it cannot be ascended.

To the west of the Glacier Point group are several nearly isolated masses, along the same drainage axis, of which one, an irregular group with a sharp, well-defined cone at its southeastern extremity, is remarkable for the great crater on its southwestern side, from which flows a tributary of the West Dolores. This cone was made a primary triangulation station, and named *Dunn Peak*, in honor of Gen. W. McK. Dunn, Judge-Advocate General of the Army. It is 13,502 feet in height, and being quite detached from the lofty ranges, is a much more marked feature of the landscape than peaks of the main group exceeding it in height. West from the Dunn's Peak Crater is a beautiful sharp cone, perfectly symmetrical, reaching but little over 12,000 feet. It is the most beautiful peak I have ever seen. It is entirely detached from the other mountains, and rises, a solitary, graceful peak, 3,000 feet above its base. It was named by me West Point. This is the last peak to the west until the Sierra Abajo and Sierra Le Sal are reached. South of these detached masses, and between the two main forks of the Dolores, is a small group of peaks, less than 13,000 feet in altitude, rising abruptly from the cañons of the Dolores.

From West Point the horizon is a succession of nearly flat sweeping plateaus from the southwest around to the north. The tributaries of the San Miguel and Dolores, which head in the San Miguel drainage-axis, all cut deep cañons in the sandstones.

The upper surfaces of the rolling country bordering the San Miguel and its tributaries, as soon as they emerge from the mountains, are most luxuriantly clothed with the richest of grasses. Wild oats and the tall seed stalks of various mountain grasses reach to a man's thigh, and the whole region resembles in the latter part of August an immense field of waving grain; clumps of quaking aspen and of yellow pine vary the surface and beautify the landscape. The region is too high for other purposes than summer grazing, except in the narrow cañons of the streams, where occasional small plats sufficiently low for cultivation may be irrigated. Game seems to be quite abundant; wild turkeys were seen upon the San Miguel South Fork. On the Dolores, bears, both black and cinnamon, are frequently seen, and sleep was nearly impossible for us on account of the continuous and multiplied cries of the California lions. Small game, such as grouse, and, on the bald-topped mountains, ptarmigan, or mountain quails, and in the streams mountain trout are easily captured.

Divide between Dolores and San Juan, Sierra La Plata, and upper drainage-area of the La Plata, etc.

The radial drainage-axis, which from the head of the South Fork of Mineral Creek trends south 35° west as far as to the Sierra La Plata, is (with the exception of a few trachytic masses, less than 13,000 feet altitude, south and west of the Dolores headwaters) a ridge of comparatively low broken hills, covered with timber, and drained by Hermosa Creek, a tributary of the Animas, and by steep rills flowing into the cañon of the Dolores, until the group of mountains known as the Sierra La Plata, twenty-six miles from the rim of the Animas Basin, is reached. Here this drainage axis attains a maximum height of 13,300 feet at the head of the North Fork of the Rio Los Mancos, where the highest of the La Plata group is situated.

The Sierra La Plata is not, properly speaking, a Sierra—a single range of serrated peaks—but is a group of peaks and mountains gradually lessening in height to the southward, and from its highest peaks, which occupy the northern part of the area covered by the group, breaking rapidly down to the northwest to the level of the Dolores. The Rio La Plata heads in the northern portion of the group and flows west of south, bordered for 13 miles of its course on either side by peaks of the La Plata group. The eastern side of this double range is drained by tributaries of the Animas River, the western slopes by the two minor forks of the Rio Los Mancos. To the northwest the water flows into the Dolores. The mountains are composed largely of sedimentary rocks in many places, altered by trap-dikes and igneous flows. This group is now attracting considerable attention from the extensive placer or auriferous bar at the gate-way of the Rio La Plata, and on account of the rich lodes of gold and silver ores lately discovered in them.

The streams flowing from the La Plata Mountains are quite insignificant. The La Plata and Mancos are about 40 feet in width and perhaps a foot and a half deep. Junction Creek is small, and the Hermosa, which has quite an extensive drainage-area, receives but little of its water from these mountains. It is, however, quite a large stream, equal in size to either the La Plata or Mancos.

Of the valleys of the streams flowing from the La Plata group those of the Mancos and La Plata only are important. On the Mancos, near the junction of the two forks, is, perhaps, 1¼ square miles of arable land, extending down the borders of the streams and along the stream at intervals; below are said to be good farm-sites.

The La Plata, within the limits of the recent purchase from the Utes, is too high for certainly-successful agriculture, but below the boundary are several thousands of acres which, if the stream furnishes sufficient water, will produce good crops. The Utes now raise corn, squashes, and melons with success, and roots and grains would, no doubt, successfully mature.

Between the Mancos and La Plata are several small parks well grassed, and the hills are all covered with mountain grasses. Quite extensive ranges for cattle can be found here. At present about four thousand head of sheep are grazed, with the permission of Ignacio and his band of Utes, who, notwithstanding the purchase of 1873, still claim the land.

The Mancos, a short distance below the junction of its forks, enters its cañon, cut through the Mesa Verde, a mesa several hundreds of square miles in extent, covered with wood, and much cut by cañons; an Indian trail leads through the cañon.

The drainage-areas of Junction and Hermosa Creeks are worthless, unless coal of good quality exists in the rocks bordering the streams. There are no valleys or flats of any extent, but the entire area, particularly the drainage of Hermosa Creek, is a mass of a small hills and cañons with vertical walls of sandstone, a chopped-up area of tangled cañons and worse-tangled fallen timber. Indians have made no trails through the Hermosa drainage-area, and whites have no occasion to.

SOUTHERN RIM OF ANIMAS BASIN, ANIMAS RIVER AND PARK.

The southern side of the rim of the Animas Basin is cut through by the Animas River, which escapes to the southward through a deep-cut gorge, which, though continuing as a close but not deep cañon below that point as far as to Old Animas City, 25 miles in air-line below its head, ends as a deep, formidable obstruction at the mouth of Cascade Creek, 15 miles below Baker's Park.

This cañon, to my mind, for reasons which will be given hereafter, is a very important element in the problem of routes of communication onward from the extensive and prospectively valuable and rich mines of the Animas, Eureka, and Uncompahgre mining districts. It is avoided now as utterly impracticable, and, unfortunately, the heavy rains and snows of the month of September so much delayed the work of the party that we could not make a detailed examination and profile of the cañon, or even devote sufficient time to it to find out whether it is passable or not. Two stations were occupied along the western rim of the gorge, but the surface of the river below was not reached.

From the foot of Baker's Park to the mouth of Cascade Creek the cañon is bordered by high peaks of quartzite, (with the exception of the eruptive rocks at Sultan Mountain,) reaching in some cases over 5,000 feet above the stream, which inclose the river in a cañon narrow at bottom, but opening out at top until from the stream to the summits of the peaks bordering it are intervals of from 1 to 2½ miles, or the mean slopes of the cañon walls vary between twenty-five and forty-five degrees. Near the stream the slopes are more nearly vertical, and are cut through metamorphic rocks as hard as flint; Cascade Creek, near its mouth, flows in a similar cañon, but not so deep.

From the mouth of Cascade Creek to the head of the Animas Valley, about 11 miles, the Animas flows in a lesser cañon, cut through the eastern edge of the floor of a valley about 1½ miles wide, bordered on the east by a ridge gradually decreasing in altitude to the south, and on the west by a vertical ledge of red sandstone, which limits the drainage area of the Hermosa to the eastward. Above the mouth of Cascade Creek the country is very rugged, and broken by the deep-cut valleys and steep ridges between of the streams draining the southern side of the Animas rim. This entire section below high-timber line is covered with fallen timber and many little bogs and springs, which, in addition to steep grades, make the trails over the high passes at the heads of Cascade and Lime Creeks difficult and unsatisfactory.

Trails of this character, steep, boggy, and obstructed with logs, lead over the heads of Cascade Creek to the Dolores and to the south fork of Mineral Creek; and over the head of Lime Creek to Baker's Park.

The mountains in the angle between Cascade Creek and the Animas are of metamorphic rocks, quartzite, and farther west and north on the southern slopes of the rim are trachytic, overlying extensive exposures of the limestones and sandstones of the Carboniferous series; which latter give, by their worn and exposed faces and small dip, mesa-like bunches or berms at the bases of many of the trachytic slopes, notably near the headwaters of Cascade Creek and at Engineer Peak, ascended and described by Lieutenant Ruffner's party in 1873.

South of Cascade Creek the rolling floor of the bench or plateau valley cut through by the Animas, along its eastern edge is heavily grassed, and quite well watered by streams flowing from cliffs in the sandstone bluffs bordering it to the west, and by lagoons and lakes on its surface near the river. It is about 8,600 feet in altitude, and is a fine summer range for a small herd of cattle. At Animas City the river emerges from the low walls of its cañon, and the valley opens out a short distance below to a width of from one-half to 1 mile and extends for 10 or 12 miles, (within the area purchased from the Utes,) its nearly flat surface broken in one or two places by low, rounded hills rising, perhaps, 25 to 50 feet above the surface of the valley. In this park or valley very fine crops of corn, wheat, oats, and barley were raised last year, and all of the usual products of the vegetable garden were growing to perfection. Up to the 1st of October, as I am informed, there had been no frost, which the green tops of growing plants further attested. Potatoes grow to great size; single bulbs weighing two pounds or more each were in several instances found among the few purchased for the use of the party, and were of fine flavor and texture, solid throughout. These were said to have been grown from the miserable little refuse bulbs obtained at Tierra Amarilla, N. Mex.

The Animas River is here about 100 feet in width and 3 feet deep, of sluggish flow, and very tortuous; bordered near the center of the valley with soft banks of alluvium, and in places in its bed obstructed with quicksands. Fords, however, are found at several places in the valley where rapids occur. Junction and Hermosa Creeks furnish sufficient water to irrigate the western side of the valley, and the eastern can be irrigated by a ditch starting below Animas City, where the banks of the stream are not too high. There are probably, in all, between 3,000 and 4,000 acres of arable land of fine quality in the Animas Valley, north of the boundary of the Ute reservation, varying in altitude from 6,400 to 6,800 feet. All of the desirable ranches have been homesteaded, and will, it is expected by the owners, prove sources of wealth more certain than the silver lodes which attract their customers to the mines above.

Old Animas City is situated on a low bunch of gravel overgrown by quite a forest of large yellow pines. Elsewhere, near the valley, there is abundant timber and fuel.

Just below old Animas City, which, by the way, is not inhabited, are several springs of carbonated water impregnated with soda-salts and with sulphur, and of quite pleasant taste. The temperature of one was taken, and found to be 107°.3 Fahrenheit. They have formed a dirty ocherish yellow deposit about them.

The Needles or Quartzite Crags.

East of the great cañon of the Animas and south of the headwaters of the Rio Grande, about the headwaters of the Rio Los Pinos and Florida, is a group of crags of quartzite rocks, reaching to an altitude above sea-level of 14,000 feet. The forms of these crags are so unusual in the mountains of Colorado, and especially here where rounded summits and large masses with sweeping curves and outlines are so frequent, that they deserve mention, although the area covered by them is very small, and in all probability is forever useless, unless unexpectedly mines of silver or gold are found. Among these there are no slopes covered with grass, no valleys, scarcely foothold for pines on their slopes, but everywhere thin crags and slabs with knife-edge or saw-teeth summits, as a rule wholly inaccessible, pierce the clouds, encircling like a crown the heads of the western tributary of the Rio Los Pinos and short tributaries of the Animas above Cascade Creek. Nowhere in Colorado can be found such steep slopes, such shapeless crags, such rocky and impassable ravines, such generally detestable characteristics and features as are here seen.

The hard metamorphic rocks are shivered along their cleavage planes for hundreds of feet, leaving here odd pinnacles, there the likeness of the shattered outspread wings of some gigantic bird, and again of the grim grinning teeth of Death.

AP. JJ—7

Under the most favorable circumstances one must be impressed with a feeling akin to terror in merely viewing these lofty crags and these deep-cut gorges between them; but, under the conditions they were visited in 1874, by Mr. Nell and myself, they cannot fail to leave behind a lasting dread of them and of everything connected with them.

On the 1st of October, after considerable difficulty, we succeeded in reaching the limit of tree-growth on the flanks of one of the most southerly of these crags, by crossing the Atlantic and Pacific divide at the head of the Ute Creek, tributary to the Rio Grande. Camping here at the head of a profound gorge 11,750 feet above sea-level, we awaited the next day to make a triangulation station upon apparently a thin slab, which from its appearance from the north we had christened "The Hunchback." In this, however, we were destined to disappointment, for it began to snow, and the next morning the ground was covered to a depth of 6 inches. Thin fleecy clouds were floating about, and cheated us with the hope that the storm was spent and the peak could still be made. We delayed, therefore, until the summit of the crags should be visible, but the clouds only thickened, and soon again the snow was flying so thick and fast that we could not see 50 yards; and to make the white gloom more visible, and our condition more amusing, perhaps, forked streams of electricity, followed by tremendous bursts, claps, and rolls of thunder—which, attending a snow-storm, was to me a special and unheard-of exhibition of heavenly wrath—leaped from cloud to crag, in this group of natural lightning-rods, in the most inexplicable and to us unassuring and unnecessary manner. It blew and blustered, stormed, crashed, and thundered as if heaven were determined to level and blow away in dust these already shattered and ruined masses of flint. All day it would storm and snow, and at night the stars would come out twinkling, cheering us with a vain hope for the morrow.

Three days we lay here unable to move, while those light, beautiful, almost impalpable flakes were gradually enveloping us, removing from us hopes of escape, and making life miserable by insinuating themselves like fog, everywhere, into our tents, into our beds, almost into the pores of our skins. We dare not move, lest, lost in this white darkness, we should be precipitated over some ledge or be lost in banks of snow drifted by the winds. Our poor mules, which, in their previous revolutionary attempts to subvert the established order of their packs by erratic excursions down hill on *side trips*, had succeeded only in crippling themselves; motionless, and covered with as much snow as would lie upon their backs, stood gloomily around, gazing at us with that dumb expression of unutterable misery, despair, and reproach which even mules can assume, and which appeals irresistibly for sympathy and aid.

On the morning of the fourth day the snow was slowly falling, and the fleecy clouds rolling hither and thither above us and in the gorges below, every now and then giving us a glimpse of sun and snow-crowned crag or deep-cut gorge, showed to us poor shivering mortals that the storm was nearly spent; our mules gained courage, and began to break their three days' fast by ravenously devouring the twigs and branches of the scattered shrubbery and the grass obtained by pawing the deep snow.

We hardly needed the additional spur to exertion given us by our cook, who, when preparing our breakfast, incidentally remarked, as a matter of no consequence, as he patted his bread into the oven, "That's the last loaf," adding, however, this additional unimportant bit of news, " We've got lots of beans."

We excavated our packs from the snow, and packing our hungry mules, commenced with many misgivings to retrace our steps. Our supplies were at Pagosa Springs, 60 miles distant, and it was necessary to first cross the continental backbone to the Rio Grande side, and then again recross it to the head of the eastern fork of the Los Pinos. The snow varied from eighteen inches to 4 and 5 feet (where drifted) in depth, and the divide where we crossed was at both places over 12,000 feet in altitude. It was necessary for us to dismount and tramp down the snow to make a passage-way for the loaded and nearly famished mules, which, as fast as the leading animal would become exhausted by the impeding snow, were changed in their order of travel. We succeeded fortunately in reaching each night places where, in lower altitudes, the snow had partly melted and left the grass sticking through it for our animals, (otherwise we would inevitably have lost them,) and in five days, tired out, we reached our camp at Pagosa, having subsisted entirely upon our cook's " plenty of beans" for that interval. This article of food had been persistently avoided by us hitherto, but most modestly asserted itself in our hour of need.

The quartzite crags are inseparably connected with this disagreeable experience, and in my condemnation of their countenances I am deeply prejudiced against them.

HEADWATERS OF THE RIO LOS PINOS, RIO FLORIDA.

Crossing the divide at the head of Hine's Fork of the Rio Grande del Norte, we find a stream flowing in a deep and rocky gorge through the midst of the crags just mentioned to the east of south. Northwest some 10 miles and within 4 miles of the cañon of the Rio Grande, heads under Simpson's Pyramid and flows through a flat, badly drained summit far below timber-line the east fork of the same stream—the Rio Los Pinos. This flat summit of the Los Pinos Pass is strewn, especially just above the head of its cañon, with immense bowlders of granite, sometimes 20 feet in diameter. The pass resembles in this respect, over very limi·ed areas, however, the areas of drift in some of our Eastern States. Along the headwaters of this stream, indeed, are the only granite rocks encountered in this whole group of mighty mountains, save a few exposures upon the Piedra headwaters. Ten miles below its head the East Fork cañons, and for 14 miles flows through a cañon with bluff-like sides, the stream choked with granite bowlders and the banks with fallen trees. Below the cañon the stream is joined by its West Fork, before mentioned, which, emerging from its quartzite gorge, flows in a beautiful little valley several miles in length above the junction, and averaging a half to two miles in width, the latter being the diameter of the valley at the junction of the streams. Below this point the valley is narrow ; in places covered with magnificent yellow pines, and at others with a few grassy valleys. The stream is alive with the finest and largest of mountain trout. In its cañon the Los Pinos is joined by two quite large tributaries flowing down from the easternmost of the quartzite crags ; but from the east side, where there is near the border of the stream a thin range of mountains reaching 12,000 feet, the tributaries are very short and insignificant. The Rio Los Pinos below the junction of the two forks is 80 feet in width—a bold, deep, and pure stream of sparkling snow-water. Quite extensive strips of land along its banks are susceptible of irrigation.

The Florida, between the Los Pinos and Animas, is of nearly the same character. It is about 35 feet in width and 3 feet maximum depth, rises in the southernmost of the "Needles," and drains the sloping ridges above timber-line between the West Los Pinos and Animas Rivers. Its valley is similar to that of the Los Pinos, but there is above the boundary of the reservation but comparatively little arable land upon its border, the upper portion of the valley above the bend of the stream being narrow and high. South of the junction of the two main forks of the Los Pinos, and drained by the two streams mentioned and those east of them, and within atlas-sheet 61C, the country is rolling, the high lands averaging 8,500 feet, eroded by water, exhibiting along the stream frequent low bluffs of sandstone of Cretaceous age. It is well timbered with large yellow pines, oftentimes 3 feet in diameter, and for 50 or 60 feet without a branch or knot. Everywhere is fine grass.

We made the first trail through the Los Pinos Cañon. The Utes, and whites after them, avoid the cañon by turning to the left above its head, and then following along the eastern slopes of the divide between the Los Pinos and Piedra, near the summit of the ridge, and come down again to the Los Pinos at the beautiful little park above the junction of its two forks. If the bogs were cut out of the trail the cañon would offer better grades, a harder surface, and a shorter route than the trail above it, which is generally execrable ; as it is, however, the cañon can only be passed by work. The stream is well peopled with beaver colonies, and on the small flats bordering it within its cañon many deer were seen.

On the Florida and elsewhere in the San Juan area, coal has been discovered, and is said to have good coking qualities. If it be as represented, it will prove of great economical importance in the smelting of the ores of the mines in this area, especially if the route via the Animas Cañon is ever opened.

The mines of this, the most important part of the area surveyed, will be briefly noticed hereafter in this paper.

3d. FROM THE LOS PINOS PASS TO THE SUMMIT MINING DISTRICT.

This division of the Rio Grande loop was only partially visited by my party, but other parties of the expedition during the past two seasons completed its survey.

From the pass mentioned to the headwaters of the west fork of the Rio San Juan the range is of less height, but more broken and cut by cañons than the opposite or first division, while at the same time exhibiting the plateau-like aspect in many of its masses.

With the exception of a few peaks near the headwaters of the western branch of the Piedra, there are no marked and dominating peaks, but a great number of small mountains, with ridge-like, rounded, or flat summits along the divide, in some instances peaked and reaching 1,000 feet above timber-line, but as a rule not exceeding 12,200 feet.

The passes from the Rio Grande to the heads of the Piedra are all below timber-line, and, except at a few places where bluffs prevent, the dividing-line between the waters of the Rio Grande and Piedra can be followed on horseback from the head of the Los Pinos as far as to the head of the eastern tributary of the Piedra. This was done by me the past season without great difficulty. On each side of the divide, vertebral ribs between the streams flowing into the Rio Grande and Piedra, separated on the southern side by very steep, rough cañons, make travel nearly impossible, and the attainment of the divide from the valleys of the Piedra area a very difficult undertaking.

The peaks retain for 6 or 7 miles south from the divide a nearly uniform height, and then in a very abrupt fashion break down in nearly vertical slopes to the level of the streams which flow through narrow gateways in these outer bluffs to the open country to the south of them. The ribs between the streams to the northward, or on the Rio Grande side, are wider and with more unbroken and rounded summits, presenting between the Los Pinos and Macomb's Peak smooth crests or 1 ench-like tops, bounded by steep walls or slopes toward the streams. Between the San Juan and Piedra Rivers is a very decided spur, or a serrated range of mountains, varying in height from 12,560 to 13,150 feet, which, trending 15° east of north, continues from the Pagosa Peak, the most southerly of this range, nearly to the Rio Grande. On the north side this range forms the divide between Thunder Creek, heading under Macomb's Peak, and Hot Spring Creek, both tributary to the Rio Grande. Its extremities are more gradual slopes than usual in these mountains, on one side sloping from below timber-line on Pagosa Peak gradually down to the oak-covered hills of the Piedra and San Juan basin, and on the northern ending in large, massive, rounded mountains, sloping with smooth slopes to the Rio Grande between Bristol Head and Wagon-wheel Gap. Around its northern end the Rio Grande describes a great curve before cutting the cañon at Wagon-wheel Gap.

Six miles east of Macomb's Peak is another quite high rib from the continental divide, serrated to the south, and drained by the San Juan west fork, but to north showing round, flat tops between the western tributaries of the south fork of the Rio Grande and Hot Spring Creek. From this rib to the north and west to the summit mines the country is mostly below timber-line and cut by steep cañon of the tributaries of the south fork of the Rio Grande, and quite heavily timbered.

Between the south fork and the San Luis Valley the mountains again rise above timber-line and extend from the rounded, dome-like mass southwest of Del Norte to the west of south, rising to a high sierra about the heads of the eastern fork of the San Juan. These mountains were not visited by me, but were surveyed by the party of Lieutenant Morrison. They were also visited by your party in 1874. From the head of the Piedra to the summit mines there is no traveled trail crossing the divide, but at the head of the west fork of the Rio Grande, which empties into this stream below Antelope Park, I crossed a trail on the summit which at one time seems to have been well traveled by the Utes. It follows up the middle fork of the Piedra and crosses the divide below timber-line. It will never be available as a wagon-route. Another trail of the same character leads across the headwaters of the south fork Rio Grande. However, all mountains can be crossed by pack-trains with a little work, and this division is not naturally as impassable as the first. The divide can be reached in many places, but the cañons are rocky and steep. The prevailing rock being still the brown trachytes met with throughout the Rio Grande loup; a coarse-grained, red feldspathic granite is occasionally seen.

As soon as they leave the mountains the tributaries of the Piedra flow through valleys, of greater or less width, similar to those on the Los Pinos and Florida, all of which are splendid cattle-ranges, and in great part sufficiently low for agricultural purposes. The Piedra, however, cañons up a few miles north of the southern boundary of atlas-sheet 61c, (37° 20'.) and the agricultural lands are limited thereby. Between the streams are low hills, covered with great pines and undergrown with scrub-oak, the acorns of which attract in the fall many deer, making this a favorite hunting-ground of the Utes at this season.

The San Juan for 15 miles of its course above Pagosa hot springs is bordered by similar valleys, ranging in altitude from 7,057 feet at Pagosa to 7,700 feet at the point the trail to the head of the south fork of the Rio Grande leaves its western tributary. On the south fork of the Rio Grande are no valleys, save one too high for agriculture on its eastern branch, and another more narrow on the upper portion of the middle fork. The western tributary is in a cañon its entire length.

The valley of the Rio Grande itself, above Del Norte, has already been well described in the reports of Lieutenant Ruffner and Prof. J. J. Stevenson, and needs no extended description here. From the head of Antelope Peak to Del Norte the river flows in a succession of narrow valleys well grassed, but too high 'or agricultural purposes, until the mouth of the south fork is reached, where potatoes succeed in maturing. In this distance the stream is inclosed by cañons for about 6 miles of its course at Wagon-wheel Gap and below. From the mouth of the south fork to Del Norte is quite a wide valley, (averaging 1½ miles,) much of which may be easily irrigated. At Del Norte the Cerritos of basalt close in, forming a comparatively narrow gateway in which Del Norte is situated, and through which the river flows out into the San Luis Valley. At Del Norte the work of the season was practically brought to a close by snow.

MINES.

No special examination was made by me in person, but the following general information was gained from miners and others interested in the mines, and from the observations of myself and party made at random while pursuing our topographical work.

Districts.—At present the mineral-bearing region is divided into six mining districts. On the *Animas* are the *Animas districts*, extending along the Animas from 2 miles above Howardsville to the south and including the southern and southwestern portion of the rim of the Upper Animas basin, and the *Eureka district*, adjoining the Animas district, and including all the country in the northern portion of this basin as far as the divide. On the *Uncompahgre*, the *Uncompahgre district* includes the drainage-area of the Uncompahgre above its gorge. The *Blaines Peak*, or *Mount Sneffels district*, the lower mineral region, of which *Ouray* is the town.

On the Lake Fork is the *Park district*, extending from the headwaters of the fork to Cottonwood Junction.

The *Lake district*, including the Hensen Creek Mines, and those upon the Lake Fork north of Cottonwood Junction.

In the Animas districts the ores are mainly argentiferous galena, with, in some cases, traces of gold, assaying from 40 to 60 per cent. of lead, and from 60 to 400 ounces of silver per ton; assays of culled ore have been made far up in the thousands of ounces. The ores are contaminated by iron and copper pyrites. The Little Giant Mine and others in Arrastra gulch are gold-bearing, the matrix being quartz with a streak of manganese in the form of ripidolite, which seems to be richest in gold. Gray copper ores are also frequent along the Upper Animas. On the Uncompahgre the veins are of immense width, with pay-streaks varying from 10 inches to 4 feet; the ores on the southern side of the district being smelting, while those on the northern fork are in great part free milling ores. Petzite, gray copper, bromide of silver, ruby-silver, native silver, and antimonial silver ore, and various sulphurets assaying high in the thousands of ounces, have been discovered. Many of the lodes here show traces of gold.

On the Lake Fork of the Gunnison and on the Hensen Creek are about 800 locations, the ores being, as a rule, of lower grade than in the Animas, Eureka, and Uncompahgre districts, but of the same general character. Several of the mines, however, give very rich ores, as, for instance, the Hotchkiss lode gold and silver assays from $200 to $12,000 per ton, averaging $1,500; *Ute lode*, $200 to $1,200; *Little Chief*, gray copper ore, $200 to $15,000; *Lone Chief*, petzite and black sulphurets, $150 to $800, and perhaps a dozen others of high-grade ore. The above veins are mentioned only for illustration. I personally know nothing about them, except from information which I consider reliable.

Below Lake City are several lodes of gold bearing quartz assaying from $20 to $1,240 per ton.

Up to October, 1874, the following assays have been made in the Uncompahgre district, and will show what is considered a fair average of all the lodes and claims in the San Juan country, near the surface. The ores are generally low grade, but an unusual number of very rich mines have been discovered, when we consider that one good lode where transportation can be secured is sufficient to assure the prosperity of a mining district. Witness Pioche and other mining districts in Nevada:

46 assays, from	5 ounces to	25 ounces per ton in silver.
19 assays, from	25 ounces to	50 ounces per ton in silver.
47 assays, from	50 ounces to	100 ounces per ton in silver.
18 assays, from	100 ounces to	300 ounces per ton in silver.
9 assays, from	300 ounces to	500 ounces per ton in silver.
11 assays, from	500 ounces to 1,000 ounces per ton in silver.	
5 assays, from 1,000 ounces to 2,000 ounces per ton in silver.		
2 assays, from 2,000 ounces to 3,000 ounces per ton in silver.		
3 assays, from 3,000 ounces to 5,000 ounces per ton in silver.		
1 assay, from	6,300 ounces per ton in silver.	

The above may, of course, be regarded as results of assays of picked ores near the outcrop. Ores worth less than $150 a ton at present prices for milling (averaging $75 per ton) and for transportation ($100 to $120 a ton to Del Norte) will not pay to work, and for this region are classed as low grades.

New discoveries are still being made almost everywhere in this group of mountains, and the known area covered by the mineral croppings and lodes is continually widening. I found float galena and cupreous ores of silver near the head of the San Miguel, and picked up several small nuggets of gold from the rocky bed of one of its tributaries. Doubtless there will be found rich lodes of gold and silver here also. The veins throughout the region are well defined, with good walls, and give, by their persistency along the surface, every outward evidence of permanency. They trend northeast to southwest, northwest to southeast, and near the head of the Lake Fork in the "American basin" nearly east and west, there seeming to have been several injections or systems of veins which upon Mineral Mountain have intersected each other, creating thereby quite a confusion, which, if the mines situated therein turn out as expected, will necessarily provoke quite lively litigation.

The veins are in great number, and in many cases of enormous width, containing wide pay-streaks of oftentimes very rich ore. All that is needed for the success of the mines is facile communication and cheaper transportation. The shortness of the summer season—if there be such a thing here—and the length of the winter is a serious drawback. The mines, as a rule, are between 11,000 to 13,000 feet above sea-level. Water freezes nearly every night. Snow may be expected any time after the 15th of September, and by the 1st of November the higher passes into the Animas Basin may be closed until about July 1. July, August, and September, then, are about the only months that outdoor work can be carried on, and almost daily rains during this interval call for the genius of a Tapley and the aquatic disposition of an Aleut to preserve an equable and serene frame of mind. If one attempts to cross a range of mountains, he is tripped by fallen logs or mired in the mud. If out of his tent or hut after half past 1 or 2 o'clock in the afternoon, he is drenched with rain, pelted with hail, or dredged with snow. If so unfortunate as to be caught upon a high peak about this time, he will be shocked by electricity and frightened out of his wits by singing rocks, stinging ears, fingers, and nose, lively hair and beard, loud crashes of thunder, and horribly-uncertain streams of lightning. The manifestations of induced electricity are very interesting, but very unpleasant. Every hair and every projecting member of one's body becomes a small discharging-point, and continued small discharges, prior to a great discharge of lightning from the clouds above, upon one's nose and ears and hair, are excessively annoying; one feels like a pincushion in demand. Every rock upon a peak becomes similarly electrified by induction, and, the tension becoming too great, minor discharges take place thereto, so that every sto nebegins to sing louder and louder, until finally a stream of lightning, seemingly as large as a man's waist, of solid fire, leaps down to earth, followed by a crash of thunder like the discharge of mighty artillery. One such great discharge generally is sufficient to determine the hasty flight from high peaks of aspiring mules and men, miners and donkeys.

The mines are as yet but little developed, but very extensive preparations are in progress now. Several smelters have been and are to be erected in the various districts, and, as soon as miners can begin to make their ores pay for work, the condition of things will be much improved. Capitalists are deterred from investing on account of the inaccessibility of the districts, and the owners have been too poor in ready money to open up their mines in places where they could not receive any immediate return. There can, however, be no reasonable doubt of the future of the districts mentioned, and the great immigration to that region in the spring of 1876 shows the increased confidence felt in these mines. In 1875 the population of these mountains was about 3,500. Thirteen towns have been founded, viz: Silverton, Howardsville, Eureka, and La Plata on the Animas; Mineral City, Park City, and Ouray on the Uncompahgre; Tellurum, Cottonwood Junction, and Lake City on the Lake Fork; Parratt City on the La Plata River, where veins of gold-bearing quartz have been located and an extensive auriferous bar is about to be worked; Hermosa and Old Animas City, or Elbert, on the Lower Animas River, in the Park.

Average cost of mining the ore is from $8 to $15 per ton. Cost of driving tunnel on main vein, $25 per foot. Cost of sinking a shaft in main vein, $25 to $30 a foot.

Points of supply Del Norte and Saguache, Colorado Springs and Cañon City. Freights, $160 a ton to Cañon City or Colorado Springs, 220 miles.

There is no grain and hay in market; the grass is good throughout the mountains.

No Indians within limits of the mineral belt except the Weeminuche Utes on the La Plata; the Tabequache Utes occasionally visit the mines, but they have now no claim to the land.

Mountain-sheep and bears are about the only large game, and these are becoming very scarce, having been driven away by the blasting in the mines. Grouse and ptarmigan are quite plentiful. The San Miguel and Dolores are filled with trout, and also the Lake Fork below the falls.

Saw-mills have been erected in nearly all of the districts, and lumber and shingles can be purchased at from $20 to $40 per thousand for the former, and $7 per thousand for the latter. Everywhere is abundant water for machinery of nearly any power required of pure snow water of constant and rapid flow.

ROUTES OF COMMUNICATION.

At present the San Juan mining region is reached from the last of three main lines of communication.

1st, From Cañon City, via Pleasant Valley, Puncho Pass, Cochetopa Pass, to Lake City, on the Lake Fork of the Gunnison; thence a road leads to the Animas Forks and the mines on the Animas and headwaters of the Uncompahgre.

2d, From Pueblo or Cucharas Station, via the Mosca, Abeyta, or Sangre de Christo Pass, to Del Norte, thence via the headwaters of the Rio Grande del Norte to the Animas mines; a branch road from Antelope Springs also leads over to Lake City, connecting with the first-mentioned line near the lower end of San Cristobal Lake.

3d. From Santa Fé and northward to Taos a road leads via Abiquiu and Tierra Amarilla to the Animas Park, thence a trail to the Animas mines.

From Colorado Springs the first route is joined by a road via the Ute Pass, Colorado Salt-Works, Trout Creek Pass, and the valley of the Upper Arkansas, near the mouth of Puncho Creek, after which it coincides with this first route. Also from Conejos and the southern portion of the San Luis Valley the third route may be reached by a road via El Rito and Ojo Caliente.

Annexed hereto are tables of distances by way of either of the roads except the third, which has not been surveyed throughout its length. The notes thereto attached are sufficient to enable one designing a visit to the mines to select the most convenient route. A detailed description of each is not necessary here, since they have already been described, or at least the eastern portion of them, in former reports.

The greatest interest attaches to the immediate approaches to the mines. Granting that it may be practicable to reach by rail the San Luis Valley from the east, and that it is from this direction that a railroad-route will be built, tapping the mines, the question of communication is narrowed to the practicability of attaining the mines from this valley. As far as possible, I caused profiles to be made of all approaches at present in use, or which seemed to me to offer possible routes, either for wagons or railroad, within the limits of the area assigned to me, the most important of which are hereto attached, with the grade per mile between the barometric stations computed. Although perhaps not sufficiently accurate for engineering purposes, they show the comparative value of the trails and roads surveyed, and may be sufficient data to save the expense on the part of those interested in such subjects of additional surveys for preliminary information.

Along the first mentioned of these routes it is practicable to reach by a railroad the valley of the Gunnison River via the Cochetopa Pass, north of the new wagon-road. Along the course of the road, however, between Cochetopa Creek and the Lake Fork of the Gunnison, the grades (as may be seen from the profiles) are too great for a railway. 1st. At the divide between the Cochetopa and Cebolla. 2d. Between the Cebolla and the Lake Fork of the Gunnison. South of this road, in the Gunnison drainage-area, there is no route as practicable as this. North of it the Cochetopa Cañon and the ridge to the west of it offer formidable obstacles, and a road will thereby be forced to follow practically the route surveyed by Capt. J. W. Gunnison, topographical engineer, in 1853, and reported upon by Lieut. E. G. Beckwith, Third United States Artillery, as far as to the Lake Fork of the Gunnison.

Here, in endeavoring to reach Lake City, an amount of blasting and rock-excavation along the cañon of this stream is greater than may be incurred, unless the products of the mines become so great as to justify the expense. So far it has not been considered advisable to gain, even at much less expense, good grades along the wagon-road above the narrowest part of the Lake Fork Cañon. Following Gunnison's wagon-road, however, to the Uncompahgre, although from his report much cutting, filling, and blasting will be necessary, it will be more easy to attain by rail the new mining town of Ouray, in the Uncompahgre Park, below the gorge of that stream. From the west possibly the same point may be reached by a road leaving the Utah Southern Railroad at Provo, in the Provo Cañon, or, better, perhaps, in the Sevier Valley, below Santaquin, and thence via Gunnison, Salt Creek, and the Wahsatch Pass, to the eastward. The difficulties along this route will be mainly in the plateau and cañon country of the Grand, Green, and Gunnison drainage-areas, for which see Captain Gunnison's report. South of the Wahsatch Pass there can be no rail-route from the west until south of the Grand Cañon of the Colorado. At best, then, the mines may be tapped by the northern route, if their products will ever demand it, at great expenditure of money and labor, by a round-about route through the Cochetopa or Wahsatch Pass.

From Del Norte a road may be built with comparative ease along the cañon of the Rio Grande, with a maximum grade of 80 feet to the mile, as far as to the mouth of Pole Creek, 13 miles from Baker's Park. From this point, however, to the summit of the range, the grade is on an average 300 feet to the mile; and from the summit to the Animas Valley at Howardsville the descent averages 900 feet per mile, and, near the summit, for 1¼ miles the fall is 1,300 feet per mile. This being a slope of 20° is not a practicable route for wagons, although there have been many wagons over it. From Del Norte to the mouth of Pole Creek the main expense will be in bridging the Rio Grande at frequent intervals to avoid cuts, in excavation between Antelope Park along the cañon of the Rio Grande for 12 miles to the mouth of Lost Trail Creek, and again for several miles between Lost Trail and Pole Creeks, the slopes of the river being followed as nearly as practicable. For some distance above the mouth of Pole Creek a road might be built, but for ordinary running the increased expense from additional power required on increased grades, the greater cost of construction, and the want of level space above this point for depot, &c., would not, perhaps, make this advisable.

From the mouth of Pole Creek pack-animals are now, and will necessarily be, the only reliable means of transportation, unless, indeed, (and this will sooner or later be done if mining industry in this section proves a success,) a tramway, or partially counterpoised railway worked by the weights of ascending and descending freights and by stationary steam-power, be built across the continental divide at the head of

Cunningham Gulch. Such a line of railway in connection with this tramway seems to me to be, from its less probable cost and near approach to the various mining districts, to be the project most likely to be remunerative. At Antelope Park it would pass within 30 miles of the mines on the Lake Fork of the Gunnison, and would also be the outlet of the mining region about the headwaters of the Uncompahgre.

From the south the most promising route to examine for a railroad route seems to be that via the valley of the Chama to Abiquiu; thence to Cañon Largo, thence to the San Juan River, thence via its cañon and that of the Animas to Animas Park, 40 miles south from the Baker's Park mines. Above this point the Animas flows in a narrow cañon for some 11 miles from Cascade Creek, which does not seem to offer a very promising route, and the benchlike valley along the west side of the Animas can only be attained by a road with gradients of 300 feet per mile. Above the mouth of Cascade Creek practicable gradients may be secured by blasting along the sides of the Animas Cañon and the road carried into Baker's Park. Such a route at best, like the northern one, must be, from the nature of the mountains and mesas of the San Juan area, very tortuous and expensive in its construction, and, for a railroad, with steep gradients at many points.

If the Denver and Rio Grande Railway should succeed in entering the valley of the Rio Grande above Taos, a branch route to the mouth of Pole Creek would certainly be the most advantageous line to secure the freights of the mining district. If the Thirty-fifth Parallel route be ever completed, the communications with the south may, perhaps, be simplified, since the drainage lines of the San Juan River being all in approximately north and south directions, may offer practicable gradients for a branch railroad along lines not interrupted by vertical walls or deep cañons. Beyond the freights of the mining-camp of the La Plata range, and the possibility of reaching the mines of the upper Animas by way of the cañon of that stream, there does not seem to be anything to induce the construction of more than a good wagon-route along or near the southern route. · In the valley of the Gunnison and Grand and Green Rivers it is different. This is one of the main possible, but at the same time one of the most difficult, of the transcontinental routes. There are here extensive agricultural areas and fields of coal, mines on both the northern and southern tributaries of the Gunnison, and prospective communication with the rich mines of Utah and Nevada. A road in this area, next to that via the Rio Grande Cañon, offers to be soonest remunerative, and would probably always be a link of through and extended communication, which the Rio Grande branch can never become.

In the immediate vicinity of the mines, and across the rim of the upper Animas Basin, the following trails and roads exist:

1st. Road via Cinnamon Gulch Pass to the Lake Fork of the Gunnison. This will be extended through the cañon of the Animas from La Plata, at the forks of the Animas, to Eureka and Howardsville.

2d. From Howardsville, in Baker's Park, over the continental divide at the head of the Rio Grande to Del Norte.

3d. From the Animas Fork over head of Animas River to Mineral City, on the head-waters of the Uncompahgre.

4th. Trails east and west of Sultan Mountain to the Animas Park.

5th. Trail over south fork of Mineral Creek to the San Miguel and Dolores.

6th. Trails over Mineral and Crescent Creeks to Red Mountain Valley, on Uncompahgre waters.

Profiles are given herewith of the mountain parts of the first, second, fourth, and fifth.

Taken in connection with the fact that the gorge of the Uncompahgre below the mines may for the present be regarded as impassable, the problem of access to and egress from the mines of the Animas and Uncompahgre districts is a difficult one. Nowhere is this rim crossed by trail or road leading outward from the mountains with a grade of less than 800 feet to the mile, which means practically no wagon transportation at cheap rates. Of the existing roads, the Lake Fork route is the best, but this is too steep for free use. The cañon of the Animas offers to those districts the only possible route with good grades for wagons, and this is rock-bound; but it may be imperative if the mines are productive, even if the stationary steam-power line be put across the divide, to get an outlet by way of this cañon at the expense of blasting a road-bed in the flint-like sides of the cañon walls. In other words, the mining districts about the heads of the Animas and Uncompahgre must look for facile wagon communication to the south, or else to the northwest via the Uncompahgre gorge, which is equally rock-bound and of three times as steep gradients.

Besides the existing routes of communication for wagons, of which the Saguache and Lake City, and the Del Norte, Antelope Park, and Lake City roads only are available, it seems to me advisable on the part of the people of Cañon City and Colorado Springs to examine the following route for a wagon-road:

From the point the Puncho Pass and South Arkansas toll-road leaves Puncho Creek, the proposed line passes over the divide at the head of this stream to the Tumichi;

thence to the Los Pinos agency; thence over the head of the Los Pinos to the Cebolla; thence via the headwaters of the western tributary of this stream, effecting a junction with the Antelope Park and Lake City road near the summit of the range, and over its road-bed to Lake City. Of this route there will be required to be built 16 miles of mountain road from Puncho Creek to the Tumichi Valley, and about 20 miles of road from the Saguache and Lake City road, above the old agency, to the summit, or in all about 36 miles of new road. The rest of the links already exist, or else the country is level gravel which requires only a wagon to be drawn over it to make a road. By this route the long detour by Puncho Pass around by Saguache and the Cochetopa Pass to the old Los Pinos agency, and also that to the Lake Fork of the Gunnison, 20 miles to the north of the point aimed at, will be avoided and a direct route gained, shorter by four days' teaming from railroad communication than any existing line. There will be one steep pitch on the west side of the range at the head of the Tumichi, a quite difficult descent to the Cebolla, and considerable work in rock-blasting in the upper cañon of the Cebolla; but taking everything into consideration, it seems to me from a cursory examination that this route is at least worthy a close survey and study on the part of the business men of the two places mentioned.

Now that it is conceded that the mines of San Juan, La Plata, and Hinsdale counties are valuable, rich, and extensive, this problem of routes of communication must touch these towns, which are at present (and who wish to preserve this status) the principal points of supply, very nearly in their business relations, and make for them and for the transportation companies the prospective traffic of this rich mining and producing region an alluring prize.

All hypsometrical data collected by my party bearing upon this subject is attached hereto.

Respectfully submitted.

W. L. MARSHALL,
First Lieutenant of Engineers.

Lieut. G. M. WHEELER,
Corps of Engineers, in charge.

APPENDIX B.

PRELIMINARY REPORT UPON THE OPERATIONS OF PARTY NO. 3, CALIFORNIA SECTION, SEASON OF 1875-'76, WITH A VIEW TO DETERMINE THE FEASIBILITY OF DIVERTING THE COLORADO RIVER FOR PURPOSES OF IRRIGATION, BY LIEUTENANT ERIC BERGLAND, CORPS OF ENGINEERS.

LOS ANGELES, CAL., *February* 1, 1876.

SIR: I have the honor to submit the following preliminary report with regard to the operations of party No. 3, California section, Geographical Surveys West of the One hundredth Meridian, during the field season of 1875-'76:

DUTIES OF THE PARTY.

The following extract from letter of instructions will explain the object and extent of the investigations with which I was charged:

"UNITED STATES ENGINEER OFFICE,
"GEOGRAPHICAL SURVEYS WEST OF THE 100TH MERIDIAN,
"*San Francisco, Cal., June* 10, 1875.

"SIR: You are hereby assigned to the charge of a party specially organized for the purpose of making a preliminary examination with a view to a further and more complete investigation as to the feasibility of the diversion of the Colorado River of the West from its present channel, for the purposes of irrigation. Your present survey will be confined to that portion of the river embraced between the foot of the Lower Grand Cañon and the vicinity of the Needles, the results of prior examinations made by parties of the survey under my charge during the past few years having shown the impracticability of turning the river from its present course at a point near the confluence of the Green and Grand, or between that point and the foot of the Grand Cañon, to direct its waters to portions of its western basin. The following route to and from the scene of your labors is suggested: From Los Angeles direct to Point of Rocks, in the Mohave Valley, thence to station at Black's Ranch, thence via Ivanpah to Cottonwood Island, on the Colorado. In returning, your line may leave the river near the Needles, running thence to the lower end of Coahuila Valley along the line of least profile. A reconnaissance-line duly checked by astronomical determinations and leveled barometrically will be measured, and as much mountain topography adjacent to this and other lines necessary to be traversed in the prosecution of your labors as time and means will permit, should be gathered.

"At mountain-peaks occupied, such instrumental observations will be taken as will permit of the introduction of such points into a scheme of triangulation, hereafter to be developed, reaching from the Coast Range near the latitude of Los Angeles to the ranges bordering on the Colorado after its great bend to the south. From present information, the most feasible, if not the only possible, points within the above limits at which the river could be taken from its present channel and carried over large alluvial areas are: 1st, at or near Cottonwood Island; 2d, mouth of Vegas Wash; 3d, foot of Virgin Cañon; 4th, Needles.

"If it is found practicable at either of these points to divert the entire body of the stream at any or all of them, you will cause surveys to be made sufficiently in detail to guide in the projection of the necessary constructions and to govern in the laying out of a canal, with an approximate estimate of the cost of such works, including embankments and cutting. In this connection the surrounding country should be scoured in search of the proper earth, rock, or other material requisite, and advantage of such examination should be taken to gather topography of the surrounding ridges in detail.

"The flow of the river and the character of its sediments will be determined at Camp Mohave and at the mouth of the Rio Virgen.

"Your attention should be especially directed to areas of marked depression along the route, and their geographical extent, with approaches thereto as far as practicable.

"Incidentally you will determine the points at which artificial reservoirs can be most easily constructed, taking advantage of the contour of the subdrainage basins; the more or less impermeable character of the soil underlying them; the value for agricultural purposes of arid tracts encountered if water can be had, and the probable amount that can be reclaimed; the analysis of alkaline, saline, and other deposits; the probable climatic changes to ensue; character of present vegetation; probable changes in the average total flow of the river in different seasons, &c. Any suggestions or recommendations growing from your examinations will be freely communicated. The physical obstacles of the section of territory to be traversed and visited are known to be great, and your operations may be materially modified and restricted thereby.
*   *   *   *   *   *   *

"Very respectfully, yours,

"GEO. M. WHEELER,
"Lieutenant of Engineers, in charge.

"Lieut. ERIC BERGLAND,
"Corps of Engineers."

ORGANIZATION.

The party was organized at Rendezvous Camp, Los Angeles, Cal., and consisted of myself as executive officer and field-astronomer, 1 chief topographer, 1 assistant topographer, 1 geologist, 1 meteorologist, 1 odometer-recorder, 3 packers, 1 cook, a guide, and 3 enlisted men belonging to Company G, Twelfth Infantry; making, in all, 14 men. The means of transportation were 28 pack-mules and 1 bell-mare.

DESCRIPTION OF THE ROUTE.

The party left Rendezvous Camp on the 21st of June, and reached Martin's Ranch, at the mouth of the Cajon Pass, on the 29th of June. At this place a side-party was detached to make the ascent of Cucamonga Peak, and one of the enlisted men was sent by stage to Camp Mohave with barometer and psychrometer, to take observations during the summer, to which might be referred the observations taken on the route.

The march to this place had been necessarily slow, as the mules were nearly all wild and unbroken when the party started, on account of which much time was lost in loading and unloading the packs, and keeping the animals on the road after the train started from camp each morning.

From Martin's the route was through the Cajon Pass, over a divide of 4,487 feet altitude, thence across a portion of the Mohave Desert to the Mohave River, at Lane's Upper Crossing. The slope is gradual from the divide to the river, a distance of 18 miles. This is a barren waste, without wood, water, or grass. The only vegetation seen was sage-brush, stink-weed, a few juniper-bushes, and several varieties of the cactus family.

The yucca trees, which are found here in great numbers, obtain a height of 30 to 40 feet, and present the appearance of a magnificent forest. Rabbits, jack-rabbits, and field-rats were the only animals seen; an occasional bird; but lizards and horned toads were noted numerously. The soil consists mostly of coarse gravel, but could be made productive if it were possible to irrigate it.

The river at Lane's Crossing was about 100 feet wide, with a maximum depth of 3 feet. A considerable volume of water flows through this portion of the river at all seasons, all of which disappears in the sand a few miles below.

From this point the route followed the river to Point of Rocks Station. My instructions were to proceed north from this place as far as Black's Ranch, and thence eastwardly to Ivanpah. From the appearance of the country already traversed, and from information gathered along the route, I concluded that the attempt would be hazardous in the extreme. But little rain had fallen during the previous ten months throughout all that section of country. Springs, which formerly had never failed, were found dry, and no grass or other feed for the animals could be obtained if we left the river; hence, I concluded to follow the course of the river as far as the Saline Flats of the Mohave, called Soda Lake by Lieutenant Whipple. This latter place was reached on the morning of July 13.

The route to the Saline Flats of the Mohave kept near the bed of the river, in which running water is seen at but few places, and in diminishing quantity as one proceeds toward Soda Lake Spring. Water can be obtained by digging a few feet in the sandy bed of the river at almost any place. Vegetation occurs at points along the river where the water comes near the surface, and at several points along the route, as Point of Rocks, Cottonwood Station, and near old Camp Cady. Extensive groves of cottonwood and mesquite trees are found. Grass was scarce along the river on account of the great number of sheep which had lately passed over that route on their way to Arizona.

Between Grapevine Station and forks of the road, the road crossed the bed of a dry lake or basin. The bed is clay, baked hard and dry, and perfectly smooth. Not a particle of vegetation is seen, not even sage-brush. It is about 2 miles wide and 4 miles long. During the rainy season, water accumulates here to a depth of a few inches, the clay becomes soft, and the road is impassable for wagons and animals.

The heat during the day kept on increasing as we descended the valley, the thermometer indicating over 100° in the shade nearly every day. The nights were generally cool, at one time the minimum temperature being as low as 44°. At Soda Lake Spring, however, no respite from the heat was obtained at night. The hot, dry wind coming from the south was almost suffocating, removing the moisture rapidly from the body, thus necessitating a great consumption of water to keep up the supply. The water here is obtained from a spring from which flows a considerable stream, clear and limpid, but strongly alkaline, and nauseating when drunk in large quantities. The Saline Flats of the Mohave constitute a basin about 20 miles long, with an average width of nearly 10 miles. The surface is composed of a white crust of saline and alkaline material deposited from the evaporated water. In the summer months nearly the whole basin is dry, but after heavy rains during the winter the surface is covered with water which is too brackish for use by animals or men.

Splendid mirages were seen here which gave the distant portions of the basin the appearance of a large lake. The illusion is perfect, and tales are told of weary and thirsty travelers who have hurried on to quench their thirst, only to be disappointed and meet a horrible death.

The drainage of the basin is toward the north, and the supposition is that it is connected with Death Valley, but this surmise has not yet been proved to be a fact.

From Soda Lake Spring to Cottonwood Island on the Colorado River, the route is northeastwardly after passing around the north end of the basin, over a range of mountains to Ivanpah, a small mining town, which contains one stamp-mill and a smelting-furnace; thence eastwardly across the dry bed of a basin without outlet, of considerable extent, over the summits of the Providence range, then down a gradual slope to the Colorado River at Cottonwood Island.

The route to Ivanpah is a trail, very heavy and sandy in some places, but passable for wagons with light loads. From Ivanpah to the summit of the river range there is an Indian trail practicable for riding and pack animals. From the summit to the river there is a gradual slope, the trail which follows the bed of the wash being practicable for wagons.

The only water found between Soda Lake Spring and Ivanpah was at Hallovan Springs, Camp 17, and France's Spring, Camp 18. At the former the supply was limited. Between Ivanpah and Cottonwood Island no water was found except at Crossman Spring, Camp 20.

At Ivanpah we were assured that an abundance of water would be found in wells on the east side of the basin; but on arriving there the wells were found dry, and it became necessary to push on to the spring. There but a small quantity of water was found, which was soon exhausted by the animals. The mud and decayed vegetable matter which nearly filled the spring were then removed, but the water trickled in so slowly during the night that there was not enough for the animals in the morning. But little grass was found on this part of the route. " Gallete " grass, very coarse and dry, was found in considerable abundance east of the basin, near Ivanpah; the latter place obtains its supply of grass here. Near the summits of the ranges some bunch-grass was found and a better quality of " gallete," but generally the grass was too far from the water to be of avail. Hallovan Springs was an exception to this rule; but here the water gave out, so we could not take advantage of the excellent bunch-grass found near this place. From the summit of the river-range we obtained the first view of the Colorado River. From this point it lay before us like a silver band, sparkling in the sunlight, a most pleasant sight, as it gave assurance of abundance of water for a time at least.

Cottonwood Island, with its majestic cottonwood trees and rich vegetation, afforded a pleasant relief to the eye, after having seen nothing but black, barren rocks and parched, sandy valleys since leaving old Camp Cady.

The island, which is 5 miles long and less than half a mile wide, is occupied by a number of Pah-Ute Indians. Others of the same tribe have rancherias along the west bank of the river. They raise a few vegetables, a little corn, melons, and wheat; but their principal food is the mesquite bean. They had no supplies to sell to our party as the products of their small gardens are consumed as fast as they ripen. They make no efforts to catch fish, but occasionally shoot a mountain-sheep, and hunt regularl for a species of large lizard and the field-rat. They visited our camp daily, begging for bread and tobacco, but otherwise did not molest us, nor did they show any propen sity for stealing.

At this place a party was sent to Camp Mohave for supplies to replenish our nearly exhausted stores, and after a few days' rest we proceeded up the river to Stone's Ferry

The trail followed the river closely as far as El Dorado Mill. From thence a wid detour had to be made to get around the Black Cañon, and we reached the river again at the mouth of Las Vegas Wash. From this point another detour was made to avoid the Bowlder Cañon. The route hence was up the Callville Wash, over the divide, and down a gradual descent to the Virgen River, which was reached about 15 miles from its mouth; thence down the Virgen to the Colorado at Stone's Ferry, which point was reached on the evening of August 4. This part of the route was almost entirely desti tute of vegetation; even the camps on the river afforded no grass for the animals. A Bitter Spring, Camp 28, some salt grass was found, but it was poor feed for the animals and harmed more than it benefited them. The water at Bitter Springs is decidedly alkaline and unpleasant to the taste. Signs of water were observed near the summi of the Callville Wash, and some green grass was found in the immediate vicinity. The water of the Virgen was unfit for use by men or animals. It was intensely saline, in color brick-red, and surface covered with floating slime of the same color.

The trail from Cottonwood Island to Eldorado Mill is very rough, in some places pre cipitous and dangerous, entirely impracticable for wagons. Thence to mouth of La Vegas Wash the trail is not so difficult, and nearly all of it could be traveled by lightly loaded wagons. Thence to Stone's Ferry the route is practicable for wagons. Call ville is entirely deserted; the wood-work has been removed from all of the houses, and nothing but crumbling walls indicate the site of the city at the "head of navigation."

In the vicinity of Stone's Ferry and along the Virgen are found extensive deposits of rock-salt, and near the ferry there is a natural curiosity called the "salt well." This is a hole in the mesa, nearly circular, with steep walls, partly filled with a strong solu tion of salt. The greatest depth of water was found by sounding to be 96 feet, the di. ameter of the water-surface 118 feet, high-water mark above present surface 4 feet and surface of water below crest of well 43 feet. The surface of the water in the well was found to be 3.9 feet above the surface of the river at the nearest point.

A considerable quantity of rock-salt is obtained in this vicinity, which is used in the reduction of ores both in Nevada and Arizona.

At Stone's Ferry we remained until the 14th of August. The time was occupied in taking soundings and current-observations. A description of the method used and results obtained will follow.

From Stone's Ferry, after crossing the river, we followed the road through a wash which heads in the almost insensible divide leading to the Sacramento Valley; then along this valley to Chloride; from this point, through Union Pass, to Hardyville and Camp Mohave. The first march, from the river to Mountain Spring, is a long stretch of over 40 miles without water. The ascent from the river is gradual and at the rate of about 80 feet to the mile. The first 10 miles is through a narrow wash, destitute of vegetation, then the wash gradually widens into a valley of magnificent proportions, bounded on the east by the Sacramento Mountains, and on the west by the river range. This valley was covered with different kinds of grasses—in the northern part princi pally coarse gallete, but farther south more nutritious grass was found in quantities sufficient to feed thousands of cattle, which is not available because water cannot be procured. Attempts have been made to dig wells in this valley, but no water has been obtained except in the vicinity of the springs near the summit of the valley. The spring at Nobman's Ranch furnishes water for about two hundred head of cattle. At Chloride we obtained good water from a well about 40 feet deep. The town was almost deserted, but half a dozen men living there. The mines were not worked, and the smelting-furnace stood idle at the time of our visit. There is a good wagon-road all the way from Stone's Ferry to Camp Mohave, a toll-road having been built through Union Pass.

The weather was pleasant and temperature moderate in the Sacramento Valley, but as soon as we crossed the summit of the river range we began to experience the dry atmosphere and excessive heat that had been felt previously in the valley of the Colo-

rado. Camp Mohave was reached August 20, and here we remained two weeks, taking soundings and current-observations.

On September 5 we left Camp Mohave, moving down the east bank of the river to Liverpool Landing, where a crossing was effected by means of a small ferry-boat. Thence down the west bank to a point opposite Ehrenberg, where supplies were obtained for the return trip. The first part of the route was through the Mohave Valley, as it is called in Lieutenant 'Ives's report, probably because it was then, as it is now, occupied by the Mohave Indians. The bottom-land fit for cultivation is on the east side of the river, and all is subject to overflow when the river is unusually high, which happens whenever there is a heavy fall of snow along the upper portion of the river.

The Mohaves cultivate small patches of ground, raising vegetables, melons, corn, and wheat. They are more industrious and provident than the Pah-Utes, and usually lay in a supply for the winter, and have enough surplus to supply the post with melons and vegetables during the summer. The valley ends at the Needles, where the mountains close in on both sides, forming a narrow cañon. Below this cañon is the Chem-e-hue-vis Valley, occupied by a portion of the Chem-e-hue-vis Indians. Thence down to Ehrenberg, the valley or bottom-land is of varying width, most of it being on the east side of the river. A few miles above Ehrenberg the river crosses over to the east side, leaving a wide bottom-land on the west side opposite this place. These bottom-lands, formed from the deposits of the river, have rich soil, and, with irrigation, could be made very productive, if the river could be confined to a constant channel and prevented from overflowing the bottom-lands. Along the river there is a rich growth of trees, principally cottonwood, and here the fuel is obtained for the river-steamers.

The road from Camp Mohave to the Needles is good. Thence, over the Needle Range, the trail along the river is in places very difficult, and practicable only for pack-animals. After crossing the river at Liverpool Landing the trail leaves the river to the left, passing around Mount Whipple, and, after crossing the Monument Range, strikes the river again nearly opposite the Indian reservation at old Camp Colorado. Thence to Ehrenberg the trail follows the river, part of it being difficult and fatiguing, as it continually crossed deep washes. After crossing the river, grass was scarce, the best supply being found, as usual, where the water was not accessible.

From Ehrenberg we followed the old stage-road to Chuckawalla, thence to Dos Palmas, Los Torros, Agua Caliente, Whitewater, through the San Gorgonio Pass to San Bernardino, and thence back to Los Angeles, which point was reached October 4.

After crossing the river-bottom opposite Ehrenberg the road ascends to the first summit, which is 934 feet above sea-level. West of Chuckawalla we cross another divide at an altitude of 2,689 feet; thence the slope is downward to Dos Palmas, beyond which place the road crosses a portion of the low basin below sea-level. From thence the ascent is slow and gradual to the summit of San Gorgonio Pass, which has an altitude of 2,743 feet. Westward from this point there is a gradual descent to Los Angeles. The road from Ehrenberg west as far as Los Torros is tolerably good. From the latter place to Agua Caliente it is very sandy and difficult for wagons. This portion of the Coahuila Valley is covered with immense sand-hills, some of which seem to be permanent, as large-sized mesquite and other brush grow on their summits. Others are constantly shifting their position, according to the direction of the prevailing winds. From Agua Caliente to San Bernardino the road is good and the grades easy. The stages having been taken off this route and transferred to the Mohave River route, most of the stations had been abandoned, hence we found it difficult to obtain water and feed for the animals at several points. At Ehrenberg I was assured that water of good quality and sufficient quantity would be found at Mule Spring Station, but on arriving there no trace of the well could be discovered, and not a drop of water obtained. A cloud-burst had occurred a few days previous to our arrival, the water which had formed a new wash over the site of the well, and it had been completely filled up and obliterated by the sand and gravel washed down from the hills. The sand in the bottom of the wash was still moist, which gave us hopes that water might be obtained by digging, but our efforts were in vain, and we were forced to start for Chuckawalla, though uncertain if water could be procured there. After a long march of over 40 miles the latter place was reached late in the evening, and, fortunately, water was obtained here. Some of my men suffered (mentally, at least) considerably from thirst before this place was reached, as the contents of the canteens had given out long before we reached it, and had we not obtained water here the consequences might have been serious. The springs of Chuckawalla had been filled up by the washing from a cloud-burst, also, but a couple of men had gone there a few days previously to open the station, and they had removed the sand and gravel from the springs. The water here is quite good, it having but a slight alkali taste. At Dos Palmas the water is very alkaline and salty. Quite a stream issues from the ground at this place. At Los Torros the water is good, and there is a plentiful supply of it. Agua Caliente the water is warm and sulphury; its temperature is about 100°. On standing, it loses most of its sulphury odor and taste. White River rises in the San Bernardino Range, flows across the valley, and disappears in the sand near the base of the San Jacinto Peak.

Green "gallete" grass was found in the washes near Mule Spring Station. These washes also contain a plentiful supply of mesquite trees and a few willows. At Dos Palmas there is a little salt grass. Los Torros and Martinez are fertile spots in the desert. Here the water rises near the surface and the soil remains moist the whole year, hence vegetation flourishes and the ground is eminently fit for cultivation. At these places and at Agua Caliente are found numerous Indian rancherias, occupied by the Coahuila Indians. These Indians cultivate small garden-plats and raise some grain and vegetables.

The temperature was high when we passed through the valley, and in the lower portions of it the heat was excessive, being about the same as had been experienced on the Colorado River, although it was now the latter part of September. The atmosphere was also very dry, and the same hot, desiccating winds were encountered here. Portions of this valley bear unmistakable evidence of having been covered with water. The beach-line is very distinct just before reaching Dos Palmas, and also in the vicinity of Los Torros. Great quantities of shells are seen in a good state of preservation near this beach-line, which would seem to indicate that the time the lake existed was not very remote.

After attaining the first summit from Ehrenberg a valley of considerable extent is seen to the west, which apparently has no outlet. This basin is not as low as that of the Coahuila Valley. (On passing this summit on my second trip I ascended a high peak near the road. The view obtained from that point leads me to believe that the valley referred to has a drainage to the river through a wide wash several miles above the Lagura.)

SECOND EXPEDITION.

By letter of instructions dated United States Engineer Office, Geographical Surveys West of the One Hundredth Meridian, Washington, D. C., January 19, 1876, I was directed to organize a party and proceed to Ehrenberg, on the Colorado River, thence down the western bank to the boundary, for the purpose of continuing the investigation as to the feasibility of the diversion of the river for purposes of irrigation, and also to determine, if practicable, the approximate area of the depression below the level of the sea. The middle route, via Temecula, was specified as the inward route from Fort Yuma to Los Angeles.

ORGANIZATION.

The party was organized as follows: 1 executive officer and field astronomer; 1 topographer; 1 assistant topographer; 1 meteorologist; 1 odometer recorder; 2 packers; 1 cook; 1 teamster; 2 enlisted men. Total, 11.

Means of transportation were: 10 riding-mules; 7 pack-mules; 3 extra mules; 6 team-mules. Total, 26.

One enlisted man, Sergeant Eugene Farnham, Company G, Twelfth Infantry, was left at Los Angeles as barometer observer during my absence. I will take occasion to remark here that he performed his specified duties to my entire satisfaction, recording the observations carefully and accurately.

DESCRIPTION OF ROUTE.

From Los Angeles City to Ehrenberg, Camp 14.—The party left Los Angeles February 13, and reached the Colorado River opposite Ehrenberg, Ariz., March 3. The outward route was the same as the inward route of last trip. Several changes had taken place since we previously passed over it. The Southern Pacific Railroad was now finished as far as Whitewater Station, and the grading completed to a point opposite Indian Wells. Two stage-lines were running from the terminus of the railroad to Ehrenberg, and thence through Arizona and New Mexico. Consequently the stations had been fitted up and wells and springs cleaned out, so that the supply of water was abundant. A few new wells had been dug and stations established by the new stage-line, (Wells's Express Company,) as at a point about half-way between Torros and Dos Palmas, and another 6 miles east of Mule Spring Station. A well was being dug on the Chuckawalla Bench, between Cañon Springs and Chuckawalla, with but little chance of striking water.

The country appeared rather more sterile and forbidding than before, as the mesquite trees had lost their leaves, and not a spear of green grass was to be seen between Whitewater and Ehrenberg. The temperature was, however, tolerably cool during day, and the nights were sometimes quite cold. Ice formed during the night at Whitewater, Chuckawalla, and one night at Ehrenberg. The day we went from Cañon Springs to Chuckawalla we had a drizzling rain, and this was accompanied by a cold, piercing wind on the high mesa near the divide. At Ehrenberg the river was found to be about 2 feet lower than it was at the time of our first visit. A long sand-bar was exposed between the ferry-landings, which rendered it necessary to tow the ferry-boat half a mile up stream before a crossing could be effected.

MUD VOLCANOS,
South-Colorado Desert.

From Ehrenberg to Fort Yuma, Camp 19.—There being no wagon-road near the western bank of the river, I was obliged to cross my wagon and team, and send it down the river on the eastern side via Tyson's Wells and Gila City. The party with the pack-train proceeded down the river, keeping as close to the bank as the conformation of the country would permit. We left Ehrenberg on the 8th day of March, and reached Fort Yuma March 12. Camp 15 was on the river, near an Indian rancheria; Camp 16, at Panchos Ranch, a deserted Mexican hut; Camp 17, at Round Hill, (Ives,) just below Light-house Rock; Camp 18, at Picacho Mill, in the Cane-brake Cañon, (Ives.) The route from Camp 14 to Camp 15 is mostly a wood road; from Camp 15 to Camp 16 a trail impracticable for wagons, but not difficult for pack-animals. Several detours have to be made to avoid deep lagunas, and the trail leads over a long sand-flat in the river-bottom. From Camp 16 to Camp 17 the trail is mostly in river-bottom, but at times on mesa, where it crosses several deep, dry washes. At time of high water the trail in the river-bottom cannot be used at all; travelers then have to take the trail along the mesa and over projecting spurs of the mountains, which is very steep and difficult in several places. Near Camp 17 was found the only patch of grass along the river.

From Camp 17 to Camp 18.—Trail near the river principally on the mesa, crossing frequently deep washes with high, steep sides. Camped at the Mill. This mill, which has five stamps, was standing idle, and had a sheriff's writ of attachment nailed to the door. The ore found at the mill is said to yield $17 per ton. It is friable and easy to crush. The mines are located up a wash near the Chimney Peak. A good wagon-road leads from the mines to the mill. No wood, except mesquite, in the vicinity.

Camp 18 to Fort Yuma, Cal.—At the Mill the trail leaves the river, and follows the road to the mines until the vicinity of Chimney Peak is reached, when it leaves the wash and ascends the mesa. This has a gentle slope up to an altitude of over 700 feet. Near the summit a number of Mexicans were working placer-diggings. Water being scarce, they used the dry process for separating the gold. The yield is small, but is enough to pay industrious men well.

The descent to the river-bottom on the south side of the divide is through a narrow wash or cañon, whose walls are almost perpendicular, and sometimes overhanging, and in some places not more than 6 or 8 feet apart, and over 50 feet high. The side washes are deep, narrow fissures, too narrow for the passage of a pack-mule. The river-bottom is several miles wide, and the soil of the same character as that observed above Ehrenberg. It is covered with mesquite trees and arrow-weeds; nearer the river, however, are found a number of large cottonwood trees and willows.

At Fort Yuma, the commanding officer, Major Mizner, Twelfth Infantry, kindly allowed me to use a vacant building near the ferry for the shelter of my party, and during my stay at the post he and the other officers willingly gave me all the assistance I required, as also did Captain Bradley, assistant quartermaster, in charge of Yuma quartermaster depot.

During our stay here, current-observations were made and soundings taken, to determine the discharge of the river. At this time the river had reached its lowest stage; no rise from melting snow in the upper portions of the Colorado or Gila Rivers had yet taken place, hence the discharge shows the minimum amount which the two rivers combined will furnish. Experiments on evaporation were also made here.

April 2 the party left Fort Yuma, following the road along the west bank of the river to Algodon Station, just below the boundary-line. At this place the party was divided.

A portion, with the pack-train, went in a southerly direction to visit the hot springs and mud volcanoes in the vicinity of Mount Purdy, while the remainder, with the wagon, followed the stage-road. The parties were to meet at New River or Indian Well Station.

Algodon, via Mount Purdy, to Mud Volcanoes.—The road taken by the side party leaves the main road a short distance below Algodon, when it leaves the river-bottom. The direction thence is southwesterly toward Mount Purdy. It crosses several of the channels of New River, which at this time were entirely dry. Water was found in three places along route, contained in natural reservoirs, these being depressions in the dry channels. The last water occurs about ten miles from Mount Purdy. Here the kegs were filled, and a dry camp made near the base of this peak. Thence the distance to the volcanoes is about 5 miles.

The country from Algodon to Mount Purdy is nearly level; the portion near the river is thickly covered with mesquite trees and willows. As the distance from the river increases the vegetation becomes more sparse, until, within 5 or 6 miles of Mount Purdy, the plain is entirely destitute of it. The ground here is covered with a crust of salt for many miles in extent. At the base of Mount Purdy there is a stream of running water, which is intensely salt. This flows toward the northwest, or in the direction of New River Station. Mount Purdy is the crater of an extinct volcano. The crest is about 600 feet above the level of the plain at its base, and the interior of the crater is filled up to within about 100 feet below the crest. From the summit a good view is obtained of the surrounding country, except toward the west, where the Cocopah range obstructs the view. From Mount Purdy to Mud Volcanoes the direction is southeast, the distance about 10 miles. Trail is good most of the way, the only difficult place being the crossing of a salt creek with marshy banks. After crossing this creek the trail is on a hard, gravelly mesa until it nears the mud volcanoes. The ground within an area of 200 by 500 yards is covered with large and small craters formed from the mud which has been thrown up into conical mounds. These mounds vary in height from 3 to 6 feet, and in diameter, at the base, from 5 to 20 feet. Some have large open craters, within which the hot mud can be seen constantly boiling and bubbling. At short intervals columns of mud are thrown up to the height of 4 to 6 feet, but no regularity in the pulsations could be discovered, nor did they occur at the same instant in the different craters. The smaller cones had small openings at the apex, from which issued sulphurous vapor with a hissing noise.[*] The center of this area was occupied by a lake of boiling mud, all parts of whose surface were constantly agitated, and from which the mud was occasionally thrown up several feet in height. A small pond of clear water is situated within the area covered by the mud volcanoes. The temperature of the water is 100°. A small spring of clear water was found near the mud lake with a temperature of 199°. A large pond or lake of clear water is situated east of the mud lake and at a lower level. The water of this lake has a temperature of about 96°, and also has a strong taste of alum. The temperature of the boiling mud was found to be 210°, and that of the vapor issuing from the smaller orifices was about the same. A large mound situated some 200 yards to the southeast of the mud lake appears to have been thrown up by this volcanic action. The crust is composed principally of sulphur, much of which occurs as pure crystals. It is not in action now, but the hollow sound heard when walking over it seems to indicate that the mound is a hardened crust with a partly-filled interior which possibly communicates with the active volcanoes. The liquid mud is black, but on drying it becomes gray, and is very pungent to the taste. A quantity of this mud was collected for analysis, and bottles of water from the lake, pond, and hot spring were obtained.

The surface of the ground between Mount Purdy and the mud volcanoes is dotted over with extinct solfataras, with here and there one from which hot vapor issues. A few were also observed east of Mount Purdy. Indians living in the vicinity and old white settlers say that at night flame is seen issuing from these volcanoes, and sometimes high columns of steam. This usually occurs during an overflow of the river.

From Mud Volcanoes to Indian Wells.—After visiting the volcanoes the side party returned to camp, and started the following day for New River station. The route follows closely one of the New River channels. These channels can be detected by the rich growth of mesquite trees, which grow in the bed and along the banks. The soil is very rich, and after an overflow the grass springs up and matures rapidly. When we passed these flows were entirely dry and destitute of grass, there having been no overflow for several years.

At New River Station water is obtained from a well in the bed of New River; water brackish and disagreeable to the taste. Between New River and Indian Wells Station we followed the stage-road, which here passes over an almost level plain, when we had the phenomena of *mirage* the whole day, the plain appearing as an immense lake of water. At Indian Wells we were joined by the main party, who followed the road. Here also we experienced one of the most disagreeable features of this desert country. Shortly after our arrival in camp it commenced to blow. The winds increased in force, and soon a hurricane was blowing. This wind carried with it clouds of fine sand, which penetrated eyes, ears, and nostrils, as well as instruments and provisions. Traveling or work of any kind was out of the question. There was nothing to do but to wait until the wind subsided. After three days of impatient waiting there was a lull in the storm, and during the night the clouds of sand settled so that we could see some distance in front of us. During this storm high mounds of sand had accumulated around the houses at the station whenever the smallest obstruction permitted a slight lodgment. The wood-pile disappeared entirely, being covered over with some 3 feet of sand. The effect of these sand-storms seems to be to cover the plain to the west of Indian Wells with sand-dunes. A few years ago the sand had not reached Indian Wells, but the station is now entirely surrounded with sand-hills of varying height.

From Indian Wells Station I ascended Signal Mountain. The aneroid indicated an altitude of 2,300 feet at the summit. Here an extensive and extended view was obtained of the greater portion of the Colorado Desert, Lake Maquata (Stretch,) a salt lake on the west side of the Cocopah Range was seen to be almost dry. The water lines on both sides of the range could be plainly discovered, and the cones of the mud volcanoes in the Coahuila Valley were seen with the telescope. The northwestern end of the sand ridge lies nearly due north from this position. The eastern end is near Algodon Station. The course of New River could be traced, as it bends to the northwest of the station, and continues on into the desert. There is no grass or other vegetation on the mountain except a few bunches of "palo verde" near the summit. No water

[*] NOTE.—On approaching these cones the rumbling noise within could be distinctly heard.

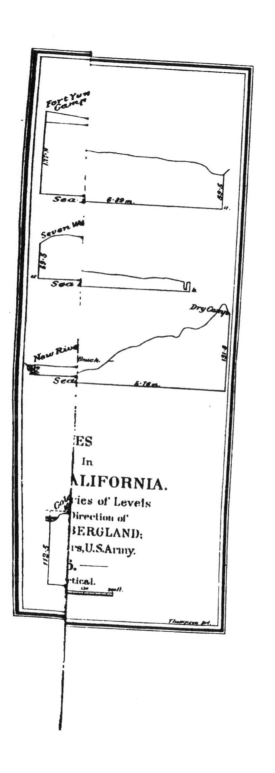

Fort Yuma Camp

Sea

6.19m.

Seven Wells

Sea

New River

Dry Canyon

Sea

6.74m.

ES
In
ALIFORNIA.
Series of Levels
Direction of
BERGLAND;
rs, U.S.Army.

rtical.

Thompson del.

was discovered in the vicinity or on the mountain. Galleto grass in considerable abundance grows on the plain at its foot.

From Indian Wells, Cal., to Los Angeles, Cal.—The party left Indian Wells on the 14th of April, and made a dry camp near Sacket's Well. This well has been entirely obliterated, and no water is found between Indian Wells and Carizo Creek. From this latter place the route leads through Vallecito, San Felipe, Warner's Valley, Oak Grove, Bergman's, Temecula, Laguna Grande, Temescal, down the Santa Ana Cañon to Anaheim, and thence to Los Angeles, which place was reached on the 7th day of May. The road is tolerably good, except over the Carizo Creek Hill, Vallecito Hill, and through Vallecito Cañon. From Indian Wells to Vallecito it is quite sandy. A little salt grass was found at Vallecito, but at San Felipe and westward good grass for the animals was obtained. Some of these valleys were very beautiful, being covered with wild flowers of different hues, which, mingling with the rich green of the flourishing grasses, made a pleasing picture for the eye. Water in abundance was found in these valleys, and of good quality, except at Vallecito. The passage through the Santa Ana Cañon was rather difficult, as the river had to be crossed seven times, and the stream was rapid, and bottom, in places, composed of dangerous quicksand.

THE DIVERSION OF THE RIVER.

This subject will now be considered. The Colorado River from the mouth of the Grand Cañon down to Chem-e-hue-vis Valley, just below the Needles, may be said to flow through a cañon which occasionally widens into a narrow valley, as at the mouth of the Virgen, Las Vegas Wash, Cottonwood Island, and the valley from Hardyville to the Needles. These valleys are separated by spurs and ranges through which the river has cut its way, forming the Bowlder, Black, and Pyramid (Ives) Cañons, and the cañon of the Needles.

Below the Needles we have the Chem-e-hue-vis Valley, which is again separated from the Great Colorado Valley (Ives) by the Monument Range.

The ranges between which the river flows are nearly 4,000 feet high at the great bend, the divide on the west side opposite Cottonwood Island being 3,900 feet, and that on the east side at Union Pass 3,800 feet; while the summit of the Sacramento Valley near Chloride is about 4,100 feet above sea-level. As we proceed southward the river ranges become sensibly lower. Thus the highest altitude of the trail over the Monument Range is 2,300 feet, and the summit of the river range on the west side opposite Ehrenberg is 934 feet. Beyond the ranges which inclose the river are other higher ranges, with valleys between, whose altitudes are greater than that of the river.

These topographical features can be plainly seen by referring to the map and profiles of the route. This being the case, it is evident that the river cannot be diverted from its present bed between the mouth of the Grand Cañon and the head of the great Colorado Valley.

In this valley there is a large area which could be made productive if irrigation were practicable. An effort in that direction has been made at the Indian reservation at old Camp Colorado, on the east bank of the river, above La Paz. Here an irrigating-canal several miles long has been tried. From the information I received about it (I did not have an opportunity to inspect it) I learned that the soil was so porous and unstable that the banks were constantly undermined, causing them to cave in and fill the canal. After repeated trials the projector had concluded to flume the entire canal, which can only be done at great cost where lumber is scarce and prices high. Even if irrigation were practicable, it would be necessary to build levees to prevent the river from overflowing the bottom-lands and destroying the irrigating-canals and ditches. The Great Colorado Valley is terminated at the south by the Chocolate Range, through which the river passes, and emerges from the Purple Hill Pass into the wide valley, which extends to its mouth.

Below the Purple Hills there are no formidable mountain ranges on the west side of the river, except a short detached range, called the Cargo Muchacho, and Pilot Knob. The divide between these mountains is about 278 feet, while the altitude of the water-surface at Fort Yuma is 120 feet. A canal through this opening would therefore require a cutting of nearly 160 feet, and, besides, would have to cut through the sand ridge west of Pilot Knob in order to gain lower ground more rapidly than by keeping to the north of the sand-hills. Even in this case the length of the canal from Fort Yuma to the point where the surface of the ground is of the same altitude as the water-surface at the latter place would be at least 30 miles long.

These conclusions are arrived at from data obtained from the Texas Pacific Railroad surveys, and also from levels run by my party from Hanlon's Ferry over the divide north of Pilot Knob. In order that the canal should be entirely within the California boundary, it must cross the divide to the north of Pilot Knob. This, as has already been shown, would necessitate a long, deep cutting, partly through rock, and a passage through the sand ridge. This passage could only be effected by means of a flume or tunnel, to protect it from filling up with sand. A canal from some point below the boundary would be more practicable and less expensive in construction and maintenance. A line of levels was run from Fort Yuma to Indian Wells, and to the dry camp beyond. The profile shows that there is a steady descent, with but few breaks, from the altitude of 137.89 feet at Fort Yuma to 388 feet below sea-level at a point 6¼ miles west of Indian Wells. From this point the rise is rapid and constant up to the mountains which border the desert on the west. One of the branches of New River leaves the Colorado near Algodon station, and this artificial channel might be utilized in the construction of a canal for diverting the water of the river into the depressed area to the northwest. The exact course of this canal cannot be determined without further surveys, but it seems probable that it would be necessary to run it below the boundary-line nearly as far as Seven Wells in order to avoid the sand ridge. From this point it could bend more or less toward the north, according to the downward slope of the surface in that direction. The amount of cutting required is difficult to estimate without further surveys, but it would doubtless be moderate, as the water flows into this area from the river when it overflows its banks. From several persons who have resided for years in this section I have obtained the following information with regard to the flow of the water through New River and changes in the bed of the Colorado. Mr. Hanlon, owner of the ferry at Hanlon's Ferry, 7 miles below Fort Yuma, says: "The channel of the Colorado opposite Pilot Knob is now about three-fourths of a mile east of where it was at the time of Mr. Wozencroft's examination. It then runs close to the point of rocks which jut out from the eastern base of Pilot Knob. In June and July, 1861, there was a great overflow. The current at Alamo Station was more rapid and four times greater in volume than at Fort Yuma now, (March, 1876,) water flowing to the northwest toward the desert."

Mr. McMasters, station-keeper at Algodon, says:

"The great sand billows southwest from Pilot Knob are constantly moving toward the south, caused by the prevailing north winds. There has been no overflow for three years. During years of unusually high water in the river the water was 2 feet deep in the station-house, which is 600 yards from the river. (Judging by the depth of water in the well at the present time, the river must have risen about 17 feet above its present stage.) Captains Poole and Polhemus, of the Colorado Steamboat Line, say that up to 1864, during summer, at high water, boats took slough at base of Pilot Knob. The highest floods occurred in 1862 and 1867."

Mr. Conners, engineer on steamboat, says:

"Crossed New River at Indian Wells in July, 1862. Water 7 feet deep and flowing north. Volume about twice that of the Los Angeles River in July."

Mr. Redondo, butcher at Yuma, says:

"Slough from Algodon runs to New River, and from there came the water of New River during the heavy flood of 1862."

Mr. Jaeger, owner of ferry at Fort Yuma, and a resident of the place since its establishment as a military post, says:

"Heavy floods in 1840, 1852, 1859, 1862, and 1867."

Tasted the water flowing in channel at New River Station in 1862, and found it fresh water. A Mr. Jones (now dead) told me that he came along the west side of the great desert basin in 1862, on his way from San Bernardino to New River, and saw in the basin a great lake some 60 miles long by 30 wide. This came from the overflow in 1862.

From information obtained and examinations made, it may be taken for granted that there is a current which sets in along the channels of New River during high floods, and that this current flows toward the north into the depressed area. At the same time, a large portion of the country between Pilot Knob and Mount Purdy is submerged, as are also portions of the plains or meadows between Signal Mountain and Indian Wells. On the subsidence of the water these plains soon become dry, and also the different channels of New River, except in the deeper portions, where reservoirs are formed, in which water remains for a year or more, depending on its depth.

Several questions of importance remain to be considered in this connection, such as probable difficulties and cost of keeping the channel free from sand blown in from the sand-hills, as well as from the settlement of the sediment in the water. These cannot be determined without closer investigation and a more detailed survey.

The area of the depressions below sea-level can be obtained approximately from data obtained from Southern Pacific and Texas Pacific Railroad surveys, together with the level-lines run by my party. The accompanying sketch shows the outlines of this area, as well as the direction and extent of the lines of profile, and altitudes at different points; these altitudes all being referred to the Southern Pacific Railroad bench-mark at Fort Yuma. From these data it appears that the northern limit of the depression is near Indian Wells, in the Coahuila Valley, and extends westwardly below the Mexican boundary. This gives an approximate area of nearly 1,600 square miles lying within the limits of California.

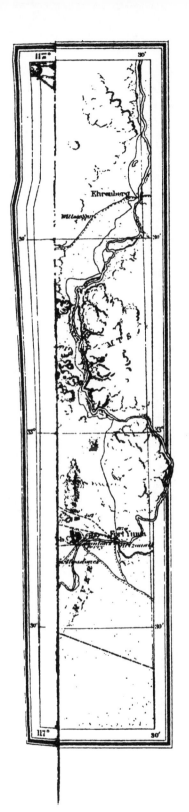

CURRENT OBSERVATIONS.

Observations for gauging the Colorado River were made at Stone's Ferry, Nev., Camp Mohave, Ariz., and Fort Yuma, Cal.

The surface-floats used were (for want of better materials) made from thin boards, obtained from packing and cigar boxes. A piece of sheet-lead was nailed to the under surface to steady the float and bring its upper surface near the surface of the water. A small flag was attached to the upper side by means of a wire about 3 inches long. As but one boat could be obtained at Stone's Ferry and Camp Mohave, a sufficient number of floats were made and allowed to run through without being picked up.

At Stone's Ferry the subsurface-float was a box 10 inches square by 14 inches high, open at both ends, and weighted with lead to sink it. This was attached by a cord to a small empty canteen, which served to keep it at a constant depth. To the box was attached a long trail-cord, which was paid out as the float left the boat, and by which it was pulled back after making the transits. On account of the rapid current and the resistance offered by the box, it was extremely difficult to draw it back to the boat; hence a large canteen, which was nearly filled with water, was substituted for the box. This canteen was 9 inches in diameter, and 3½ inches thick. Even this required great exertion in pulling it back to the boat, and the trail-cord finally broke, allowing the whole arrangement to float down the river.

The boat used here was large and unwieldy, extremely difficult to manage in the rapid current. As no anchor could be obtained, a substitute was used in the shape of a sack filled with stones. This had to be very heavy in order to hold the boat against the strong current, making it very tedious and laborious to handle. A base-line of 280 feet was carefully measured on the south bank and transits placed at the extremities. The times of transit were taken by a separate observer. The boat was pulled into position and the anchor dropped. It was then located by means of the transits, sounding taken, and at a signal from the observer at the upper transit a surface or subsurface float was sent out from the boat, the time and position of the float at each transit being noted by the observers. As material for surface-floats was scarce, chips were occasionally thrown in and their transits noted.

A sufficient number of soundings were taken to determine the cross-section of the river opposite the base-lines, and meander-lines were run and topographical stations made to furnish material for a complete topographical sketch of the locality.

For plotting, the section was divided into divisions of 50 feet each, numbered consecutively from the base-line.

OBSERVATIONS AT STONE'S FERRY, NEV.

The following values were obtained by measurement and computation.

The fall of the river was found by leveling to be 2.13 feet per mile, or slope = $0.000403 = s$.

Area of section = 5723 square feet = a.
Width = 480 feet = W.
Wetted perimeter of section = 481.4 = p.

Mean radius = $\frac{a}{P} = r = 11.89$ feet B $= \frac{169}{(r+1.5)\frac{1}{2}} = 0.4618$.

Under the circumstances it was impracticable to take a full set of mid-depth observations; hence, reliance has to be placed on the results obtained from the surface-floats. Sections were only measured between the transit-lines, consequently the formula for the slope cannot be used. The determination of the mean from the surface-velocity has been made according to the method indicated in chapter IV, Humphreys and Abbot's Physics and Hydraulics of the Mississippi River.

No correction has been made for wind.

The mean velocity as deduced, is $v = 3.217$ feet, and the discharge, $Q = va = 18410.38$ cubic feet per second.

CURRENT OBSERVATIONS AT CAMP MOHAVE.

The method of observation and the character of surface-floats were the same as at Stone's Ferry.

The subsurface-floats used here consisted of a square tin can with top and bottom covers removed. This was kept at the required depth by means of a cord attached to a small cork float which carried a small flag. A trail-cord was attached to pull the float back to the boat, after making the transits, as but one boat could be obtained near the Post. The base-line was 300 feet long, measured with a compensated steel tape. All the water passed through the section, except a small portion which flows through a shallow channel near the west bank. This had a cross-sectional area of 10 square feet, and the water flowing through it a mean velocity of 1.25 feet per second.

The fall per mile obtained by leveling was, 1.2 feet or $s = 0.000227$.
For the section at this place we have the following values:
Area of section = $a = 4628$ square feet.
Width = W = 1116 feet.

Wetted perimeter $= p = 1116.7$ feet.
$r = 4.144$ feet.
$b = .713$ foot.

For the same reasons as at Stone's Ferry, but few subsurface-floats were observed, and the mean velocity has been computed from the mean surface velocity, as at that place, but here a correction was made, for wind. The results obtained are as follows:
Force of wind $= f = + 1.5$.
$V = 2508.64$ feet, and Q $ra = 11610.93 + 12.5 = 11623.43$ cubic feet per second; 12.5 cubic feet added represents the discharge through the channel not included in the section.

The river fell 5.11 feet from July 9 to September 5, according to observations taken at Camp Mohave. During the interval between August 11, when current observations were taken at Stone's Ferry, and September 3, when taken at Camp Mohave, the river at the latter place fell 1.84 feet.

This will account for the greater portion of the difference between the discharge at Stone's Ferry and that at Camp Mohave, as a rise of 1.84 feet will cause a corresponding increase of section at the latter place of over 2,000 square feet. A small portion is lost by evaporation, but a considerable quantity flows below the bed of the river, in sections where the bottom is sandy, as at Camp Mohave.

CURRENT OBSERVATIONS AT FORT YUMA, CAL.

Here the conditions were more favorable. Two boats were obtained, one of which was used to pick up the subsurface-floats after passing through the sections. The sections were also more regular than at the former places, and mid-depth floats could be sent out from all boat-positions without the annoyance of having them drag on the bottom at any point. The floats were of the same character as those already described, except that no trail-line was used with the subsurface-floats. The observations were taken March 15 to March 20, 1876. The river was then at its lowest stage, and at a constant height (or nearly so) during the time of observation. The base-line was 300 feet long, measured along the right bank below the ferry. The following values were obtained:

$$A = 2726.5 \text{ square feet,}$$
$$W = 461.0 \text{ feet,}$$
$$p = 466.2 \text{ feet,}$$
$$r = 5.848 \text{ feet,}$$
$$b = 0.6234 \text{ foot,}$$

Fall per mile as determined by leveling $= 1.21$ feet.

With the above values, and the formula for mid-depth velocities (formula (25) H and A), we have:
$$V = 2.809 \text{ feet, and}$$
$$2 = Va = 7658.74 \text{ cubic feet per second.}$$

The following table gives the number of floats from which the calculations were made, and the mean surface on mid-depth velocity in each division at the three places where observations were made:

Mean division velocities.

Divisions.	No. of floats.	I.	II.	III.	IV.	V.	VI.	VII.	VIII.	IX.	X.	XI.	Remarks.
Stone's Ferry......	98	.94	1.68	3.12	3.11	3.94	4.41	4.04	3.35	2.26	1.08	..	Surface-floats calm.
Camp Mohave	173	.39	.90	1.65	2.73	3.19	3.13	3.06	2.92	2.76	2.41	1.67	Surface-floats wind-up=1.5.
Fort Yuma	80	1.73	2.77	2.60	2.69	2.7?	2.78	2.74	2.90	3.22	2.89	....	Mid-depth floats.

Divisions.	No. of floats.	XII.	XIII.	XIV.	XV.	XVI.	XVII.	XVIII.	XIX.	XX.	XXI.	XXII.	Remarks.
Stone s Ferry	98												Surface-floats calm.
Camp Mohave	173	1.51	1.64	1.92	2.21	2.77	3.06	2.86	2.00	1.35	.94	.50	Surface-floats wind up=1.5.
Fort Yuma	80												Mid-depth floats.

The accompanying sketches and profiles show the location of the bases and the topography in the immediate vicinity, as well as the form and area of the sections.

EVAPORATION.

Experiments on the amount of evaporation were made at Stone's Ferry and Camp Mohave. The vessel containing the water was placed in the river so that it would retain the temperature of the river-water. The results were as follows:

At Stone's Ferry several experiments, each one lasting from 2 to 3 days, showed the evaporation to be 0.23 inch per 24 hours. The atmosphere was dry and either calm or stirred by gentle breezes; a few sand-storms of short duration occurred during the

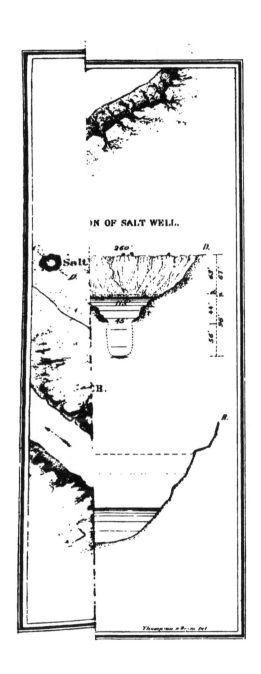

experiments. The thermometer ranged during these days from 75° at sunrise to 114° at 3 p. m., while the temperature of the river-water varied from 79° to 81°.5 per day. The average relative humidity was, at sunrise, 0.459, and 3 p. m. 0.173.

One experiment, made August 10, after heavy showers over a great extent of the surrounding country, showed a considerable decrease in the amount of evaporation, being 0.18 inch for the 24 hours. The temperature of the air ranged from 71° at sunrise to 99° at 3 p. m., and the relative humidity from 0.645 to 0.310. The temperature of the river-water was the same as in the previous experiments.

During the latter part of August, the experiments were continued at Camp Mohave. A strong hot wind blew during the day, commencing about 9 a. m., and lasting with but little interruption until sunset; hence the amount of evaporation was much increased, as will be seen by the following table:

Date.	Hours.	Temperature of air.	Mean relative humidity.	Evaporation in inches.	Remarks.
August 29 and 30 ..	7 p. m. to 7 a. m	95. 1–74. 4	0. 359	0. 15	Temperature of water, 80°, calm.
August 30	7 a. m. to 7 p. m	74. 4–92. 0	0. 399	0. 69	Temperature of water, 82°; strong wind; sand-storm after noon.
August 30 and 31....	7 p. m to 7 a. m	92. 0–72. 1	0. 351	0. 11	Calm.
August 31	7 a. m. to 7 p. m	72. 1–92. 6	0. 352	0. 62	Strong wind.

From these few experiments it is impossible to form any correct estimate of th yearly amount of evaporation in the vicinity of Camp Mohave, since this amount is subject to so many variations, depending on the temperature of the air and water, relative humidity, and force of the wind. It is evident, however, that the monthly evaporation will be much greater during the four hot months, June, July, August, and September, which have an average temperature of 94°.8, than that during the cooler months. The mean temperature of May is 79°.47, while the mean temperature of the remaining months is only 65°.1 F., as shown by the meteorological observations taken at this place.

If we estimate the daily evaporation during June, July, August, and September at one-half of that observed August 29–31, or equal to 0.392 inch, and that the total amount of evaporation during the remaining months is equal to that in the four hot months, we get a total yearly evaporation of 95.77 inches, which compares well with the observed evaporations in other localities.

Thus the yearly evaporation is—

	Inches.
At Cumana...	130
At Dead Sea...	96
At Marseilles..	73. 2
At Palermo...	58. 4
At Manchester...	41. 0
At London..	28. 8
At Rotterdam..	23
At Breslau..	14. 8

If we take the mean of the observed daily evaporation at Camp Mohave = .784 inch, we find that the daily evaporation from 556 square miles of lake-surface will equal the discharge of water in the river for 24 hours at the same place.

At Fort Yuma experiments were made from March 19 to April 2. The water-pan placed on the roof of the commanding-officer's quarters was exposed to the direct action of the sun, and the temperature of the water taken when the depth of water in the pan was measured. Owing to the exposed position of the vessel, and the absence of any large body of water in the immediate vicinity, the results obtained are probably in excess of what would have obtained under circumstances similar to those at Camp Mohave.

At Indian Wells Station, Cal., observations were made April 12 to 14. Here the vessel was placed in the shade of a tree.

In 1868 Dr. Lauderdale, post surgeon at Fort Yuma, made some observations, using the rain-gauge at the post.

In August, 1853, some experiments were made by Lieutenant Williamson, at Ocoya Creek, Tulare Valley, lasting four days. The water-vessel (an ordinary milk-pan) was placed on a stand 2 feet above ground, and a cover of brush built above it. All of these results are tabulated below:

FORT YUMA, CAL.

Date.	Daily evaporation, inches.	Temperature of water.			Temperature of air.						Mean daily relative humidity.	Remarks.
					Maximum.			Minimum.				
		Date.	Hour.	Degree.	Date.	Hour.	Degree.	Date.	Hour.	Degree.		
Aug., 1868	.409											Maximum daily evaporation, 0.583 inch; minimum, 0.020 inch; Dr. Lauderdale's rain-gauge.
Sept., 1868	.408											Maximum daily evaporation, 0.666 inch; minimum, 0.009 inch; Dr. Lauderdale's rain-gauge.
Mar., 1876, 19 and 20	.560	20	11.30 a. m.	91	19	2.00 p. m.	90	20	5.00 a. m.	58.4	.304	Pan showed a little leakage; wind light, clear.
20 and 21	.300	21	12.15 p. m.	68	20	4.00 p. m.	83	21	4.00 a. m.	56.7	.301	Wind very light, cloudy.
21 and 22	.330	22	12.30 p. m.	78.5	21	4.00 p. m.	77.7	22	4.00 a. m.	56.8	.389	Moderate breeze; cloudy during day; night clear.
23 and 24	.440	24	4.30 p. m.	69.5	23	4.30 p. m.	77	24	5.30 a. m.	54.0	.449	Light breeze, clear.
24 and 25	.450	25	9.00 a. m.	70.5	24	5.00 p. m.	82.8	25	5.00 a. m.	55.0	.449	Do.
25 and 26	.430	26	9.30 a. m.	71.5	25	5.00 p. m.	82.3	26	6.00 a. m.	57.0	.363	Very light breeze, clear.
28 and 29	.430	29	9.54 a. m.	69.3	28	4.00 p. m.	82.0	29	6.00 a. m.	57.0	.301	Light breeze, clear.
29 and 30	1.02				29	4.00 p. m.	86.4	30	7.00 a. m.	52.0	.294	Moderate breeze during day; sand-storm during night.
30 and 31	.57	31–1	6.30 p. m., and 9.30 a. m.	52–78	30	3.00 p. m.	68.0	31	7.00 a. m.	58.4	.229	Sand-storm day and night.
31 and April 1	.39				31	3.00 p. m.	72.0	1	7.00 a. m.	53.0	.331	Afternoon strong breeze; clear, cool night.
1 and 2	.57				1	3.00 p. m.	76.5	2	7.00 a. m.	57	.394	Afternoon strong west wind.

INDIAN WELLS STATION, CAL.

April 12–14	.26	13	12.00 m.	73	12	2.00 p. m.	78	13	7.00 a. m.	50.5		Light breeze, clear.
April 12–14		13	7.00 a. m.	41	13	2.00 p. m.	80	14	6.00 a. m.	42.8	.320	Do.

OCOYA CREEK, TULARE VALLEY, (Lieutenant Williamson.)

August, 1853, 4 days.	.25					6.00 a. m.	52		3.00 p. m.	100		Pan 2 feet from ground in the shade.

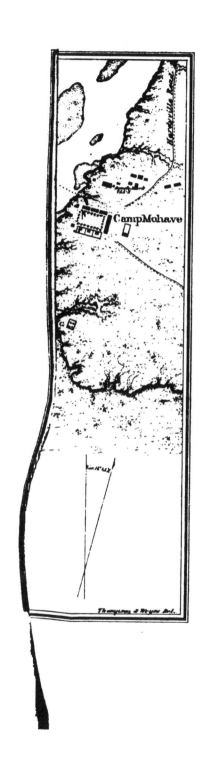

The mean daily evaporation at Fort Yuma from March 19 to April 2, 1876, was 0.5 inch, and the discharge of the river per second, as found above, was equal to 7,659 cubic feet. Hence the evaporation from a lake-surface of 570 square miles would be equal to the quantity of water which the river could supply. Since the river was at its lowest stage, this gives minimum amount supplied by the Colorado and Gila combined.

A few observations were made to discover the extent of the influence of small bodies of water, as the surface of a river, upon the increase of humidity in the air. Observations made on the river-bank, with the psychrometer, show not only a decrease of temperature, but also an increase in the relative humidity. A distance of a few hundred feet from the river this influence is hardly perceptible, as a great bulk of air is mingling continuously with the stratum of cooler and moister air, which rests on the surface of the river.

Where the river-bank is lined with trees, the cooler air is longer retained and does not mix so quickly with the neighboring strata; hence in such localities the difference in temperature and humidity is considerable. Thus it was repeatedly observed at Stone's Ferry that the air under the trees on the river-bank had a temperature of 96° to 98°, when that upon the mesa, but a few yards distant, showed a temperature of 105° to 108° F.

In the following table a few observations are given, which show the extent of the influence of the river on the air:

Date.	Time.	Dry bulb.	Wet bulb.	Relative humidity.	Remarks.
		°	°		
July 24	Sunrise	90	73.2	0.399	On river-bank.
July 24	...do	92	71.5	0.339	400 yards from river.
August 31	...do	72.3	61.0	0.510	River-bank.
August 31	...do	74.4	60.2	0.42?	On mesa, 400 feet from river.
August 31	2 p. m	92 6	69.1	0.280	River-bank.
August 31	.. do	95.0	66.1	0.209	On mesa, near river.

TOPOGRAPHICAL WORK.

This was carried on by my efficient topographer, Mr. Gilbert Thompson, assisted by Mr. George H. Birnie, on the first trip, and by topographer F. A. Clark and G. H. Birnie on the second. The route throughout was meandered with the Casella theodolite, and distances measured with the odometer. Bearings were taken to prominent peaks and the topography of the country adjacent was noted. Whenever time and opportunity permitted peaks were ascended, and topographical and triangulation stations established. Owing to the excessive heat experienced while in the valley of the Colorado, during the summer-months, it was impossible to make as many ascents as might have been made in a cooler climate. The mountains being destitute of water and grass, rendered it necessary to carry these supplies up the mountains, where an ascent was made, and this it was not often practicable to do.

In order to obtain the profile of the route the aneroid barometer was read at each meander-station, and at camps near the river the altitude of the ◯ of the barometer above the level of the water was determined by leveling. At Stone's Ferry, Camp Mohave, and Fort Yuna a daily record was kept of the fall and rise of the water in the river, and at Camp Mohave a permanent bench-mark was placed, to which the height of the river can be referred in future. This bench-mark is a stout iron pin driven down into the ground at the east end of the hospital. Its head, which is flush with the ground, is at an altitude of 755.2 feet above sea-level, as determined by our barometer observations. An arrangement for observing the rise and fall of the river was placed a short distance from the bank.

A piece of iron water-pipe was taken and cross-wires placed at each end. The pipe was then firmly fastened to two stout posts in such a position that the line joining the intersections of the cross-wires was horizontal. A long graduated rod is placed upright at the edge of the water, when the observer looks through the pipe and takes the reading. The axis of the pipe is 42.52 feet below the bench-mark, and the surface of the water on September 2 was 50.24 feet below, or 704.96 feet above sea-level.

At Fort Yuna a bench-mark was also established by driving an iron pin into the ground near the southeast corner of the platform which surrounds the flag-staff. The altitude of the top of this pin, as obtained from Southern Pacific Railroad levels, is 204.56 feet above sea-level, referred to the Fort Point tide-gauge. The altitude of pin above surface of water in the river near engine-house, March 18, 1876, is 84.41 feet. A record of these altitudes was placed in a cavity in the upper end of the pin.

The distance of surface below high-water mark was measured wherever the point could be accurately determined, and found to be as follows: At Stone's Ferry, 17 feet, high-water mark of 1871; south end of Cottonwood Island, 14 feet; Camp Mohave, 8 feet, high-water mark of 1872; proposed railroad crossing, north of the Needles, 20 feet; Needles, 18 feet; Camp 40, 15 feet; opposite Ehrenberg, 8 feet; Fort Yuma, 10.19 feet, high-water mark of 1862; Fort Yuma, 9.16 feet, high-water mark of 1873; Fort Yuma, 6.07 feet, high-water mark of 1875.

When the observations were taken at Camp Mohave the river had nearly reached its owest stage, although it usually continues to fall slowly until after Christmas.

METEOROLOGY.

As previously mentioned, one enlisted man, Private Charles Lengert, Company G, Twelfth Infantry, was sent by stage to Camp Mohave to take barometric and psychrometric observations during the summer. This duty he executed faithfully and well, for which he deserves a great deal of credit.

Observations were taken in camp at regular hours, and comparisons of instruments and observers were frequently had. At Fort Yuma our barometers were compared with that of the signal-office, and a series of hourly readings were taken, which was also done at other camps where we remained two or three days. Cistern-barometer readings were also taken on mountain-peaks and on the principal divides. In this way a great number of observations were obtained from which to calculate the altitudes of the different camps.

The observations taken on the first trip were reduced and altitudes computed in the temporary office at Los Angeles during the winter months. Those taken on the last trip have not yet been computed.

The method of computation used differs from that used in the office at Washington for this reason: It was found that the horary curves at Los Angeles and Camp Mohave differ considerably; that of the latter place being much sharper and having a wider range. This was to be expected, since the climate differs so greatly from that at the former place. For this reason the proper correction was applied to the readings before computations were made.

The horary correction for Camp Mohave was used for all places included within the desert area whose climate was nearly the same, while at places west of the Cajon and San Gorgonia Passes the horary correction for Los Angeles was used. The daily means of synchronous observations were computed separately, after which the results were examined, and those which were unmistakably bad were thrown out. The mean of the remaining ones were then taken as the true altitudes. To check the results the altitudes of several camps were computed with both Los Angeles and Camp Mohave as reference-station, and in every case the coincidence was as close as could be expected.

The computed altitude of Camp Mohave, 755.8 feet, I was disposed to consider too great, as it exceeded by 100 feet all previously determined or estimated altitudes. Hence, every precaution was taken to determine this accurately. The daily means of the synchronous readings for over 70 days were separately computed. The wandering from the mean was next obtained and the observations thrown out which showed an abnormal variation, and a new mean obtained. This altitude was then used to compute the altitudes of camps already referred to Los Angeles, giving results differing but little from those previously deduced.

GEOLOGY, MINERALOGY, AND NATURAL HISTORY.

Dr. Oscar Loew, chemist and geologist, had charge of these branches during the first trip, and was indefatigable in the prosecution of his duties.

He took copious notes and sketches of the geological formations, collected numerous mineralogical specimens, visited all the mines in the vicinity of our route, and obtained many specimens of plants, insects, and reptiles.

Mineral water was taken at several points, and specimens of soils and saline deposits were obtained for analysis. He also made some original investigations on the influence of extreme heat on the rapidity of the pulse, inhalation and exhalation, and the absorption of water. He also obtained hundreds of new words from different Indian tribes, and conducted the experiments on evaporation.

It would be desirable to have his report accompany mine. On the second trip collections of the flora and fauna of the country traversed were made, specimens of soil from various localities collected, and bottles of water brought in from all thermal and mineral springs within our reach.

CLIMATE.

The climate in the Colorado Valley during the hot months is not one which a sane person would select in which to spend the summer. From the middle of June to the 1st of October panting humanity finds no relief from the heat. As soon as the sun appears above the horizon its heat is felt, and this continues to increase until a maximum is reached about 3 o'clock in the afternoon, after which the temperature falls slowly, and oftentimes very slowly, until sunrise. During the hottest part of the day

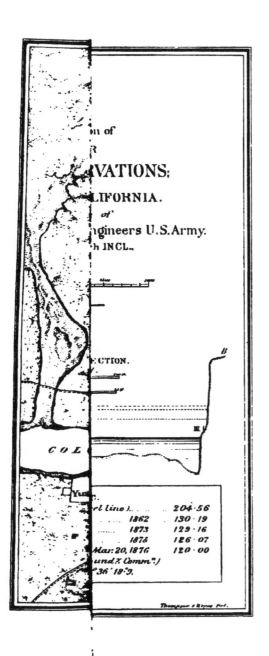

on of

R

VATIONS;

LIFORNIA.

of

gineers U.S.Army.

h INCL.

CTION.

B

COL

Yum

el line).		204·56
	1862	130·19
	1873	129·16
	1875	126·07
Mar: 20, 1876		120·00
und ? Comm?.)		
36′ 18″9.		

Thompson & Knowy Prt.

exertion of any kind is impossible; even while lying perfectly quiet the perspiration oozes from the skin and runs from the body in numerous streams. Everything feels hot to the touch, and metallic objects cannot be handled without producing blisters on the skin. The white sand reflects the heat and blinds the traveler by its glare. Rain scarcely ever falls during the summer months, and not more than 3 or 4 inches the year round. Cloud-bursts frequently occur in the mountains, and at times we saw heavy showers all around us, but not a drop fell along the river.

The atmosphere is so dry and evaporation so rapid that the water in our canteens, if the cover was kept moist, kept at a temperature of 30° below that of the air; a most fortunate circumstance, as it obviates the necessity of using ice-water where it would be impossible to preserve ice.

Great quantities of water are drank during these hot days, and no uncomfortable fullness is experienced. One gallon per man, and sometimes two, was the daily consumption.

Notwithstanding this excessive heat, no sunstrokes occurred, although we were at one time exposed in a narrow cañon to a temperature of 120°. All of the party preserved good health during the summer.

There is no danger of catching cold in this climate, even if wet to the skin three or four times during the day or night. No dew or moisture is deposited during the night, hence no covering is required.

The hot wind which blows frequently from the south is the most disagreeable feature of the climate. No matter where you go, it is sure to find you out and give you the full benefit of a gust that feels as if it issued from a blast-furnace, and parches the skin and tongue in an instant. Then there is no recourse but to take copious draughts from the canteens to keep up the supply of moisture in the body. If water cannot be obtained, the delirium of thirst soon overpowers the unfortunate traveler, and he dies a horrible death.

To illustrate the difference in the climates of Los Angeles and Camp Mohave, the following observations are tabulated:

Hour.	Los Angeles, July.		Camp Mohave, July.	
	Thermometer.	Relative humidity.	Thermometer.	Relative humidity.
	°		°	
7 a. m.	71	.695	88.8	.400
3 p. m.	77	.576	104.5	.216
9 p. m.	75.5	.604	94.6	.302
Mean	74.5	.625	95.96	.306

Hour.	August.		August.	
	Thermometer.	Relative humidity.	Thermometer.	Relative humidity.
	°		°	
7 a. m.	75.6	....	86.5	.429
3 p. m.	86.8	....	107.5	.206
9 p. m.	77.2	....	94.8	.275
Mean	79.9	....	96.3	.303

In conclusion, I wish to express my thanks to Mr. Thompson, topographer, and Dr. Loew, geologist, and the other assistants, for their cheerful co-operation and attention to their duties under trying circumstances.

I am also under many obligations to the officers stationed at Camp Mohave and Fort Yuma for their uniform courtesy and assistance.

Respectfully submitted.

ERIC BERGLAND,
First Lieutenant Engineers.

Lieut. GEO. M. WHEELER,
Corps of Engineers, in charge.

APPENDIX C.

EXECUTIVE AND DESCRIPTIVE REPORT OF LIEUT. W. L. CARPENTER, NINTH INFANTRY
ON THE OPERATIONS OF PARTY NO. 3, COLORADO SECTION, FIELD-SEASON OF 1875.

UNITED STATES ENGINEER OFFICE,
GEOGRAPHICAL SURVEYS WEST OF ONE HUNDREDTH MERIDIAN,
Washington, D. C., March 14, 1876.

SIR: I have the honor to submit the following executive report of party No. 3, Colorado division, under my charge during the field-season of 1875.

The party rendezvoused at Pueblo, Colo., early in June, and on the 8th day of that month the organization was completed—the party being well equipped and ready to take the field for six months' service. The *personnel* was as follows: First Lieut. W. L. Carpenter, Ninth Infantry, executive officer and naturalist; F. O. Maxson, topographer; A. R. Conkling, geological assistant; Allston Ladd, recorder; Private J. F. Kirkpatrick, Company C, Battalion Engineers, recorder; 2 packers, 1 laborer, and 1 cook.

On the 9th of June the party left Pueblo, with orders from First Lieut. W. L. Marshall, Corps of Engineers, commanding the Colorado section, to survey the southern portion of Colorado and northern part of New Mexico, this being an area of about 7,000 square miles, comprised between longitude 104° 7' 30'' and 106° 30' west, and latitude 35° 30' and 37° 20' north.

The month of June was spent in working down the eastern side of the mountain-range which separates the water-shed of the Mississippi from the drainage of the Rio Grande; in occupying East Spanish Peak and other less prominent points as triangulation-stations, and in meandering and measuring important roads.

On the 7th of July the party crossed the Sangre de Cristo Mountains through Taos Pass and descended into the valley of the Rio Grande. Drawing supplies at Taos, the party turned northward, following the range to Fort Garland, examining en route all passes which appeared likely to afford means of communication with the eastern slope. The rainy season commenced July 4, and continued, with slight intermission, until October 1, covering the country with a luxuriant growth of grass and filling all the streams with an abundant supply of water, but proving a source of great annoyance to the party by the persistence with which clouds and fogs covered the mountains for many successive days, sadly interfering with topographical work. Taos Peak was occupied eight days without obtaining complete results; and on many other mountain-stations did the topographer spend weary hours watching a rift in the clouds, vainly hoping that it would even for a few moments disclose some distant and important station. I doubt if ever similar work was carried on under more disadvantageous circumstances than by this party during the month of July, 1875, said to have been the *rainiest* summer ever known in New Mexico.

After drawing supplies at Fort Garland, the party proceeded southward, keeping close to the foot of the mountains. By great exertion this area was completed, notwithstanding the continuance of unpropitious weather, and reached Elizabethtown August 24. Here I left the party in charge of Mr. Maxson, and, taking Conkling and one packer, proceeded to the Gallinas *mauvaises terres*, for the purpose of exploring that region and collecting vertebrate fossils from the locality visited by Prof. E. D. Cope in the season of 1874. I did not rejoin the party until October 11, at Fort Union, it meanwhile working in the range from Taos Pass to the southern extremity, below Santa Fé, and completing unfinished work in atlas-sheet 69*d*.

At Fort Union supplies were drawn for the last time at a military post, and advantage taken of the kind proffer of assistance from Capt. A. S. Kimball, assistant quartermaster, to make some much-needed repairs on pack-train before starting out for the fall work.

While at this post Mr. Conkling, in obedience to orders, made a careful examination into the geology of the vicinity, with a view to determine the feasibility of sinking an artesian well at the garrison. He reported adversely upon the project, finding that a true hydrographic basin at a practicable depth was wanting, owing to the existence of a synclinal depression in a basaltic formation of great thickness which underlies this locality. His detailed report on this subject was forwarded to you from the field.

The subject of artesian wells is one of great importance to New Mexico. The Territory possesses a large area of excellent grazing-land now almost worthless for want of water, but which only awaits the means of furnishing a small artificial supply to render it of great value. There are many localities where the basaltic formation does not occur, which appear to offer advantageous sites, with a reasonable prospect of obtaining a flowing well. A considerable sum of public money intelligently expended in this work would be a wise disbursement. If successful, it would stimulate private enterprise to sink other wells, which would operate to reclaim hundreds of square miles of an arid region now uninhabitable; while, if the experiment resulted in failure, none could so well afford to sustain the loss as a Government whose policy has always wisely been to aid States and Territories in matters of public welfare.

The rest of October was spent on the plain country east of the Canadian River, making rapid progress with the topographical work, as the weather was now perfectly clear. Trinidad was reached November 2, and after connecting the system of triangles with the established base-line at that place, started for Wet Mountain Valley. On the 5th of November it commenced to snow, and continued stormy and very cold for ten days, making it impossible to complete the survey of the valley. Camped two days in Rosita, Colo., where the party was made very comfortable by the kindness of Mr. Livingstone, who offered the use of a vacant log house, which was very acceptable, as the thermometer stood at 6° above zero and the ground was covered with snow. During the stay in the valley the weather was so stormy that the astronomical observations necessary for the establishment of a meridian-line could not be taken, and accurate topographical work was rendered impossible. The trail leading from the Huerfano River up Williams Creek into Wet Mountain Valley was meandered; also the wagon-road from Rosita to the Huerfano by way of the Muddy Creek Pass, and such other topographical results obtained as the intensely cold and stormy weather would permit. The principal mines at Rosita were inspected by Mr. Conkling, and collections of their valuable ores secured. This little mining-town enjoys a well-merited prosperity seldom to be observed in the West. Its mines are wonderfully rich, and, being situated only 30 miles from the Cañon City Railroad, possess facilities for shipping their products, which give it superior advantages over other richer and more extensive mining-districts, but which are farther removed from railroad transportation.

Leaving Wet Mountain Valley November 17, the party started for West Las Animas, Colo., meandering the Apishpa River on the way down, from the stage-road to its mouth. Arrived at West Las Animas, the terminus of the Atchison, Topeka and Sante Fé Railroad, November 25, and disbanded next day.

The result of the season's operations is very satisfactory. The topographical work planned for the party was faithfully carried out; connection made with the bases previously measured at Trinidad, Cimarron, Fort Union, and Santa Fé; also sufficient accurate detailed geodetic data obtained to construct a reliable map of that region. During the season a total of 2,742 miles were marched; and of this number 1,199 miles accurately meandered. The number of triangulation-stations occupied was 24, the topographical stations made being 133.

Especial care was taken to determine the best routes of communication between the plains and the valley of the Rio Grande. With this end in view, the Sangre de Cristo range, which is the most formidable obstacle, was examined at all accessible places from Morino Valley northward to the Sangre de Cristo Pass. These mountains were crossed four times and a series of barometric profiles secured, which establish the fact that the Taos Pass, starting from the southwest end of Morino Valley, is the best route for either wagons or a railroad. Traveling from this place north, the next pass which can be made available for a road is called Red River Pass, and is situated at the northern end of Morino Valley. It follows up a small tributary of the Cimarron River, and crossing the range at an elevation of 9,460 feet above the sea, with an easy grade descends a little more abruptly by the Red or Colorado River, a small branch of the Rio Grande. The grade throughout the entire distance, although quite steep in some places, is perfectly practicable for wagons or railroads. But on account of the narrow walls of the Red River Cañon and the tortuous course of the stream, a considerable outlay would be required to build a wagon-road, because of the great number of bridges which would be necessary, while the cost of a railroad would be too great, for the same reason.

The next pass to the northward, known as the Costilla Pass, is impracticable, even for horsemen; while the fourth and last, called Trinchera Pass, has an exceedingly rough mountain-trail over it, which answers very well for pack-animals, but which cannot be utilized for wagons. The proximity of the Sangre de Cristo Pass, through which there is a good wagon-road, answers all purposes of travel at present, and it will probably be many years before any other means of communication for this section will be considered necessary.

Taos Pass has an elevation of about 8,625 feet, and offers many advantages over any other. There is an abundance of wood, water, and grass, until the valley of the Rio Grande is reached, when the country changes abruptly from a pastoral to an agricultural region. A proposed railroad (the Arkansas Valley and Cimarron) has been already projected along the Cimarron River into Morino Valley, and it should have the Rio Grande Valley for its objective point, as there is no intervening obstacle if constructed via Taos Pass.

Large collections in natural history were made, the movements of the party being so conducted that a collection embracing a deep vertical range could be secured during its progress from the plains, at an altitude of 5,000 feet above the level of the sea, to the summit of the loftiest peaks. And as this movement was continued in a sort of winding course down the mountains from Fort Garland to Santa Fé, it afforded opportunities for securing a valuable collection, which will undoubtedly prove extremely interesting in its bearing upon the geographical distribution of the fauna of the western mountains. In making my collections above timber-line, I received much assistance from Mr. F. O. Maxson, whose duties as topographer often required his presence at extreme altitudes.

The collection of vertebrate remains obtained from the *mauvaises terres* of New Mexico is of importance, as this locality is a new field, visited but once before, by Prof. E. D. Cope. It was a matter of regret that more time could not be spent in its exploration; but being dependent upon rain for water, movements were necessarily restricted to the vicinity of a few pools which were found; and as there were but three persons in the party, the exposed condition was a subject of anxiety during the sojourn here, on account of the number of semi-hostile Indians about. It required the constant presence of one person in camp to guard the property, while the other two went out each morning and, with rifle in hand, hastily collected the petrified bones and teeth of the huge animals which once inhabited this remarkable region. I fully appreciate the efficient aid rendered by my two companions under many trying circumstances which occurred during an absence of six weeks from the main party, without which success would have been impossible.

The geological and mineralogical collections were in charge of Mr. A. R. Conkling, who, under many difficulties, visited every mining-region passed through, and not only made valuable collections in his specialty, but also found time to render me some assistance in the department of natural history.

The past year afforded a good opportunity to see Southern Colorado and Northern New Mexico under the different conditions of a wet and dry season. The spring was said to have been the dryest ever known, and consequently the country appeared under every disadvantage. From Pueblo to the Cimarron River it was parched with the drought; there was barely enough water in some streams for live-stock, and the crops suffered for want of irrigation, although every drop was utilized economically. New Mexico, like Colorado, is a region requiring irrigation to produce crops. Although in a favorable season indifferent crops might be raised without resorting to that process, yet no dependence should be placed entirely upon natural moisture in this peculiar climate, where the rain does not fail in the season when most required by the farmer. Agriculture on a large scale must always be precarious when the means of irrigation do not exist. Most of the natives, however, are content to eke out a bare existence by cultivating a small patch of ground with wooden ploughs, and thrashing out their grain by driving herds of sheep over it. In no part of New Mexico can anything approaching to eastern thrift and industry be found. The picture which the writer saw, of a thrifty field of corn in which a herd of cattle was peacefully browsing upon the tender grain while the owners were in-doors taking their noon-day siesta, is but a type of native shiftlessness too prevalent for the common weal.

The corn raised throughout New Mexico is a short, stunted kind, well adapted to a quick growth and moderate yield. It is probably the same grain which has been cultivated on the upper Missouri River by the Indians since the days of Lewis and Clarke, where it is known as "Ree" corn, so called from the fact that the Ree Indians were the first people in that region to raise it extensively. If brought originally from Mexico or the Isthmus, its introduction to the northern tribes becomes apparent through the agency of traders and trappers who years ago were in the custom of traveling from the Missouri region to Sante Fé to dispose of their furs. But as it is a kind especially adapted to a cold climate, on account of its rapid growth, it might also have been first imported from the East, and be a variety of the original James River maize, considerably modified by climatic conditions.

The soil of Northern New Mexico, along the eastern slope from the boundary-line south to the end of the Rocky Mountain chain, is very fertile and easy to cultivate The entire valley of the Rio Grande is also a fertile tract, capable of supporting ten times its present population. This comprises about all the arable land of that region. Westward of the Rio Grande the bad-lands occur, rendering an area larger than the State of Connecticut worthless for any purpose except an indifferent grazing section during the winter-months. A short distance east of the mountains the plains appear covered with a light alkaline soil, and, being furnished with a scanty supply of wood and water, present a very uninviting appearance to the agriculturalist. The farming-land, then, of Northern New Mexico is a narrow belt on the eastern and western sides of the main range, considerably contracted toward the mountains during the prevalence of an unusually dry season; and although this region is sufficiently. large to render the Territory self-supporting for all time, yet its agricultural importance will not bear comparison with other more favored Territories. Its chief wealth must always be found in the valuable grazing advantages which it affords. The writer has visited nearly all the Western States and Territories, and having had good opportunities for judging of their relative merits for agricultural and grazing purposes, has no hesitation in declaring that as a live-stock country New Mexico is far superior to any other west of the Mississippi. Morino Valley and the mountains just north of there is a region especially adapted to the raising of cattle and horses during the summer and fall, but it is an inferior winter-range, owing to its great elevation. The adjacent foot-hills, however, afford shelter and abundant food until the storms of winter have passed. The section drained by the Canadian River and its tributaries, the Mora and Pecos, is a fine tract for all kinds of live-stock, and is *par excellence* the future great wool-growing center of the West. Far to the north, where severe winter-storms and scarcity of food cause an annual loss of from 10 to 20 per cent., and even occasionally as high as one-half, the business is considered a paying one, and if made profitable under such disadvantages, what degree of success may not be expected in a region where the winter-storms are mild and of short duration; where the nutritious native grasses furnish all the food necessary during summer and winter; and where sheep thrive without shelter, and are free from the diseases so prevalent in the East. The native animal is a descendant of the old Spanish Merino stock, considerably deteriorated by years of interbreeding; it, however, breeds back to pure Merino with wonderful facility, and may be readily improved in this way. It is quite small, and shears but about 3½ pounds of wool of a good quality; the custom of shearing twice a year is prevalent, the mild winter appearing to make this practice desirable. The Cotswold sheep in the few instances where it has been introduced has done well, and could not fail to prove a good investment on a large scale, either to raise pure, or for the purpose of improving the native stock. Sheep may be purchased by the thousand after the fall-shearing for the sum of $1 per head, and may generally be obtained at any time for $1,500 per thousand. A competent shepherd may be hired for $15 per month, and the necessary corral and buildings erected with a maximum outlay of $200. With care and attention, it appears feasible that, aided by so many local conditions peculiarly favorable to the business, the emigrant to this region who turns his attention to sheep-raising will be certain to achieve success.

The hills and mesas in the vicinity of Tierra Amarilla are covered with the finest growth of white-pine timber to be found between the Mississippi Valley and the Sierra Nevadas. Here are countless thousands of lofty trees, valueless at present, and which may remain so during our generation, but which the advent of the locomotive will one day utilize for the benefit of mankind.

The mineral wealth of New Mexico is not great, and has been generally overestimated. There is not at present a thrifty mining-district in the Territory. The gold mines at Elizabethtown still yield a moderate profit, but the gold-bearing area is of very limited extent, and must soon cease to be remunerative. Red River Cañon is a promising locality for gold, which, if present there in paying quantities, could be mined to advantage, on account of the abundant supply of water, wood, and other natural facilities which tend to cheapen mining operations and render them profitable. Gold in small quantities, with silver and copper, have already been found here, and further prospecting will probably disclose these metals in greater richness.

The mines of Trinidad, Colo., must exercise an important influence over the development of Southern Colorado and Northern New Mexico. Although the precious metals do not exist here, there are coal and iron mines of vast extent, capable of supplying the western railroads with excellent material whenever the time shall come that the price of labor will allow of competition with eastern manufacturers.

With all its natural advantages, its bright skies, pure atmosphere, and healthful climate, it may be pertinent to inquire why it is that the emigrant-wagon is so seldom seen bringing pioneers to new homes in this desirable region, while other Territories, far less inviting, are being populated so rdwidly. The reason will be found to be a peculiar state of affairs, affecting the price and ownership of land, which does not exist in any other Territory. New Mexico has the misfortune to have a large part of her lands under the control of so-called land-grant companies, which have obtained grants from individuals whose title generally came from the Mexican government, before the cession of New Mexico. There are nine of these land-grants, claiming in the aggregate over 8,000 square miles of the best agricultural and grazing land in the Territory. As certain of these grants have been confirmed by Congress, their title may be regarded as secure, but others yet lack ratification. The claims of some companies conflict with others, causing uncertainty regarding the titles and sometimes annoying litigation. Where there is so much doubt concerning the validity of these land-grants, and so little desirable Government land available for pre-emption, it should not be a matter of surprise that the population of the country is not rapidly increasing. Under these circumstances, people will not settle down to till the soil, erect buildings, and fence in farms, but choose rather to engage in the less expensive business of stock-raising, which is more suitable to a nomadic life in the event of an ejectment by some more lawful claimant.

It would be economy for Government to buy up all these claims at a good round sum and throw the land open to settlement under the homestead laws. The speedy increase in the population and amount of taxable property, and general prosperity of the people, would soon more than repay the original outlay, and change this conservative, inert Territory into a thriving rival of her less-favored neighbors.

Very respectfully, your obedient servant,

W. L. CARPENTER,
First Lieutenant Ninth Infantry.

Lieut. GEO. M. WHEELER,
Corps of Engineers, in charge.

APPENDIX D.

EXECUTIVE REPORT OF LIEUTENANT R. BIRNIE, JR., THIRTEENTH UNITED STATES INFANTRY, ON THE OPERATIONS OF PARTY NO. 2, CALIFORNIA SECTION, FIELD-SEASON OF 1875.

UNITED STATES ENGINEER-OFFICE,
GEOGRAPHICAL SURVEYS WEST OF ONE HUNDREDTH MERIDIAN,
Washington, D. C., May 1, 1876.

SIR: I have the honor to submit herewith report of operations of party No. 2, California section, for the field-season of 1875.

The country to be surveyed by the party comprised the eastern portion of atlas-sheet 65, of which Owen's Lake, California, forms nearly the central figure. The main topographical features of this eastern portion, lying approximately between longitude 116° 30′ and 118° 15′ west from Greenwich and north latitude 35° 35′ and 37° 20′, proved exceedingly simple. Here was found most markedly a type of that large section of country lying east of the Sierras, in Nevada and California, where the general trend of the mountain-ranges is very nearly parallel to the Sierras, and, so far as our area extended, uniformly decreasing in altitude to the eastward, and becoming more barren and sterile as they decrease in height, seeming to endeavor to assimilate themselves to the alkaline sandy and desert valleys that separate them. It is also to be observed that these valleys decrease in altitude with the ranges, that is to say, considering the main divisions of ranges and valleys. First the Owens River Valley separates the Sierras from the Inyo range; and second, considering the Argus as but the southern extension of the Inyo range, we find this separated from the Panamint range by the Salinas and Panamint Valleys, these valleys being separated by an east and west spur that tends to connect the Panamint and Inyo ranges; third, Death Valley separates the Panamint range from the Amargosa on the east. The order of the altitude of the ranges has been found as stated, and of the valleys, Owens River is the most elevated. Salinas and Panamint are next, and Death Valley the lowest, being below the level of the sea. Eastward of the Amargosa range is the valley of the Amargosa River or the Amargosa Desert, which rising a thousand feet or more above the level of Death Valley, broke the uniformity heretofore observed and, moreover, formed the limit of our survey to the eastward.

On the 23d of June, 1875, the party left your rendezvous camp near Los Angeles, Cal., composed as follows: Louis Nell, chief topographer; F. Brockdorff and W. S. Waters, recorders; Benjamin P. French and Lenardo Aguilar, packers; Frank Reyer, cook; E. S. Stevens, private Company G, Twelfth Infantry, and myself as ex-officer and field-astronomer. After two days' march the cook was discharged, and Andrew Hoos employed in his stead.

At Los Angeles we were distant about 120 miles from Pilot Knob, a mountain-peak situated nearly on the southern line of our area. Up to this point, however, there was to be traversed the main road from Los Angeles to Panamint, much traveled before the establishment of the Southern Pacific Railroad terminus at Caliente, when a road from that point to Panamint and the surrounding country offered better facilities for the receiving of freight from San Francisco.

The route of the party was along the road first mentioned, following nearly the line of the Southern Pacific Railroad as far as Spadra, Cal., about 30 miles from Los Angeles, and then passing through Cucamanga.

We were detained at Martin's, Cajon Pass, until July 2, awaiting the return of Mr. Nell, who, with a small party, left us at Lytle Creek, June 29, to make the ascent of and the necessary observations upon San Antonio Peak. From Martin's, through the Cajon Pass, the road crossing the divide a few miles east of the crossing of the proposed Los Angeles and Independence Railroad into the Mohave Desert, we crossed the Mohave River at Huntington's, and again near the cottonwoods, after it has made its remarkable turn to the eastward. Its bed here is dry most of the year, and the prevailing west and southwest winds, blowing uniformly from noon to midnight in the summer season, pile up ridges of the sand athwart its course, and one wonders that its waters could flow here at all, rather than that the volume of water seen 20 miles higher up the stream should here be nowhere visible. A few settlers are found along the river, groves of cottonwood, and a little grazing, but the soil is too sandy to be productive, and the difficulty of obtaining water for irrigation, except in early spring, is almost insurmountable. Water for drinking purposes is obtained from wells. From the Mohave we cross a low divide into a valley, in which is Black's ranch; the soil of the valley incapable of cultivation, but sand-grass and a coarse grass somewhat used for hay in the country are found, and afford sustenance for several hundred head of cattle. Good water is obtained from wells.

July 7 we reached Granite Springs, near the southwestern base of Pilot Knob, and this peak was ascended for triangulation and topographical purposes. In our last day's march to this camp, marching northward, we came obliquely upon a line of volcanic rock, that could be clearly traced in a slightly northeast direction, having completely shattered the group of hills forming the Cosco Mountains, and, in the southern extension of Owens River Valley, showing a probable elevation of the land, now a low divide, to the south of what now forms Owens Lake and the cutting off of the river, for there are indications of an old river-bed south of this divide; an immense old crater and lava-flow is found again just south of Cerro Gordo Peak, and in the same general direction, in the northern part of Owens River Valley, is a great group of old craters, extending into the foot-hills of the Sierras. The earthquake of 1872 took place along this line, and slight shocks, not felt at any great distance to the east or west of it, are of frequent occurrence.

From Granite Springs the party moved westerly, by way of Surveyors' Well and Willow-Tree Spring, to Panamint Station, (a station of the Cerro Gordo Freighting Company, about 75 miles from Panamint,) over an old and little-used wagon-road, passing also by the El Paso (silver) mines, that have not been worked for several years. From Panamint Station a small party ascended Owen's Peak, in the Sierra Nevada range, near by, and then made its way northward through the mountains, principally along the south branch of the Kern River, to Olancha Peak, and after ascending it, met the remainder of the party at Olancha post-office, (near the eastern base of the peak and the southern end of Owens Lake as well,) whence they had come by the direct wagon-road along the base of the Sierras from Panamint Station.

In the great longitudinal basin in the Sierras, through which flows the Kern River and its south fork, the two streams only being separated by a subsidiary ridge, and especially about the headwaters of the streams, fine meadows abound and excellent pasturage-ground extending up the mountain-slopes. These are utilized in the pasturage of cattle and sheep, principally the latter, thousands of head being brought in the warm season from the settled country of the valleys nearer the sea-coast, the pasture-grounds there being insufficient to sustain them. Many of these lovely meadows are entirely laid bare and their sod almost ground into the dust, and nearly to the tops of the highest peaks these flocks of sheep find their way.

The party remained in camp at Olancha two days, and then proceeded, skirting the foot-hills of the Coso Mountains on the north by way of Arab Spring to Darwin, a new mining-town, only established in the early part of 1875. A topographical station was made near at hand, and the town and some of the principal mines located in position. [For a report of the mineral resources, &c., of this mining-district, see report of Dr. Loew, who visited this place in October, 1875.] The town was found in a growing and most flourishing condition. Water had been conducted by pipes from springs distant about 7 miles, and in sufficient quantity for the use of the town and furnaces. The party camped on the night of the 23d of July in Darwin Cañon, about 7 miles from the town, where a magnificent spring bursts from the side of the cañon. Its waters, however, sink after flowing a few miles and before reaching the valley; and such is found to be the case with all the other little streams of this range (the Argus) and those of the Panamint and Amargosa ranges to the east of it.

Returning through Darwin, we moved a short distance westerly to Coso, an old but now abandoned gold-quartz mining-town (except by a few Mexicans) in the Coso Mountains. From these hills at present is obtained the supply of wood and charcoal for Darwin; this supply, with any great demand, as seems probable, will, however, soon be exhausted. From Coso, following a wagon-road, we crossed the Argus range into Shepperd's Cañon, and thence directly across Panamint Valley to Cañon Station, (of the Cerro Gordo Freighting Company,) in Panamint or Surprise Cañon, on the eastern slope of the range, a few miles from the valley. We remained in this camp about a month, (until August 28,) occupied in measuring a check base-line in Panamint Valley; the ascent of Telescope Peak; a trip to Lookout and Rose Spring mining-districts; another to Panamint; and another to Borax Lake and factory, in which the Slate range was crossed; Argus Peak occupied for topography, a route meandered from this peak northward to develop the topography of the angle formed by the junction of the Argus and Slate ranges, (Borax Lake lying between these ranges,) and to occupy Maturango Peak of the Argus range.

Panamint Valley, a little more than 1,000 feet above the sea-level, is exceedingly desert and alkaline, and, together with the other low and desert valleys in this country, very hot in summer. Hot springs are found at nearly the lowest point, which is on the east side of the valley, at the base of the Panamint range. A valuable salt marsh is near them.

The mining-town is situated near the summit of the range, in a cañon of the same name, (Panamint,) and is reached from Panamint Valley by a very steep but excellent wagon-road. (Dr. Loew also visited this place for report.)

While at this camp, all our animals, except a few required for short trips, were sent a distance of about 30 miles, where pasture could be obtained; we were thus enabled to recruit materially those that had developed signs of weakness.

From this place, with a portion of the party, I crossed the head of the cañon in which the town is situated, and from the summit of the pass we overlooked Death Valley and the Amargosa range beyond on the east and Panamint Valley on the west.

The first portion of the descent to Death Valley by trail was very steep. In the cañon through which we passed grass and a short running stream were found, also a small cultivated piece of ground, where vegetables were raised with facility by irrigation. Entering Death Valley, we turned northward and camped two days at Bennett's Wells nearly on the edge of the great alkaline deposit of the valley; but permanent and very good water, only a little alkaline, was found in the wells—three or four holes scarcely more than 6 feet deep—while a coarse green bunched grass furnished pasture for the animals and mesquite trees grew sparsely around. Our barometric observations show this camp to have been within a few feet of the sea-level, (below.) Two trips were made to determine the lowest point of the valley. On the first day, moving out into the valley nearly in an easterly direction, I was induced to turn northward from the washes met with, showing a flow of water had been in that direction. I believe, however, the slight difference of level throughout allowed the water (in the wet season) to be driven by the prevailing winds from the south, and at least greatly assisted in the formation of these channels.

This valley on the south receives the flow of the Amargosa, and on the north that of Furnace Creek and of the more northern and higher portions of the valley, with much of the drainage of the Panamint and Amargosa ranges. Water collects in these drains only in the wet season. The southern and nearly flat portion of the valley seems to extend north and south for 20 to 30 miles. From what I could observe, I am inclined to think there are two low portions situated on either side of the center of this flat portion. Two trips across the valley, made after this first, the one to the south and the other to the north, showed a less elevation.

It is to be remarked, however, that our first and central trip was not entirely across the valley, as in the other cases, but was far enough to support the opinion expressed. The second day we crossed the valley to the south of the first line, and, as before remarked, our observations showed a greater depression, which was near the eastern edge. In this vicinity, where water had evidently stood not long before, the surface was an unbroken crust of salt, probably an inch thick; the appearance of a beach and water-waves was thoroughly impressed upon the white surface of the salt as the water dried up, and the effect was very fine.

Even at this time, nearly the driest season of the year, great difficulty was experienced in traveling about the valley; the marshy, soft ground in one place, and the dry honey-combed loose surface in another, made traveling with animals exceedingly dangerous, while again the ground was smooth and baked hard enough to readily support them. On both these days we were compelled to leave our animals and make the rest of our journey on foot, sometimes sinking nearly to the knees in ash-colored mud beneath the salt. This flat portion of the valley is about 5 miles wide, (opposite Bennett's Wells,) and while the lowest point may not have been discovered, enough has been developed to show that there is a large tract nearly flat, and a great portion of this approaching 100 feet depression from the sea-level. I may here remark that probably the only way to thoroughly solve this problem would be to run several instrumental level-lines through the length of the valley, and this would be almost impracticable from the marshy character of the ground; many places are, I am sure, impassable.

Two fresh-water springs are found on the western edge of the flat north of the wells, but the waters of both seem somewhat medicinal, and are said to be unwholesome. We did not suffer as much from the heat as had been anticipated, although the thermometer was noted at one time at 145° F. in the sun; yet, as at this camp we could receive all the benefit of the breeze from the south, which blew a great portion of the time, the oppressiveness of the heat was obviated, except when there came, as at intervals, blasts of hot air.

It is claimed by intelligent persons who have visited the valley that water remains in a body in portions of the flat throughout the year. I was unable to discover any such body of water. (1 September, 1875.)

From Bennett's Wells we moved northwardly along the western border, and crossed to the northeast, over the old emigrant road to Furnace Creek, (noting a second portion at Salt Springs lower than on the central trip,) where we awaited the arrival of the rest of the party that had separated from us at Cañon Station, and had moved along the western foot-hills of the Panamint range to Willow-Tree Station, (on the Los Angeles and Panamint road via Pilot Knob;) there occupied Brown's Peak, in the southern portion of the range, for triangulation and topography, and passed to the south of the range by Leach's point and Owl Springs into Death Valley at Saratoga Springs; thence along the dry bed of the Amargosa by Resting Spring and Clark's Fork, and left the Amargosa to pass to the east into Water Valley and Pah Spring, this last the rendezvous of a numerous band of Indians; thence by Ash Meadows to our camp at Furnace Creek, which they reached September 8, having been absent thirteen days.

Furnace Creek is formed by numerous warm springs that have their rise (all about of the same altitude) in the low slopes on the north side of the cañon, the warmest of which is a little over 90° F. The stream, 3 or 4 miles in length, sinks where it reaches Death Valley. The south walls of the cañon are a curious strongly-cemented conglomerate, and a small amount of calcareous matter seems to be the only deposit from the springs.

After occupying a peak to the south, in the Amargosa range, for topography, we moved to the northeast, crossing this range by a pass heretofore unexplored, and then north across the Armagosa Desert to Oasis Valley. This march of 38 miles was entirely destitute of water or grass, and having first to cross the range, but 20 miles were accomplished at sundown, a night march was made until 1 a. m.; and, starting early the next morning, a good camping-place was found at 10 a. m.

Bare Peak, near by, was occupied, and a trip made to occupy Toe-li-cha Peak, in the northeastern corner of atlas-sheet 65. While this was being done, a reconnaissance was made with a light party, to gain the necessary information for recovering the Amargosa range into Death Valley. Several springs were discovered along the eastern slope of the range, but Boundary Cañon was found the first practicable pass south of Grape-Vine Cañon. Between these passes is the highest portion of the range, and some very pretty mountain country on the eastern slopes, which are gentle, and there is good grazing. Piñon is almost the only wood found. The place is a resort for Indians, who gather the piñon-nuts in quantities in the fall. Here was also first observed the peculiar *blinds* made by the Indians just beside the springs for killing the birds—quail and some very small varieties—that come in numbers to the only water to be had. These blinds have the general appearance of a bee-hive, are made of rushes and small boughs interlaced, with an opening for entrance on the side away from the spring. The interior is large enough to seat one person, a small opening being toward the water through which to shoot the arrow, and with string attached for recoving it no alarm is produced. In Oasis Valley duck-blinds were found constructed in the same way near little artificial ponds.

The ascent of this range on the east side is easy, but, crossing to descend into Death Valley, the faces of the strata are broken into abrupt descents, and *boxed* cañons are almost universal and often impassable.

From Oasis Valley we passed the Armagosa range by Boundary Cañon, stopping to occupy Wah-guy-he Peak, and, entering Death Valley a second time, camped at Salt Wells. This portion of the valley is, if possible, more dreary than the southern part, for, except this salt marsh, where grows a tall green grass, but of poor quality for grazing, its general appearance is a great stretch of white sand, interspersed, it is true, with mesquite trees, but these are perched upon hillocks sometimes 15 or 20 feet high, their roots having protected the sand about, while the wind has carried away the intermediate portions. We found the water at the Salt Wells unfit for use, (during a portion of the year it can be used,) but in crossing the valley and at a short distance from our camp we found several holes with a small supply of very good water. The excavations showed a stratum of saturated sand but a little below the surface, and the indications are that it would not be difficult to obtain water by a little digging almost at any place in this vicinity. The valley is here about twelve miles wide. Our march was through Cottonwood Cañon, in the Panamint range. We reached water and a camping-place at sundown. On this day's march our animals were much exhausted, two having to be left, unable to travel; one died almost immediately, the other was brought into camp the next day. The water met with had been sufficient only for ourselves, and riding animals and some of the pack-mules had been without for forty-eight hours; and in this one encounters the difficulty of traveling with more than a few animals. Besides the scantiness of the grazing, the springs are too small to furnish sufficient water.

We passed along the stream of several miles in length, lined with cottonwoods, and following a good trail, camped at springs near the summit of the range and the head of the cañon.

The uniformity of the ranges is broken at this place, and a spur that separates Panamint from Salinas Valley juts out from the Panamint range. The highest portion of this spur, somewhat plateau in character, is interspersed with springs, and good grazing is found. A mule-ranch has been established. Occupying a triangulation-station here, we passed along the spur and crossed the Cerro Gordo range by a very steep trail, and came to the mining-town of Cerro Gordo on the eastern slope and near the summit, (for

AP. JJ—9

report of mines, &c., see Dr. Loew's report,) and thence by excellent wagon-road down the mountain to Cerro Gordo landing, on Owens Lake. The country from Panamint to Cerro Gordo was without settlers, except the small garden in the cañon east of Panamint.

Skirting the northeastern shore of Owens Lake, we crossed the river by a bridge a few miles above the lake, and thence through Lone Pine to Camp Independence, arriving the 3d October, and remained until November, small parties being sent out from time to time to prosecute the work in the vicinity.

October 6, Mr. Nell occupied a triangulation-station on the eastern divide of the Sierras and nearly west of Camp Independence.

Careful observations were made to connect this post with primary triangulation points, and a series of sextant-observations taken for latitude.

October 9 we started to occupy Mount Whitney, the highest peak in this portion of the Sierras, and probably the highest in the range. The most feasible route was found by returning to Lone Pine and ascending the very steep eastern slope of the Sierras by the way of the Hockett trail, (from Lone Pine to Visalia, Cal.,) passing the divide by the headwaters of Cottonwood Creek, and then turning northward from the trail when in the basin of Kern River, which drains directly the western slope of the peak which we occupied October 13. The view from this peak is most grand and comprehensive, more than 11,000 feet above the Owens River Valley, overlooking it and the ranges to the east, and including almost in one view the two great ridges on either side of the basin of the Kern River and the rocky barrier at its head that separates it from King's River to the north and west.

In the Sierras we find the strata dipping to the west, just the opposite of what is found in the Amargosa range, (the eastern one of the uniform system heretofore remarked.) A close study of these ranges in connection with the intermediate ones, the Inyo, Argus, and Panamint, would I think, in this particular alone be very interesting. In the Inyo (next to the Sierras) the strata seem but little inclined from the perpendicular. This range has been considered remarkable from its height compared to its base, rising 7,000 to 8,000 feet from a base of scarcely more than 8 miles, both sides being at this point (Inyo Peak) about equally precipitous.

The glacier action to be seen on the western slope of Whitney's Peak is very grand; the immense size of the cañons and the smooth and polished surface of the rocks extending up their sides attest the action that once has taken place; the eastern slopes of the range observed near Camp Independence also bear evidence of this action, notably on Glacier Cañon, just north of Kearsarge Pass, and almost at any place along this slope may be found broken rocks worn and polished on one side, showing they were fixed in position when this abrasion took place.

Returning from Whitney's Peak, we were compelled to leave our pack animals at Lone Pine to recruit, while Mr. Nell and myself made a trip to occupy Cerro Gordo Peak and New York Butte, both of the Inyo range; then collecting the party, we returned to Camp Independence. Next a trip was made to occupy Wau-co-ba Peak, (Inyo range,) returning by the Eclipse mill and mine and crossing the river at the mill. A triangulation-station was also made upon one of the group of craters just south of Big Pine.

The Owens River is scarcly fordable within 20 miles of its mouth, but is crossed by three bridges. The banks, though low, are steep, and the river-bed soft or the approaches swampy. The sediment brought down by the river is deposited in bars in the lake, perpendicular to the thread of the current; behind them are forming lagoons. The bed of the river near its mouth is a hard-pan formation, in which deep and shallow places alternate, seemingly without reason.

In the valley is distinctly observable a fissure formed by the earthquake in 1872, when the ground sank in the valley and along the west bank of the river; and next to the foot-hills is observed the exposure of the west side of the fissure. There seemed a striking resemblance between this and a terrace formation, its position facing the valley also, and I think in a short time, when the bank shall have become rounded by the weather, it will be hard to distinguish from an old terrace.

On November 3, having completed as far as practicable the area assigned us the party moved toward Caliente, Cal. Passing along the western shore of Owens Lake the road keeps along the eastern base of the Sierras to Tehacapai Pass, and through this to Caliente, where we arrived November 12.

At Little Lake a halt was made to occupy a station in the Coso Mountain, and for me to visit interesting boiling springs that had been reported in the vicinity. Dr. Loew has analyzed the waters from these springs and reports upon them.

Supplies for the party were received at Los Angeles for forty days, at Panamint for sixty days, and at Camp Independence for forty-five days.

Lieutenant Whipple and party arrived at Caliente also on the 12th, and Mr. Klett and party on the 15th. The parties were disbanded as soon as practicable, and the property inventoried and packed in cases. Lieutenant Whipple, in charge of train and property, left for Los Angeles on the 19th. Mr. Klett and myself left the last, on the night of the 19th, for Washington, D. C.

The party was in the field one hundred and forty-three days. A system of triangles was closed over an area of about 8,000 square miles of territory, and the topography of this area, with about a third as much more obtained by running meanders exterior to the system, was, as far as practicable, carefully studied and can be mapped.

The parallel ranges of mountains, with but 15 or 20 miles between their crests, and these marked with well-defined points, afforded an excellent opportunity for carrying on the triangulation.

A base, 6.78 miles in length, was measured in Panamint Valley, where we were able to find a nearly level and favorable surface, and well-conditioned triangles enabled us to include two prominent mountain stations (Telescope and Malurango Peaks) in the first extension. This system was definitely connected with that of the Los Angeles base by the two parties occupying in common Whitney's, Olancha, and Owens Peaks.

In the measurement of the base an instrumental line was laid out between the extremities and marked by small white flags planted at intervals of fifty paces, obstructions and irregularities of surface were removed, and the line carefully measured with a spring-steel tape-line, compensated for temperature. The line runs a little west of north from the south end, which is near the center of the valley, and about one-fourth of a mile north of the direct road from Shepperds Cañon to Panamint Cañon.

This end is marked by a rough-hewn stone monument, projecting 15 inches above the surface, rectangular in shape, 6 by 8 inches, and 1875 inscribed on the side away from the base. In this valley our survey was also connected with the land-survey, a three-point station being made at section-corner of ranges 42 and 43, township 20, sections 7, 18, 12, and 13. This section-corner is situated 5 miles, and bears north 70° west, from the northern extremity of the base, which is a monument of rough stones.

A series of observations was made to determine the azimuth of the base, and of sextant observations for latitude at the south end. In addition to this, sextant observations were taken at sixteen other points, which were also connected by instrumental bearings with the triangulation. In all, thirty-six sextant latitude stations were made, and complete observations for azimuth of sides of triangles at four different points.

Magnetic variation was determined by observations on Polaris; thirty-nine results have been recorded, including those determined at the apices of triangles, where the difference between the magnetic and true azimuth of one of the sides was, whenever practicable, carefully determined, the true azimuth resulting from computations afterwards made.

Eighteen triangulation-stations and thirty-two topographical stations, including the eighteen triangulation-stations, were made.

At starting, the party was well supplied with instruments for taking the meteorological observations required ; both our barometers were out of order by July 12, but new tubes being received at Panamint in August enabled us to use this instrument for the remainder of the season, good comparisons for error being made on our return.

The party was unfortunate in that, from breakage or loss, the full set of meteorological instruments was not carried throughout the season ; and the results of the observations taken have been curtailed on account of this ; the meteorological observations were taken throughout to conform to your written instructions, and, together with the odometer record, duplicated and compared in the field. Eighty-two cistern-barometer stations were made, and four hundred and sixty-two aneroid stations in addition.

The principal roads and trails were meandered, and a table of distances for these, with remarks as to wood, water, &c, prepared in the field.

Twelve hundred and sixty-one miles were meandered and 611 traversed but not meandered ; thirty-three point-stations were made as checks, and six hundred and seventy-two other stations upon meander-lines.

The instruments used were the sextant, Stackpole & Brother's, for latitude ; Wurdemann theodolite, graduated to read to 10″ of arc, for triangulation ; Wurdemann graduated and Young & Son's small transit, for topography and meander work ; cistern and aneroid barometer, with wet and dry bulb, and maximum and minimum and pocket thermometers, for meteorological observations ; the compensated steel tape for accurate and the odometer for road measurements. Short meanders were sometimes made with the pocket-compass.

A number of mining-camps were visited, and everywhere the party met with the kindest treatment from the people of the country and a willingness to impart information that proved of great value to us in traveling. Especially are we indebted to Capt. A. B. MacGowan and the officers at Camp Independence, and Mr. Maclean, the superintendent of the Cerro Gordo Freighting Company, for assistance rendered us.

The members of the party, with scarcely an exception at any time, worked with a unison and cheerfulness that is a pleasure for me to record.

Very respectfully, your obedient servant,

R. BIRNIE, JR.,
First Lieutenant Thirteenth Infantry.

Lieut. GEO. M. WHEELER,
Corps of Engineers, in charge.

EXECUTIVE AND DESCRIPTIVE REPORT OF LIEUTENANT C. C. MORRISON, SIXTH CAV-
ALRY, ON THE OPERATIONS OF PARTY NO. 2, COLORADO SECTION, FIELD-SEASON OF
1875.

UNITED STATES ENGINEER OFFICE,
GEOGRAPHICAL SURVEYS WEST OF THE 100TH MERIDIAN,
Washington, D. C., May 1, 1876.

SIR: I have the honor to render executive and descriptive report as follows of the
operations of party No. 2, Colorado section, of the survey for field-season of 1875.

Reporting in person to Lieut. W. L. Marshall, Corps of Engineers, in charge of the
Colorado section, at the camp of organization, at Pueblo, Colo., on June 5, I found the
personnel of the party completed as follows: First Lieut. Charles C. Morrison, Sixth
Cavalry, executive officer and field-astronomer; topographical assistant, Mr. Fred. A.
Clark; recorders, Messrs. W. C. Niblack and Anton Karl; general assistant, Mr. W. H.
Rideing; packers, Alex. Harbeson and Samuel Abbey; laborer, N. Bascom; cook,
Green Terrell.

The necessary supplies, record-books, &c., being received later, the party took the
field June 12. Moving southward upon the Santa Fé stage-road, our objective point
being Fort Garland, we crossed the Sangre de Cristo Pass June 15, and reached the post
the following day. It is a military post for four companies, built of adobe or sun-dried
brick, not very prepossessing in appearance, although very prettily located on the
eastern edge of San Luis Valley, near the junction of Sangre de Cristo and Ute Creeks.
From the latter of these the post draws its supply of water by *acequias*, or ditches.

My instructions were to carry the survey from Fort Garland westward, to cross the
San Juan range north of the San Antonio Creek, to extend the triangulation on a por-
tion of this range unfinished the previous year. Having completed this, to go to Fort
Wingate by way of Pueblo Pintado, surveying a large section in that vicinity; to
examine Washington Pass as to its possible use as a wagon-pass; thence to proceed
down the Bonito Creek to the Puerco of the West; thence to Wingate, surveying the Zuni
Mountains southeast of Wingate, and large areas on each side of the Wingate and
Albuquerque wagon-road; to carry the survey from the Rio Puerco of the East to longitude 104° 7' 30", between latitudes 34° 50' and 35° 40'. Having finished this belt, to
proceed to West Las Animas, Colo., the disbanding camp, reaching there by November
15, the return route being up Ute Creek to its head. The route followed was essentially as directed, and incurred traveling by this party of 4,627 miles. The area surveyed during the season was 11,300 square miles.

It became necessary to divide the party for a portion of the time to accomplish the
survey of this great area. During such time Mr. Karl acted as assistant topographer
to the main party, and did so very satisfactorily indeed. The ration-points need during the season were: Pueblo, Tierra Amarilla, Fort Wingate, Albuquerque, Santa Fé,
Anton Chico, and Stone Ranch, rations being distributed to those points from Santa Fé
and Fort Wingate.

The triangulation was carried over nearly the whole area and profile lines run.

Hypsometrical data were gathered by careful observations taken regularly throughout the season. Latitude-observations were made, when possible, at all camps not
located by triangulation.

The examination of the country between Tierra Amarilla, N. Mex., and Fort Defiance, Ariz., developed two possible wagon-routes of communication to the latter
place by way of Washington Pass; thence into Southern Utah or Northern Arizona.

The one, striking west just south of Nacimiento, crosses the main divide not far from
the eastern head of Chaco Creek, follows down this drain past Pueblo Pintado to
where it passes out of the cañon opposite Mesa Teohada; thence, a little north of
west, crossing the Vaca Creek, nearly opposite the pass; thence over Washington
Pass; thence, bearing to the south, skirting the Black Lakes, down the drain of the
Bonito past Fort Defiance, nearly to the mouth of the creek; then crossing over to the
main wagon-road from Fort Wingate to Prescott, Ariz.

The second route crosses the Atlantic and Pacific divide at the head of Cañon Largo,
where the main wagon-trail crosses north of San José; thence running down Cañon
Largo about 17 miles, it leaves the main trail, crosses over to the southwest to Ojo Nuestra Señora; thence running nearly due south it strikes the Chaco in the vicinity of
Pueblo Pintado; thence it follows the same route as above.

On either route there would be a scarcity of water.

Little work would be necessary on the first route from Nacimiento as far as Pueblo
Pintado; from there there would be some blasting, cutting, and filling in Cañon Chaco, as also in Washington Pass, particularly near the summit on the eastern slope;
here the pass narrows, and a short, sharp rise around a point of rocks would require
much blasting. A very careful selection of route would be necessary on the whole eastern approach; the rise is almost 3,000 feet. On the west little difficulty would be met

with; a little blasting would be needed about 200 feet below the summit, and some cutting and filling, with two or three small bridges, would be all required.

On the second route little additional difficulty would be experienced, except south of Ojo Nuestra Señora. A line of bluffs running nearly east and west would necessitate a very careful selection of route in this vicinity. One of these, in connection with a proposed wagon-road from Fort Garland, crossing the Rio Grande at Colonas or Myers Ferry, thence to San Antonio, from there by the southern or Brazos branch of the Chama to Tierra Amarilla, would make a very direct route to Northern Arizona.

The distances on these routes will be found in tables of distances to be published from this office. It is not intended to convey the idea that these routes are now practicable, for they are far from being such; but that they are possible routes, necessitating work. There is at present a good road as far as San Antonio, Colo.; from there but a horse-trail; in many places not even that.

No traverse-line was run from Pueblo to Garland, as that portion of the country had already been surveyed in previous years by your parties.

From the summit of Sangre de Cristo Pass the San Juan range was seen enveloped in smoke of burning forests; and upon leaving Fort Garland there was still no more than a few snow-capped points visible. It certainly was not encouraging for triangulation, as we were dependent upon Banded Peak in the center of this section for the developing of the work to the south and west, the connection with Mount Taylor being very necessary, and only to be well obtained with a clear horizon.

Two miles out from Garland, June 19, we crossed Trinchera Creek, a clear mountain-stream, as yet but little utilized for irrigating purposes, but capable of supplying water for quite a tract of land now almost barren. Further on, the Culebra was reached, with its Mexican ranches scattered along its banks. Flowing through low ground, with several channels, it makes much marshy land in its broad valley which could be easily reclaimed and cultivated.

As we approach the Rio Grande the bare basalt crops in places, and a few hundred yards below Myers or Colonas Ferry the cañon of the Rio Grande heads, but a few feet deep, gradually increasing till it reaches 800 or 1,000 feet in depth—a narrow black gorge in the eruptive rocks—the great outlet of what has once been a large lake, gradually drained by its outlet cutting deeper, till finally the sandy plain of disintegrated lavas was left, rich in soil but unproductive for the want of rains or economic distribution of its present waters. The ferry is a small flat-boat swung from a cable stretched from bank to bank, the boat being propelled by the action of the current. At this point, or possibly a little farther north, would be the best railroad-crossing for a route to follow the Rio Grande, forced as it would be, in order to be built at a reasonable cost, to follow the plateau on the west bank, to come down the river again by the Ojo Caliente Creek and Rio Chama.

The route of the party from Myers Ferry was up a short valley, thence passing over a low divide down into the drainage of the San Antonio Creek. Here the country had been overflowed by the San Antonio and Conejos Creeks, which unite a few miles below where we made our camp, having between them a large marshy flat which could easily be drained, and quadruple the land now under cultivation could be devoted to agriculture had the people the necessary enterprise. At the southern end of this valley the San Antonio Mountain, a great wooded dome of eruptive rock, rises high above the surrounding country. Isolated as it is it would be a fine natural triangulation-point were it not for its wooded crest. A built station on the tree-tops would make it a fine central point to carry the triangulation from the eastern or main rocky range to the west.

The Conejos, up which lay our route for the morrow, is a broad mountain-stream, at this season difficult of crossing, swollen as it was by the melting snows of the mountain. Several promising Mexican plazas attest the appreciation of the agricultural advantages of the section, and the many sheep in the valleys near the mouth of the cañon speak well for the grazing. The Prospect Peaks, standing as sentinels over the entrance to the country beyond, rise up from the plain just north of the mouth of the cañon. The forests were still smoking, giving promise of much waiting. We left the main road 4 miles above the town of Guadalupe, and commenced our work in the mountains. Without the serrated outlines of granite-formations the range is none the less beautiful in its many inclosed valleys and cañons. Originally a great plateau of eruptive rock over the Cretaceous sandstones, it falls gradually, with trend to the southeast, extending from the main San Juan range in the northwest till it disappears in the foot-hills and low mesas in the vicinity of Ojo Caliente. Broken up by many cañons and small valleys which have been carved out in long ages, there is left in this section but a system of narrow tables, here and there a half-mile across, generally but a few hundred feet, frequently but a ragged edge along which a man cannot pass. It is not a mountain-system, but a succession of steppes, narrow flats, and small basins with vertical rims. With summit above timber-line, bare of vegetation, in the day-time warm and pleasant, at night freezing, the year round, with slopes covered with snow, stands Banded Peak in the center of this region, the southern prominent point of the range. Dimly outlined on the southern horizon, 158 miles away, is Mount Taylor. Nearly as far away to the southwest and west are the Tunicha range, the Carrizo Peaks, and the Sierra Lata; to the northwest and north, white with snow, the innumerable peaks of the San Juan range, like the spires of the churches of a wide-spreading city, are seen. Springing from its sides flow the Los Pinos Creek with its heads on its northeast slope, the Navajoe from its northwest slope, and the two northern heads of the Chama from its southern sides. Divided here by the narrow, nearly vertical wall, but thirty feet across its top, waters of the Atlantic and Pacific take rise, the one to find outlet by the Rio Grande in the Gulf of Mexico, the other to seek the Pacific by the Colorado River and Gulf of California.

The nearest prominent points to the north are Meigs and Monument Peaks. The country between is broken into small flats and basins, in which cluster many little lakes from the melting snows. The more protected of these are bordered by a dense growth of quaking aspen, which, with its silvery bark and bright green leaves, contrasting with the somber foliage of the pine, is far more beautiful to the eye than enjoyable to him who has to pick his way through these groves, dense as cane-brakes. At times crowded into the lake by the impenetrable growth, the pack-mule mires. There is no escape. Standing knee-deep in water of melted snow, you must take off his pack and pull him out by ropes. The delay has so belated you that in the approaching darkness you can no longer trail the mules that have gone on before, and are compelled to make camp without shelter and nothing to eat. At this season of the year it was impossible to travel over the country, on account of snow and water. The drifts to be crossed had not a crust strong enough to bear the mules, and they were in constant danger of being wedged in between hidden masses of rock. Around the entire horizon the forests were still burning as they had been for over a week; there seemed no prospect of its abating, and we were constrained, by our rations running short, to postpone the occupation of Meigs and Monument Peaks till later in the season.

In September the topographer returned, and accomplished this. Although loath to do so, (for it would have required but four more days had the horizon been good to finish the work here,) we turned our backs on Banded Peak and followed down the middle branch of the Chama. Taking its head immediately under this peak in two lakes, which find their outlet in falls some 600 feet high, the Chama gathers these waters into a bright stream 10 feet wide and 2 deep. With quite a volume of water where they flow over the rim, these falls are but spray before they reach the basin or amphitheater, inclosed on all sides by nearly vertical walls, excepting the narrow cañon through which the stream finds outlet.

This whole region abounds in game. The elk, black and white tailed deer, grizzly, black, and cinnamon bears are numerous. The banks of the stream are lined with aspen, spruce, and fir, while higher up the bare rocks give silent testimony of the colder air where even the grass is refused a living.

Following the stream down we emerged from the mountains and entered the flats of the Upper Chama, in the section known as Tierra Amarilla of Northern New Mexico. On the east the spur continues to the southeast; on the west we see the mesa-country bordering the great barren Bad Lands of the Territory. Near the point the trail leaves the mountains the middle fork makes junction with the western fork of the Chama, and together they flow to the south, bearing slightly east. We soon passed through the towns of Los Ojos and Los Brazos, and camped a little above the town of Nutritas, the agency for the Jicarilla Apaches. It being issue-day the Indians had come in for their supplies and were encamped about the town. Nearly all of them were Utes and Apaches, although some few Navajoes were easily picked out from among them by their straight, slight figures and intelligent faces; they'had come over to trade with the other tribes and were willingly taking advantage of issue-day to obtain supplies.

Our camp is in a forest of pine trees which grow in this flat as upon the mountains. Such is the elevation of Tierra Amarilla that in winter the snows are very heavy, but in summer the grazing is fine and the country cannot be excelled in that season as a cattle-range. A very undemonstrative Fourth of July was here passed, posting records and recuperating our animals for the hard stretch of plain country to be surveyed.

With no guide but our instruments and the information that we shall probably find but one spring of water permanent in its flow, we start for Fort Defiance. About 4 miles from Las Nutritas the Chama is crossed at the Mexican town of La Puente, so named from the fact that formerly there was a bridge across the Chama near this point. Moving nearly due south we strike the Chama again and camp for the night. Cretaceous sandstones crop along our route during the day. The grazing is good, but after we passed La Puente no cultivated land was seen, although numerous small flats along the Chama might have been reclaimed, and much land on the right bank might be devoted to agriculture did the people partially drain the Chama and throw its never-failing water on the sage-brush plain extending towards the Gallinas. Our camp was near the crossing of the old Santa Fé and California trail, now entirely unused.

From here we pass to the west of the Gallinas Mountains, a low range much broken by short cañons, but having no marked distinctive features other than the point at its

northern end, and the deep rugged cañon through which the Gallinas Creek breaks
its way, cutting the range asunder to its very base. Frequently dry, the stream, from
the recent rain-falls, now runs its banks full. Just west of the stream are the Bad
Lands, or dry, barren Cretaceous deposits, wherein are found the fine fossil-beds. A succession of hog-backs, or uplifts with cliff-faces to the east and gradual slopes conform-
able to the dip of the rock on the west, extend to the divide between the Atlantic and
Pacific drainage. They form, with the Gallinas Mountains, a valley by no means attractive to the eye, scant in vegetation, even the sage-brush and greasewood becoming
scarce, the scattered piñon-trees alone breaking the forbidding hue of hot sandy
drifts.

Pushing west from the Gallinas we cross the divide, here but 7,561 feet elevation,
and descend upon the Pacific slope by even grade, easily to be made favorable to
wagons, and follow the heavy Navajo trail which runs down Cañon Largo from Ojo San
José. The rain falling in torrents brings but the consolation that, if we have to make
a camp where there is no permanent water, we can probably find enough in pools to
partially satisfy the animals.

The day's march was 26 miles through a hard rain, and we camped with only such
water as we had carried in our water-kegs and that soaked up by our clothes, tents,
and packs.

The following day we passed Ojo de Nuestra Señora, a fine bubbling spring, situated
in a drain running into Cañon Largo. The ground round about is marshy; large
bowlders are scattered over the ground low mesas of sandstone inclose the drain, and
numerous trails concentrate at this one volcanic spring in the desert. From here we
traveled southward, over a gently rolling country, here and there broken by low
mesas, with little to distinguish them one from the other, the drainage-lines all converging toward Cañon Largo. After some 18 miles' travel a sudden descent or cliff is
reached. A deer-trail gives us means of descending some 400 feet into Cañon Blanco
drainage, which drain unites farther north with Cañon Largo near its outlet, San Juan
River. The grazing is very poor, and the country would be almost worthless, but for
the lignite which crops at the head of these cañons.

Our route lay to the south, our objective point being the Pueblo Pintado, an ancient
ruin, before referred to in our reports as Pueblo Bonito. It is situated on the south
side of the Chaco Creek; creek simply because it flows water in the rainy season, but
perfectly dry nine months of the year. Its southern and western walls are still standing, showing in its present state at least four stories; the outlines of one hundred and
three rooms are easily traced on the ground-floor. The walls on the east, south, and
west sides have been at right angles to each other; that on the northern front facing
the water has been an arc of an arch, with three large towers built so as to defile all
the ground between the building and the stream. In the interior has been a court
with several circular rooms, like the present *estufas* or assembly-rooms of the Pueblo
Indians of New Mexico.

The whole structure is of stone and wood; no evidence of iron is found. The
masonry consists of thin plates of sandstone, dressed on the edges, laid in a coarse mortar, now nearly as hard as the stone itself. Every chink is filled. The usual stone is
from half inch to an inch thick, with occasional layers of stone 2 or 3 inches thick
occurring regularly every 15 to 18 inches' interval, evidently to strengthen the masonry.
The exterior face of the walls is as smooth as one built of brick and beautifully
plumbed. At the base, 2¼ feet through, the wall at each story decreases in thickness
by the width of a slight beam, on which rest the girders of the floor, the larger ones
setting in the wall. There are no doors opening on the side away from the court, and
the only means of light seem to have been through the inner rooms and through some
small port-holes opening outward on the stories above the first. There are no perfect
arches found in the building; the only approach to such being in having the successive
layers over the windows extend one beyond the other till one stone can span the space.
Usually the doors and windows were capped by lintels of wood, which were but slight
round poles, with their ends, as were those of all the girders, hammered off, apparently
by some stone implement. In one of the circular rooms was found what appeared to
be an altar, built out from the side of the wall in the very center of the building; it
was probably here that their worship, since lost or perpetuated in an altered form by
the present Pueblos, was carried on.

The most striking peculiarities of the buildings were the wonderfully perfect angles
of the walls, the care with which each stone had been placed, the perfection of the
circular rooms as to their cross-section, and the great preservation of the wood. With
an architecture so advanced in other respects, their glaring inability to tie joints in
corners, each wall being built up against and not united with the others, makes it
comparatively weak; indeed, it is to be wondered at that the walls are still standing,
depending as they do each upon its own base, without abutments.

Usually the Chaco is dry; doubtless at one time there was plenty of water, for an
apparent difference in the weeds and grass just above the building indicates that the
ground was once cultivated. We found no implements other than a section of a

metate, or hollow trough of stone, similar to those now used by the Indians and Mexicans, in which they grind corn and coffee. Innumerable fragments of pottery were found very similar, although none perfect, to that made by the present Pueblo Indians.

A few hundred yards down the stream, as also above the buildings, are found traces of other buildings, with, in some cases, the outlines of the walls easily distinguishable. In the cañon, which commences less than three miles below, are seven or eight other ruins equally well-preserved on the cliffs above; these are what have been apparently watch-towers.

South of the Chaco the country rises to a table-land, presenting on its southern and western slope for about 30 miles but two places to descend the cliffs, which are about 300 feet nearly vertical. On the southern face, probably 120 feet above the valley, with no visible way of getting up, nor could we reach them from above, we found several smaller buildings, probably coeval with the larger ruins, presumably used by the shepherds. They are built under the overhanging walls of the cliff-rocks.

On the level surfaces above were found numerous cisterns from $2\frac{1}{4}$ to 8 feet deep, hollowed out in the rock by the action of water, possibly aided by the hand of man.

Descending from the table-land, which we were a day in doing, as a ramp had to be built for each mule down certain vertical places, we camped on a small drain, tributary to the Chaco from the south. A mile north of us was the Mesa Fachada, an isolated mass, which looks like a grand old church and marks the outlet of the cañon Chaco. On another drain just west of this we found another ruin similar in the main features to the others, but differing in that it had a tower-like room running clear to the top, inclosed in rectangular walls, so that the perimeter of cross-section was a square on the outside and circle internally, the segments where the wall was thickest being filled up by rubble-masonry. The ruin was on a slight elevation above the valley. From opposite the face of the former ran a built wall of earth, with stone revetment across the drain, possibly a roadway with bridge, more probably a dam, 10 feet across the top, 5 feet high, and 15 feet across the base. Here, as at the other ruins, was found much broken pottery. In one of the ruins on the main Chaco drain the topographer entered a room now almost under ground from *débris* of the falling walls. It was entirely destitute of furniture or tools of any sort, but was very interesting in that it showed the manner of making the floors; also that the interior walls were plastered with a mortar containing but little lime. In the walls were small square recesses, as if for shelves. The ceiling, which was the floor of the room above, consisted, first, of heavy poles about 5 inches in diameter and at intervals of about 3 or 4 feet; on these transversely were placed smaller poles, and again across these in juxtaposition were laid small square poles, all held down by withes. Nothing bore evidence of the people leaving suddenly, for, though the hole was barely big enough to admit of a man crawling through and had only lately been unearthed by the rains, there was no sign or trace of anything manufactured by man left behind; nothing but the bare walls. In this we were much disappointed, for it was but reasonable to suppose, if we could find a room in fair state of preservation, that some articles of household-furniture might remain.

Throughout this whole section the grazing was very indifferent, and we should have suffered much for water were it not for the rains, and even that which we obtained while camped on the Chaco, where the water was most abundant, held in suspension fully half its volume of silicates.

Nothing could have been more welcome than the refreshing draught of cold water obtained when the small branch of the Vaca Creek, which flows from the eastern side of Washington Pass, was reached. Here were camped several bands of Navajoes, who pushed into camp and endeavored to be extremely friendly, rather too much so, for it proved to be but a cloak for the opportunity of the better making small peculations. They were, however, generally well disposed to whites, but seemed fearful that the object of surveying in their country was either to establish new lines to their reservation or to run a railroad through it. The latter possibility seemed particularly distasteful to them; they were shrewd enough to have seen the civilizing effect of railroads elsewhere. There is possibly no Indian of the plains as intelligent as these Navajoes; of straight, lithe figures, wonderfully square shoulders, the average man tall, quick of movement, with bright, intelligent, rather pointed faces, they are easily distinguished when mingling with other Indians. While still having all the characteristics of the nomadic tribes, they are better able to support themselves. Even now they raise corn and beans and have very large herds of sheep and horses; they have many four-horned sheep. The head of an old buck presents a very strange appearance: in addition to the heavy horns turning up, there is still a second pair equally as large, which turns down and back; and one old fellow had what appeared to be a third set of short horns growing from between them, directly to the front. As I saw this latter, however, at a distance, I could not vouch for their being, as the Indian told me, a third set. They claimed to have quite a number of six-horned bucks; the four-horned were certainly numerous.

At our third camp in the Tunicha Mountains, near the western mouth of the Washington Pass, we saw the squaws making blankets It is a slow process. Between two upright poles are three horizontal ones, two of them placed apart a little more than the length of the blanket, running vertically from one to the other; the warp-threads are stretched; then those of the woof are put in, one at a time, by hand; a narrow thin board is put in after each of these, and they are hammered down by striking the board with a cleet, which they handle so skillfully as to rarely break a thread. Some of the yarn is furnished by the Government, but the best white yarn they make from the wool of their sheep, and the finest red they make by picking an English cloth and spinning the yarn. These blankets are a perfect protection against rain, and are wonderfully warm.

The Tunicha section is a beautiful country and abounds in game. The grazing is particularly good, and there is much arable land in the reservation. The tribe is said to number 9,000. It is fortunate they are peaceably inclined. They rarely do more than steal a few sheep or horses. Although there was, at the time we were there, some little reason to fear an outbreak, some turbulent spirits grown up since the last Navajo war were anxious to show their prowess, and we heard dark rumors of how they were going to kill all the whites in the country. The better counsel of the older men, who had large herds to risk and little to gain, prevailed against such attempt.

On the evening of the 20th of July we reached Fort Defiance, the agency for these Indians, now unoccupied by troops. It is situated in a little valley at the mouth of Cañon Bonito. The masses of sandstone rock tower up above it several hundred feet. The narrow gorge in the red sandstone through which flows one branch of the Bonito is very impressive, but not so striking as the cañons farther north. The Tunicha range, which breaks down into low mesas a little to the south of this, is or has been a large plateau, which is now broken up into many beautiful valleys. There is much fine timber and plenty of water. From Defiance the party traveled by two routes, the main party following the road to Wingate, the topographer going down the Bonito some distance and crossing over to the Rio Puerco of the West, which is a dirty little arroyo or wash emptying into the Little Colorado. The first night out from Defiance the main party camped at Rock Spring, a small pool formed by a dripping spring under the rocks a little to the east of the road and about half way between Defiance and Wingate.

At Wingate we spent several days recuperating the animals and posting records. We are much indebted to the kindness of the officers there, who offered us every private and official assistance possible. Wingate is a very pretty four-company post, built of adobe, situated some 15 miles west of the divide between the Atlantic and Pacific waters. It is dependent on a large spring, the Ojo del Oso, for its supply of water, which is carried by ditches and pipes through the whole post. It stands on the northern slope of the Zuni Mountains, which rise about 1,000 feet behind it. The adobe, or sun-dried brick here used, has as main constituent a pale-blue clay, which is very pleasing to the eye as well as possessing the more practical advantage of being particularly durable for that class of masonry. The roofs are shingle; not the ordinary earth ones used throughout New Mexico.

After our rest at Wingate we started into the Zuni Mountains with renewed energy, and found them far different from what we had expected. Through what has always been represented as the crest of the mountains, we found an anticlinal axis. Following the axial line was a wide valley, running nearly the entire length of the range, abounding in the most beautiful glades, with bunch-grass 18 inches high standing as thick as it could grow, here and there rooted out in the damper places by red and white clover. These natural meadows were a rare treat to our animals, which had so long been living on but scanty provender. Ten or twelve sheep-ranches already located and one large cattle-ranch showed that these advantages were not unappreciated. Here we replenished our larder by mutton as well as by venison killed by one of the party. The game is very abundant. Bear, elk, deer, and turkeys were seen in numbers, and had we had time there would have been rare sport. From Gallinas Spring, in the eastern part of the range, 4½ miles from Agua Fria, the party broke up into three parties, the topographer, with two packers, going to the famous Inscription Rock, with instructions to proceed eastward from there; the main party going to Agua Fria, thence to old Fort Wingate; I myself going down the Cañon de Zuni to Blue Water, up through the Blue Water Cañon, and back to old Fort Wingate. From old Fort Wingate, which is on the waters of the Ojo Galle and about 3 miles south of it, with the assistant topographer we ran two more lines through the Zuni Mountains, occupied several topographical stations, and from there went to McCarty's ranch, which is on the San José Creek just under Mount Taylor, Gallo Spring furnishing most of the water for the San José, its other heads being Agua Azul, which is the outlet of the Valles Ovagones of the Zuni Mountains. Here we waited the arrival of the topographer. The Zuni Mountains, which we had finally covered by the survey, are but a low range, reaching in no place much over 9,000 feet. Heavily timbered, with but few prominent points, its northern slope is very gradual, but it is much cut up by small cañons. Parallel to its axis runs a valley having on the north a line of hills or hog-backs nearly perpendicular on their southern face. To the south, the gradual rise of the main slope, presenting a mass of timber to the eye, runs up to the west. Here it drops down nearly vertically, 500 to 600 feet into a broad valley, extending through the range, varying in breadth from ½ mile to 4 miles. Beyond the valley again rises the crest, vertically, or nearly so. From there it slopes off again into the plains through which runs the Zuni Creek, on whose banks is found the old town of Zuni, one of the ancient "seven cities" discovered by the old Spanish explorers in the sixteenth century. The whole range is very finely grassed and, at the time we were there, well watered, although some of the springs are said not to be permanent in very dry seasons. To the southeast from the range are a number of extinct craters, and it is but sorry work to the surveyor to go through them. Although beautiful natural triangulation-points, they were only beautiful as such, for in traveling over them the animals had almost no grass and were without water three days. The shoes were torn from their feet by the sharp points of lava. Leather ones were substituted. Everything was done to try to save their hoofs, but without success. Two of the five died. The other three were nearly worthless for some time. One poor brute had his hoofs worn away till the bone protruded below the hoof. Another had his leg broken by breaking through the shell over a large lava-bubble and becoming wedged in the crack. The men were too much exhausted, nor had they torches, to explore a cave discovered in the lava, which seemed in its black depths to be of immense size, running in the direction of the great flow which extends for miles around. Hungry and utterly worn-out, they reached the main camp August 16, leading their animals.

From here, after one day's rest, the party was again divided ; the topographer with one man was sent north to occupy Mount Taylor, thence to cross over to the Puerco at Casa Salazar, from there into the Jemez Mountains, thence into Santa Fé. Refitting there, he was to return to Banded Peak to do the little work left undone earlier in the season. The main party proceeded to Sheep Spring. From here work was done to the southwest, in the Sierra Verde and Mesa Lucera, by the assistant topographer and myself. We suffered some from want of water ; at two camps in succession we had but a pint each, and that strongly alkaline.

After a hot day's work, nothing is more necessary than water ; food can much more readily be dispensed with. In a perfectly-dry climate, where evaporation from the body is so great, it needs but a short abstinence from water to produce suffering.

Returning from the Mesa Lucera, we found the main party encamped at Juelites, a forsaken little Mexican town in a broad, brownish-red flat, where everything looks burned up ; the adobe-houses, the parched soil, the gaunt goats, the lean dogs, the very men, look as if they were being slowly consumed by hunger and heat. Few towns in New Mexico will compare with Juelites. Our first camp after leaving this place was at Isleta, on the Rio Grande, which we were glad enough to see. It is an Indian town, some 15 miles south of Albuquerque, on the west bank of the river. The fruits, grapes, and melons were just ripening, and were a great treat to us. The grapes of the Rio Grande Valley are particularly good, being in flavor very similar to the California grape. The country traversed from McCarty's ranch was mainly a succession of sandstone mesas. The valley of the San José, through which runs the principal road, has on its north mesas of red sandstone, capped with white, rising in successive steps about 800 feet above the valley. South of the valley the Mesa Lucera, some 600 feet above the valley, stretches to the southwest, till it unites with the foot-hills of Cerro Verde to the southeast, till it reaches the foot-hills of the Ladrones Mountains. It is gently rolling on top, gradually rising to the south in a line of small hills. The surface-rock is basalt.

Near the town of El Rito, which is some 12 or 14 miles west of Sheep Spring, is a deposit of gypsum. The water is strongly impregnated with the salts of lime and soda, making it almost undrinkable. The grass has been fair, but was well eaten off throughout the whole valley of the San José. The herds of the Mexican town of Cubero, the Indian town of Laguna, and the Mexican and Indian towns of El Rito, fill the valley to its full grazing capacity. This stream finds its outlet in the Puerco, just to the north of Quelites. The latter stream, flowing between high banks of sand and clay, mainly the former, is quickly drunk up by the porous thirsty earth, and, though a fine stream at its head, it barely runs water enough at this point to indifferently support two or three ranches, and a large part of the year is perfectly dry. Between Quelites and Isleta, on the narrow neck of land, are several craters, which form very pretty cones. Being entirely bare of timber, they were a valuable acquisition in the secondary triangulation.

The valley of the Rio Grande, up which our route lay, is a broad open valley, in which not more than a fourth of the land has been utilized for agricultural purposes that could be taken advantage of. Passing through the towns Pajarito, Padillas, and Atrisco, we reached Albuquerque, the county seat, and second only to Santa Fé in size in the Territory. Formerly a military post, there remains but the falling walls to show where once was the garrison. In a low bottom but little above the river-bed, the location does not strike one favorably, and yet it is quite a thriving place. From Albuquerque to Santo Domingo, indeed to Peña Blanca, the fields were looking well and the fruits were everywhere to be found. The grapes of Bernalillo were very fine. Here the valley is lined with vineyards. In this latter place, one stock-owner alone has some 200,000 sheep. Algodones, Corrales, Sandia, are all well-to-do towns on the river; and Santo Domingo, the Indian pueblo, is the picture of peaceful content. The Pueblos were thrashing their grain by driving goats and horses round in a circular inclosure. These Indians who have, ever since known, lived in towns, cultivating the soil and raising small herds, are to be wondered at for having retained their manners and customs so distinct from those of the Mexicans. Contented with little, they certainly have not much.

From the valley of the Rio Grande we went up the Galisteo Creek to the town of Galisteo, covering the country known as the Cerillos, and also the plateau running out to the north of the Sandia Mountains and east to the Placers. In this section are to be found large anthracite-coal beds, which will with the age of railroads become valuable. In the Cerillos are silver and galena deposits and an old mine of turquoise. Several of these mines, known to have been worked by Indian labor by the old Spaniards, are found through the country. In some cases the shafts are filled with brush; others are illy concealed with earth. Just west of the Cerillos is La Bajada Mesa, a volcanic plateau, through which the Santa Fé Creek has cut a narrow cañon. The principal point on the mesa is the Tetilla, a sharp cone, which many centuries in the past has overflowed the surrounding country with the molten rock, now covered with turf from the easily-decomposing feldspathic lava. There are two main routes from Santa Fé to the lower country in the vicinity of Bernalillo, Algodones, and Albuquerque. The one now little used, but formerly the stage-route, follows the general direction of the Santa Fé Creek, south of west to the edge of La Bajada Mesa; thence rising the mesa, it crosses it by a good road and descends by what is known as the La Bajada Hill, a steep, dangerous descent, particularly in its present condition. It comes to the river at Santo Domingo. The other route, known as the Pinos Ranch road, runs to the east of the Santa Fé Creek, strikes over on to the Galisteo Creek drainage, crosses the Galisteo Creek beyond Pinos Ranch several miles, and comes to the river in the vicinity of San Felipe. There are also two main routes to Fort Wingate and Northern Arizona from Santa Fé. Following the La Bajada Hill route to the foot of the hill, the right fork of the road leads to Peña Blanca, situated on the Rio Grande just south of the mouth of Santa Fé Creek; crossing the river here by a good ford, it runs through the foot-hills of the Valles Mountains to San Ysidro; thence, over a rolling country, crossing the Puerco Creek of the East near Cerro Cabezon, going north of Mount Taylor, passing Cross Spring, Willow Spring, and the Mexican town of San Mateo, it comes into Blue Water Ranch; from there to Bacon Spring and Fort Wingate; thence westward down the cañon of the Puerco of the West to the Little Colorado into Northern Arizona. This is one route to Prescott. From Wingate there is a second road running south over the Zuni Mountains to the Puebla of Zuni, down the creek of the same name; thence, crossing over to the White Mountains at Summit Spring, the road runs to Camp Apache. The second route to Wingate follows either route from Santa Fé to Albuquerque, crossing the Rio Grande there, the Puerco of the East at the bridge 6 miles above Quelites, running by Sheep Spring, El Rito, Laguna, Cubero, and McCarty's ranch, south of Mount Taylor, it joins at Agua Azul with the first-mentioned road. From just beyond McCarty's ranch a fork of the road leads off to old Fort Wingate, thence to Agua Fria, Ojo Pescado, and Zuni, thence to Camp Apache, or, branching at Deer Spring, to Prescott.

Santa Fé has been so often described that it is useless to refer to it further. While there a new triangulation-station, well weighted, was built on the hill near the meridian-mark and the points re-occupied for the development of the triangulation to the south and east.

The country assigned to the party to be surveyed after leaving Santa Fé lies south of the main stage-road as far as Vegas, thence north to latitude 35° 40', thence east to longitude 104° 07' 30'', thence south to latitude 34° 50', thence west to the Rio Grande, an area containing approximately 6,600 square miles. This work was accomplished before our return, but, of necessity, with a delay of nine days in arriving at Las Animas.

The main rocky range, which breaks up just northwest of Mora into two ranges, the Santa Fé range and the Las Vegas range, with the Pecos Valley between them, loses its ridge-like characteristic at the break through which runs the present stage-road, built by Colonel Macomb, United States Engineers, in 1859. South of the road and east of the main Galisteo branch is a high plateau, more or less cut up by cañons extending to Cañon Blanco on the south and to the Pecos River Cañon on the east. The grazing on it is very fine. Just east of the Pecos rises another table-land, much broken up, extending between that stream and the Tecalote. From beyond this stream rises the Mesa Chupaines, through which the Gallinas cuts a deep cañon. Beyond the Gallinas it is a grand level plateau, extending to Ute Creek, cut through by the Canadian River and its tributary Mora Creek. Evidences of volcanic or eruptive action are found on nearly all of it.

The principal streams of this section of the country are the Pecos and the Canadian. The former drains an immense area. Taking its head in the numerous small lakes on the east slope of the Santa Fé range, it gathers these waters together with those of the Las Vegas range, the eastern slope of this latter range being drained by the Gallinas, which empties into Pecos beyond the mountains.

The foot-hills of the Santa Fé range run down just southeast of Santa Fé nearly to the Galisteo. The Arroyo La Java, the southern fork of the Galisteo, heads opposite the head of Cañon Blanco; the two constitute the terminal drainage-lines of the range west of the Pecos. The Conchas, a fork of the Canadian, rises entirely without the range in the plateau and, cutting a cañon down through it, debouches on the plain at the Rincon de las Conchas; thence flowing nearly due east it empties into the Canadian about six miles above Stone Ranch, known to the Mexicans as the plaza of Don Francisco Lopez.

The subdrainage-basin of the Cañon Blanco and Galisteo is very limited. To the south the greater portion of the water south of Cañon Blanco sinks in the plain, being cut off by the run running south of the Arroyo La Java and the above-named cañon. To the south of Galisteo Creek rises a new formation or succession of uplifts, not a continuous range, but rather spasmodic uplifts, which usually abound in mineral veins. The Placer, San Ysidro, San Pedro, Sandia, and Manzana Mountains rise from the Placers, each distinct in itself and together forming a group. The Sandia, running for 10 or 12 miles with the dip of the rock to the east, the western face being nearly vertical, the eastern slope smooth, have at their southern end the Cañon Tijeras. East of the Sandia, and separated from them by a broad, beautiful valley, lie the Placer Mountains on the north, then the Tuerto or New Placers south of them, then the San Ysidro, and on the southwest of these the San Pedro, a low line of wooded mountains. All of these have gold, copper, and iron in veins and deposits, and traces of silver and lead. Coal is also found in the vicinity. The placer-diggings are very fine. I saw several nuggets found in the Tuerto or new placer-diggings, weighing 2 or 3 ounces of very fine quality gold. The little mining now done there is by a very rude rocker or *batea*. No steps have been taken to form a reservoir of water; indeed, everything shows but the individual efforts of men who cared but to make a livelihood or had not the capital to properly work the placers. There was no work being done on the veins or deposits, and never had been much more than prospect-shafts. It is said, however, that prior to the Mexican war, at the little town of New Placer, there were some 4,000 souls living on the products of these placers; frightened away by the approach of the American Army, they never returned. One Mexican, the principal man of the town, stated, and his story was corroborated by the other old men of the place, that he once found a lump of gold which weighed 11 pounds 9 ounces avoirdupois. I give the story for what it is worth. He certainly had done so, or else had told the story so often that he, as well as everybody else in the town, believed it. The Ramirez lead or deposit—it is hard to tell which in its present unworked state—is a wonder in size. Prospect-shafts show it to be well defined by a wall-rock on at least one side. It runs with very slight dip to the east nearly 30 feet thick at the points of exposure. The country-rock is porphyrite. The ore is easily quarried; mainly of copper-carbonates, it carries quite a proportion of gold.

The San Ysidro Mountains should have mineral wealth from their formation, although no leads have been as yet worked. In every respect the section was the most promising we had seen during the summer. With the San Antonita, San Pedro, Cañon del Agua, and Agua Limpia Valleys running through it, furnishing the best grass we had during the season, with water enough for large sheep-herds, plenty of timber on the mountains, which will be of great value when the railroad is built, which will be forced to pass through this region or within 10 or 15 miles of it, as it is the only timber near fit for ties, the section is a valuable one, and it seems unfortunate that it is covered by land-grants, the Cañon del Agua and San Pedro, whose owners have so far left its wealth undeveloped.

In the San Pedro Valley and to the south of here, in the Manzana or Monte Largo range, we found fields which did not depend on irrigation, which is quite unusual. To the east of the Tuerto, San Ysidro, and Manzana Mountains extends the great dry basin, finely grassed for 8 or 10 miles from the mountains, but dependent upon the water therein to be utilized for grazing. At the little town of La Madera, at the northeast base of Sandia Mountains, was a saw-mill, at which they were making lumber of the timber from along the base and sides of these mountains. Four or five large sheep-herds were seen grazing in the San Pedro Valley. Nearly all the hay used in the towns Bernalillo, Algodones, Sandia, and Santo Domingo, and much of that for the Government corrals at Santa Fé, is drawn from here

The most prominent point of the Manzanas is Mosca Peak, rising from the range with two sharp points, the southeastern one of which was used as triangulation-station. It will prove of great value in carrying the triangulation from Mount Taylor, Santa Fé range, and Las Vegas range to the south and east. The mountains are finely grassed, and game is very abundant. The range north of Mosca, to the Tijeras and Gutierres

Cañon has little strong character, being quite low and heavily timbered, broken up by branches of these cañons, as well as by Hell Cañon and Cedar Cañon, with their branches. Traces of copper are found throughout the range. Chilili is the only town of the range south of Tijeras within the area to be surveyed by the party, Tesuque lying just south of it. The equinoctial storm, which struck us while in this section on the 21st September, commenced with a cold driving rain, which turned into a snow-storm. East of these mountains extends the plain of Guadalupe, or, as it is sometimes called, the plains of Galisteo. It is a dry expanse of country, with no running streams, and but three springs, these near the mountains. Antelope Spring, the largest, is about 15 miles east of Mosca Creek. Here, as also at Buffalo Spring, 8½ miles to the north, there is a ranch. The other spring is but a black hole, from which the water does not run. It is known as Stinking Spring, and is situated about 10 miles north of east of Buffalo. The numerous bones of mules and horses bleaching on the plain around the spring show how bad the water is. The plain is unbroken from the mountains to the Pecos divide, excepting by the Cerros de Pedernal, the Cerrito del Lobo, and the Cerrito del Cuervo. Beyond these, indeed commencing with the Cerros de Pedernal, which are sharp, bare points of porphyritic rock, the divide runs to the north-east, connecting with the low hills south of Cañon Blanco. From Pedernal the Cañon Piedra Pintada runs eastward to the Pecos River, with a very uniform grade of 37 feet to the mile. It would be much preferable to the Cañon Blanco route, as a connection with a road running up the Canadian as far as Fort Bascom, thence up Pajarito Creek, crossing over the heads of Arroyo Cuervo and down the Tanques to an easily-bridged crossing, thence up Piedra Pintada, thence over to Abo Pass, thence to the Rio Grande. There is, to be sure, but little water in the cañon, but it could easily be obtained by wells. Nor is there more on Cañon Blanco. This for a connection between the proposed road of the thirty-fifth parallel and the Southern Pacific would be a very short and easily-graded route. The grade in Cañon Blanco is much heavier and not so uniform, the outcome at the head being quite steep. Between the two cañons the country is gently rolling. The divide west of the Pecos waters is covered with small timber, mainly piñon. At the Lagunas Coloradas, at the head of Cañon Blanco, water can nearly always be found, so also at the water-hole on the south side of the principal point of the Cerro Pedernal, just east of the Anton Chico and Fort Stanton road, about 100 yards and slightly above it. In Cañon Piedra Pintada in the wet season water is always found.

Near where the road from Pedernal crosses the Cañon Blanco, at the head of an arroyo running into it, is Aguaje Guajolotes, which almost always has water in it. Near the mouth of Cañon Piedra Pintada is the Agua Negra spring of permanent water. In the cañon the grazing is very fine; 15,000 sheep were seen between the Pecos and the divide. From there to within about 10 miles of the mountains the grass is very poor. Just below the mouth of the Piedra Pintada Cañon, or, as it is here called, the Cañon del Agua Negra Grande, is the town of Puerto de Luna, the most southern of the towns visited on the Pecos River. With its clean white houses nestling in the cañon of the Pecos, it is strikingly pretty. Just north of here, at a little opening in the cañon, down which flowed the Agua Negra Chiquita, are the scattered houses of the town of Santa Rosa. Above this the river passes through a cañon impassable to animals, opening at the deserted plaza of Esteros. Still farther north, where it opens again, are the towns of Colonias and Plaza Abaja, and at the mouth of Gallinas Creek is the town of La Junta. Farther up are La Cuera and Esteritos; immediately above the mouth of Cañon Blanco is Montosa, and 2 miles from there Anton Chico, the most important and largest town south of Las Vegas and east of Santa Fé. This, as Galisteo, is a radiating point for roads. From the north there is the main road from Las Vegas, which is the heavy freight-road. West of the Pecos is the San Miguel road, which crosses the stream at La Cuesta. Forking from this road is the Albuquerque road, which runs up over the mesa to the west, coming into Cañon Blanco near the lakes and town of Los Griegos, a small, scattered town on the north of the cañon. From this road, about half-way between Anton Chico and Los Griegos, branches the route via Pedernal to Fort Stanton, known as the dry route. Running to the southwest of the stream is the road to Puerto de Luna. East of the stream is the road to La Junta; another to Fort Sumner, branching from which, near Whittemore's ranch, is the route to Fort Bascom. Still farther north is the road over to the towns of Los Torres, Chaparrito, and Rincon de las Conchas. From Galiesto run the roads to Santa Fé to the north, and Fort Stanton to the south, there being two to the latter place, one by way of Buffalo Spring, the other by Stinking Spring; the former, branching before reaching Buffalo Spring, runs through the Cañon Gutierres to Albuquerque and the Rio Abajo country, at Buffalo Spring again branching, the western fork running in the foot-hills past Chilili to Fort Stanton, the other going to Antelope Spring or Ojo Berenda, from there to Stanton. The routes to the east and west from Galisteo are the road up the Arroyo La Java past the Lagunas Coloradas and Los Griegos to Anton Chico, the route up the Arroyo Cristobal over the mesa to San Miguel, thence to San José, 3 miles above. Following a branch-drain of the Cristobal past Los Fuertes and Aguage Abrego is the route to Koslowskeys. Farther west is the route to Rock Corral and that to Santa Fé. To the west run the roads to Real de Dolores and Real San Francisco, and the route down the Galisteo to the Vaca, thence to Santo Domingo. The only roads not already mentioned between the Pecos and Rio Grande of importance within the section to be surveyed by the party and south of Santa Fé are the routes from Santa Fé, via San Marcos Spring, Old and New Placers, San Pedro, San Antonio, San Antonito, and Tijeras to Albuquerque; or, turning off at the mouth of the cañon to Peralta, the route through Cedar Cañon from Tijeras to Chilili, that through Hell Cannon, from Peralta and Isleta to the same place, that from Tijeras through Gutierres Cañon, past the Lagunas Coloradas, joining with the Galisteo and Anton Chico road to the latter place, and the route from Antelope Spring past Pedernal down Cañon Piedra Pintada to the Pecos. For distances on these routes see tables of distances to be published.

East of the Pecos, between the two large plateaux, at one time connected, lies the broad rolling plain broken by occasional mesas extending down the Canadian drainage. Along the Conchas are the Corazon, a high wooded dome, the Bareadero, a flat rectangular mesa, and other minor mesas. To the south of Corazon is El Cabre Mesa; southeast from there, Mesa del Pino; still farther south, Barejon, Churisco, Cuerro, and Cuerrito; these latter are at the head of Arroyo Cuerro, which rises in Churisco and Cuerro Springs and other small springs; running to the northeast, it enters a cañon, the mesas of which are much broken by other minor drains; sallying from here, it empties into the Conchas, rarely running water—a supply can always be had in the water-holes. Beyond the mesas of the Cuerro is the Mesa Riceo, the highest in the section; this divides the drainage of the Conchas from that of the Pajarito, an arroyo running from the south to the east of Mesa Riceo, and emptying into the Canadian near Fort Bascom. Of these isolated mesas, the Rica is the largest; rolling on top, it is 10 or 12 miles long. Three points of it were occupied for secondary triangulation, as were also Mesa del Pino, Mesa Cuerro, Cuerrito, and Corazon. The Cuerro just at the head of Cuerro Arroyo is the sharpest, best-defined point. South of this opening extends the great wall of the Llano Estacado, across the northern portion of which we worked for some days getting the drainage-lines of the Arroyo Juan de Dios and Alamo Gordo.

The main routes of communication in this section east of the Pecos are, first, the road from Los Vegas via Taylor's Ranch and Gallinas Spring to Fort Sumner; the route from Anton Chico past Gallinas Spring to Stone Ranch and Fort Bascom; the Buffalo road, nearly parallel to the northern edge of the Llano Estacado, running from La Junta; the route from Fort Union to Bascom, across the head of the Cañon Largo; and the route from Vegas down the Gallinas Creek. On the road to Fort Sumner, about 8 miles from the crossing of Gallinas Creek, is the Laguna del Alto de Los Esteros, at which there is usually water. At the Esteros, given on maps as Hurrah Creek, from the thirty-fifth parallel surveys, there is water in holes, as also at Los Tanos. Throughout this whole region the grass is wonderfully good, and numerous herds of sheep were seen, although not 1 per cent. of what the country would support. There is very little large timber in the whole of this region—almost none. The party reached Stone Ranch November 10. From there they proceeded up the Canadian for two days to the northern limit of the atlas-sheet, or rather a little beyond, to the mouth of the heavy cañon of the Canadian. There is a cañon of about 300 feet deep extending from the mouth of the main cañon. Beyond this is the main cañon, 800 to 1,000 feet deep. From the mouth of this the mesa wall runs westward to Mesa Chupainis and eastward to Ute Creek. Just north of here, indeed very near the mouth of the cañon, the country is covered with basalt. Below the mouth of the cañon, on the first slope or plateau, is another basaltic formation. About 2½ miles north of the mouth of the cañon comes in the Mora, a stream supplying nearly all the water of the Canadian. Also coming from the west, between the Mora and the face of the mesa, is the Cañon Largo, a narrow gorge, impassable in places to men on horseback.

Rising the mesa at the mouth of the cañon, we pushed north of east for 58 miles to Tequesquite, on a small drain, branch of Ute Creek from the west. Here we found encamped the topographer, who had not been with the main party since August 17.

The following day we started on our homeward march, running but a traverse line. Following up the Ute Creek 22 miles above Tequesquite, we came to where it emerges from the cañon. Entering the same, we still continued up it, till, forced by the undergrowth, we ascended with difficulty to the plateau above. Many sharp volcanic cones dotted the plain, which, gently rolling, extended far to the east. We pushed on to the head of the cañon, which is 20 miles long. The Cerritos del Aire, in which heads the Ute Creek, were in sight. They are a cluster of volcanic cones, with but little running water in the drains. Just to the east of these runs the old Kansas City freight-road, now unused. A short distance beyond them we reached the head of the Dry Cimarron, striking the Fort Union and Las Animas road, which we followed to Las Animas, reaching there November 24. All the country traversed between the Pecos and the head of the Dry Cimarron was very fine grazing-land. Although still considered dangerous on account of roaming bands of Indians, we saw none. From Dry Cimarron to Las Animas the grass was very poor, probably from the herds of cattle driven from Texas on this route.

At the head of Ute Creek, as well as on the Dry Cimarron, we found ranches. In and about the head of the latter stream is a very pretty section of grazing-land and some little arable land.

We had been in the field 166 days; during that time the party traveled 4,627 miles, surveying an area of 11,300 square miles; in this, 24 sextant latitude-stations, 94 mountain and mesa points for triangulation and topography, 160 three-point stations for the location of important points were occupied, 2,968 miles traverse line were run with stations or meander, 147 cistern-barometer, 1,767 aneroid barometer stations were occupied. The highest altitude reached was 13,393 feet, on Meigs Peak, Colorado; the lowest point was Las Animas, Colorado, a little less than 4,000 feet.

Too much credit cannot be given to Mr. Anton Karl and Mr. Niblack for the energy and ability with which both carried out the duties assigned them.

I desire to tender my thanks to the officers at Santa Fé and Fort Wingate, N. Mex., particularly Capt. Charles P. Egan, C. S., U. S. A, and Lieut. L. II. Walker, Fifteenth Infantry, for their prompt official kindness.

I am, sir, very respectfully, your obedient servant,

CHAS. C. MORRISON,
First Lieutenant Sixth Cavalry.

Lieut. GEO. M. WHEELER,
Corps of Engineers, in charge.

APPENDIX F.

EXECUTIVE REPORT OF LIEUTENANT C. W. WHIPPLE, ORDNANCE CORPS, ON THE OPERATIONS OF SPECIAL PARTY, CALIFORNIA SECTION, FIELD-SEASON OF 1875.

PHILADELPHIA, PA., *April* 25, 1875.

SIR : I have the honor to report the following outline of the operations of the party of which I was in charge during the field-season of 1875. It was composed as follows : Frank Carpenter, topographer; Frank M. Lee, meteorologist; Sergeant Eugene Farnham, Twelfth Infantry, odometer recorder; 2 packers, and 1 cook.

Leaving Los Angeles, Cola., on the 28th of June, several days were spent in the vicinity of Wilmington to enable Mr. Carpenter to connect by a line of levels the southwest end of the Bates line with the bench-mark of the San Pedro Breakwater. This being completed, the party moved past the little settlements of La Buyonna and Santa Monica, and at the foot of the range of the same names, along the beach to the Malaga Ranch, at which point an ascent and station were made on a somewhat prominent peak. The great difficulty so frequently encountered afterward, and which was so serious an obstacle to the progress of my work, met me here. Excepting a small, scant stubble-field of barley, there was literally not a particle of feed to be obtained for my animals, which were already suffering severely from unaccustomed work. I wished to stay at this ranch a couple of days, but I could not buy the permission for more than a single night. The following one, after a most trying march through soft, shifting sand, and over deep, dry gulches, we were even worse off, for we had, in spite of representations to the contrary, neither water nor grass. Finally leaving this shore, we succeeded in forcing our way through the chaparral, and camped on the summit, in the vicinity of which we made a topographical station.

Throughout my stay in these mountains, work was very much retarded by the almost constant existence of heavy fogs. The ocean was scarcely ever visible, but far away, clear and distinct, we could see the peaks of the various islands rising above the clouds. So intense was this fog that even the ordinary meander-work was accomplished with difficulty, and sextant observations were impossible.

Descending into the valley of the Simi, we occupied the extreme western point of the Santa Monica ranges and connected with several Coast Survey stations. Following the dry wash of the Simi, (called in this vicinity " Lost Palos,") we crossed the divide of the Susana range; made a station on its most prominent peak, and, keeping along the Conejo and Triunfo Creeks, finally crossed the San Fernando Plains, and camped, July 14, with your party at the Mission. A week well occupied in meandering the Little and Big Tujunga Creeks, in visiting the Charlotte mines, situated at the head of the latter, and collecting topographical details of this vicinity. Crossing the Arroyo Seco, which your own topographers had already meandered, we kept along close under the San Gabriel range, with the intention of crossing it by what is known as the Wilson trail. This plan was changed in consequence of meeting Mr. Cowles, of Dr. Kamfp's party, who informed us that they had just returned from a triangulation station by that trail, which only ran to the top of the mountains, that the country beyond was impassable, and that Mr. Joy, of their party, was lost, and might possibly be found by going up the San Gabriel Cañon. With the intention of making such an attempt, I moved on with all haste, and had the fortune to meet Mr. Joy at the entrance to the cañon. The information I succeeded in obtaining in regard to the possibility of finding grass was so exceedingly discouraging that I decided not to risk my animals by taking the San Gabriel trail, the details of this vicinity having already been collected.

Crossing by the Cajon Pass, we followed down the Swarthows Cañon on to the Mojave Desert, and camped near the mouth of Rock Creek. This stream meandered, we struck for Soledad Pass, and, occupying stations on the south side, marched straight down the Santa Clara Valley to San Buenaventura, on the sea-shore.

The Santa Clara rises in some springs and small mountain creeks near the mouth of the Soledad Pass, and, a few years ago, it is said, was a continuous running stream. Now it is dry for much of its length. At Riley's Station, a few miles above the stage-crossing, the present occupant told me that in 1869 there were large lakes in this valley, both above and below him, and an abundance of water in the river. He then ran profitably placer-diggings in the range to the north. That year the water began to fall at the rate of an inch and a half per day, and he commenced to sink his well, which is now 80 feet deep, with a very scant supply of water. I was shown a great variety of ore specimens, including graphites, blue and green carbonates, magnetic iron; but no mines were then worked nearer than the Charlotte at the head of the Big Tujunga Cañon. With the exception of the Camules Ranch, which consumes the waters of the Penn Creek in irrigation, but little cultivation is attempted in the valley above Santa Paula, which is some 16 miles from the sea.

We followed the San Buenaventura River to its junction with the San Antonio, and, passing through the pretty little village of Nordhoff, crossed the Topa Topa Mountains by a very rough trail, and reached the Cespe Creek. The scarcity of grass and the critical condition of my animals hurried my departure from a locality where the single man brave enough to strive there for an existence did all in his power to assist me; and I crossed the White Sandstone range, which incloses the north side of the cañon, by a trail which he indicated, making a station on one of its highest points from a camp on the summit, and, following Ray's Creek, moved round the base of Cuddy's Mountain, and joined you at Tejon August 15. Here my party was increased by the addition of Mr. Douglass Joy, assistant geologist. After your departure, and upon deliberations with Dr. Kamfp, chief of triangulation, I made arrangements for the erection of a monument at Mount Pinis, and returned to the White Sandstone Mountains to complete from a point in this range, adjacent to the one already occupied by me, the triangle between that station, Mount Pinos, and Tehachipi Mountain.

As an example of the difficulty attending an instrumental work during the season, t may be said that after one day had been spent in cutting down trees, another followed, during which the fog and mist did not permit us to catch a glimpse of Tehachipi, and barely of Mount Pinos, though scarcely 10 miles away. Sleeping on the mountains, Mr. Carpenter and myself were fortunate enough to see both of our points, which soon after sunrise, however, were covered by the mist which rose from this valley, not again to be seen for the day.

With the intention of collecting the topography of the range north of the Santa Clara, and completing unfinished work, we then crossed Motor's Flat, where in a number of cavities in an immense rock we found some very odd Indian hieroglyphics and paintings, and following down the Hot Spring trail camped on the Cespe in the vicinity of the springs themselves. These were exceedingly interesting, and it is a source of regret that the bottles of water collected there were broken during their shipment to Washington. The flow of water comes from the base of the mountains, forms a very considerable stream, possesses a temperature of 195° at its point of emergence, and probably contains sulphur and iron.

Thanks to a multiplicity of stock trails, I was misled into attempting the cañon, and had the satisfaction of spending six days there, during which I crippled some of my animals, smashed one wheel of my odometer vehicle, and met with other mishaps. I succeeded in getting within about 5 or 6 miles of the mouth of the cañon, far enough indeed to see from the mountain-side into the open valley of the Santa Clara, only to find that farther progress was absolutely impossible and that I must retrace my steps.

Over immense bowlders, through pools of water deep enough to carry the mules off their feet, through a country where in the most accessible spots the soil had been stripped of the meager supply of herbage it perhaps once possessed, we struggled on and finally crossed the Topa Topa by the same trail we had used before, and, leaving Nordhoff on our right, camped at the Ojai. Our road from here to the Santa Paula took us around the base of the Sulphur Mountains, out of which run the Tar Springs, which frequently flow across the road, and in the heat of the day are soft enough to make travel difficult.

Some days were spent in Santa Paula, meandering the creek and occupying a very difficult point at its headwaters. The Peru then having been meandered for some distance from its mouth, I sent Mr. Carpenter and Sergeant Farnham to San Fernando, to collect some topographical details in the vicinity of the Pacoima Creek, and awaited their return on the Santa Clara.

Moving up the Castac Creek I then followed an excellent trail across the mountains to a point on the Peru near where I had terminated my meander before, and keeping to its headwater, passed the Los Alemos and Gorman's ranches, and reached old Fort Tejon, where I had left my supplies, on September 19. From here our march took us past Roee's Station, and along the foot-hills to the San Emidio, then round the edge of Kern Lake, through Bakerfield and across Sage-brush plains, which numbers of little pre-emption houses indicated a hope on the part of the inhabitants to reclaim, to a point on the sluice at the western extremity of this area allotted me.

I moved north from here under a broiling sun toward Posa Creek, and though this country was so barren that I begged from a humane ranchman at my first camp the corn-stalks which formed his shed-cover to prevent my animals from starving, he was sanguine enough to herd sheep there in large numbers.

When I arrived at his camp he was driving around an ill-conditioned horse, pumping up water from a well 90 feet deep, while his thousands of sheep were huddled together in bands under care of dogs, waiting their turns to drink; and when I left him in the morning he was similarly employed.

Following Posa Creek, in which water stands in occasional pools, we climbed the back-bone of the Green Horn Mountains, and, camping on the summit, from a station looked down into the valley of Kern River. Here, as everywhere, was apparent the over-stocked condition of the country; every vestige of grass seemed trodden out, even in spots almost inaccessible. By the most perfect road I have ever seen, we moved past the lumber-mill into the valley and up the river to Kernville, where a day was employed in visiting the Sumner mines, and where every courtesy was shown us by Mr. Burke, the superintendent.

We camped over night at the hot springs a few miles below, and then, following up Erskine's Creek, took a trail which leads us to the summit of the Pah-Ute Mountain. The mines and work here were carefully inspected, thanks to the extreme kindness of the proprietors, and a station made on the peak to the eastward. Passing through Clarasillo, we visited the John's mine and made a station on a beautiful point we called the "Crowned Butte," on the edge of this desert, and, keeping down Kelso Valley and along Cottonwood Creek, we crossed a divide and camped at the warm springs of Caliente Creek. Moving from here into Walker's basin, a station was made on Mount Breckenridge and an abandoned mine visited, and I then took my party into Caliente for supplies. From here our road lay along the line of the Southern Pacific Railroad in the Tehachipi Pass, and making a camp for a few days in the valley, we meandered Sand Cañon, Charlie Morris's Cañon, and Caché Creek, and made two stations in the range between Tejachipi and Kelso Valleys. The difficulties in executing nice instru-mental work had been increased. Besides the mist and fogs, the wind blew constantly almost a gale, filling the air with dust and sand.

Crossing Little Oak Creek, we then passed by Desert Springs and Elizabeth Lake, and camped at Lopez Ranch, on the edge of the Desert, long enough to permit us to make an excellent station in the range north of Polvadera Pass and near its mouth. From Elizabeth Lake, I sent Mr. Lee into Los Angeles to assist Lieutenant Bergland in his meteorological computations. We then passed the head of Castac Creek, and, having made two stations in this range north of the Santa Clara, we crossed the hills, and following down a valley, passed Cow Springs and Gorman's Ranch and over on to the San Emidio Creek.

Here, November 2, for the first time during the season's work, we were detained by a very severe storm, which covered the mountain-side with snow. Moving due west, we made a station near the Paleta Ranch, and, descending into the valley of the Cuyana, followed it for miles through as barren a country as I had yet seen, and by very hard marches crossed to Lockwood Creek, and reaching the San Emidio again, moved on to the Plato. Here we visited the antimony mines, recently established, and of which great hopes are entertained.

For some miles the bed of the Plato ran in an exceedingly pretty valley, where we startled deer and quail from cover at every bend; but it gradually narrowed, until it became an absolutely impassable cañon, and we were forced to climb for hours with jaded, worn-out animals over immense hills. Scattering with pistol-shots bands of mountain-sheep, we climbed to the highest point, made a station, and descended over rolling foot-hills to the valley.

Striking from here across the plains, we made a direct march toward Caliente, and, climbing the hills by the old Bakersfield Road, reached that place and joined Lieuten-ant Birnie on November 13. After Mr. Klett's arrival with the other party, a week in Caliente sufficed to disband this expedition. I then conducted the trains to Los Angeles and stored the material.

It is a source of consolation to me that, before leaving this part of Southern Cali-fornia, I had the satisfaction of seeing grass springing up in the valleys and along the hills. Ocular demonstration was necessary to convince me of the fact that such could be the case, for in the six months my party had been in the field my animals had but once found feed (this on the summit of a high mountain) that I did not pay for. That this country was formerly noted for its grazing is undeniable, and I took pains to gather from every source of information some explanation of this change. Mr. Cuddy, pointing to the hills opposite his house, near Fort Tejon, said that the bunch-grass which once covered them no longer grew there at all.

Another source of disquietude to those interested in the welfare of this section of our country, is the diminution in the supply of water. It is true that in many localities where the attempt has been made, notably on the plains of Los Angeles and San Bernardino, artesian wells have been successfully sunk and furnish a plentiful supply of water. Even at Bakersfield, at a depth of 260 feet, water has been struck. But the grazing sections can hardly depend upon these for their wants, and everywhere I found evidences that the creeks and springs were failing. The Santa Clara River and Posa Creek, and the springs which feed them, may be cited as illustrations; their gradual, steady subsidence indicating something more serious than a failure in a winter's snows.

As regards the springs, I was told that the sheep "trampled them out;" but acquitting them of this responsibility, they are undoubtedly inflicting an injury on this country of a very serious nature.

Very respectfully, your obedient servant,

C. W. WHIPPLE,
Lieutenant of Ordnance, in Charge of Party.

Lieut. GEO. M. WHEELER,
Corps of Engineers, in Charge.

APPENDIX G 1.

METEOROLOGY AND HYPSOMETRY.—FIELD-SEASON OF 1875.—BY LIEUTENANT W. L. MARSHALL, CORPS OF ENGINEERS.

UNITED STATES ENGINEER OFFICE,
GEOGRAPHICAL SURVEYS WEST OF THE 100TH MERIDIAN,
Washington, D. C., July 8, 1876.

SIR: I have the honor to submit the following summary of the barometric work of the past field season, and of the office-work in connection therewith. The observations and work performed have been of the same character as in 1874, and reference is made to the report upon this subject in the annual report of that year for information other than is contained herein.

The various parties of the California and Colorado sections of the expedition of 1875 were provided each with duplicate sets of instruments, comprising cistern and aneroid barometers, psychrometers, and pocket thermometers, and with the necessary books of record and printed instructions with reference to observations, instruments, and books. The barometers, psychrometers, and thermometers were made by James Green, of New York; the aneroids with attached thermometers by Casella, of London.

Comparisons of all instruments were made at Washington, D. C., with the standards of the United States Signal Service, and again at Pueblo, Colo., and Los Angeles, Cal., before final distribution among the various parties, and their initial errors of indices determined. Daily comparisons among the instruments of each party served as checks upon their indications and afforded the means of detecting changes in instrumental errors, and the errors of such instruments as were fitted with new tubes during the season. The corrections for capillarity in barometers were included in the errors determined, having first been corrected for approximately, by shifting the scale to agree with the computed effects.

The observations taken during the field season were mainly for psychrometrical purposes, the time spent in the field being too short, and the observations at any one place too few to be of much service in general meteorological investigations in a region where general climatological features are as well known as may be given by cursory observations.

For the California section two reference stations were established by the survey, one at Los Angeles, Cal., and another at Fort Mohave, Ariz. The altitude of the cistern of the barometer at Los Angeles was determined by a line of levels run from the United States Coast Survey tide-gauge at the Wilmington breakwater, by Mr. Carpenter, of this survey; that of Fort Mohave, Ariz., was carefully computed by Lieutenant Bergland, by referring barometrical observations there by daily and monthly means to corresponding observations at Los Angeles, and this determination was afterward checked and verified by referring to the Signal Service barometers at San Francisco and San Diego, Cal., and Santa Fé, N. Mex. Camps, where parties remained several days, were taken as reference stations for all observations made in their vicinities; the camps themselves being referred to the stations above mentioned.

151

In Colorado and New Mexico the cisterns of the United States Signal Service barometers at Santa Fé, N. Mex., and Colorado Springs, C. T., were taken as distance-points, and the temporary camps of the parties, as in California, taken as reference stations for all observations made in their vicinity during occupation.

Horary tables were secured for Los Angeles, Cal., and Fort Mohave, Ariz. The horary tables published by Colonel Williamson in his paper on the "use of the barometer, &c.," for stations in California, and that for Camp Independence, Cal., secured by your expedition of 1871, were also used as far as they were deemed applicable. In Colorado and New Mexico no additional horary corrections were secured, but those determined in 1873 and 1874 being applicable, were used.

The observations taken in the field were copied by the observers on forms adapted to their complete preparation for final reduction, avoiding errors in copying and transcribing again and again upon special computation sheets. When filled, these forms were forwarded, when practicable, to the office in Washington, D. C., for revision and reduction, the certificate of the chief of party that they contain true copies of the original observations being required.

In following out the system of observation and record required by the printed instructions published by the survey, a great number of observations at various points were secured and have been reduced.

Upon the receipt of the observations from the field, the records of the various instruments were carefully examined, and their relative indices of error determined at every comparison, and therefrom their absolute errors upon the standard of the United States Signal Service in Washington.

The hourly observations were then corrected for instrumental errors, reduced to 32° F. and to level, and after correcting erroneous or erratic observations revealed by plotting. Horary tables were formed for the reduction of isolated cistern barometer observations and for use in reducing the aneroid profiles. In forming the tables of horary corrections I have had the observations reduced to second level plotted, and have selected those days only when well-marked diurnal oscillations are exhibited.

The relative humidities were not computed for each observation separately, but the means of the temperatures and wet-bulb indications were taken at each hour for the entire series, and the resulting computed humidities for the differences of these means taken for use in the barometric formula. Single results are liable to be so erratic and unreliable, that it is supposed that this course would give a sufficiently near approximation with less labor of computation for use in the barometric formula.

In computing the altitudes of camps, &c., from cistern barometer observations, the full formula of Plantamour containing a term for humidity, and as represented by the tables in Williamson, has been used. In the reduction of aneroid observations the smaller terms of this formula have been omitted, since the aneroid differences of altitude are generally small, and from the construction of the instrument some of these terms necessarily do not apply.

The observations have been referred by divers means to corresponding observations at the base station. When a long series of observations have been taken, the observations have been referred by a mean of the daily means, or of the observations at 7 a. m., 2, and 9 p. m.

Isolated observations have been corrected for horary oscillations when suitable tables were known, and referred to the corresponding daily mean at the reference station; otherwise simultaneous barometric observations have been obtained by interpolation and an approximate mean daily temperature used.

Observations taken upon peaks have been referred, (save in a few instances where they have been compared direct with the initial reference stations) to temporary stations at their bases, the approximate daily mean temperature being used in the temperature term, this temperature being observed at the base station, and found approximately as follows for the mountain station: The camps of the ascending party (as a rule) were established at the limit of tree growth, where observations of the thermometer were taken at 7 a. m. and 9 p. m. on the day of the ascent before and after the occupation of the station. These observed temperatures have been reduced 3° F. for every 1,000 feet difference in elevation between this camp and summit, and taken in connection with the 2 p. m. observation on the peak for an approximate daily mean. Where a peak is high above the reference station, and for lack of fuel, cold will prevent observation at night upon its summit, some such device must be inaugurated for securing approximate daily mean temperatures. The temperature will necessarily always be observed near the hottest part of the day, and will give results the more erroneous the higher the peak is above the reference station. The changes in temperature must take place principally near the heating body, i. e., the earthy surface; and especially in the excessively dry atmosphere of the West, where a conservator of heat in the shape of aqueous vapor is wanting, the air directly in contact with the earth undergoes great fluctuations in temperature, during the day giving ranges of climate temperature not at all commensurate in effects with the less horary oscillations of the barometer, whereas, if there be quite a thick and wide stratum of air between the summit of the peak and the reference station not in contact with the earth, the actual mean temperature of the entire stratum varies but slightly, if at all, during the 24 hours, unless disturbed by abnormal warm or cold currents ; so that while the observed temperatures near the surface would when plotted, during say a month or a year, show well-marked daily maximum and minimum, the mean temperature of the thick stratum would show nearly uniform changes of temperature from season to season, but slightly affected by diurnal oscillations. This would seem but a natural consequence of the fact that in a dry climate, where but little heat is made sensible by the condensation of aqueous vapor in the formation of clouds and rain, and still less is absorbed by the dry air, the entire diurnal oscillation in temperature is due to the actual heating of the particles of air by contact with the heated earth, which heat, taken by the air, may disappear by radiation or be expended in work in expanding the particle and overcoming the resistance to its upward motion, leaving but comparatively little sensible heat, carried by connection, to be distributed throughout the great body of this stratum. If this surmise be correct, not only is it necessary to obtain the daily mean temperatures for use in the barometric formula, but the means for a longer time, if practicable.

It would be interesting to find by experiment the limit of time for which the mean observed temperature, taken in connection with the observed height of the barometer corrected for horary oscillation, would give the best results.

Corrections for difference in *phase* of abnormal waves or for abnormal oscillations have not been directly attempted. As far as practicable, however, observations at points where altitudes are required, especially at semi-permanent stations, have been referred to two or more reference stations and the results weighed in accordance with the known general directions of atmospheric waves and the distance of the point whose altitude is sought from the reference stations estimated in the direction of this wave motion. This supposes that the position of the wave between the two reference stations is an inclined plane, which is true only within small limits compared with the area covered by an entire wave at any given instant, but this seems the only means at our disposal by which we may readily and approximately eliminate the effects of abnormal disturbances from the resulting altitudes, when, as must necessarily be the case in the thinly-settled west, observations must be referred to distant points in other phases of disturbance, and when the observations are not at points sufficiently numerous to determine the barometric gradients.

Since the middle of December the following work has been performed :

Cistern-barometer stations computed.. 930
Aneroid-barometer stations computed.. 5, 013
Hourly observations recomputed and plotted, days...... 270
Horary tables deduced.. 14
Cistern-barometer altitudes copied into permanent record-books, and indexed.. 2, 424
Aneroid-barometer altitudes copied into permanent record-books, and indexed.. 8, 696
Total number of altitudes computed from barometric readings since 1871...... 11, 125

All of the barometric work has been re-examined and such results as are considered most useful have been arranged for publication with Vol. II.

The computations have been made by Assistants F. M. Lee and George M. Dunn, and privates William Looram and John F. Kirkpatrick, Battalion of Engineers, except Lieutenant Bergland's observations, which were computed under his own supervision at Los Angeles, Cal.

Respectfully submitted.

W. L. MARSHALL,
First Lieutenant of Engineers.

Lieut. GEO. M. WHEELER,
Corps of Engineers, in charge.

APPENDIX G 2.

ON THE METEOROLOGICAL CONDITIONS OF THE MOHAVE DESERT.

UNITED STATES ENGINEER OFFICE,
GEOGRAPHICAL SURVEYS WEST OF THE 100TH MERIDIAN,
Washington, D. C., February 10, 1876.

DEAR SIR : I have the honor to submit herewith a report upon the meteorological conditions of the Mohave Desert.

During the past five years observations upon the daily temperatures were made at the two military posts of Fort Yuma and Fort Mohave, both in the Colorado Valley ; and to these military authorities thanks are due for information furnished. I hope that this report may furnish some interesting points, as the greater portion of the Mohave Desert was until a comparatively recent date a *terra incognita*.

This report treats of the winds, rain-fall, cloud-bursts, sand-storms, electrical phenomena, hot winds, temperature of air, river, and soil, relative and absolute humidity, and ozone. Numerous comparisons with other countries have been added, in order to elucidate more clearly the specific character of this region.

Very respectfully, your obedient servant,

OSCAR LOEW.

Lieut. GEO. M. WHEELER,
  Corps of Engineers, in charge.

There exist probably but few regions on earth where two totally different climates are separated by a single mountain range, as in Middle and Southern California, where the uniform sea-climate of the coast-strip, with comparatively small variations in temperature, forms a most remarkable contrast with the climate on the other side of the gigantic chains of mountains, the Sierra Nevada and its southern continuations, the San Bernardino and Jacinto ranges, which represent the backbone of the State of California.

The state of humidity, the direction of wind, the electrical condition, the daily oscillations in temperature, how far different are all these conditions on the west and on the east side of those mighty elevations which separate the Mohave Desert from the fertile coast!

In regard to the winds of the Mohave Desert, it is a striking fact that southeast winds are by far the most prevalent in summer time. It is also easily observed that the clouds and summer rains come from that direction. This observation was confirmed, on our arrival at Fort Mohave, by the hospital-steward of that post, who attended to the meteorological observations for the past three years. The great prevalence of these winds in Northern Texas, Southern New Mexico, and Arizona, was recalled to my memory; these same winds producing there the summer rains.

That this moisture-carrying current comes from the Gulf of Mexico, and is the continuation of the great equatorial current which, like the equatorial marine current, is a consequence of the earth's rotation, and flows from east to west in a width from 28° south latitude to 28° north latitude, cannot be subject to doubt.

By the development of the high summer temperature of those countries, the air-current deviates, upon its arrival in the Mexican Gulf, from its original westerly direction to the northwest, and appears therefore as a southeast wind. That this stream of air must contain a great deal of moisture, may be inferred from its passage across the warm and extensive oceanic area between Africa and Central America. Although deprived of the greater part of moisture after its flowing across Texas, New Mexico, and Arizona, it still carries enough moisture as far west as the Mohave Desert, to produce occasional summer rains there; but at the backbone of California this influence is completely broken; the wind, now deprived of its moisture, is forced into a different direction, and no summer rains fall on the California coast. Here, on the contrary, prevails the northwest wind, which saturated at a comparatively low temperature with moisture, and passing over the much warmer continent, becomes drier; its relative humidity sinks rapidly; hence no rain can fall. However, when in the winter months the power of insolation is weakened, the land cools off, then the aqueous precipitates become extensive and the coast country covers itself with verdure.

In winter time the northwest winds also prevail over the entire Mohave Desert, but having lost most of their moisture on the passage across the high ranges, bring but rarely rain or snow in the desert. The absolute humidity, or rain-producing powers of the northwest wind, must naturally fall far below that of the southeast monsoons, on account of their respective temperatures.

While the summer rains in the Mohave Desert fall chiefly in great cloud-bursts that are always accompanied by electrical phenomena, neither is the case on the coast; thunder is at least there a very rare occurrence. The phenomenon of cloud-bursts is doubtless produced by the alternating position of hot, deep, and sandy valleys and steep, high ranges; the reflected heat of the former shifts the tendency of the clouds to a condensation to rain upon the extreme point, and these, approaching the somewhat cooler mountain-chain, must discharge their moisture at once, and sheets of water come down with such force that fragments of still uncondensed clouds are carried along. We witnessed this interesting phenomenon in more than one case; the clouds had the singular appearance of a lake-surface reflecting the light like liquid water.

The rains falling principally upon the mountains, rarely in valleys, it is not surprising that the rain-fall at Forts Mohave and Yuma is exceedingly small, both these posts being situated in the low and hot valley of the Colorado River. The following table, taken from the records of these posts with permission of the authorities, shows the rain-fall in four years:

Fort Yuma.		Fort Mokare.	
	Inches.		Inches.
1870	2.71	1870	3.07
1871	0.78	1871	2.01
1872	3.34	1872	3.02
1873	3.84	1873	3.04

If the rain-fall in the neighboring ranges had been measured, the results obtained would doubtless be three to four times as high. More than once did we see extensive rains in the mountains when not a drop fell in the adjacent valley, and were surprised on three occasions by heavy rains; once falling 1.06 inches of water in four hours.

Sand-storms, hot winds, and whirlwinds, the latter producing columns of sand 100 to 200 feet high, are not infrequent, while they are quite rare in the coast counties. The hot wind (simoon) is a most disagreeable feature; it continues sometimes until late at evening, with a temperature of 108° F. The sand-storms produce certain electrical phenomena; sparks were drawn by us one night from a blanket that was covered with a stratum of sand deposited by a sand-wind.

Analogous to other deserts, the climate of the Mohave Desert is in summer time generally very hot and of great dryness. In connection with the latter, stand the great daily variations of temperature.

In the annexed table are placed side by side some of our observations with those of Rohlfs made at Rhademes, an oasis in the Sahara Desert, and a comparison will show how closely both climates agree. The highest temperature Rohlfs observed at Rhademes was 43°.9 C. = 111° F., August 11, 1865. This was next to Shimmedru in the Oasis Kauar, the hottest place he struck in his travels through Africa; and at this locality he observed on a series of afternoons 53° C. = 127° F. At Mursuk, where he remained from November, 1865, to April, 1866, he observed as the highest temperature 36°.1 C. = 96°.9 F., (March 12, 1866.)

According to Humboldt, the temperature of the Llanos of Caracas seldom rises above 98°.6 F., although these plains are situated near the equator.

The highest temperature our party observed in the Mohave Desert was 116° F., on the 6th of August, at Stone's Ferry, in the valley of the Colorado River in Southern Nevada. However, the temperature reaches sometimes still higher limits. Mr. Jennings, who lived for nearly eight years at Saint Thomas, a small settlement on the Muddy, 24 miles from the Colorado, stated that in 1871 the thermometer rose in the shade every afternoon for three weeks in succession to 123° F., and on one occasion to 127° F. A temperature of 122° F. was also repeatedly observed at Fort Mohave. The temperature at night during the summer-months sinks frequently to 70° F., but does not fall often below 90° F., particularly when the air is in motion.

The hottest portions of the Mohave Desert are the Coahuila Valley, (altitude—100 feet,) the Lower Virgin River Valley, (altitude +1,300 feet,) Death Valley, (altitude −200 feet,) Panamint Valley, (altitude +1,400 feet,) the Saline Flats of the Mohave, (altitude +1,200 feet,) and the Colorado River Valley up to 1,200 feet altitude. While the four months June, July, August, and September are prominent by a very high temperature, the remaining eight months have a very moderate climate, and ice and snow have been repeatedly seen on the Saline Flats of the Mohave in January. Warm winters, however, exist on the Lower Colorado and in Coahuila Valley, in which latter wild palms form an ornament of the oasis.

As regards the mean temperature of the hottest months, it was calculated for July, 1873, in Fort Mohave to be 100°.9 F., but for this calculation three observations served, taken at 7 a. m., 2 p. m., and 9 p. m. It is evident that the 7 a. m. observation is not a correct element in this calculation, as the temperature shortly before sunrise is considerably lower than at 7 a. m. We arrive clearly much nearer to the truth if the lowest temperature in 24 hours be introduced into the calculations of the mean. In July, 1873, were observed as the monthly means the following figures:

7 A. M.	12 M.	9 P. M.
91°.4 F.	100°.2 F.	98°.7 F.

If we take as the mean of the lowest temperatures admissible in July = 75° F., and calculate then the total monthly mean, we obtain 93°.6 F., which still is a very high degree.

The mean temperature of the hottest month in—

	Degrees C.	Degrees F.
Cairo, Egypt	29.9	85.8
Madras, India	31.8	89.2
Abushar, Persian Gulf	34	93.2
Llanos of Caracas	31.5	88.7
Shimmedru, Sahara	35 to 36	95 to 96.8
Rhademes, Sahara	32.1	89.6
Ghadames, Sahara	32.4	90.3
Fort Mohave, Colorado River, Arizona	34.2	93.6

We see from this comparison that July at Fort Mohave counts among the hottest periods of the globe.

A table taken from the records of the post by the kind permission of the authorities may find here a suitable place. It shows the monthly mean of all the 7 a. m. observations, that of the 12 m. and of 9 p. m. in 1873.

	7 A. M.	12 M.	9 P. M.
	F.	F.	F.
January	47°.9	63°.8	56°.7
February	47.6	69.3	53.6
March	58.6	81.9	68.4
April	64.0	83.0	69.7
May	69.1	88.3	78.2
June	81.7	104.1	89.3
July	91.4	110.2	96.7
August	85.6	101.4	88.2
September	78.2	102.0	88.2
October	64.0	86.3	74.0
November	54.1	77.0	67.1
December	45.1	58.2	51.2

As in barren countries the development of heat by insolation on one hand is just as powerful as the loss of heat by radiation in absence of sunlight on the other, December is here observed to be the coolest month, and the sudden transitions from May to June and from September to October.

How insignificant, compared with these conditions, appear the changes of temperature between winter and summer on the coast at Santa Barbara! The following table showing this is taken from the Santa Barbara Press:

Month.	Monthly mean.	Month.	Monthly mean.
1872.	Fahrenheit.	1872.	Fahrenheit.
January	51°.00	July	68°.00
February	54.00	August	69.00
March	57.00	September	67.00
April	60.00	October	64.00
May	63.00	November	60.00
June	69.00	December	63.00

In Los Angeles, some 30 miles from the coast, the contrasts are already increased, July reaching a mean temperature of 75° F., January 52° F.

The temperature of the soil is generally above that of the air in sunshine, but this is more the case in barren stretches of the desert than anywhere else; indeed, temperatures of 150° F. of the soil can easily be observed when the air has but 112°. While, however, the surface loses much by radiation during the night, and has at sunrise that of the air, the earth in a foot depth retains still a high degree of heat. I found it one morning still 94° F., when at the same time the surface and the air had but 73° F.

In regard to the temperature of the Colorado River, it may be mentioned that from middle of July to middle of September it was never found below 78° F., and not above 82°.5. How small this variation compared with that of the air!

The temperature of the Rio Grande in Colorado and New Mexico was found in 1874 to be, within small variations, 63° F. from July to October. Humboldt found the average temperatures of Rio Apure and Orinoco during the hottest months to be 27° C., or 80°.6 F.

In regard to the relative humidity of the air, great variations may, a priori, be expected where the daily extremes of temperature are great. Thus, the relative humidity August 6 was, at sunrise, 0.526, (saturation = 1.0,) while at 3 p. m. = 0.093.

Rohlfs[*] observed as the mean relative humidity in August at Ghadames (Sahara) = 0.330, and in July = 0.275. The lowest relative humidity was found in November, 1865, at Murzuk, (Sahara,) at 0.07, (dry bulb = 82° F., wet bulb = 55° F.)

But not only the relative humidity is subject to a great range, but also the absolute. While this was observed in the Colorado Valley after a heavy shower to be increased to 15 grams per cubic meter, it amounted at dry weather only to 6, sometimes 3. Still lower figures, however, were obtained in the fall of 1874 in Northern New Mexico, (see Annual Report of Explorations and Surveys west of one hundredth Meridian for 1874.)

* His extensive meteorological reports are published in Peterman's Goegr., Mittthe-lungen, 1872.

As a general rule, we see, with the *decrease* of the absolute amount of moisture, an *increase* of the daily difference of extremes, as shown in the following table:

Locality.	Date.	Absolute humidity.	Difference of extremes.
			Fahrenheit.
Cottonwood Islands	July 12	7.8	44°.00
	July 27	12.9	15.00
Stone's Ferry	Aug. 6	6.0	39.00
	Aug. 12	15.1	18.00
Rhadames in the desert Sahara, (Rholfs)	Aug. 15	11.7	27.00
	Aug. 18	13.6	20.00

In some cases the differences of daily extremes stood nearly in inverted proportions to the *square* of absolute humidities; in other instances, especially when two cases with widely differing absolute humidities were compared, the differences of the daily extremes stood nearly in inverted proportion to the *simple* absolute humidities. A law expressible by a mathematical formula can only be found, however, after a long series of very careful observations. It is natural that only observations made at one and the same locality and in the same season can be compared with each other.

In regard to the ozone, none, or but very weak reactions, were obtained when the temperature rose above 106° F., which failure is either due to the volatilization of the iodine from the test-papers prepared with iodide-of-potassium starch, or to the real absence of ozone; however, during cool nights moderately strong reactions were obtained in midst of the desert.

Thermometer and psychrometer observations in the Mohave Desert.

Locality.	Date.	Sunrise. Dry bulb.	Sunrise. Wet bulb.	12 m. Dry bulb.	12 m. Wet bulb.	3 p.m. Dry bulb.	3 p.m. Wet bulb.	Sunset. Dry bulb.	Sunset. Wet bulb.	Remarks.
Cottonwood Island, Colorado River Valley, 40 miles above Fort Mohave.	July 21	69.0	64.4	106.5	72.2	104.0	72.1	85.5	72.0	A hot east wind blew in the night 23d–24th of July.
	July 22	63.0	56.5	106.0	73.4	107.0	73.0	91.2	70.1	A heavy thunder-storm occurred in the night 25th–26th of July.
	July 24	90.2	72.0	99.2	74.5	104.1	76.2	90.0	74.0	
	July 26	73.0	70.2			98.0	75.3			
El Dorado Cañon, Colorado River Valley.	July 27	81.0	68.5	102.3	69.0	96.0	74.5	93.5	70.2	Few clouds; moderate wind.
	July 31	78.2	61.5			106.0	78.5	105.0	68.3	A hot wind blowing until 9 p. m.
Stone's Ferry, on the Colorado River, Southern Nevada.	Aug. 1	81.2	58.0	112.0	75.3	114.0	74.1	107.0	68.8	Hot wind in the afternoon.
	Aug. 6	75.0	68.0	105.0	75.3	112.0	74.2	106.2	68.5	Moderate southeast wind; partly cloudy.
	Aug. 7	83.5	63.8	106.0	77.4	113.1	74.9	100.0	73.5	Clouds in southeast; hot wind in the night.
	Aug. 8	88.0	67.2			104.2	76.8	97.0	73.5	Rain on the 9th, and in the night from 10th to 11th of August.
	Aug. 10	80.6	70.0							
	Aug. 11	73.2	73.2	84.5	75.0	93.0	76.9	90.0	78.5	Cool breeze all day.
	Aug. 12	82.0	74.0	97.8	77.2	98.0	77.1	89.0	76.2	Cool southeast wind; clear sky; sand-storm in evening.
	Aug. 13	79.0	67.5	98.0	75.0	101.0	72.0			Sky clear.
Rhadames, in North Africa, (Sahara).	Aug. 2	73.1	56.1			102.0	72.0	84.0	64.9	Observations by G. Rohlfs in 1865.
	Aug. 3	75.9	57.7			100.0	82.0	89.9	70.7	Do.
	Aug. 4	79.0	60.9			106.2	75.0	91.9	70.8	Do.
	Aug. 5	77.0	59.0			108.0	77.7	93.0	70.8	Do.

REPORT ON THE GEOLOGY OF A PORTION OF SOUTHERN CALIFORNIA, BY PROF. JULES
MARCOU.

LOS ANGELES.—The hills which surround Los Angeles, forming a vast amphitheater
open only on the west, are composed principally of friable sandstone of a yellowish-
gray color, with some intercalated layers of sandy clay and limestone of the same color.
These rocks, which have all the appearance of real *molasse*, can be studied to the best
advantage in the trenches which have been dug on the side of the old *presidio*, or fort,
which overlooks the town on the north ; and also by following the trenches of the
aqueduct and the cañon of the Rio de los Angeles or Rio de Porciuncula, after leaving
the town, as far as the valley of San Fernando. The strata are all much elevated,
and dip in a south-southeasterly direction, at an angle varying from 30 to 35 degrees.
The thickness of the sandstone-beds varies from 1 to 5 feet, but beds are nowhere
found of sufficient hardness to furnish good building stone, on account of the friability
of the sandstone, which is rapidly decomposed by the action of the atmosphere.

Fossils are very rare, with the exception of the bones of *Cetacea*, which are in a bad
state of preservation, and it is difficult to obtain good specimens. On the hill of
the *presidio* fragments of the cast of a small gasteropod are sometimes found in the
sandy clay ; of this not even the *genus* can be determined.

On the road from Los Angeles to Cahunga Pass, descending the northern declivity
of the hills, before reaching the plain of the Balonas or Bayona, we find several scanty
petroleum-springs, and the limestone in the vicinity of the springs is strongly impregnated with asphaltum. The petroleum, or bitumen, when dried in the sun, becomes hard, and is then known by the name of " brea " to the Mexicans, who use it as
a covering for the roofs of their houses and as a pavement for sidewalks.

The hills of Los Angeles begin at the southeast of the lagoon, and of the Balona or
Bayona ranch, rising at the east-northeast toward the cañon of the Rio de los Angeles
or Rio de Porciuncula, which cross them ; they then rise in gentle acclivities against the
granite of the Sierra Madre, not far from the ranches of the San Gabriel Mission ;
finally, they again descend toward the south, and next turn once more to the east,
and form the small basin of El Monte. Their height is small, and varies from 50 to
300 feet above the bottom of the valley of Los Angeles.

PLIOCENE ROCKS OF LOS ANGELES.

The sandstone rocks of Los Angeles may be considered as a brackish deposit of an
estuary, or of lagoons, subsequent to the formation of the valley of San Fernando,
where they do not exist. The sierras of Santa Monica and San Fernando had already
undergone a movement which had obliged the sea to recede more to the south and to
the west. The formation of this sandstone is evidently quite recent, and, as it is placed
between the Miocene rocks of San Fernando and the Quaternary alluvial rocks of the
bottoms of the valleys, it must be regarded as being of the age of the Pliocene Tertiary.

Fish-bearing limestone rocks of Los Encinos.—Below the molassic sandstone rocks of
Los Angeles we find a group of limestone rocks whitish, chalky, and containing a
small quantity of magnesia, stratified in rather thin layers, the entire thickness of
which is from 100 to 150 feet. These white limestone rocks are found in the hills
which face San Gabriel and along the road between San Gabriel and the San Fernando
Valley.

In these limestone rocks, especially in the upper portions, we meet here and there
with remains of fishes, such as scales and vertebræ, generally in a bad state of preservation.

Denudations have carried away these limestone rocks in which fish are found in the
cañon of the Rio de los Angeles or Rio de Porciuncula, and likewise at the beginning
of the San Fernando Plains ; they reappear, however, in the San Fernando Valley,
where they form a large portion of the bottom of the valley, and a part of the first
" counterfort " of the mountains which surround it, beginning a mile and a half to the
west of Cahunga Pass, as one goes toward Los Encinos. At the ranch of Los Encinos
these limestone rocks attain their greatest thickness, which is nearly 200 feet ; and
they contain in their upper portions, close by the houses of Mr. Eugéne Garnier's
ranch, a large quantity of fossil-fishes, large bones, vertebræ, and ribs of *Cetacea*, fossil-
plants, *Crustacea*, and, it is said, even birds ; also a quite common bivalve mollusk—the
Pecten Peckanii Gabb.

These fish-bearing limestone rocks of Los Encinos belong to the Miocene Tertiary
rocks, of which they form the upper portion. As regards their age, they may be compared to those at Oeningen, on the banks of the Rhine, near Schaffhausen, Switzerland,
or to those of the quarries of Aix, in Provence, France. They are a deposit of brackish,
almost fresh, water, notwitstanding the presence of a small *Pecten* and of *Cetacea*. It
has not, as yet, been possible to collect the fishes in sufficiently large numbers, or to
examine them in such a manner as to be enabled to obtain correct ideas with regard
to the ichthyological fauna of Los Encinos.

At the ranch of Los Encinos the direction of the dip of the strata is north-northeast, at an average angle of 15°. These strata consist chiefly of light, grayish, pale-yellow, or chalky-white limestone rocks, in layers of a thickness varying from 2 to 3 feet; they are easily divided into slabs, with numerous dendritic impressions. The fossils are found chiefly in the upper portion of the group. These fish-bearing limestone rocks have evidently formed the entire bottom of the beautiful valley of San Fernando, as is shown by the hillocks or isolated mounds which have been left here and there in the plain by erosions, and on which remains of fossil-fishes and the *Pecten Peckanii* are found. I refer particularly to the mound by the side of the stage-route which leads to Santa Barbara.

THE SIERRA OF SANTA MONICA.

The San Fernando Valley is closed on the south by the Sierra de Santa Monica. The center of this sierra is between the ranches of Los Encinos and of Santa Monica, or San Vicente, is formed of a gray syenite, composed of fine particles, passing to a true granite, and easily decomposed by atmospheric action. Dioritic rocks here and there intersect the syenite, and from massive dikes toward the crest of the sierra, on the south side of the chain. On these syenites rest large-grained sandstone rocks, scarcely ever more than 250 feet in thickness, and covered in concordance of stratification by the fish-bearing limestone rocks of Los Encinos. These sandstone rocks are sometimes asphaltic, and contain some fossils, especially the beautiful *Pecten Veatchii* Gabb, and *Pecten Whipplei*, new pieces. Beautiful specimens of these two large *Pectines* have been collected a little to the west of Los Encinos, at the Malaga ranch, by Lieut. C. W. Whipple, of the Ordnance Corps, who had charge of one of our exploring parties.

At the side of the principal house of the Encinos ranch, at the top of an isolated eminence, which overlooks the basin of mineral-water, numerous remains of vertebræ, ribs, and head-bones of an enormous fossil cetacean are found. There also are found the upper beds of this formation of white fish-bearing limestone. These beds are arranged in the form of calcareous lenticular nodules inclosed in silex. These nodules are from 2 to 3 feet in diameter, and, when opened, are found to contain fossil-fishes, fragments of bones of *Cetacea*, or plants.

The Sierra de Santa Monica runs from west to east, and strikes perpendicularly against the granite and pegmatite of the Sierra Madre, on the other side of the cañon of the Rio de los Angeles or Rio de Porciuncula. This cañon itself has been formed through this sierra, which rises rapidly in the eastern part, after leaving Cahunga Pass. From a height of from 500 to 600 feet the

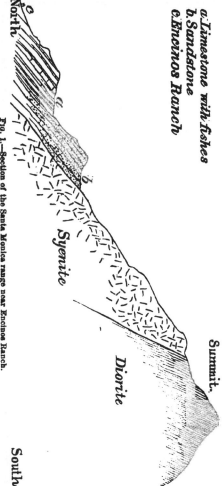

Fig. 1.—Section of the Santa Monica range near Encinos Ranch.

North

South

Syenite

Diorite

Summit.

a. Limestone with fishes
b. Sandstone
c. Encinos Ranch

crests of the sierra suddenly rise to an altitude of from 800 to 1,000 feet on either side of the cañon of Los Angeles.

Hills of sandstone and of fish-bearing limestone, identical with the beds of the section near Los Encinos, fill the eastern extremity of the San Fernando Valley, and rise against the granite of the chains of the Sierra Madre at the entrance of the large cañon of the Big Tuhunja or Tujunga. Towards the point of contact with the granite and the crystalline metaphoric rocks the Miocene Tertiary strata are much raised and contorted; the direction of their dip is easterly, opposite to that in which the Sierra Madre runs. We follow these mountains or Tertiary hills which form the counterforts of the Sierra Madre, from which they are separated by a kind of large ditch all along the eastern and northeastern extremities of the San Fernando Valley as far as the entrance of the Pacoima or Pacoña Cañon. There they end abruptly; erosions and denudations have destroyed them for a distance of 2 miles along the foot of the Sierra Madre, which here bears the name of the Sierra Pacoña or Pacoma; afterward these same Miocene strata are again met with before one reaches the Grapevine Cañon.

SIERRA MADRE, PACOÑA OR PACOIMA CAÑON.

The sides of the Pacoima Cañon are perpendicular, and it runs through gray granite rocks, gneissoid in some parts, with serpentinous metamorphic rocks and crystalline limestone at the entrance of the cañon. These rocks are the same throughout the length of the Sierra Madre chain, from the Cajon Pass Cañon, where I observed them in the spring of 1854, to the San Gabriel Cañon, in the Big and Little Tujunga Cañons, at La Soledad, or Williamson's Pass, and at the San Francisquito Pass. This gray granite, composed of fine particles, is the same granite as that of the Sierra Nevada, of which the Sierra Madre is a prolonged spur in the shape of a bayonet-like fault. This fault, or rather this bayonet-like rupture, of the primordial crystalline rocks, occurs between Tehachipi Pass and the Cañada de las Uvas or Cañon of Fort Tejon. The Pacoima Cañon, 4 miles to the east of the San Fernando mission, is narrower, deeper, and more zigzag than the San Gabriel and Tujunga Cañons. Four or five miles from its issue in the San Fernando Plain it becomes quite impassable, in consequence of deep, perpendicular cataracts which bar it entirely.

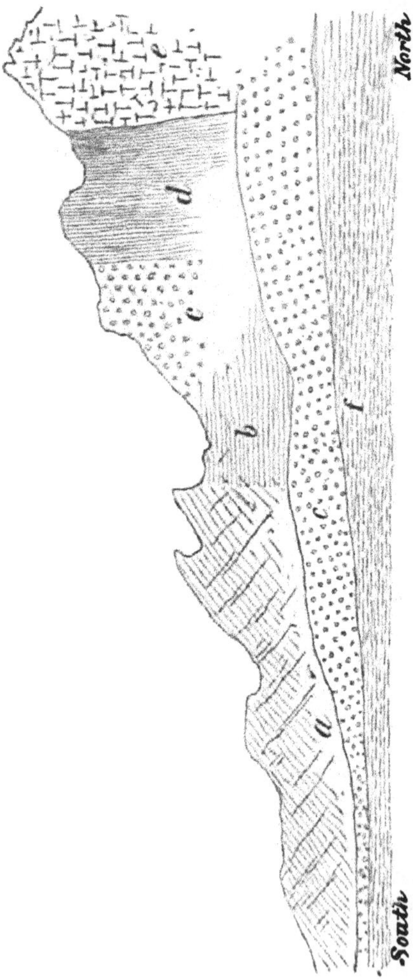

I give below a drawing of the geological section to the left of the arroyo, at the place where the Pacoima Cañon issues into the San Fernando Valley.

The crystalline metamorphic rocks are from 100 to 120 feet in thickness; they form the entrance to the cañon, and are covered halfway up by a very heavy Quaternary drift. The arroyo of Pacoima has carried away much of this drift, which is very coarse, with rounded and quite heavy blocks; it may be examined all along the banks of the arroyo in cliffs or bluffs which have an elevation of 60 feet, quite perpendicular to the bed of the arroyo. This Quaternary drift evidently indicates a very large *cone of dejection** at the issue of a cañon which existed here during the Quaternary period. This cone is very well marked, more to the northwest, at a distance of a mile from the

FIG. 2.—Plains of San Fernando, (near the village.) *a*, Miocene sandstones; *b*, Gray clay; *c c*, Quaternary; *d*, Metamorphic; *e*, Granitic rock; *f*, Modern rocks, Arroyo and Cañon de Pacoima.

North

South

---

* See torrents of the Upper Alps, by Surrell, page 11; "Torrents des Hautes Alpes, par Surrell, page 11."

cañon, where superb remains of it are seen plastered, as it were, against the foot of the Sierra Madre.

The Tertiary-molassic strata dip towards the foot of the Sierra Madre—that is to say, to the north-northeast, which is a proof that the Tertiary strata have not been dislocated and uplifted by the Sierra Madre, but by another system of upheaval prior to the last upward movement of the Sierra Madre, which took place during the Quaternary period. The mass of Tertiary rocks which is nearest to the Sierra Madre is composed of a gray clay, which fills the bottom of a valley or kind of ditch, extending parallel to the Sierra Madre, all along the foot of the metamorphic rocks.

GEOLOGY OF THE VICINITY OF THE SAN FERNANDO MISSION.

Just north of the San Fernando mission there is a line of hills, having an elevation of from 150 to 200 feet above the mission; they are composed of limestone and sandstone identical with those found at Los Encinos, on the other side of the San Fernando Plains. The fish-bearing limestone rocks, however, are less thick than at Los Encinos, but they retain the same lithological characteristics, such as nodules of silex with chalcedony, and, as at Los Encinos, scales, vertebræ, and other fragments of fossilfishes are found here, together with fossil-plants, in a bad state of preservation. The sandstone is coarse and hard, of a gray color, and in certain places it becomes a true conglomerate.

The direction of the dip of all the strata is to the north-northwest, at an angle of from forty to forty-five degrees; the heads of the strata run from northeast to southwest. Near the Lopez and Bernardi ranches, numerous specimens of the *Pecten Cerroensis* Gabb are found in the sandstone, together with a new species similar to the *Pecten Deserti* Conrad, which I call the *Pecten missionis*, in honor of the San Fernando mission.

Grapevine Cañon.—In the eastern part of the Sierra of San Fernando, which is called Monte de Pinos on account of the pine trees which cover it, 3 miles to the north of San Fernando, on the right-hand side of the road which leads to the tunnel, the Tertiary rocks reappear toward the foot of the Sierra Madre; they rise higher as we approach the Sierra of San Fernando, which is entirely formed by them. At the points of contact of the molassic Miocene strata with the granite and pegmatite of the Sierra Madre in the Grapevine Cañon, the sandstone beds are very asphaltic in certain parts; and in consequence of folds and ruptures of the strata small springs of petroleum or mineral oil are found. The asphaltum flows over rocks, and in certain parts we find a real covering of pure asphaltum or Mexican "brea," which extends over several square yards around the springs. The rock has been bored unsuccessfully in the hope of finding petroleum, and the following is the geological section at the spot where the bore has been made. After reaching a depth of 50 feet the lead entered the pegmatite and remained there.

The sandstone is very friable, nearly of a bluish-gray color, like the molassic rocks of Switzerland, and often becomes a true conglomerate, or "nagelfluh ;" it is impregnated with a considerable quantity of asphaltum. The strata are often very much dislocated, raised almost perpendicularly, inverted and folded, while close by they are almost horizontal. Yellow spots of oxidized iron are often seen in some of these beds of sandstone and conglomerate.

Fossils are quite frequently found, although generally in a bad state of preservation. Fine specimens of sharks' teeth are found, of the genus *Carcharodon*, probably the *Carcharodon rectus* Agassiz; also specimens of the *Fusus*, *Pecten*, *Tellina*, *Lucina*, &c., all of which indicates a Miocene-Tertiary fauna. In general, in the Grapevine Cañon the direction of the dip of the strata of the Californian "molasse" is south southwest, at an angle of 45°.

THE SAN FERNANDO SIERRA.

Following the San Fernando road to Lyon's Station or Petroleopolis, almost as soon as one enters the cañon to the right of the road in the direction of the Grapevine Cañon, superb folds in the strata of the Miocene sandstone rocks are seen. The direction of the dip of these molassic sandstone rocks is south, at an angle varying from 10° to 16°; the following is the geological section:

As we go higher up in the cañon a repetition is seen of the same strata of sandstone, conglomerate and fish-bearing limestone, in consequence of the numerous foldings to which all this Miocene formation was subjected at the time when the dislocations and elevations of the San Fernando Sierra took place.

The San Fernando tunnel.—When we reach the southern extremity of the tunnel the sandstone rocks are found still more massive; they here appear in the form of superb blue molassic rocks, with indistinct stratification, in enormous beds of from 15 to 20 feet in thickness. Here and there, scattered at different heights throughout these rocks, calcareo-sandy nodules are found, containing numerous fossil-shells, and forming the genuine shelly sandstone or *muschelsandstein* of the Swiss geologists. The following is a list of the principal fossils which I have found in the *muschelsandstein*, the

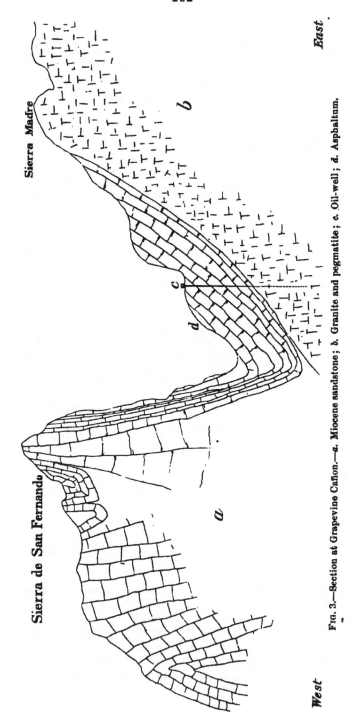

Fig. 3.—Section at Grapevine Cañon.—a. Miocene sandstone; b. Granite and pegmatite; c. Oil-well; d. Asphaltum.

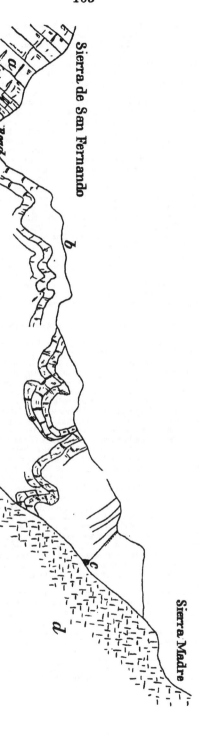

FIG. 4.—Entrance of Grapevine Cañon.—a. Limestone with fossil fishes; b. Yellow and blue sandstone; c. Oil-well; d. Granite and pegmatite.

appearance of which calls to mind the marine Miocene fauna of the *Helvetian* group of Switzerland and Suabia: *Neptunæ Humerosa* Gabb, *Turritella in—od, Venus pertenuis* Gabb, *Callista Voyi*(?) Gabb, *Schizothorus Californicus*, n. sp.; *Pecten corrosensis* Gabb, *Pecten Fernandii*, n. sp., very large and beautiful; *Cardium Californianum* (?) Conrad, and *Lucina Richthoferi* Gabb. In consequence of the extreme dryness of the climate and of the dust which results therefrom, the Miocene sandstone rocks of the entire San Fernando Sierra seem to be of a gray color. If a quarry is opened, however, it is seen that the real color of these molassic rocks is a pale blue; and the whole formation presents a general lithological, stratigraphical, and paleontological aspect which reminds one to such an extent of the Miocene "molasse" of Switzerland, as seen from Lausanne to Freiburg, Berne, Lucerne, Zurich, and the Lake of Constance, that I have often thought myself in Switzerland while really in California.

The San Fernando Pass.—As we approach the summit of the pass, after passing the Tall House, the strata become more and more inclined, until at the summit the inclination attains an angle of 60°, the direction of the dip being always southerly. It is also remarked that the molasse, the beds of which are from 10 to 15 feet thick, contains pebbles disseminated here and there in the sandstone. These pebbles finally become so numerous that the strata become a true conglomerate or *"nagelfluh,"* similar to the *"nagelfluh"* of Righi, Rosenberg, the Emmenthal, and Appenzell, in Switzerland. Half a mile from the summit of the pass, descending toward Lyon's Station, we find a considerable mass of sandstone and conglomerate rocks of a dingy yellow color, 200 feet in thickness. The direction of the dip is southerly, or rather a little west of south. The direction of the dip of the strata does not change until we reach the secondary chain, only a mile this side of Petroleopolis, or Lyon's Station. The direction of the dip then changes to the north, or rather a little east of north, at an angle of 20°. We must remark that the road from San Fernando to Petroleopolis is very near the point of contact of the San Fernando Sierra with the Sierra Madre, and hence it follows that the inclinations of the strata are slightly different from the normal inclinations of the most western portions of the Sierra, in consequence of numerous plications and setbacks caused by the obstacle of the Sierra Madre.

ASPHALTUM AND MINERAL OIL NEAR SAN FRANCISQUITO RANCH.

Six miles to the west by south (ouest-sud) of Lyon's ranch, or San Francisquito ranch, precisely on the crest or arête* at the point of division from the San Fernando Sierra, there are two springs of petroleum-oil on each side of the summit of the mountain. On the north side Mr. Lyon's petroleum-spring is also known by the name of Pico's Spring, while that on the south side bears the name of Temple's Oil-Wells.

Before reaching Pico's Spring we cross two ridges of mountains, very abrupt on the south side, as are the greater part of the mountains of the San Fernando Sierra. These mountains are formed of huge masses of molassic sandstone, of conglomerate, (*nagelfluh*,) numerous strata of which are thoroughly impregnated with asphaltum, and whose entire thickness varies between 1,500 and 2,000 feet. The direction of the dip of all the strata is toward the bottom of the Santa Clara Valley; that is to say, to the north, a little west of north, at very variable angles, between 20, 85 degrees, and even the perpendicular. Under these asphaltic sandstones, strata of arenaceo, marly schist mingled with rather thin beds of sandstone and limestone, intercalated here and there in these schistous clays, are seen to crop toward the center of the Sierra, and the petroleum-springs bubble up at the points of contact of these clays with the masses of sandstone and asphaltic conglomerate. These springs are not abundant, for two very simple reasons: first, they are at the very summit of the Sierra, and there are no basins to fill them; and secondly, the climate is an extremely dry one. During the very short rainy season these springs increase in volume and the petroleum is much more abundant. If these beds of sandstone and asphaltic conglomerate could be flooded, a large quantity of petroleum could be obtained. The only hope of obtaining a good and sufficient flow of this oil is by means of artesian wells bored at the bottom of the Santa Clara Valley. There, at a great depth—between 2,000 and 3,000 feet—there is some prospect of reaching the supply of petroleum and of its flowing in a rich and abundant stream.

Hitherto all the boring has been done in the worst localities that could possibly have been selected. Wherever a few insignificant petroleum-springs gave speculators the hope of making a sudden fortune, they went to boring at once, without thinking of the future. Whoever wishes to succeed must leave the vicinity of the Sierra Madre and the summits of the San Fernando Sierra, and go to the valley of Santa Clara or to that of San Fernando. There will be a better prospect, however, at the bottom of the Santa Clara Valley.

---

*The arête of a mountain is defined by Bescherelle as a curved line which usually separates the principal declivities of a chain of mountains where the highest peaks are found.

(The height above the level of the sea, measured by my friend, Mr. Francis Klett, is 2,260 feet.)

On the other side of the Sierra, looking then toward the south, are the springs and the bored well of Mr. Temple. The rocks present the same folds and contortions as at Pico's Spring; only the direction of the dip of the strata is to the south, at an angle of from 15° to 25°.

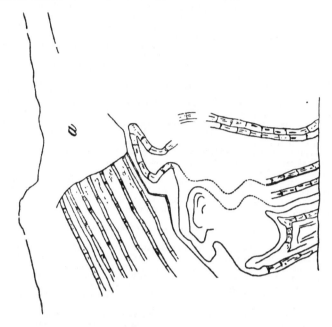

FIG. 5.—Pico's Spring.—*a* Asphaltic sandstone covered with vegetation.

If there is no hope of ever finding abundant springs of mineral-oil in the San Fernando Sierra itself, there is at least a prospect of getting magnificent and very rich quarries of asphaltum. These quarries will certainly one day be worked with the most satisfactory results. In the vicinity of the San Francisquito ranch the sandstone contains fossils, the most abundant of which is the *Pecten Cerrosensis* Gabb.

San Francisquito Pass.—The plain from the San Francisquito ranch, as far as Moore's Station, is composed of fine sand, of great thickness, which forms the entire bottom of the valley, and indicates that large rivers must have descended from the passes of Soledad and San Francisquito; and there is some reason to suppose that at the commencement of the modern period the Santa Clara Valley was the outlet of a part of the Western Rio Colorado.

A mile to the east of Moore's Station we again meet with the Miocene sandstone or molassic rocks, more or less inclined in directions varying from north to west and south. However, the westerly direction seems to prevail.

Before reaching the stage-station, doleritic trap-rocks, interstratified in the blue molasse, are found on both sides of the road, but particularly on the right. The molasse is then somewhat metamorphic, and its argillo-arenaceous schist becomes a lustrous black schist, similar to the fishy schists of Glarus or to the "flisch" of the Alpine geologists. The reader must not be surprised if I constantly remind him of the great lithological resemblance between the Tertiary rocks of California and those of Switzerland, Vorarlberg, and Bavaria. It is unusual to find so many points of comparison and similarity at such great distances from each other on the terrestrial globe. The conditions of the deposits must evidently have been the same in California and in Switzerland. The Sierra Nevada, or its prolongation, the Sierra Madre, performs the same part as the Alps in furnishing the materials which were deposited in the Californian and Helvetic Tertiary seas. The Tertiary rocks, both sandstone and schist, come to an end at Humphrey's ranch, where they are very much raised, some even to

AP. J J—11

the perpendicular, and some strata are even inverted. Suddenly, immediately after leaving the sandstone schist and dolerite, we find ourselves before a granite wall, which rises abruptly from 100 to 150 feet above the Tertiary rocks. This is the Sierra Madre, with its gray granite, consisting of fine particles, its pegmatite, its veins of quartz, its green and micaceous schist, and its old diorites. All at once the San Francisquito Cañon, which until now has been pretty wide, except near the dike of dolerite, becomes narrow and very sinuous, like the granitic cañons of San Gabriel, Tujunga, Pacoima, and Soledad. The directions of the crests or summits of the mountains also change abruptly, and instead of running from east to west, as in the San Fernando and Santa Monica Sierras, they run from south to north. Between Humphrey's ranch and the summit of the pass there are three parallel ranges of mountains running from south to north. The Delano ranch is situated in a little valley between the first and second of these ranges. After passing the valley of Lake Elizabeth we come to a fourth granitic range, known by the name of the Sierra de Liebre, which borders on the Californian Desert. These four ranges of the Sierra Madre are of unequal heights, the highest being in the locality where the "col" of the pass is found. The valleys which separate them, moreover, are of very different widths. The first two are the narrowest, while the valley of Lake Elizabeth is comparatively wide and open.

Origin of the name California.—Lake Elizabeth extends from east to west, and an extremely violent west wind blows there night and day, scarcely ceasing for a single instant; it is a dry wind, and its heat, which reminds one strongly of the air of a hot oven, has evidently given rise to the name California. The Mexicans or Spanish-Americans, who first came to this country by land *via* Sonora or New Mexico, soon remarked this temperature and this very peculiar climate, which has nothing in common with the three divisions which they had adopted, until that time, for Mexico, viz, *tierra fria,* *tierra templada,* and *tierra caliente, i. e.,* cold country, temperate country, and warm country; and they very properly called this country, so different from the three others, *tierra california,* that is to say, country hot as an oven.

SIERRA LIEBRE AND CALIFORNIA DESERT.

After crossing the Sierra Liebre, which separates the valley of Lake Elizabeth from the great basin or California Desert, if we follow the east foot of this granitic sierra, we several times cross wide river-beds, which are hollowed out in the sand, and through which not even small brooks now run. These dry beds indicate that at no very distant day—for one would suppose that the water in them gave out only a few days ago—large rivers descended from the Sierra Madre, or rather from that part of the Sierra Madre which is called the Sierra Liebre, discharging their waters into the great valley of the Dry Lake, where is now Willow's Station. The climate of Southern California has evidently undergone great changes in modern times. The sand of the desert covers this entire basin, which is a genuine American Sahara. Nevertheless, two miles to the east, before reaching Liebre's ranch, we again meet with the Tertiary sandstone, with the conglomerate, which here forms a counterfort to the Sierra de Liebre.

The strata dip to the south-southeast and north-northwest, at an angle varying from 15° to 30°, showing here and there the synclinal line of the strata. These sandstone rocks, which are of a gray and sometimes of a reddish color, are seen all along the road from Liebre ranch as far as Gorman's ranch, at the entrance of the cañon of Fort Tejon or Cañada de las Uvas. I was unable to determine the exact age of these Tertiary rocks, in consequence of a very long and painful journey in the month of July, when the weather was excessively hot. It is possible that some of the strata are of the Miocene epoch; I am, however, rather inclined to regard them as Eocene rocks. Where was the strait or passage through which the Tertiary sea communicated from west to east across the Sierra Madre and the Sierra Nevada? I have not been able to recognize it. It was evidently neither through the San Francisquito Pass, nor the Cañada de las Uvas or cañon of Fort Tejon, nor through the Tehachipi Pass. The Tejon Pass is the only one which I have not visited, but toward this all the Tertiary rocks to the east and west of the pass seem to run; and it was probably through this pass that the Tertiary sea, which covered the entire western portion of California, penetrated across the granitic chains into the great basin of the California Desert. This communication, however, possibly took place through the valley of the Rio de Peru, the water passing by the south foot of Mount Pinos and flowing into the great basin in the comparatively low mountains which are found between Liebre ranch and Gorman's ranch.

Cañada de las Uvas.—The Tertiary rocks in the vicinity of Gorman's ranch, which are very much uplifted and which dip to the west toward the Sierra de Liebre at an angle of 60°, are crossed by dikes of dolerites and euphotides, the real *gabbro* of Tuscany. The summit of the pass of the Cañada is on these euphotides, with yellow sandstones, and Tertiary, whitish, limestones. As soon as we descend we come to the granite of the range, which runs to the east of Lake Castac; and then the entire cañon of Fort Tejon runs through the granitic and crystalline rocks, gneiss; and a mile and a half this side of Fort Tejon we find a large dike of grayish-white crystalline limestone, more than 100 feet thick. This is real marble, and runs from north-northwest to south-southeast. All this cañon is formed of gray granite exactly like the granite of the Sierra Madre and the Sierra Nevada, intersected here and there by dikes of milk-white quartz, dikes of rose-colored feldspath, and dikes of diorite. Sometimes, in place of the granite, we find gneiss and mica-schist, very much bent and twisted in all directions.

Tertiary rocks in the vicinity of Fort Tejon.—At the outlet of the cañon into the plain of Tejon's ranch, we find Tertiary rocks composed of a heavy mass of clay, more or less sandy, with some thin strata of sandstone and limestone intercalated at different heights of the mass. The Tertiary rocks stand against the Sierra, and rise only to one-half the height of the mountains which overlook the cañon. Some fossils are found in the sandstone, at the very entrance of the Cañada de las Uvas; but, as these fossils are much more numerous and in a better state of preservation three miles farther east, as one follows the foot of the mountains, at the *Arroyo de los Alisos*, (Alder brook,) or *Arroyo del Rancho Viejo*, (Old Ranch brook,) I am going to give a detailed section, with a list of the fossils collected in this locality.

The corral and the ruins of the Rancho de los Alisos are situated on the Quaternary drift, which is not very thick. Immediately afterward we come to hills of trachyte, trachytic conglomerate, and dolerite, which form the basis or first counter-fort of the mass of mountains. These eruptive rocks are the same as those of the Cañada de las Uvas near Gorman's ranch. They seem to have raised the Tertiary strata very much. These latter dip to the south-southeast, toward the foot of the mountains, at an angle varying from 60° to 80° and 85°, almost perpendicularly at the spot where there is a fault in the bottom of the principal lateral ravine to the east of the arroyo. At the place where the section was taken, on the right bank of the arroyo, the eruptive rocks form four small ranges of hills, which gradually rise to a height of 350 feet above the plain of Tejon ranch.

Immediately after passing the fourth hill, going up toward the mountain, we come to an abrupt bluff, which at first sight seems to be formed of drift. One soon sees, however, that the numerous rolled and much rounded pebbles that are met with are much flatter, and rocks whose composition are very different from that of those which constitute the ordinary Quaternary in the cañons, and the cones of dejection of the cañons of Tejon Pass, Las Tunas, La Pastorina, and Las Uvas. These rolled pebbles, moreover, are imbedded in a gray sandy clay of great hardness; finally, the whole is very much raised and dips in a south-southeasterly direction. This kind of argillous conglomerate forms the basis of the Tertiary rocks of the region about Fort Tejon. Toward the top of the hill, instead of this conglomerate, we find gray clay, containing here and there intercalated strata of calcareo-arenaceous sandstone; some are entirely calcareous, while some other strata consist of pure sandstone. These calcareo-arenaceous strata vary from 1 foot to 2½ feet in thickness; they dip to the south, at an angle which is very much inclined, and which even reaches the perpendicular.

At the bottom of the hill which separates the fifth hill or eminence from the sixth, we find a well-defined fault; the strata of fossiliferous sandstone now dip to the north, at an angle of from 45° to 50°, and the beds on each side of the ravine do not correspond. Having reached the summit of the sixth hill, we see a small valley, with a gentle declivity, which lies at the point where the stratified Tertiary rocks form a junction with the granite and other crystalline rocks of the group of the Fort Tejon Mountains. These Tertiary rocks are, approximately, from 250 to 300 feet thick. All the strata of sandstone and arenaceous limestone contain fossils in greater or less abundance; some are even full of them, and form a real *Lumachella*; thus, there are strata which might be called turrilitic beds, to such an extent do the *Turrilites* abound in them, reminding one of the ceritic limestone of the basin of Paris.

In consequence of the hardness of the rock, it is difficult to obtain complete specimens of fossils; although they are in general well preserved, it is very difficult to collect anything more than fragments. Dr. Horn, the surgeon at Fort Tejon, has made a good and numerous collection of these fossils, which has been described by Mr. Gabb in the first two volumes of the "Paleontology of California," published by that State. After having visited the locality explored by Dr. Horn, and having studied the geology of a portion of Southern California, I cannot adopt the opinion of Mr. Gabb, who considers this formation as cretaceous, and particularly resembling the Maastricht beds of Central Europe. I was not able to find a single cretaceous fossil, nor even any true cretaceous generic forms, in this entire formation; and I am altogether of the opinion expressed by Mr. Conrad, many years before Mr. Gabb, in volume 5 of Pacific Railroad Explorations, pages 318, 320, *et seq.*, who, judging from certain fossils found in an isolated block at the entrance of the Cañada de las Uvas, has very judiciously referred these rocks to the Eocene-Tertiary formation. I go even further. I think that the rocks of the Arroyo del Rancho Viejo of Fort Tejon belong to the superior Eocene epoch, and that they are of about the age of the "coarse limestone or calcaire grossier" of Paris. The fauna of Tejon reminds one very much of the fauna of the sands of Anvers, near Pontoise, and of the sands of Gregnon, near Versailles. The following is a list of the

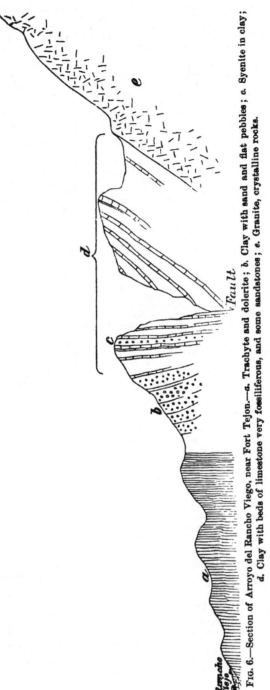

FIG. 6.—Section of Arroyo del Rancho Viego, near Fort Tejon.—*a*. Trachyte and dolerite; *b*. Clay with sand and flat pebbles; *c*. Syenite in clay; *d*. Clay with beds of limestone very fossiliferous, and some sandstones; *e*. Granite, crystalline rocks.

fossils which I collected at the Arroyo de los Alisos, near Fort Tejon : *Fusus* (*Hemirfusus*) *Remondii*, Gabb; *Tritonium Whitneyi*, Gabb; *Trit. Californicum*, Gabb; *Trachytriton Tejonensis*, Gabb; *Olivella Mathewsonii*, Gabb; *Fasciolaria ficus*, Gabb; *Mitica Uvasana*, Gabb; *Lunatia* (*gyrodes*) *Conradiana*, Gabb; *Nererita globosa*, Gabb; *Never secta*, Gabb; *Turritella Uvasana*, Gabb; *Galerus Excentricus*, Gabb; *Ringinella Pinguis*, Gabb; *Tellina ovides*, Gabb; *Tell. aqualis*, Gabb; *Meretrix Uvasana*, Conrad; *Meret Hornii*, Gabb; *Dosinia elevata*, Gabb; *Tapes Conradiana*, Gabb; *Cardium Cooperii*, Gabb; *Card. Brewerii*; *Linia multiradiata*, Gabb, &c. This fauna does not contain a single one of the cretaceous genera of Central Europe, while it abounds in European and South American Tertiary genera. There is an almost entire absence of polyparia or corals, and of echinoderms or radiata. This absence of corals is particularly remarkable, in view of the fact that it extends throughout the whole formation of Tertiary rocks of California, in which only two or three specimens of zoöphytes have as yet been discovered, and they were found in the vicinity of Mount Diablo; not a single one has been found in Southern California.

Tertiary rocks of California.—The middle and lower portions of the Eocene Tertiary rocks are well represented in California by the lignitic strata of Mount Diablo, which are found immediately beneath the strata of Fort Tejon, after which some beds are found, which bear the name of the Martinez and Chico groups, which succeed each other without discordance of stratification, and the fauna of which (chiefly Tertiary) contains some cephalopods badly developed, small, and stunted, as are often the last representatives of a great family or genus which is on the point of becoming extinct. Dr. John B. Trask, who, with the late Dr. Randall, the discoverer of the fossils at Chico Creek, first called attention to these beds at Chico Creek containing cephalopods, correctly referred them to tqe Eocene Tertiary formation, notwithstanding the presence of the few cephalopods, which he called *Ammonites Chicoensis* and *Baculites Chicoensis*. (See Description of New Species ef Ammonite and Baculite from the Tertiary Rocks of Chico Creek, in Proceedings of the California Academy of Natural Sciences, vol. L., p. 85, San Francisco, 1854–'58.) There is absolutely no law in geological science which makes it necessary for cephalopods of the genera known as *Ammonites, Baculites, Hamites, Heliocera, Turrilites, Ancylocera*, &c. to disappear entirely from the surface of the terrestrial globe with the rocks of the Secondary epochs. In Central Europe these cephalopods have hitherto disappeared with the cretaceous rocks; that, however, is but an empirical law, a negative fact, applicable only to Central Europe, and to extend this fact to the whole earth as a well-established law would be contrary to all that we know of the laws of the geographical distribution of animals.

The general zoölogical, lithological, and stratigraphical characteristics of the California rocks which extend from Chico Creek and San Francisco to San Diego and San Bernardino are such as to render the whole analogous to the great Tertiary rocks of the Americo-European basin of the Atlantic, of which it is the representative and the equivalent in the great basin of the Pacific; only some *Ammonites, Baculites, Heliocera,* &c., continued to live in the Tertiary seas of the Pacific regions when these genera no longer existed in the Atlantic hemisphere, a fact which is observed in all the geological periods through which the terrestrial globe has passed until the present time. This is not an exception; it is, on the contrary, a law of the geographical distribution of beings. Real cretaceous rocks are only found, in California, in the northeastern portion of the State, around Mount Shasta. But all the southern and central parts, the real coast range, from San Luis Obispo to Cape Mendocino, the city and peninsula of San Francisco, Mount Diablo, and Fort Tejon, are formed exclusively of stratified rocks of the Tertiary period.

Glacial rocks of Southern California and Pike's Peak.—It cannot be doubted that the Sierra Madre, from Mount San Bernardino to Tejon Pass and Tehachipi Pass, was covered with glaciers during the Quaternary period. The long cañons of San Gabriel, of the Little and Big Tujunga, of Soledad and San Francisquito, served as receptacles or beds (*lits*) for the glaciers which descended from the many high peaks of this chain of mountains, still so little known geographically. In the San Gabriel Cañon we see in several places before reaching Dr. Winsor's silver-mines traces of ancient lateral "moraines" on each side of the cañon, and toward the entrance of the cañon, a mile from Buell's Hole, heavy blocks of granite or true bowlders are seen, arranged in the form of a crescent, as if for a frontal "moraine." My exploration of this cañon was too superficial to enable me to speak with confidence of the existence of an ancient glacier; it seemed to me, however, that there were some traces of a large glacier which descended from the peak of San Antonio.

The traces of glaciers are more visible and striking in the San Francisquito Pass, especially near Jesus Gallejo's ranch, where diorites appear in great masses, and not long after reaching a large lateral valley on the right, as one ascends the pass, we see a very heavy Quaternary drift, with erratic blocks and indications of glacial striæ, on the dioritic and Miocene sandstone rocks which form the bottom of the valley.

I however met with really unmistakable "moraines" only at a single point of my explorations in Southern California, and that was on the road which runs along the foot of the mountains between Tejon Pass and Caliente. There, just as one goes up from the great plain of Lake Tulare to the plateau which overlooks Caliente, one sees enormous blocks of granite with an entirely northern exposure, piled the one upon the other, with glacial drift mixed among them, which forms, without the slightest doubt, a beautiful frontal "moraine." The road which leads from Bakersfield up to Cerro Gordo, Panamint, and the California Desert passes close by these remains of "moraines."

Although outside of my exploratio·., I cannot forego the pleasure of referring to the magnificent frontal "moraine" of the great glacier which must have come down from Pike's Peak, have covered the entire valley of the Manitou Springs, and have come to a stand-still midway between Manitou and Colorado City. The house or inn, called the "Half-way House," on the road is built on this frontal "moraine" itself, and the road crosses it at this place. Lateral "moraines," moreover, are seen on each side, especially in the direction of the Garden of the Gods.

Mountain chains and their ages.—Our knowledge of the mountains of California to the south of San Francisco is so limited and such strange confusion has arisen that, notwithstanding the small number of my observations, and the meagerness of the results obtained, I do not hesitate to present them, in the hope that they may lead to a more rational and systematic study of this subject. In general, everything in California that is not in the Sierra Nevada, properly so called, is thrown into the coast range. This coast range is naturally known by various local names. Great confusion has evidently arisen from the placing together of chains of mountains running in totally different directions, of different relative ages and different geological constitutions.

The Sierra Madre.—The Sierra Madre is but an uninterrupted continuation of the Sierra Nevada, which deviates abruptly from its general direction, that is to say, from north to south, with slight tendencies toward the west and east, to form an elbow at Walker's Pass; this elbow runs through the Tehachipi Pass, the Tejon Pass, and stops at the Cañada de las Uvas, in order to resume its principal direction from north to south. This double elbow may be compared to the point of junction of a bayonet on a gun. Geologists have long been familiar with breaks of this kind in the rocks of the earth's crust, and faults of this shape are often met with, which are very appropriately called "bayonet-shaped faults."

What is the extent of the Sierra Madre, which, like the Sierra Nevada, is formed of several parallel ranges, (*chaînons?*) I cannot tell. I may say, however, of my own knowledge, that the Cajon Pass crosses it, and Mount San Bernardino forms a part of it. In 1854, in my exploration on the thirty-fifth parallel, in company with my lamented friend, the late Gen. A. W. Whipple, Engineer Corps, I was struck with the identity of the crystalline rocks of the Cajon Pass with those which I saw in the vicinity of Nevada City and Grass Valley, in the Sierra Nevada, and I did not hesitate at that time to regard them as being of the same geological age, and as containing, in all probability, the same valuable minerals. (See Pacific Railroad Explorations, vol. 3, 4to, Geological Reports, page 171, Washington, 1856.) Since then, my explorations of the San Gabriel, Pacoima, and San Francisquito Cañons have confirmed me in this opinion; Winston's silver-mine, moreover, in the San Gabriel Cañon, reminds one in every respect of the silver-deposits of Virginia City and Washoe. After crossing an enormous granitic mass, on a width of at least 4 miles between the entrance of the San Gabriel Cañon and the Winston mine, one meets with serpentinous dioritic rocks, very hard, and which, near their points of contact with the granite, contain silver and copper in abundance.

Throughout the length of the Sierra Madre one meets with more or less auriferous drift, identical in all its characteristics with that of the placers of the Sierra Nevada; the want of water alone has prevented its being washed to advantage.

The Sierra Madre being only a bayonet-like prolongation of the Sierra Nevada, is of course of the same geological age. This age has hitherto been a problematical one, notwithstanding that it has been proclaimed with some ostentation to be of the Jurassic epoch. There is no doubt that the Sierra Nevada and all the ranges or sierras of Colonel Frémont's "great basin" are much more ancient than Jura.

Like all complicated chains of mountains which extend over large surfaces, the Sierra Nevada was not made at once, but at various times and at different geological dates. The Sierra Nevada and the Sierra Madre have been *terra firma* from the most ancient paleozoic times: and it is certain that the existence of gold dates from those remote periods, like the gold of Australia, the Ural, Wales, Canada, the Carolinas, and British Columbia. Elevations and ruptures of crystalline and stratified rocks took place in the Sierra Nevada, the Sierra Madre, and others of the great basin, toward the close of the Carboniferous, Triassic, Jurassic, and Cretaceous eras, and left very perceptible traces in certain places, especially near Mount Shasta, Bass's ranch, Plumas County; El Dorado Cañon, Humboldt ranges, Mariposa County; Inyo range, Cerro Gordo, and Panamint.

The Tertiary sea washed the western sides of the Sierra Nevada and Sierra Madre, and in some places even penetrated these mountains especially at the bayonet-like elbow which unites them. Nevertheless the dislocations which broke the Tertiary strata, and raised them to form the Coast Range, the Mount Diablo range, the sierras of San Rafael, of San Fernando, &c., did not appreciably affect the Sierra Madre and still less the Sierra Nevada.

Meeting with a powerful barrier, the Tertiary strata were pressed back against the obstacle of the Sierra Madre and were in some places folded back (*repliées*) upon themselves, becoming contorted, and their beds being turned in an opposite direction, perpendicular to that of these granitic and crystalline mountain-chains. I have not seen any indications which prove that the Sierra Madre was ever subjected to the uplifts of the Tertiary epochs. But, then, at the close of the Quaternary epoch, and perhaps even in the Modern epoch, there are proofs of uplifts and dislocations, which are particularly perceptible on the eastern side of the chains of the Sierra Madre throughout the whole length of the California Desert. The principal proofs of great movements on the sides of the Sierra Madre toward the end of the Post Pliocene age, and during the Modern epoch, are: First, the eastern counterfort of the Sierra Madre at the Cajon Pass, at the very place where the pass attains its greatest elevation above the level of the sea, which, according to Gen. A. W. Whipple, is 4,559 feet; here there are very heavy strata of white sand with rolled pebbles, arranged in beds or scattered through the sand; these strata have been greatly uplifted by the Sierra Madre. Although the stratification of this kind of drift is quite indistinct and confused, it is clearly seen that the whole of this formation dips in an easterly direction at an angle which even attains 45°, which is a very great inclination for rocks which are almost friable, (*meubles.*) The Sierra Madre evidently forms the anticlinal or uplifting mass. The thickness of this formation may be estimated at 1,500 or 2,000 feet. What is its age? It is scarcely possible to tell, because no fossils have as yet been found in it. Nevertheless many of the rolled pebbles contained in this sandy mass are fragments of trachytes and basalts, which shows that this formation is very recent, certainly much more recent than the Pliocene in the vicinity of Los Angeles, and that it can only be referred to the Post Pliocene or Quaternary formation.

On the other hand, the great California Desert, between the Sierra Madre, the Sierra Nevada, Death Valley, and the valley of the Mohave River, indicates the existence of an ancient lake, now dry, the banks and old beaches of which are still seen in many places, notwithstanding the continual movement of the sand, which is violently driven by the west wind. As extinct volcanoes are met with all along the Colorado River, in the basins of the Mohave and the Amargosa, and as very violent earthquakes still take place throughout a great portion of this region, it is natural to suppose that the last uplift and elevation of the Sierra Madre, and of a portion of the Sierra Nevada, took place at the close of the Quaternary period, or even in modern times.

This is the conclusion which I reached in 1854, and which I stated in my "Sketch of a geological classification of the mountains of a part of North America," (see *Geology of North America*, Sierra Nevada system, page 79; Zurich, 4to, 1858,) only I expressed no opinion with regard to the dislocations which had previously taken place in this system of the Great American Desert, which I called a *second meridian system* in North America. The discoveries made since, principally by miners, in the Sierra Nevada and the ranges of the Great Basin, of the primordial or Taconic fauna, and of the Carboniferous, Triassic, and Jurassic faunas, show that from the earliest times of the Upper Taconic, there have been *terra firma* in that region which had emerged from the sea; that the granitic *arêtes* of the ranges which cross this entire country date from the beginning of the Paleozoic ages; that the Carboniferous, the Trias, and the Jura penetrate only into narrow valleys of this system; and that during the Tertiary ages there was an enormous mass of *terra firma* here, which, like the Alps, furnished the arenaceous and pebbly materials for the deposits of the marine and of fresh-water Tertiary rocks, which are now found in California, Utah, Wyoming, and Colorado

"*En résumé,*" the Sierra Madre is altogether the most ancient and the most modern mountain chain of this region of Southern California; that is to say, that the granite, pegmatite, gneiss, dioritic, and metamorphic rocks which form its principal mass date from times anterior to the Paleozoic epochs, *ou tout au plus paléozoïques mêmes;* and that the counterforts of sand, sandstone, and conglomerate, which form the summit of Cajon Pass and of other portions of the eastern region of this chain, date from the Post Pliocene or Quaternary epoch.

COAST RANGE.—The name of *Coast Range* signifies a chain of mountains which follows the line of the coast of the Pacific Ocean. By way of extension, all mountains which are near the coast have been comprised in California under this elastic designation, no regard being had to the direction in which the chains run, whether parallel to the coast or perpendicular to it. After leaving Point Conception and Santa Barbara, and even the Sierra de San Rafael, we find that the mountain chains, instead of running from north-northwest to south-southeast, run from west to east, being perpendicular both to the Sierra Madre and to the sea-coast; so that, properly speaking, the Coast Range, which is so well defined at Monterey and San Francisco, terminates in the southern part of San Luis Obispo County, as was correctly observed by Dr. Trask in the first geological survey of the State of California. (See *Report on the Geology of the Coast Mountains*, &c., p. 10 ; 1855 ; 8vo : Sacramento.) It may be said in general that the Coast Range divides the Pacific Ocean from the valley of the San Joaquin River and of the Tulares beyond San Emidio, not far from Fort Tejon, and that it comprises all the parallel chains which reach the great bay of San Francisco. These mountains continue beyond the bays of Suisun, San Pablo, and the Golden Gates, in a north-westerly direction. What is the principal age of this system of mountains? In a word, at what geological epoch did it make its appearance? I now think, as I did in 1854, when I saw it for the first time with my friend Whipple, that it should be referred to the end of the Eocene Tertiary deposits.

Sierras of San Fernando and Santa Monica.—These mountains, which, with the Sierra Madre on the east and the Santa Susana range on the west, inclose the charming valley or plain of the ancient mission of San Fernando Rey de España, run from west to east. The strata are much broken, uplifted, and inclined, and they are all Miocene Tertiary or molassic rocks, which fixes the age of the appearance and formation of these mountains at the end of the Miocene epoch.

The Santa Susana range, between the Triunfo and Simi Valleys, is but a counterfort of and an appendage to the Sierra de Santa Monica.

The Sierra of San Rafael, which runs right up to the Sierra Madre, at the foot of Mount Pinos, the highest peak of this region, with the peak of San Antonio, seems to belong to this west-easterly system of the San Fernando and Santa Monica Sierras. The same is the case with the Sierra de Santa Ines, back of Santa Barbara. The Santa Clara Valley, with its prolongation almost to Soledad, forms a part of this system, as do the San Francisquito and Castac Valleys.

In consequence of their directions being from west to east, this system of mountains enters the Pacific Ocean on one side and the Sierra Madre on the other, and intersects and completely isolates Southern California from the central and northern portions of the State. The separation is even so great that the railway from San Francisco to Los Angeles is obliged to cross the Sierra Madre twice, viz, at Tehachipi Pass and at Soledad, through several long and costly tunnels, for the sole purpose of avoiding the barrier placed in its way by the Sierra San Rafael; and having reached the Santa Clara Valley, the railroad is still obliged to cross the San Fernando Sierra through a very long and deep tunnel, issuing at last into the San Fernando plains, whence it reaches Los Angeles by following the valley of erosion, which fortunately crosses the Santa Monica sierra, intersecting it perpendicularly almost to its eastern extremity. In reality, Southern California is more disconnected and isolated from California proper than is the latter from the States of Nevada and Oregon.

Hills of Los Angeles.—The hills which surround the city of Los Angeles and separate it from the valley of Bayona or Ballona, from Monte and from Anaheim, are of the Pliocene Tertiary epoch. It is possible, and in my opinion highly probable, that their age is identical with that of the mountains of Cajon Pass, and that they represent on the west side of the Sierra Madre the uplifts and elevations on the east side of that chain to which I have already referred. The sands and conglomerates of the summit of Cajon Pass must, then, be of the same age as the molassic rocks and sandy clays of the old Presidio de Los Angeles. In that case, the deposits of Cajon Pass, instead of being of the Post Pliocene, must be of the Pliocene epoch.

At all events, these two elevations and dislocations of the hills of Los Angeles and of the summit of Cajon Pass, if they did not take place simultaneously, did so at periods by no means remote from each other.

As to the various mountain-chains to the south of Los Angeles, as far as San Diego, and even farther, I can say nothing as to their relative ages, not having visited them. To sum up, we have the following systems of mountains for a portion of Southern California :

I. Sierra Madre, of the Primordial epoch, or Laconic, anterior to the Silurian.

II. Coast Range, of the close of the Eocene epoch.

III. Sierras of San Fernando and Santa Monica, of the close of the Miocene epoch.

IV. Hills of Los Angeles, of the close of the Pliocene epoch.

V. Mountains of Cajon Pass, (east side of the Sierra Madre,) of the close of the Post Pliocene or Quaternary epoch, or, perhaps, even of modern times.

JULES MARCOU.

CAMBRIDGE, MASS., *December* 30, 1875.

APPENDIX H 2.

REPORT ON THE GEOLOGICAL AND MINERALOGICAL CHARACTER OF SOUTHEASTERN
CALIFORNIA AND ADJACENT REGIONS.

UNITED STATES ENGINEER OFFICE,
GEOGRAPHICAL SURVEYS WEST OF THE 100TH MERIDIAN,
Washington, May 13, 1876.

DEAR SIR: I have the honor to submit herewith a report upon the geological and
mineralogical conditions of Southeastern California and adjacent regions. The country
traversed by the party in command of Lieut. Eric Bergland, Corps of Engineers, our
able and beloved leader, along the Mohave River, across the Opal, Payute, and Dead
Mountains, to Cottonwood Island, in Southern Nevada, thence to the mouth of the
Virgin River, across the Colorado to the Cerbat and Blue Ridge Mountains in Northwestern Arizona, from there to Fort Mohave and along the Colorado River to the Mohave
range, Monument, and Riverside Mountains, thence to San Bernardino, via Chuckavalla Peak, Coahuila Valley, and San Gorgonia Pass, was of unusual interest in geological, mineralogical, and chemical respects. Nor of less interest was the trip made,
in accordance with your orders, by myself alone (after the return of the expedition to
Los Angeles) to Panamint, Darwin, Owens Lake, Cerro Gordo, Benton, Aurora, and
Virginia City, Nev. An immense area was thus visited from June to November, and
valuable collections made.

The chemical analyses of rocks, soils, and mineral springs necessary in connection
with my reports were made in the laboratory of the Smithsonian Institution; and I
cannot but express my deepest thanks to the Secretary, Professor Henry, for the liberal
spirit he has shown in giving me the free use of the laboratory and all necessary utensils and chemicals whenever I desired.

Very respectfully, your obedient servant,

OSCAR LOEW.

Lieut. GEO. M. WHEELER,
Corps of Engineers, in charge.

CONTENTS.

Taking a bird's-eye view of the great area between the Lower Colorado and the
gigantic mountain-chains traversing California from north to south, (the Sierra Nevada
and its southern prolongations, the San Bernardino and Jacinto Mountains,) one cannot fail to recognize at once a considerable predominance of three formations, viz: the
Primitive, the Eruptive, and the Quaternary. A subordinate position is occupied by
the Palæozoic, the Mesozoic, and Tertiary.

While the numerous ranges of hills and mountains mainly consist of granite or
trachytic rocks, the intervening valleys are filled by Quaternary deposits. The topographical features of the Mohave Desert differ, therefore, vastly from those of the Painted
Desert in Northeastern Arizona, a realm of sandstone, with the characteristic mesa-
type of the table-lands.

THE PRIMITIVE FORMATION.

Nearly all the prominent ranges of Southern California belong to this formation,
although in a number of instances the axial rock is concealed, as at the Opal Mountains,
by Palæozoic strata, at the Blue Ridge Mountains by volcanic flows.

The San Jacinto Mountains, the most southern range of California, consist chiefly of
granite, well exhibited in the precipitous faces turned toward the east. It is of dense
structure, hard, and a splendid building stone, resembling closely the New England
granite. Of mica, it is the black variety (biotite) that dominates in the rock, and
among the accessory constituents titanite, turmaline, and garnet deserve especial notice. The titanite* is disseminated in small yellowish crystals through the rock for
over 30 miles, (Whitewater to Los Toros,) while the latter two minerals are confined
to narrower limits, (near Whitewater.)

---

* Titanite is also found in the granite of the Sierra Nevada, especially on the Moke-
lumne and American Rivers, according to Prof. W. P. Blake.

In the San Bernardino Mountains the main mass is granite, accompanied by syenite, gneiss, mica-schist, talcose-schist, and primitive clay-slate. Occasionally, as for instance between Martin's ranch and Cajon Pass, the granite gave rise to the formation of beds of arcose, a rock in which granite *debris* has been recemented, forming a sort of granitic sandstone resembling to some extent granite; but the uniform grain, friability, and rusty surface of the fragments elucidate its true nature.

The western slopes of the San Bernardino Mountains are covered by extensive beds of detritus and *débris*, which in some portions are auriferous, as in Lyttle Creek Cañon, where gold-washing is carried on on a large scale. In the vicinity of the Cajon Pass the Azoic rocks are covered by broken strata of a conglomerate exhibiting a changing dip, being at first about 30° to the south, and finally, near the Pass, 20° to the north. These inclined strata are overlaid by horizontal beds of bowlder and *débris* drift 4 to 5 feet in thickness. While dikes of dyorite and trachyte occasionally appear in the northern portion of the mountain-range, the spurs south of the Gray and San Bernardino Peaks consist largely of volcanic materials.

One of the isolated peaks east of the Jaciuto and southeast of the San Bernardino Mountains is Chukawalla Peak, with about 4,700 feet altitude, towering conspicuously above the plain, and whose shape resembles a trapezoid surmounted by a pyramid. The primitive rocks of this mountain are traversed by volcanic dikes, (chiefly trachyte and basalt,) while at the base beds of bowlder and conglomerate have accumulated, the latter showing a dip of from 20° to 25° to the south. Among the primitive rocks a mica-schist of porphyritic structure deserves mention; it consists of a fine-grained mixture of quartz and biotite, containing muscovite in plates of one-sixteenth square inch imbedded. There are said to occur also lead and copper ores at Chukawalla Mountain, but they are not worked.

The primitive rocks of the River-Side and Half-Way Mountains consist of granite and gneiss, the latter garnetiferous. While here we find the feldspar of a green color, it is pink with that of the Monument Mountains north of the former ranges.

At the Mohave range, 20 miles farther north, a series of Azoic rocks is met with, viz, a fine-grained granite, containing simultaneously biotite and muscovite, a syenite with veins of a coarse aplite, then hornblende-schist and quartzite. As this range consists largely of volcanic material, it will be again mentioned hereafter.

Among the mineral occurrences in the primitive rocks of the Cerbat range, muscovite in large plates may be mentioned. In this range are found extensive lodes of metalliferous quartz; also the neighboring Black Cañon range, Payute range, Providence Mountains, and Opal range, contain metalliferous lodes in the primitive rocks.

At the Panamint range we find primitive limestone and slate-clay as accompaniments of the granite.

THE PALÆOZOIC FORMATION.

This formation, chiefly represented by limestone and quartzite, becomes conspicuous at the Riverside Mountains, Opal range, the saline flats of the Mohave, and the Inyo range. On the eastern slopes of the Riverside Mountains, on the Lower Colorado, are exposed, for a distance of over 5 miles, layers of a gray siliceous highly-crystalline limestone, that must be referred to this formation; a view first expressed by Dr. Newberry.[*] The rock-surface is very uneven, full of little cavities, caused by the sand-winds that attacked the calcareous particles of the rock sooner than the siliceous ones and carried them off. In the rock itself paleontological evidence is in vain searched for, but the accompanying quartzite exhibits crinoidal forms.

Farther north, near Fort Mohave, is another Palæozoic region. The post stands upon a terrace 40 feet in height, consisting of rounded, water-worn bowlders washed down from the mountain-ranges on either side. These large traces, testifying of the activity of the river in former ages, when its bed was far above the present level, contain, among bowlders of trachyte, basalt, and granite, such of a peculiar quartzite resembling semi-opal, and with organic forms (chiefly of crinoids) in a state of astonishing perfection. Treatment with hydrochloric acid removes the last traces of adhering carbonate of lime and brings the forms still better to light. Proceeding farther up the river, Palæozoic limestone is again found at Bowlder Cañon. In the Opal Mountains, at the boundary of Nevada and California, this rock plays an important part, occupying portions of the very crest, exhibiting there frequently inclined strata; thus 3 miles southwest of Ivanpah dipping 60° to the west. In this limestone fossil remains are scarce, a single athyris and few crinoidal stems having been the poor result of a long search.

The flanks of the range are covered by a conglomerate consisting of pebbles of carboniferous limestone, granite, and quartzite, through which numerous gullies and arroyos have been washed. It is a striking fact that while the Opal Mountains are covered largely by Palæozoic strata, these were not met in the opposite Payute range, hardly 30 miles east and of about equal height.

Farther north this formation is met with in the Argus and Inyo ranges, where it

---

[*] See Lieutenant Ives's reconnaissance upon the navigability of the Colorado River.

acquires an especial interest on account of rich metalliferous lodes being associated with it.

At the saline flats of the Mohave (Soda Lake, Mohave Sink) the Palæozoic limestone is highly crystalline and siliceous, composes hills and mesas 30 to 40 feet in height, and its strata dip at an angle of 30° to the west.

THE MESOZOIC FORMATION.

On the entire trip through Eastern California cretaceous fossils were not seen, but a deep-red sandstone, that probably has to be referred to the Upper Carboniferous or Trias, as it forms close connection with the Palæozoic strata, was encountered in a few places, viz, the Opal Mountains and the northern spurs of the Black Cañon range, north of the deserted Mormon town Callville, in Southern Nevada. This sandstone, accompanied by gypsum, is exposed again opposite the mouth of the Virgin River, and appears to continue beneath the surface of the great detrital valley to the Cerbat range.

THE TERTIARY FORMATION.

This formation probably is wide-spread in the Mohave Desert, but hidden by the extensive Quaternary deposits. In the valley of the Mohave River thick beds of sandstone, clay, and conglomerate are exposed, whose Tertiary age was recognized by Prof. Jules Marcou.* Granite and quartzite contributed the pebbles to these conglomerates, and as the volcanic rocks, notwithstanding their great abundance all along this valley, do not participate, it is evident that their protrusion took place after the deposition of the Tertiary beds, which latter exhibit uplifts and dislocations caused by the volcanic protrusions. These inclined strata are again overlaid by horizontal undisturbed strata, of moderate thickness; as, for instance, 3 miles east of Grape-vine ranch; also a singular bend noticed in the sandstone strata 10 miles east of Camp Cady may be due to volcanic perturbations.

At Cañon Springs, on the eastern slopes of the southernmost spurs of the San Bernardino range, occur strata evidently of equal age with those of the Mohave River valley, and consisting of light-red and gray sandstone, and indurated bluish clay with seams of gypsum. These strata show a highly-inclined dip to the south and southwest, and exhibit singularly twisted and warped folds. In the horizontal strata of conglomerate overlying them pebbles of basalt and trachyte occur, while none are seen in the inclined strata beneath them.

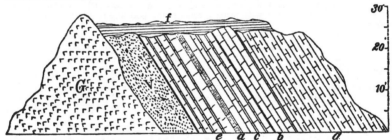

FIG. 1.—Section at Cañon Springs, San Diego County, California.

a. Dark pink-colored sandstone, 80 feet. ⎫
b. Yellow sandstone, 24 feet. ⎪
c. Light-pink sandstone, 32 feet. ⎬ Tertiary strata.
d. Yellow sandstone, 2 feet. ⎪
e. Gray sandstone interstratified with clay, 50 feet. ⎭
f. Post-volcanic strata.
g. Granite.
r. Intrusive rock, (trachyte.)

THE QUATERNARY FORMATION.

The Glacial epoch, by which the Quaternary in high and moderate latitudes was introduced, has left but few markings in Southern California. At that time the country was submerged in the Quaternary ocean, and the highest peaks projected as merely small cliffs above the surface. Still the period of cold was not entirely gone when the gradually-rising country acquired considerable dimensions, and the peaks commenced to bear glaciers to a small extent, as the traces of moraines in the San Bernardino and Jacinto Mountains indicate.

* Pacific Railroad Reports, volume 3.

The deposits of the Quaternary are formed by sandstone, conglomerate, clays, and gravel-beds, with occasional saline masses, and fill not only the valleys, but also occur at considerable altitudes in the uplifted mountain-ranges. To the saline efflorescences less attention was hitherto paid than they deserve. Their occurrence appears to indicate that since the recession of the ocean not enough rain has fallen to dissolve them and carry them off. It is true, in several instances, they are of so recent an origin that our conclusions must naturally be much restricted, but, in other cases again, they are of a considerable age. Their origin may be due either to the drying up of salt-lakes left in natural basins after the rising of the country above the sea, or to the desiccation of fresh-water lakes, containing generally, although a very small, amount of mineral salts in solution, or they are the residue of mineral springs.

As the chemical composition of these saline masses may give in many places a clew as to the origin, a number of specimens were analyzed. In the following three cases, the composition is that of table-salt of average quality:

Constituents. *	A.	B.	C.
Sodium chloride	95. 51	94. 02	95. 49
Sodium sulphate	2. 44	4. 35	2. 78
Calcium sulphate	1. 01	1. 24	0. 27
Magnesium chloride	0. 60	Trace.	Trace.
	99. 56	99. 61	98. 54

* In every analysis mentioned here, the substance was freed of its moisture in order to facilitate the comparison of the results.

A. Salt of Panamint Valley, deposit 1 to 2 feet deep, covering a number of square miles.

B. Salt covering the plains of Columbus, Nevada.

C. Salt of Death Valley, deposit 1 to 3 inches deep and covering many square miles. Analogous deposits exist in Salinas Valley. (For information and specimens regarding A and C, I am indebted to Lieut. R. Birnie, jr., of your survey.)

Other specimens have a very different composition, as the following table shows; the amount of sulphate of soda is increased and forms in one case (C) the whole soluble portion. Also, carbonate of soda forms a constituent pointing to another source than sea-water.

Constituents.	A.	B.	C.	D.
Insoluble (sand and clay)	42. 32	48. 68	44. 30	51. 57
Sodium chloride	38. 02	28. 08	Trace.	29. 00
Sodium sulphate	10. 81	8. 73	54. 08	19. 51
Sodium carbonate	7. 51	15. 06	None.	Trace.
	98. 66	100. 55	98. 38	100. 08

A. From the saline flats of the Mohave, thickness 1 to 3 inches; area, about half a square mile.

B. From Point of Rocks, forming isolated saline spots.

C. From shore of Kern Lake. }
D. From Tehachapi Pass. } Collected by D. A. Joy, geologist of party No. 2.

Another deposit of small extent, however, derived from a spring near Stone's Ferry, gave the following result:

Insoluble (clay)	57. 40
Sodium sulphate	32. 30
Magnesium sulphate	6. 32
Calcium sulphate	3. 51
	99. 53

An alkaline deposit near Benton, (Mono County,) gave the following result:

Insoluble (sand)	3. 70
Sodium carbonate	70. 31
Sodium sulphate	11. 25
Sodium chloride	13. 76
	99 02

Saline efflorescences composed of chloride and sulphate of sodium are the commonest; less frequent is the occurrence of sodium carbonate; still rarer are borates. At one place, Coyote Hole, Inyo County, the deposit of borate of soda is of considerable dimensions.

Extensive deposits of nearly pure rock-salt are found in the valley of the Virgin River, about 10 miles above its junction with the Colorado. Geological circumstances point toward their Quaternary age and indicate that Bowlder Cañon is older than these deposits. Before the cutting through of the Black Cañon range, i. e., before the completion of Bowlder Cañon by the Great Colorado, the water must have accumulated in a large lake 40 miles in width, and bounded in the east by the plateaux through which now the Grand Cañon extends. It is upon the bottom of this ancient fresh-water lake that the large rock-salt deposits are found. These conditions render it highly probable that their formation took place after or by the receding of the Quaternary ocean, and that Bowlder Cañon was in existence *before* the submergence of the country at this period, which becomes evident also by the characteristic position of the Quaternary conglomerates within the head of Bowlder Cañon and in the neighboring washes. On the other hand, evidence is furnished, by erosion, that the cañon is not older than the Tertiary epoch, for there are dikes of basalt cut through so perfectly even with the inclosing syenite at quite a considerable distance above the river, that we must conclude the basalt-dikes existed *before* the erosion of the cañon; that is, the age of the cañon has to be referred to the later Tertiary.

The lapse of time required for the completion of the great cañons of the Colorado is, after all, not so great as one would be inclined to estimate at first sight. If we take the erosion of Niagara Cañon, that is, the retrocession of Niagara Falls, as an example, which amounts, according to Hall and Lyell, to 1 foot a year, a period of one and a half million years would be required to complete the Grand Cañon of two hundred and eighty miles in length. Not less than by the saline efflorescences and salt-deposits is attention arrested by the conditions of the Quaternary conglomerate, it being exceedingly coarse and occupying frequently localities far distant from the mountains that contributed the pebbles. This fact can hardly find any other explanation than that powerful currents agitated the waters of the comparatively shallow Quaternary ocean and distributed the coarse material over large areas.

These conglomerates acquire a considerable thickness, as elucidated by the artesian borings of San Bernardino, which penetrate them for 140 feet. Neither limestone nor clay beds are struck by these borings, and the water rushes up as soon as the loose sand beneath the conglomerates is reached.

FRESH-WATER LAKES OF THE QUATERNARY PERIOD.

After the receding of the waves the country enjoyed a climate moister than that of the present day, as is indicated by the formation of a number of fresh-water lakes that left nothing but the barren clay bottom, whose numerous imbedded shells give evidence of a more numerous animated life.

Owens Lake, now charged to such a degree with salts* that molluscous or fish life is impossible, was formerly an extensive fresh-water lake, as the recent shells in its immediate vicinity indicate.

Mono Lake had gigantic dimensions, to judge from the well-defined shore-lines, 15 miles from its present margin.

Soda Lake, now a barren clay flat of forbidding appearance and desolating sight, (the saline flats of the Mohave River,) was a large water basin, but the largest lake of Southern California existed in the Coahuila Valley, and to all appearances up to a period less than a thousand years before the present day.† Among the coarse granite-sand of its former western shores formed by the eastern slopes of the Jacinto Mountains are found millions of minute fresh-water shells, (*Amnicola thryonia*,) most delicate structures, that would long ago have been crushed between the rolling sand and have disappeared, if the period elapsed since the desiccation had been a considerable one. A calcareous crust, several inches in thickness and of quite fresh appearance, covers the granite of the slopes, and marks exceedingly well, by a far-stretching horizontal line, the shores of the former lake,‡ whose depth was about 125 feet and surface probably over 1,000 square miles. This porous crust, of the structure of a sponge, contains numerous fresh-water shells. The now dry clay bottom is covered with patches of white salt efflorescences, and nothing save an occasional stunted Halostachys inter-

---

* Should this be due to the bursting forth of mineral and thermal springs opened by volcanic forces? Lava of recent origin occurs in the vicinity and mineral springs are numerous in the neighboring Coso Range.

† The Kauvuya Indians of that region have still a tradition of the lost lake.

‡ This region has been described before by Prof. W. Blake. See Pacific Railroad Reports, vol. 5.

rupts the evenness of the flat. A specimen of this clayey soil gave, on analysis, the following result :

Clay with fine silt	64.70
Calcium carbonate	3.36
Calcium sulphate	0.43
Sodium chloride	10.52
Sodium sulphate	3.90
Magnesium sulphate	Trace.
Calcium phosphate	Trace.
Potassa, lithia	Traces.
Chemically-bound water	15.68
	98.59

This soil is unfit for agriculture, containing too much clay and salts.

If, on the one hand, desiccation of former lakes proves that the amount of evaporation exceeded that of the aqueous precipitates, there exist, on the other hand, facts tending to prove that the dryness of the climate *is still on the increase*, namely, the disappearance of forests within the last three centuries and the drying up of springs within the last fifty years.*

These phenomena recall our observations in New Mexico and Arizona, where indications of increasing dryness are numerous; to mention only the forests of dead cedar-trees standing mummy-like, the occurrence of shells of land-snails (*Planorbis*) in localities where not a single snail is found at the present day, the deserted ant-hills, the dry arroyos, and the ruined towns of now barren tracts.†

In close connection with the decrease of aqueous precipitates in New Mexico, Arizona, and Eastern California, appears to stand the *increase* in Utah, the Great Salt Lake having risen 15 feet in the last twenty-five years, and will, if this accumulation continues, submerge the capital of the Latter Day Saints at no very remote period.

The most satisfactory explanation of these phenomena appears to me can be given by the assumption of changes in the country-level. A gradually rising country will experience a *decrease* in the annual mean of temperature, consequently, also, in the amount of evaporation; while, on the other hand, the aqueous precipitates will *increase*, the distance of the clouds becoming smaller. As the attraction grows with the square of proximity, a lifting of 100 feet of an extensive mountainous country will suffice to change the climate perceptibly. Two causes, therefore, co-operate to increase number and volume of springs and to swell creeks and lakes. My hypothesis is this, that New Mexico, Arizona, and Eastern California are undergoing a gradual subsidence, while Utah is, like the coast of California, slowly upheaved.

A characteristic feature of the Mohave Desert is the black coating of rocks and of the bowlders and gravel‡ that cover the barren plains, adding perceptibly to the dismal impression of the scene, the blackening appearing like a mourning garb for departed flora. Miners call these rocks "sunburnt," and as curious as this expression sounds to the ear of the naturalist, there is a grain of truth in it, the black coloration being of the deepest shade where the surface of the bowlder is exposed to the direct sunlight, while at the under side it is much less developed and sometimes replaced by a reddish color.

Analogous phenomena were observed by Littel in the Libyan Desert, by others at Syene, at the Congo River, and by Humboldt on the Orinoco. Berzelius, who examined the coating from the latter locality, upon the request of Humboldt, declared it due to mixture of oxide of iron and manganese. A chemist can hardly be long in doubt in regard to it, as hydrochloric acid dissolves the coating with liberation of chlorine, and manganese easily can be shown in the solution thus obtained. It appeared to me a matter of some interest to ascertain the quantities necessary for the production of the black coating, and for this purpose a blackened small bowlder, weighing 80 grams, was treated with hydrochloric acid until the natural color of the rock (granite) developed. The acid poured off gave:

Sesquioxide of iron	0.078 gram.
Binoxide of manganese	0.038 gram.
Oxide of nickel	traces.

* The miners of El Dorado Cañon, in the Black Cañon range, stated to me the ceasing of a large spring in the vicinity within the past fifteen years.

† See annual report U. S. Geographical Surveys West of the One Hundredth Meridian. 1875.

‡ On the eastern slopes of the Payute range, some 20 miles east of Cottonwood Island, a bed of coarse conglomerate, 10 feet in thickness, was observed, of which each pebble was provided with the black coating.

The questions arise:

1st. Whence was the manganese derived?

2d. How was the coating produced?

Chemistry points to the widely-spread *pink* and amethyst colored granites, gneisses, limestones, and sandstones, and leads to the conclusion that the manganese present in the state of proto and sesqui oxide in these rocks furnished also the binoxide for the black coating of the Quaternary drift, by becoming dissolved as carbonate of manganese upon the disintegration of the rocks, covered partly with the water of the slowly-receding, shallow ocean, and by being deposited afterward upon the rocks of the ground, where it changed gradually from the state of proto and sesqui oxide to the binoxide by the influence of air and sunlight.* This fact forms an analagon to the production of coatings of oxide of iron by waters that contain ferrous carbonate; and can be observed on a grand scale at localities where earthy carbonate of manganese is found. B. von Cotta mentions one instance of this kind in his "Treatise on ore deposits" occurring in the gold-bearing regions of the Rhine: "Widely-extended strata of those slates consist largely of white or reddish rhodonite, (carbonate of manganese,) which, when exposed to the air, turns *as black as coal, since it becomes incrusted by a very thin layer of manganite.*"

There can hardly exist a doubt that the pink color of the lime and sand stones of the Mohave Desert also is due to the manganese derived from the granites,† while, on the other hand, the volcanic rocks may have contributed their share in coloring the Quaternary drift.

At Mountain Springs, in the Cerbat range, a pink chalcedony was observed that most probably has derived its manganese from the trachydolerite of the vicinity.

That the pink color of the palæozoic limestone of those regions is due to manganese, was easily proven by dissolving a piece in hydrochloric acid, adding excess of ammonia, filtering off the hydrated oxide of iron, precipitating the filtrate by sulphide of ammonium, and fusing the precipitate with soda and a little saltpeter.

THE ERUPTIVE FORMATION.

The eruptive activity, once so extensive in the countries west of the Rocky Mountains, appears to have reached its summit in California, or, more generally mentioned, in the countries of the Pacific coast. Here appear to exist unusually favorite regions of the operations of the Plutonic powers; for the most ancient as well as the most modern geological records tell of convulsions and outbursts of molten masses, and even at the present day Pluto manifests his subterranean energy by frequent shocks and earthquakes.

There is no mountain-range of Southern California free from eruptive material, which either occurs in injections and dikes, or forms entire hill-ranges and mountains, nor are there any known rocks that do not find a representative here. From the oldest erupted gneiss, syenite, diorite, up to the porphyries, trachytes, and basalts, the series is complete. In coloration, in structural character, no greater variation could be found, and in the bursting, uplifting, and dislocating of sedimentary rocks, a geological genius could not be more productive in furnishing designs.

Eruptive gneiss.—Near the northern end of Coahuila Valley, rise abruptly above the sandy plain two hills 20 to 30 feet in height, of highly crystalline limestone, of either primitive or palæozoic age, whose strata dip at an angle of 36° to 40° to the eastward, traversed by a dike of a highly micaceous gneiss 1 to 2 feet in thickness.

The rocks are at the contact-surfaces very friable and metamorphosed. The intrusive gneiss shows by the position of its mica-plates a stratification parallel to the limestone layers, indicating the effect of pressure during the consolidation of the injected rock mass.

Eruptive syenite.—A great portion of the eastern slopes of the Buena Vista Mountains, the portion of the Inyo range in which the mining-town Cerro Gordo is situated, is composed of syenite, to whose eruption is not only due the remarkable displacement of strata of palæozoic limestone on the eastern slopes, (see Fig. 2,) but also the disturbances produced on the west side of the mountains, the strata of slate, sandstone, and limestone standing on end for a distance of several miles.

Section for the eastern slopes of Buena Vista Mountain 3 miles east of the mining-town of Cerro Gordo. Strata of palæozoic limestone, P, standing upon their vertex, and forming with each other an angle of about 80°. E, eruptive syenite, entering the strata like a wedge.

---

* Mr. Joy, geological assistant of Division No. 2, mentions in his notes a hot spring near Montan's ranch, California, whose waters deposit a black coating over rocks, probably due to the formation of binoxide of manganese from the protocarbonate eld in solution by the spring-water.

† Pink granites were observed at the Riverside Mountains, Monument Mountains, Chukawalla Peak, Mohave River Valley, Payute range, and other localities.

Eruptive granite.—At the Saline Flats of the Mohave River, at Dead Mountain, and in the Opal ranges, granite appears under circumstances indicating the eruptive character of the rock.

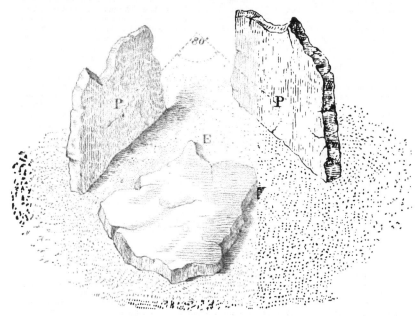

FIG. 2.—From Buena Vista Mountain.

Eruptive diorite.—Two narrow dioritic dikes, 2 to 3 feet thick and but about 15 distant from each other, traverse the calcareous sandstone of palæozoic age near Duckweiler Station, at Owens Lake, in such a manner that the phenomenon strikes even those who never were devoted to observations in nature most impressively, (see Fig. 3.) At the contact-surface of the eruptive and the sedimentary rock, the latter is fritted and very hard, (so-called porcelain jasper,) due to the metamorphosis produced by the heat of the erupted rock.

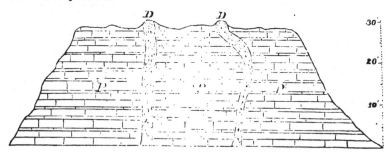

FIG. 3.—Section in vicinity of Owen's Lake.

P, sandstone wall, 30 feet high; strata dipping at an angle of 24° toward the east; D, dikes of diorite.

In the vicinity of the mining-town Ivanpah, in the Opal Mountains, and farther east, at Mount Newberry, on the Colorado River, dikes of diorite in granite are very extensive, and the rock assumes in part a porphyritic structure.

At Chuckavalla Peak, in San Diego County, occurs an eruptive rock of the appearance of diorite, but has to be probably referred to the propylite of Richthofen, although it is difficult to draw a line of distinction between these two rocks, propylite being in many cases nothing but Tertiary diorite.

Andesite.—A rock with a black vesicular matrix, containing imbedded well-formed crystals of orthoclase and hornblende, composes several peaks of the Payute range about 28 miles east of the Colorado River, at Cottonwood Island. I have no doubt that it corresponds to Richthofen's andesite, a volcanic rock so prominent in the South American Andes; whence its name.

* *Trachyte.*—This is by far the most extensive of all erupted materials in California, and the varieties which it exhibits are numerous. Basalt, covering immense tracts in Arizona and New Mexico, dwindles here into insignificance. The following list gives an approximate idea of the varieties met in various places of Southeastern California and adjacent regions :

Locality.	Matrix.	Imbedded minerals.
Mohave Range	Pink	Sanidine, biotite.
	Brown, vesicular	Sanidine, (sparingly.)
	Gray-brown	Oligoclase and zeolite.
	Brownish-lilac	Sanidine and hornblende.
	Gray	Hornblende, (sparingly.)
	Reddish-pink	Quartz, muscovite, orthoclase, and sparingly hornblende.
Riverside Mountains	Red	Orthoclase.
	Lilac	Oligoclase and biotite.
Cottonwood Island, (Colorado River.)	Gray, vitreous	Many large orthoclase crystals, with little biotite.
	Brown	Red orthoclase and white sanidine.
	Pink brown	Biotite and sanidine.
Dead Mountain, on the Colorado River.	Gray pink	Oligoclase and hornblende.
	Brown, vesicular	Zeolite, with (sparingly) hornblende.
	Light gray	Large orthoclase crystals, sparingly hornblende.
	Gray, amygdaloid	None.
Black Cañon range, (north of the former.)	Brown, amygdaloid	Sanidine and hornblende.
		Zeolites in the amygdaloid spaces.
	Green, vitreous	Oligoclase and biotite.
Payute range	Reddish	Orthoclase and biotite.
Callville	Slate color, vesicular	Orthoclase.
Cucamungo Peak	Slate color	Do.
El Dorado Cañon	Pink brown	White orthoclase.
	Gray pink	Densely crowded with small crystals of feldspar, biotite, and hornblende.
Blue Ridge Mountains, Northwest Arizona.	Nearly white, vitreous	Much sanidine and little biotite.
	Light green	Quartz, orthoclase, and biotite.
	Reddish, very siliceous	None.
Detrital Valley, Northwest Arizona.	Gray pink	Oligoclase and biotite.
	Gray, vesicular	Large number of small sanidine crystals.
Caves, Mohave River Valley.	Red, vesicular	Orthoclase.
	Gray	Biotite and sparingly oligoclase.
	Gray, vitreous	Sanidine in large crystals.
	Dark gray	Little oligoclase and hornblende.
Saline Flats of the Mohave River.	Slate color	Orthoclase.
	Gray, vesicular	Quartz, hornblende, and orthoclase.
Grapevine Ranch, Mohave River Valley.	Reddish	Sparingly sanidine and hornblende.
Camp Cady, Mohave River Valley.	Light gray	Sanidine, (sparingly.)
Chuckawalla Peak	Pink	Hornblende, oligoclase.
Cañon Springs	Red, with black stripes	None.

That these trachytic outbursts did not take place at one and the same time, but were repeated at long intervals, becomes evident after even a cursory examination. Trachytic masses repeatedly forced their way up through the identical fissures, and examples of this kind are well exhibited at the Needles † in the vicinity of the cañon which the Colorado has cut through the Mohave range. Here the strata of volcanic tufa show

*All the different varieties of volcanic rocks mentioned in this report have been collected. Of a large number microscopical sections will be studied by a specialist in this branch.

† These gigantic pinnacles, consisting of trachyte and trachytic breccia, have most probably obtained their singular shape by the erosive action of the river, when it was formerly running at a level of 1,000 feet and higher above the present bed.

manifold dislocations by later outbursts, and one of the instances is represented by the following section:

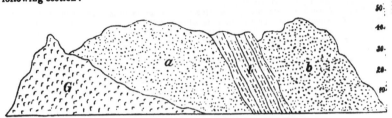

FIG. 4.—Section from the cañon at the Needles.

G. Granite.
a. Trachyte with pink matrix. containing imbedded quartz, orthoclase, and biotite.
b. Trachyte with brown matrix, imbedded sparingly oligoclase and hornblende.
t. Trachytic tufa containing pebbles of a trachyte with a reddish matrix, in which is imbedded sparingly oligoclase.

Here we have evidence of three distinct outbursts of trachytic masses; the trachyte that gave rise to the formation of the tufa, *t*, being the oldest, then following *a*, uplifting and dislocating the tufa strata, and finally the trachyte, *b*, protruded and overflowed the tufa on the opposite side from *a*.

In the Blue Ridge Mountains and Mohave range the trachytic eruptions appear to have reached their maximum, the masses measuring thousands of feet in height. At Union Pass, in the Blue Ridge Mountains, rhyolite, felsite, pitchstone, and globular porphyry, or pyromerid, form the accompaniments of the trachyte, while in the Mohave range basalt is associated with it. The volcanic tufas, conglomerates, and breccias of the latter range contain only material derived from trachyte and none of the neighboring basalt; thus giving evidence of the more recent age of the latter. The primitive rocks of the Mohave range are not so extensively covered by volcanic material as in the Blue Ridge Mountains. However, in the singular outlines of their peaks and crests, assuming fantastic shapes of towers, domes, and castles, the more sharply defined as they are devoid of forests and verdure, both rival each other.

In the Riverside Mountains obsidian accompanies the trachyte; at Chuckawalla Peak, basalt and a phonolite rich in zeolite; in the southern spurs of the San Bernardino Mountains, pumice; and in the Mohave River Valley, rhyolite and recent lavas. In this valley a fine section of volcanic rocks is exposed at the Cañon of the Caves, where, for a stretch of 5 miles, a series of trachytes of most vividly contrasting colors impress the traveler with the grandeur of inorganic nature. The spaces and fissures between the dikes are filled by Quaternary clays and conglomerate.

Trachydolerite.—This rock forms dikes and hills in the Mohave, Payute, and Cerbat ranges, at Mountain Springs. It also is largely spread upon the island of Santa Cruz.

Basalt.—The most notable localities are the Virgin River Valley and vicinity, and Black Mountain range, in Southwestern Nevada; the eastern portion of Detrital Valley, in Northwestern Arizona; the Mohave range, and the Mohave River,* in Eastern California. The displacements due to its protrusion are on a grand scale in the mountains north of Callville, in Southern Nevada, from which region the accompanying section is taken.

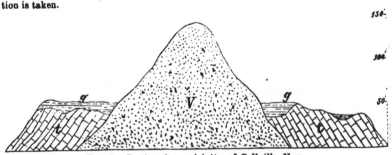

FIG. 5.—Section from vicinity of Callville, Nev.

V, Basaltic protrusion; *t*, Triassic strata; *q*. Quaternary conglomerate.

* The Mohave range has nothing to do with the Mohave River, being situated much farther east, and crossing the valley of the Colorado.

183

As I have demonstrated the presence of cobalt and nickel in basalts of New Mexico, (see survey reports Vol. III,) I searched for these elements in the California basalts, and was surprised in finding small quantities of them, widely distributed.

At first one might form the conclusion that their presence is due to the olivine, a mineral which contains nickel as a normal constituent, as shown by Stromeyer, but a closer examination will refute this hypothesis, as the basalts of Southern California contain but rarely olivine as an accessory constituent. I am of the opinion that these elements are connected with the magnetic iron of the basalts, and this idea is sustained by the fact that upon digestion of the finely pulverized rock with concentrated hydrochloric acid for from six to twelve hours the whole of the nickel and cobalt is with the iron in the solution, while the labradoritic and pyroxonic particles are but little attacked. If the acid solution thus obtained be freed from most of its acid, then supersaturated with ammonia, filtered, evaporated, ignited, and the residue treated with dilute hydrochloric acid, the presence of cobalt and nickel may be easily demonstrated by the common methods. Among the rocks and tufas tested I found the respectively largest proportion in the *tufas* of the trachydolerites of the island of Santa Cruz.

Lava.—Vesicular lava, of apparently very modern date, was observed at Camp Cady, in the Mohave River Valley, and at the southeastern margin of Owens Lake.

Are we now justified in assuming that with the lava outbursts the plutonic powers left their favorite theater, or have retired to seek another field of activity? The circumstances do not testify to the affirmative of such an assumption, for frequent rumblings indicate the existence of latent volcanic powers. In the vicinity of Owens Lake hardly three weeks pass without a slight shock being felt. The most violent shock remembered by the settlers in the Owens River Valley took place at 3 p. m. of the 26th of March, 1872, reaching from Owens Lake north as far as Aurora, Nev., a distance of about 130 miles. The changes thereby produced at Owens Lake are still visible. I saw the fractures of the earth, now partly filled with sand, the stratum of sandstone raised 14 feet, and a spring produced by newly-formed fissures. Huge masses of rock became detached on the east side of the Buena Vista Mountain and hurried thundering down the slopes. At the village of Lone Pine resulted most terrible effects; the dwellings were thrown down and sixty persons killed in a twinkling of an eye. Similar effects, but on a smaller scale, were experienced at Independence. The vibratious lasted several seconds, and the noise resembled that of a train of heavy wagons hurrying over a stone pavement.

At San Diego and Fort Yuma, in Southern California, the shocks average nine per annum. The most violent in the memory of the whites was that of November 9, 1852, with a great frequency of subsequent shocks, lasting for about two months. Professor Blake reports that a portion of Chimney Peak was thrown down, and cracks and fissures opened in wide circumference of Fort Yuma. Forty miles south from this post a number of mud volcanoes were formed, whose waters had a temperature of 170° F.

The frequency of earthquakes did not fail to be noticed by the early Spanish visitors. Sebastian Viscaino said, (1596:) "No es el mar peligroso es su costa tierra muy templada." (It is not the sea that is dangerous, it is the trembling coast.)

GENERAL GEOLOGICAL HISTORY OF SOUTHERN CALIFORNIA.

This is, in a few words, as follows: The country was partially submerged during the Azoic period, and, probably, with repeated changes of level, partly during the Palæozoic, began to rise and remained above water until toward the close of the Cretaceous, when it sank to a great extent below the ocean and remained there up to the middle Tertiary; began then to rise, and reached finally a level so far above the present one that the islands now separated from the coast by the Santa Barbara Channel formed part of the continent, as is indicated by the elephants' teeth* found on them, and by the occurrence of a small fox derived from the continental fox by gradual transmutations.

Toward the close of the Tertiary the country was the home of the buffalo, horse, rhinoceros, llama, tiger, and mastodon.† That brilliant age experienced, however, a great series of volcanic disturbances that had commenced early in the Tertiary, and continued up to quite modern date, in longer or shorter intervals.

With the decline of the Tertiary, the country sank again until the mountains projected as mere cliffs above the ocean. Mighty currents agitated the waters and assisted in forming the Quaternary conglomerates. The Quaternary period was so far advanced that the glacial epoch had passed its zenith, when the country emerged once more from the watery grave to salute the vivifying sunbeams. The coast is still rising at the present day, at a rate of about 5 feet per century, and, as the Santa Barbara Channel has 60 fathoms depth, the islands off the coast will, in seven thousand two hundred years, be again united with the main land, if the rise thus continues. It is, however,

* See Proceedings of California Academy of Sciences, 1873.
† See article of Professor Leidy in the Proceedings of California Academy, 1873.

probable there will be interruptions, cessations, for a longer or shorter time, as may be inferred from the terraces all along the coast of Southern California, indicating unmistakably the shore-lines of the ocean during the cessations in preceding periods.

ON THE METALLIFEROUS VEINS OF CALIFORNIA AND SEVERAL PECULIAR ORES.

In regard to mineral wealth, California occupies a prominent position, not only with reference to the quantities of the precious metals, but also to those metallic elements that occur but sparingly in other countries. Gold, silver, platinum, mercury, copper, lead, antimony, arsenic, tungsten, tellurium, molybdenum, bismuth, chromium, manganese, iron, nickel, cobalt, zinc, are found in various ores.

A series of the most important ore-veins, among them some discovered quite recently, lie in Inyo and Mono Counties, for description of which the reader is referred to the general report. Attention is only invited here to some peculiarities connected with them; above all, the walls slickensided to such a degree of perfection and to an extent rarely witnessed in other mining regions. The Hemlock lode, in the Panamint Mountains, shows this phenomenon on a grand scale and of extraordinary beauty. Another peculiarity are the so-called "horses," large bowlders of wall-rock that became detached and dropped into the fissure during the process of vein-making. Further mention must be made of the "breaks," large hollow fissures encountered occasionally in the walls. The "horses," as well as the "breaks," are doubtless due to volcanic disturbances; the former to earthquakes *before* the vein was filled, the latter long after the perfection of it.

If we look at the fissure-veins, on one hand, and observe on the other the immense masses of erupted rocks in those mountains, one cannot fail to suspect a connection, and to ascribe, with Baron von Beust, the production of fissures to volcanic forces, or earthquakes that accompanied the eruption of molten rocks.

But how, may we ask, were these cracks filled with vein-matter? Neither the lateral infiltration nor the injection hypothesis are sufficient to account for all peculiarities met here, but if we consider—with Elie de Beaumont—the veins as the product of hot waters which entered charged with mineral salts these immense fissures from beneath and filled them to the brim, we may easily explain the presence of the quartzite, the carbonate of lime, carbonates of lead and copper,* as well as the occasional *banded structure* of the veins, the deposits being made at first upon the walls of the fissures, and gradually filling up to the center. If the character of the mineral water changed after a certain period, a change of the nature of the deposit would of course be the result, and thus the banded structure seen occasionally in veins accounted for.

The deposition of vein-matter from the hot waters is partially due to a loss of temperature of the water, partly to a loss of carbonic acid in contact with the air at the fissure surface. The formation of the metallic sulphurets might be explained by the subsequent entering of waters charged with sulphureted hydrogen, converting the metallic carbonates into the sulphurets.†

No country abounds to such a degree with hot springs as California and Nevada, and here it is especially the vicinity of mining districts where they are encountered. There are thermal springs in Death and Panamint Valleys, adjoining the Panamint Mountains with their veins; there are others in the Coso range, near the mines of Darwin, another near Blind Spring mining district, and again between Carson City and the famous Virginia City mines. In one of the latter, the Imperial Mine, the workingmen struck, a few years ago, hot water emitted with such force that they could not escape a thorough scalding of their feet. Who would deny that the system of thermal springs was formerly much more extensive in California than at present, if one sees the glaring and decisive marks they have left, the cones they had built, the coatings produced? It may be that thermal springs are forming at the present day far beneath the surface mineral veins, and their final filling may be the result of many springs dying out after certain periods, the channels becoming closed up.

It may also be, that if large quantities of water of certain hot springs be analyzed, traces of metals, as lead, copper, or silver, could be discovered. I add here an interesting passage quoted from Cotta's "Ore Deposits," page 531: "It appears from Daubrée's researches, that the mineral water of Plombières still deposits minerals which are characteristic for the variety of lodes mentioned, and it is by no means impossible that there, at a corresponding depth below the surface, such lodes are still forming."

* There is not a single metallic carbonate that would not be a little soluble in water charged with free carbonic acid.

† A singular fact worth recording is that, as a rule, the lead-mines of Mono and Inyo Counties are in limestone, while the copper-ores occur as an impregnation of quartzite ledges in primitive or erupted rocks.

‡ A fine specimen of a snow-white coating over paleozoic limestone is seen on the eastern slopes of the Buena Vista Mountains, three miles east of Cerra Gordo.

ANALYSIS OF THE TIN ORE OF TIMESCAL, SAN DIEGO COUNTY, CALIFORNIA.

The tin mine of Timescal became famous by the continuous litigations it caused, not by the metal produced. No work was done for seventeen years, litigation preventing it. I was unable myself to visit the mine, hence can express no opinion as to the prospects, but received specimens of the peculiar ore from two persons, one in San Bernardino, the other in Los Angeles; the specimens resembled each other perfectly, and were said to represent the average ore. It is difficult to recognize in this black rock a tin ore; nobody would suspect the presence of tin; still chemical analysis reveals it, although the quantities, at least in the specimens at my disposition, are small.

Luster, dull; streak, grayish; hardness $= 3$; specific gravity $= 3.40$. In the uniform black mass brighter particles of a crystalline structure are recognizable. The ore contains no water of hydration, and is but with great difficulty attacked by acids, whereby some oxide of iron is obtained in solution. Fusing potassa decomposes the ore rapidly, and if the mass be treated with water a yellow powder, containing oxide of iron, silica, potassa, and antimonic oxide, remains insoluble, while tin is found in the alkaline solution. The former is easily decomposed by hydrochloric acid.

The quantitative analysis gave—

Silica	39.70
Oxide of iron*	35.85
Alumina	5.01
Oxide of tin	3.52
Oxide of antimony	3.98
Lime	5.81
Potassa	} 4.32
Soda	
Titanic acid	traces.
	98.19

ANALYSIS OF PARTZITE, OF BLIND SPRING DISTRICT, CALIFORNIA.

This peculiar silver-ore is thus far known to occur only at one locality, the Blind Spring Mountains, near Benton, in Mono County, California. I visited myself the mine in which it used to be found in abundance, (the Comanche lode;) but at present, with the deepening of the shaft, it becomes very rare, the sulphurets[†] predominating over the oxidized ores and carbonates. The color of the partzite varies from yellowish-green to black; no crystallization is perceptible; luster, dull; fracture, conchoidal; streak, gray to brown; hardness $= 3$; specific gravity $= 5.324$.

Nitric acid dissolves it partially, with disengagement of red fumes, due to the presence of protoxide of iron in the ore becoming peroxodized. The quantitative analysis revealed the presence of copper, lead, iron, silver, and antimony. The latter element is present in the state of antimonic acid, as I have proved by treatment of the finely-pulverized ore with potash and addition of chloride of sodium. The ore has been previously analyzed by A. Arents,[‡] who calculates the antimony as the oxide, and finds the specific gravity $= 3.8$, a number which is utterly impossible in consideration of the composition, as antimonial ores have never less than 5.2 specific gravity. The result below given would correspond approximately to the general formula :

$$S bo_5, 4 Ro, 4 Ho,$$

but it is very doubtful whether this ore can be designated as a mineral species, as the necessary requisites are wanting. It is probably a mixture of antimoniate of iron and silver with the hydrated oxides of copper and lead.

The great deviations in the composition of this ore are shown by the following comparison of the analysis made by Mr. A. Arents (2) and myself, (1.)

	(1.)	(2.)
Antimonic acid	34.03	47.65
Oxide of copper	18.39	32.11
Oxide of silver	3.28	6.12
Oxide of lead	23.51	2.01
Protoxide of iron	10.04	2.33
Water	11.30	8.29
	100.55	98.51

* A small portion of the iron is present as magnetic oxide.
† Argentiferous antimonial lead and copper sulphurets.
‡ American Journal of Science and Arts, 1867.

List of minerals and rocks collected and observed by A. B. Conkling, geologist, assistant of Division 3, in Southern Colorado and Northern New Mexico, and by O. Loew, mineralogist and chemist of Division 4, in Southeastern California and adjacent parts of Nevada and Arizona.

COLLECTED BY OSCAR LOEW.

Minerals.	Locality in California, &c.	Locality in Colorado, &c.	Remarks.
1. ELEMENTS.			
Gold, native	Little Creek; Cerro Gordo	Ute Creek; Elizabethtown.	In small quantities in most of the California silver-ores.
Silver, native	Cerbat range, Ariz.; Panamint.	Rosita	
Sulphur	Coso range		
Graphite		Santa Fé, N. Mex.; Clifton.	In small quantities.
2. SULPHIDES.			
Pyrite	Cerbat range, Ariz.	Rosita; Golconda and Aztec mines.	
Sphalerite	Cerbat and Inyo ranges	Golconda mine	
Stibnite	Tejon Pass		
Molybdenite	Payute district; Benton		
Chalcocite	Ivanpah, Cal.; Chukavalla Peak; Panamint; Cerro Gordo; Hualapai district; Cerbat range; Ord district; Benton.	Rosita	Often argentiferous.
Galenite	Darwin; Ivanpah; Lyttle Creek Cañon; Cerro Gordo; Chukawalla Peak.	Rosita; Golconda and Aztec mines.	Mostly argentiferous.
Cinnabar	New Almaden		
Pyrargyrite	Hualapai district, Ariz.	Rosita	
Tetrahedrite	Benton	do	Argentiferous.
Argentite	Cerro Gordo	do	
Stephanite	Virginia City, Nev.	do	
Sternbergite	Camp El Dorado, Nev		
Strohmegerite		Rosita.	
Stetefeldite	Panamint; Ivanpah		
Realgar	Cerro Gordo		
Chalcopyrite	Hualapai district, Ariz	Rosita; Aztec mine	
3. CHLORIDES.			
Cerargyrite	Hualapai district, Ariz.; Ivanpah; Panamint; New York district.	Rosita	
Rock salt	Valley of the Lower Virgin; Death Valley; Salinas Valley; Panamint Valley.		
4. OXIDES.			
Limonite	San Bernardino Mountains	Trinidad; Gallinas Creek.	
Hæmatite	Cucamonga Peak; Cerbat range, Ariz.	Elizabeth; Baldy Mountain.	
Partzite	Benton		
Cassiterite	Timescal		
Massicot	Darwin; Cerro Gordo		Generally mixed with minium and cerusite.
Cuprite	Ord district; Hualapai district		
5. SILICA AND SILICATES.			
Quartz crystals	Cerbat range; San Bernardino Mountains.	Spanish Peaks	
Flint and hornstone	Camp Cady; Caves		
Jasper	Riverside Mountains		
Chalcedony	Cerbat range; Mohave River Valley.		Of fine amethyst-color, at Mountain Spring, in the Cerbat range.
Garnets	Riverside Mountains; Jacinto Mountains.		
Steatite	Santa Barbara County		
Zeolite	Mohave Mountains; Chukawalla Peak.		Accessory constituent of basalts.
Amphibole	Gorgonio Pass; Chukavalla Peak		
Turmaline	San Jacinto Mountains		
Muscovite	Hualapai district		In large plates.
Titanite	San Jacinto Mountains		

List of minerals and rocks collected, &c.—Continued.

COLLECTED BY OSCAR LOEW—Continued.

Minerals.	Locality in California, &c.	Locality in Colorado, &c.	Remarks.
6. SULPHATES.			
Anglesite	Cerro Gordo	....	
Barite		Rosita	
Gypsum	Callville; Saint Thomas, Nev ...	Saint Charles; Colorado Cañon.	.
Soda-sulphate	Mohave River Valley; Kern Lake.	....	As afflorescences.
7. CARBONATES, BORATES, &C.		.	
Malachite	Ord district; Ivanpah; Panamint.	Coyote; Golconda and Aztec mines.	
Siderite	....	Rosita	
Borax	Coyote hole	....	
Cerusah	Darwin ; Cerro Gordo		
Trona	Benton		
Calcite	....	Aztec mine. N. Mex.; Nutritas; Cimmarron.	
8. ORGANIC MATTERS.			
Asphaltum	Santa Cruz Island; Santa Barbara.	....	
Mineral-oil	San Fernando	....	
Lignite	....	Trinidad ; Fort Union.	

Rocks.	Locality in Southern California and adjacent portions of Nevada and Arizona.
Granite	Riverside, Mohave, Monument, San Bernardino, San Jacinto, Opal, Payute, Cerbat and Panamint ranges.
Syenite	Mohave, Riverside, and Inyo ranges.
Mica-schist	Chukawalla and Riverside Mountains.
Gneiss	Cerbat, Riverside, San Bernardino, and Inyo ranges.
Hornblende-schist	Cerbat range, Riverside Mountains.
Diorite	Opal Mountains, Mount Newberry, Chukawalla Peak, Inyo range.
Trachyte	San Bernardino, Riverside, Mohave, Monument, Blue Ridge, Cerbat, Payute and Opal ranges, Chukawalla Peak, Mohave River Valley.
Rhyolite	Mohave range and Blue Ridge Mountains, valley of the Mohave River, Riverside Mountains.
Pyromerid	Union Pass, in the Blue Ridge Mountains.
Andesite	Payute range.
Basalt	Cerbat, Black Cañon, and Mohave ranges, Providence Mountains, Virgin River Valley, Chukavalla Peak.
Lava	Owens Lake Valley, Mohave River Valley, Black Cañon range.
Quartzite	Riverside, San Bernardino, and Panamint ranges.
Limestone	Riverside, Panamint, Inyo, and Argus ranges, Saline flats of the Mohave, Bowlder Cañon.
Conglomerates	Black Cañon range, Colorado River Valley, Providence Mountains, Mohave River Valley, San Bernardino Mountains, Opal range.
Tufa	Santa Cruz Island, Mohave range.
Sandstone	San Bernardino, Inyo, Opal and Black Cañon range, Cañon Springs, Detrital Valley.

Rocks.	Locality in Southern Colorado and Northern New Mexico.
	New Mexico.
Granite	Ute Creek. Cimarron Creek. Cieneguilla Creek. Rio Colorado Cañon. Taos range, west side. Moreno Valley. Upper Cimarron Creek. Taos range. Head of Cimmarron Creek. Colorado Cañon. Golconda mine. Costilla Mountain. Taos range, east side. Elisabeth. Baldy Mountain. Aztec mine. Rayado Cañon.
	Colorado.
	Indian Creek. Top of Culebra Peak. Top of Cerro Blanco. Head of Purgatoire River.
	New Mexico.
Granulite	Comanche Creek. Head of Cimarron Creek. Aztec mine. Uraca Mountain. Rayado Cañon.
Basalt	Hole in the Rock. Rio Hondo. Uraca Creek. Cañon of San Antonio Creek. Fort Union. Rio Grande Cañon. Rio Colorado Cañon. Ocaté Crater. Near Laughlin's Peak. Costilla Cañon.

List of minerals and rocks collected, &c.—Continued.

COLLECTED BY OSCAR LOEW—Continued.

Minerals.	Locality in Southern Colorado and Northern New Mexico.
	Colorado.
	Culebra Creek. San Luis Valley. Huerfano Butte. Head of Cuchara River. Fort Garland.
	New Mexico.
Trachyte ...,..........	San Antonio Creek. Ridge of East Costilla Peak. Uraca Creek. Comanche Creek.
	Colorado.
	Near East Spanish Peak. Cerro Blanco. Gardner. Colorado Cañon. Upper Cucharas River.
	New Mexico.
Sandstone	Aztec mine. Vermijo River. Upper Cimarron Creek. Ridge north of Costilla Peak. Turkey Mountains. Costilla Creek. Moreno Valley.
	Colorado.
Feldspar-porphyry.....	Indian Creek. East side of Spanish Peaks. Rosita.
	New Mexico.
Limestone	Aztec mine. Taos range. Rio Colorado Cañon. Comanche Creek. Six miles from Taos. Fort Union. Elisabethtown. Collier's ranch.
	Colorado.
Diorite	Indian Creek. Saint Charles River. Huerfano Butte. Cerro Blanco. Cucharas River. Indian Creek. Near Walsenberg.
	New Mexico.
Hornblende-schist ..... Chlorite-schist .... Conglomerate.......... Clay-slate........... Mica-slate .... Quartzite .... Syenite	Costilla Peak. Rayado Cañon. Elisabeth Baldy. Rayado Cañon. Fernandez Creek. Vermijo River. Elisabethtown. Top of Taos Peak. Luero Cañon. Taos range. Rio Colorado Cañon. Taos Range, west side.
	Colorado.
Hornblende-porphyry.. Conglomerate..........	Santa Clara. Cerro Blanco.
	The following rocks were observed but not collected.
Sandstone	Wet Mountain Valley. Ryder's Cañon. Long's Cañon. Spring Vale. Huerfano Park. Walsenberg. Trinidad, Colorado. Purgatoire River. Santa Clara Creek. Bear Creek. Bodito. Cerro Blanco. Apishpa River. Arkansas River. Trinchera Pass. Cuchara Pass. Raton Plateau, Colorado. Canadian River. Rock Ranch. Cimarron River. Poril Pass. Coyote. Mora Mountain. Tierra Amarilla. Dillon's Cañon. Hole in the Rock. Aqua Negra Creek. Apache Creek. Flecha's Cañon. Ponil Park.' Van Bremmer Park. Salinas Creek. Chama River, Mora River. Taos Range. Fort Union. Clifton. Santa Fé. Pecos and Abiquiu, New Mexico.
Limestone	Las Vegas. Crow Creek. Santa Fé. Rio de la Cuera. Rio de las Vacas. Pecos River. Apishpa River. And south of Arkansas River.
Basalt	Tinaja. Eagle Tail. Abiquiu. Chama River. South side Torquillo. Mesa. west of Turkey Mountains. Raton Mesa. Bragg's Cañon. Fisher's Peak. Huerfano Park.
Shale	Near Taos, New Mexico. Hole in the Rock. Aqua Negra Creek. Cuchara Pass, Colorado. Tinaja Creek.

APPENDIX H 3.

REPORT ON THE ALKALINE LAKES, THERMAL SPRINGS, MINERAL SPRINGS, AND BRACKISH WATERS OF SOUTHERN CALIFORNIA AND ADJACENT COUNTRY.

UNITED STATES ENGINEER OFFICE,
GEOGRAPHICAL SURVEYS WEST OF THE 100TH MERIDIAN,
Washington, February 7, 1876.

SIR: I have the honor to submit herewith the report on the chemical composition of the alkaline lakes and thermal and mineral springs of Southern California, visited during the field-season of 1875 by parties of your expedition.

Thus far very little had been known of the nature of these waters, and the information will, next to the scientific value, prove also very acceptable to those interested in this region.

This chapter may be considered as the continuation of the former chapters on the mineral springs of Colorado and New Mexico, published in Vol. III of the Survey Reports.

Very respectfully, your obedient servant,

OSCAR LOEW,
Chemist and Mineralogist.

Lieut. GEO. M. WHEELER,
Corps of Engineers, in Charge.

CONTENTS.

THE OWENS LAKE, INYO COUNTY, CALIFORNIA.

This lake is, next to Mono Lake[*] in Mono County, California, certainly the most interesting lake on the North American Continent. Situated in a basin of about 4,000 feet above sea-level, its shores are bounded on the west side by the majestic Sierra Nevada, rising abruptly to towering peaks of 14,000 to 15,000 feet; and on the east side by the precipitous Inyo range, with the famous mines of Cerro Gordo and an altitude of 10,000 feet. Standing on the summit of this range, the panorama spread out in all directions is one of the grandest, most overwhelming views to behold, although there is no verdure to delight the eye and to support the ornamentation of the scenery. How far beneath us lies the Salinas Valley on one side, the Owens Valley on the other! How perpendicular the mountains, how diminutive the lake! How are we deluded by the optic refraction of the superposed strata of air of different temperature! Truly, to observe the setting sun on these heights, the changing tints of the sky, the spreading of darkness over peaks and valley, is a spectacle never to be forgotten.

The Owens Lake has no outlet and is fed by the Owens River, a stream about 30 feet wide, 2 feet deep, and having a velocity of about 5 miles per hour. As the level of the lake remains constant, there must be a perfect equilibrium between the amount of evaporation and the incoming water. The lake having 110 square miles surface, an evaporation of 4.6 feet per year would suffice to swallow up the annual volume of Owens River. Those who cannot appreciate the amount of evaporation, have invented the hypothesis of a subterranean outlet, as in the case of Great Salt Lake in Utah. The water has a strong saline and alkaline taste, and is far-famed in Mono and Inyo Counties for its cleansing properties, surpassing those of soap. Neither fish nor mollusks can exist, but some forms of lower animal life are plentiful, as infusoriæ, copepoda, and larvæ of insects.

While around the lake the vegetation consists of two salt plants, Bryzopyrum and Halostachys, the vegetation in the lake is confined to an algous or fungoid plant, floating in small globular masses, of whitish or yellowish-green color in the water. These accumulate on certain localities of the lake-bottom and near the shore and undergo decay, emitting a feces-like odor, as observed also in the treatment of albuminous matters with caustic alkalies.

One of the most striking phenomena is the occurrence of a singular fly, that covers the shore of the lake in a stratum 2 feet in width and 2 inches in thickness, and occurs nowhere else in the county; only at Mono Lake, another alkaline lake, it is seen again. The insect is inseparable from the alkaline water, and feeds upon the organic matter of the above-named alga that is washed in masses upon the shore. In the larva state it inhabits the alkaline lake, in especially great numbers in August and September, and the squaws congregate here to fish with baskets for them. Dried in the sun and mixed with flour, they serve as a sort of bread of great delicacy for the Indians.

[*] This lake is said to resemble the Owens Lake in all particulars.

Humboldt relates that, in Mexico, the dried larvæ of an insect from the lake Tescuco, form an article of commerce among the Indian population of the province.

Notwithstanding the alkalinity of Owens Lake, numerous ducks are occasionally seen swimming on it. The great numbers of dead ducks and other aquatic birds seen here and there on shore seem to indicate that they tried to satisfy their thirst with this water.

In regard to the fly-species, an interesting letter from Baron Osten-Saken may be referred to here, to whom were sent a few specimens for examination. He says: "They belong to the genus Ephydra, but differ in some respects from the usual type of the genus. Usually, Ephydræ are of a metallic blackish-green, while the species in question is dead gray, and not metallic at all. Other differences may be revealed on a more careful examination. Several species of Ephydræ have been observed to live in the larva state in salt water, either in the brackish waters near the shore, or in the brine of salt-works. A species of Ephydra occurs in enormous numbers on the shore of the Great Salt Lake, Utah. I am pretty sure that it is a species different from yours; next summer I hope to ascertain the fact positively."

I had suspected the peculiar algæ of Owen's Lake a new species, and sent a specimen to Professor Wood in Philadelphia for examination; unfortunately, however, I was not favored with an answer.

The lake was evidently at one time much larger than at present, and its waters pure and fresh, so as to permit the life of various fresh-water mollusks, as indicated by numerous shells of recent species found in the sand of the vicinity.* Upturned strata of limestone and slate, containing numerous dikes of intrusive rock, as diorite and porphyry, skirt the valley of the lake on the east side, while the main mass of the opposite Sierra Nevada consists of granite and gneiss.

The taste of the water reveals at once the presence of carbonate of soda. The specific gravity is 1.051. In 100 liters (= 26.42 gallons) are contained :

	Grams.
Potassium sulphate	644. 87
Sodium sulphate	929. 07
Sodium carbonate	2, 440. 80
Sodium chloride	2, 328. 30
Silicic acid	17. 21
Boric acid	traces.
Phosphoric acid	traces.
Nitric acid	traces.
Lithiat	traces.
Lime	traces.
Magnesia	traces.
Aluminia	traces.
Organic matter	traces.
	6, 360. 25

The proportion of saline substances is therefore over double as large as in sea-water' and about one-third of that of the Great Salt Lake. But while these contain chloride of sodium, the former contains a large amount of carbonate of soda. To undertake to calculate this for the whole lake is to appreciate its value. The greatest length is 17 miles; width, 9 miles; depth, 51 feet. The total surface of the lake is very closely 110 square miles = 284.9 square kilometers, the average depth 3 meters = 9 feet 10 inches. The cubical contents are therefore = 0.8547 cubic kilometers.

One cubic kilometer = 1,000 × 1,000 × 1,000 cubic meters.

One cubic meter = 1,000 liters.

Hence the volume of the lake = 854,700,000,000 liters or = 8,547,000,000 hectoliters.

As the analysis shows, one hectoliter contains 2.44 kilograms carbonate of soda, hence the total amount in the lake = 20,854,680,000 kilograms, or 22,000,000 tons.‡ Evaporation and concentration works might be erected at a moderate cost, and the carbonate of soda separated by crystallization from the remainder of the salts. If by this process also the considerable amount of potassa, which is a valuable fertilizer, were separated and saved, the outlay for fuel might be compensated. Wood is scarce in the vicinity, and transportation at present expensive, which would be modified with the completion of the projected Independence Railroad. The analysis showed that there is one-twentieth more carbonic acid than required to form with the available soda the monocarbonate;‑hence there is a small amount of bicarbonate present, which, however, was calculated in the analysis as monocarbonate. The amount of bicarbonate formed depends upon the quality and tension of the carbonic acid coming in contact with the monocarbonate, also upon the quantity of water. Another fact to be men-

* I find this fact also mentioned by G. K. Gilbert, vol. III, Survey Reports.

† Careful search was made for traces of rubidium and cæsium, according to the method of Bunsen, but no trace was revealed by the spectroscope.

‡ The amount of sulphate of potassa would be 5,000,000 tons.

tioned is the formation of a considerable quantity of sulphide of potassium and sodium, if the water is kept in well-corked bottles for several months, in the dark. 100cc of such water were mixed with ammoniacal cuproammonium sulphate and 0.024 grams sulphide of copper obtained, corresponding to 0.019 grams sulphide of sodium. Of course there was no longer any free oxygen present in this bottle, but still the animal life was not extinct, and the minute copepoda were soon in agitation, when the water was poured in a large airy bottle and exposed to sunlight. It is astonishing under what circumstances life sometimes can exist, and a world of mysteries still lies here before us, waiting to be solved by future investigators.

THE BLACK LAKE OF BENTON, MONO COUNTY, CALIFORNIA.

About one mile west of Benton, a little mining town in Mono County, California, rises a low granite hill-chain, running from north to south, and bordering a long-stretched valley with its western slopes. In this depression a number of springs and grassy spots are met with.

In the northern portion of this valley is situated a lake of about 1 mile in length, varying in width from 100 to 500 feet, and reaching a depth of 60 to 70 in some places. It is called the Black Lake, on account of the dark color of its water—a coloration due to its containing organic matter in solution. The taste is strongly alkaline, and white efflorescences are seen all around the lake. The amount of saline constituents is much smaller than at Owen's Lake, and to this fact it is due that Black Lake does not prevent several species of juncus and gramineae from growing in the shallow places, nor permits the existence of the peculiar black fly found at Owen's Lake. Like the latter was also Black Lake formerly much larger, and formed, doubtless, the drainage-basin of a number of hot alkaline springs, like one still in existence in the vicinity. Many hot springs also exist in Long Valley, 25 to 30 miles west of Benton. Another small alkaline lake is the Slough, near Bishop's Creek.

The water of Black Lake gave the following composition. In 100 liters are contained—

	Grams.
Sodium carbonate	1,233.1
Sodium sulphate	294.0
Potassium sulphate	83.5
Sodium chloride	234.2
Silicic acid	5.2
Formic acid	traces.
Humic and crenic acid	traces.
Phosphoric acid	traces.
Boric acid	traces.
Iodine	traces.
Bromine	traces.
Lithia	traces.
	1,850.0

THE WATER OF THE VIRGIN RIVER, (SOUTHERN NEVADA.)

This tributary of the great Colorado penetrates a wide valley, of over 25 miles in length, before it empties into the latter. Not a single settlement exists in this apparently very fertile bottom—a fact that is understood as soon as the water of this river or of any well sunk in the valley is tasted, it being not only very disagreeable, resembling in taste glauber and epsom salts, but its effects are of an alarming nature, diarrhœa and vomiting being the immediate consequences of having tried to satisfy the burning thirst by means of this water. Cattle using it for a few days in succession invariably die; also many human beings become sacrificed to it. Hence the water is pronounced poisonous, and people avoid settling in this valley.

The chemical analysis did not reveal a trace of mineral poisons, but the presence of a large amount of sulphates, to which the diarrhœa must be ascribed. It is easily understood how deleterious in such a hot climate, when the body is weakened considerably, a diarrhœa must prove.

The following was found to be the composition:

In 100 liters of the water are contained—

	Grams.
Potassium sulphate	4.16
Sodium sulphate	94.71
Calcium sulphate	73.60
Magnesium sulphate	75.66
Sodium chloride	189.00
Total	437.13

Alumina, iron, and phosphoric acid were present in small traces.

While these salts are contained in solution, a reddish mud is kept in suspension, its average quantity being 170 grams per 100 liters.

THE SALT-WELL NEAR STONE'S FERRY, (SOUTHERN NEVADA.)

About 1 mile north of Stone's Ferry, and some 3 miles west of the junction of the Virgin with the Colorado, exists a singular sink, with a surface of 600 square feet and a depth of 96 feet.* It is situated in a funnel-shaped depression in the mesa, and the sink can only be reached after a careful descent of about 30 feet on the steep slopes of this funnel. As there are deposits of rock-salt in the vicinity, this singular natural well obtains its salty constituents probably from them.

The temperature on the surface was 89°.5 F., that of the air being 105° F. at the same time, August 5. The composition of the water was found to be the following:

Sodium chloride	1,813.50
Sodium sulphate	294.71
Calcium sulphate	172.04
Magnesium chloride	48.27
Aluminium chloride	trace.
Silicic acid	trace.
	2,328.52

THE THERMAL SPRINGS OF SANTA BARBARA, CALIFORNIA.

These springs are situated 6 miles west of the town of Santa Barbara, in two cañons of the Santa Inez Mountains, and in an altitude of 1,415 feet above sea-level. The steep slopes of the mountains and the fine view of the ocean and the islands near the coast render the locality very picturesque. A hotel and bath-houses exist near the springs for the accommodation of guests.

The springs number in all twenty-two, and their temperature ranges from 112° to 122° Fahrenheit. A taste is hardly perceptible; the odor very faint after sulphureted hydrogen. Issuing from fissures of a fine-grained, very hard and dense sandstone, they form pools and rills, covered with an alga of a vivid green; also, a small violet fungus can be noticed. Two specimens of springs were analyzed; No. 1 from Hot Spring Cañon main spring; No. 2 from main spring in the side cañon. In 100,000 parts of water are contained, parts—

	No. 1.	No. 2.
Sodium carbonate	29.6	24.8
Sodium chloride	8.7	7.6
Sodium sulphate	5.0	trace.
Silicic acid	4.2	6.0
Calcium, potassium	traces.	traces.
Sulphureted hydrogen	traces.	traces.
Free carbonic acid	traces.	traces.
	47.5	38.4

THE HOT SPRINGS OF SAN BERNARDINO, SOUTHERN CALIFORNIA.

Seven miles north of the town of San Bernardino, in the foot-hills of the mountains, quite a number of thermal springs take their rise. The place can be easily recognized from afar by a peculiarly-shaped large barren spot on a steep hill-side, bearing resemblance to the "ace of spades," by which term the hill is known to the people of the town. The spot was probably produced by a land-slide. A homeopathic physician, who keeps his guests and patients on vegetable diet, has established a small hotel at the springs for accommodation of about a dozen people, and erected a few bathing-houses. He also formed a large basin for the reception of the hot water of the spring, next to the house.

The formation of the vicinity consists of granite and gneiss, and from fissures in these rocks issue the hot springs, about a dozen in number, and of a temperature from 154° to 210° F. They emit no peculiar odor. Around the rim of the basins efflorescences both ochry and white are formed. The taste of the springs is very weak, being nearly that of plain water.

Two specimens of waters were procured; No. 1 from the large spring in front of the hotel, and No. 2 from the spring 200 yards west of the hotel, near the bath-house.

---

* The depth was ascertained by the topographers of our party, Thompson and Birnie, who swam with a twenty-pound weight to the center of the well and sunk it to the bottom. These gentlemen also determined the level of the well to be the same as the Colorado River.

In one hundred thousand parts of water are contained—

	No. 1.	No. 2.
Sodium sulphate	81.7	80.2
Potassium sulphate	2.3	trace.
Sodium chloride	12.8	13.4
Carbonate of lime	10.7	11.0
Carbonate of iron	trace.	trace.
Carbonate of magnesia	trace.	trace.
Silica	20.5	22.4
Total	128.0	127.0

It will be noticed how closely both springs agree in their composition. Carbonate of soda, a usual compound met with in the hot springs of California, is absent, but silica is there in increased proportion. Indeed, on evaporating the water I noticed quite a jelly-like separation of the silicic acid before the residue became perfectly dry.

THE THERMAL SPRINGS OF SAN JUAN CAPISTRANO, SAN DIEGO, CALIFORNIA.

In the cañon of the Rito de la Mission Vieja, 12 miles east of the village San Juan Capistrano, are situated six thermal springs, issuing from fissures of the Azoic rocks. The locality looks like an inviting hidden mountain summer resort.

The Tertiary and Quaternary formations so conspicuous around the town of San Juan Capistrano, cease as we ascend the foot-hills of the mountains. The springs belong to an old grant, whose owner refuses permission to build a hotel at the springs, although sums of money have been offered to him. He hates monopoly, and wants free access for every one, even the poorest, to the springs. Hence any person who wants to stay here for some time has to provide himself with blankets and a cooking outfit; and many a diseased man of means learned here for the first time, how much healthier it is to sleep under a tree than in the best furnished hotels. There are sometimes twenty to thirty persons camping in the valley; some have tents, others a few branches under which to sleep. Some of the springs form pools that are used for bathing purposes. At the time of my visit I found six individuals there shut out from the rest of the world, and living upon bread and bacon.

There exists one cold spring of 3 to 4 square feet surface, a temperature of 75° F., and a faint odor of sulphureted hydrogen. The hot springs form no deposit worth mentioning, have but a very faint odor, and are nearly insipid. The temperature of these six springs was found 120° to 123° F. A specimen of water of the main spring was procured for analysis. There are three other hot springs half a mile north of the lower group, and one of them contains a trace of iron.

The main spring furnished the following result. In one hundred thousand parts of water are contained—

Sodium carbonate	11.10
Sodium chloride	10.53
Sodium sulphate	trace.
Silicic acid	7.66
Lime	trace.
Magnesia	trace.
Potassa	trace.
Lithia	trace.
	29.29

The healing properties of this water can only be ascribed to their temperature or to a sympathetic belief.

HOT SPRING OF THE CABEZON VALLEY.[*]

Some 10 miles south of White River exists *Aqua caliente*, a little oasis with a hot spring. This spring has 1½ to 2 square feet surface, and discharges its water into a large pool, which is used by Indians as a bathing-place. The water has 100.4° F., of a weak smell of sulphureted hydrogen, and is almost tasteless. On chemical analysis, the amount of mineral matter was found indeed but very small.

In one hundred thousand parts of water are contained—

Sodium chloride	31.0
Sodium sulphate	trace.
Sodium carbonate	8.3
Lithia	trace.

[*] This valley, also called Coahuilo Valley, is a long-stretched basin in Southeastern California, south from the San Bernardino Mountains.

Lime.. trace
Magnesia.. trace.
Silicic acid... trace.
Organic matter... trace.
Sulphureted hydrogen .... trace.

39.3

WARM SPRING, NEAR LITTLE OWENS LAKE, INYO COUNTY, CALIFORNIA.

A sample of this water was collected by Lieut. R. Birnie, jr., of division No. 2, of your expedition, who kindly furnished me with the following notes:

The spring is situated 300 yards from Little Lake, near a basaltic bluff 20 to 30 feet in height, and extending north to south several miles. The water is lukewarm, odorless, and forms no deposits or incrustations. A mineral taste is hardly perceptible.

The following is the composition:

In one hundred thousand parts are contained parts—

Sodium carbonate 45. 2
Sodium sulphate 8. 0
Sodium chloride.. 26. 9
Calcium carbonate, with trace magnesium carbonate............................... 12. 0
Organic matter... trace.
Potassium, } ....traces.
Silicic acid, }

Total 92. 1

Here may be the proper place to mention that the water of Little Owens Lake (also called "Little Lake") is not brackish, nor charged with mineral salts, like Owens Lake, Black Lake, and Mono Lake, according to information I obtained from Lieut. R. Birnie, jr.

THERMAL ACID SPRINGS, IN THE COSO RANGE, INYO COUNTY, CALIFORNIA.

These singular springs, situated in the Coso range, 12 miles east of Little Owens Lake, are the discovery of Lieut. R. Birnie, jr., of your survey, through whose kindness I was provided with information regarding them and with several bottles of water for analysis. They have but a limited flow, and form pools through which steam is continually ejected. Large deposits of sulphur (the specimen I received from Lieutenant Birnie was a solid chunk of nearly pure sulphur) cover the surroundings, and hundreds of tons of this material are said to exist in the neighboring mountains, where extinct and living thermal springs are numerous.

The taste of the water is intensely *sour*, making it perfectly unfit for drinking purposes. It has no smell, but formerly there must have been large quantities of sulphureted hydrogen contained in it, as the sulphur deposits indicate. It is true, I found a large proportion of *free sulphuric acid* in these waters, but the sulphur deposits cannot be derived from this source. Chemists are, at least, unacquainted with a process by which free sulphuric acid would turn under the circumstances, as the above, into sulphur.

The composition is certainly a remarkable one, as will be seen from the following analysis:

In one hundred thousand parts of water are contained parts—

Free sulphuric acid... 78. 4
Potassium sulphate... 2. 5
Sodium sulphate 15. 1
Calcium sulphate... 15. 3
Magnesium sulphate.. 1. 2
Aluminium persulphate.. 127. 0
Iron persulphate 33. 2
Nitric acid... traces.
Phosphoric acid .... traces.
Chlorine.. traces.
Ammonia... traces.
Lithium .... traces.

272. 7

Springs or lakes of a chemical composition like this are very rare. I know only of one instance analogous to it; that is the "Sour Lake" in Texas. Singular is also the small trace of chlorides in a water so strongly charged with mineral matters.

P. S.—Quite recently Boussingault has discovered and analyzed quite analogous springs in the vicinity of the volcanoes of the South American Cordilleras de los Andes. (Annales de chimie et de physique, 1875.)

THE SULPHUR SPRING ON THE SOUTH SIDE OF SAN FERNANDO MOUNTAIN, LOS ANGELES, CALIFORNIA.

For the water of this spring collected by yourself, and for the following information I am indebted to you.

The spring issues upon the eastern side of a little cañon, coming out of the southern portion of the San Fernando Mountains. Its first superficial reservoir in the surface-rock, which is of Tertiary age, is surrounded by recent deposits formed by the spring. The water is higly charged with sulphureted hydrogen. Its healing properties are well known, and made use of by the Indians, who, led by their priests, used to make pilgrimages from the old San Fernando Mission to this spring.

In one hundred thousand parts of water are contained parts—

Sodium carbonate	6.21
Sodium sulphate	23.87
Sodium chloride	trace.
Calcium carbonate, magnesium carbonate	*50.60
Silicic acid	trace.
Phosphoric acid	trace.
Carbonic acid	in excess.
Sulpureted hydrogen	5.
Potassium, lithium, manganese, iron, alumina	traces.
Organic matter	trace.
Total	85.68

LITTLE YOSEMITE SODA SPRING, KERN COUNTY, CALIFORNIA.

Water of this spring was collected by Mr. F. Klett, to whom I am indebted, also, for the following information :

The spring is situated in the valley of the north fork of Kern River, whose calcareous ocher-colored deposits of great extent, indicate the former activity of a large number of mineral springs, of which those in existence at the present day form apparently only a small remnant. The Little Yosemite Soda Spring bubbles continually, emits no odor, has an exceedingly agreeable taste, and a temperature of 52° F. Animals prefer this mineral water to that of the Kern River, which is but 50 feet distant. The vicinity has a luxuriant vegetation. To all appearances, this will become a popular summer resort in the future.

In one hundred thousand parts of water are contained parts—

Sodium carbonate	20.97
Sodium sulphate	trace.
Sodium chloride	4.68
Calcium carbonate, with some magnesium carbonate	16.02
Iron carbonate	0.92
Silicic acid	7.31
Free carbonic acid	in excess.
	49.90

MINERAL SPRING OF ENCINO RANCHO, SAN FERNANDO VALLEY, LOS ANGELES COUNTY, CALIFORNIA.

Water of this spring was obtained by you for chemical investigation, and the following notes by Mr. F. Klett :

The spring is situated fourteen and a half miles from Los Angeles, on the Santa Barbara and San Buenaventura stage-road, and is the property of Mr. Garnier, the owner of Encino Rancho. The temperature was found 85°.5 F. at 7 a. m., and 87° F. in the afternoon, a change due to the effects of the exposure to the sun. The flow is about five gallons a minute. A basin of masonry about 12 feet in diameter contains the water, which rises at uneven intervals, as indicated by a great number of bubbles thrown up. A bathing reservoir is connected with the spring. The water is used for irrigating purposes on the ranch.

In one hundred thousand parts of water are contained parts—

Sodium carbonate	24.31
Sodium sulphate	54.46
Sodium chloride	2.93
Calcium carbonate, with some magnesium carbonate	32.17
Silicic acid	11.50
Phosphoric acid	trace.

* Present as bicarbonates in solution, but deposited on ebullition and concentration of the liquid.

Sulphureted hydrogen	trace.
Potassium	trace.
Lithium	trace.
Carbonic acid	in excess.

125. 37

THE THERMAL SPRING OF BENTON, MONO COUNTY, CALIFORNIA.

A few rods west of the small mining settlement of Benton issues a hot stream of considerable size, from the fissures of the granite. This rock is overlaid by volcanic tufa. The basin has 3 to 4 feet in diameter, and a depth of half a foot, and the water comes up with such force and quantity that I think a yield of 200 gallons per minute is not an overestimate. From the good-sized creek thus formed, a ditch carries the water to the town, where it is used for drinking and household purposes, as it is almost tasteless. That, however, some mineral matter must exist in solution, may be inferred from the fact that the rocks near the spring show white incrustations and efflorescences. No medical use is made of the spring; its temperature is 138° F., but it is said that variations up to five degrees have been noticed. It contains in one hundred thousand parts but 26 parts mineral matter, consisting of sodium carbonate, sodium sulphate, sodium chloride, with traces of potassium and calcium salts.

MINERAL SPRING OF LYTTLE CREEK CAÑON, SAN BERNARDINO COUNTY, CALIFORNIA.

The Lyttle Creek Cañon is situated on the eastern slopes of the San Antonio Peak in the San Bernardino Mountain range, and runs parallel with the road across the Cajon Pass for some distance. Three miles above the mouth of the cañon we have another side cañon, which comes in from the west. In it is situated a spring, with a large basin of about 25 square feet, but with a very moderate flow. The taste is very faint, the odor slightly after sulphureted hydrogen; bubbles of carbonic acid are continually rising from the bottom. The temperature was found to be 92° F., that of the air being 94° F. (June 30, 2 p. m.)

While the cañon is filled with trees and shrubbery, there are but few patches of arable area. Besides the few men engaged in the hydraulic-mining works in the vicinity, only two men, living like solitary hermits, inhabit the cañon.

In one hundred thousand parts of water are contained 56.8 parts solid material, consisting chiefly of sodium carbonate, sodium sulphate, sodium chloride, with traces of calcium carbonate and silicic acid.

SPRING OF THE DOS PALMS OASIS, COAHUILA VALLEY, DRY LAKE, SAN DIEGO COUNTY, CALIFORNIA.

Where the road from Ehrenberg, Ariz., to San Bernardino, Cal., enters the Big Dry Lake, a small oasis exists, prominent by two palm-trees that grow here in a wild state. The spring that furnishes the water for the pond at which these palms are growing, is but a few yards distant from the hut of the single inhabitant of this oasis. As far as the water of this oasis spreads, the salt grass—*Bryzopyrum spicatum*—grows abundantly, but farther on nothing is seen but barren clay, covered with salt efflorescences.

The whole basin was a great lake until quite a recent period.

The spring of Dos Palms has a disagreeable, brackish taste, a temperature of 82° F., and is odorless.

In one hundred thousand parts of water are contained parts—

Sodium chloride	230. 8
Calcium sulphate	32. 6
Magnesium sulphate	31. 0
Calcium carbonate	traces.
Manganese	traces.
Phosphoric acid	traces.
Silicic acid	traces.

294. 4

WATER FROM THE SALINE FLATS OF THE MOHAVE RIVER OR SODA LAKE, SAN BERNARDINO COUNTY, CALIFORNIA.

These names are given to a flat basin 14 miles long and 4 to 5 miles wide, and surrounded by chiefly volcanic ranges. The bottom is composed of clay, devoid of vegetation, but covered with patches of saline efflorescences. Here the Mohave River rises for the last time in its remarkable course, which is repeatedly a subterranean one, and forms here but a small rill of water.

While this river in its upper course has no saline taste of any consequence, it has acquired here a disagreeable taste of Glauber salts; still it is the only water that may be drank in that whole region. While it has no smell in the fresh state, the sample taken developed on boiling a disagreeable odor of crude hippuric acid. The presence of organic matter in the water cannot be a cause of surprise to any one who saw the great number of dead sheep lying in the stream, 12 miles farther up, where the river gradually sinks to re-appear at Soda Lake. But, strange to say, putrefaction does not set in so long as this water, after its subterranean course of 12 miles, is exposed to the desert air, and is not brought into too close vicinity with the decaying animal-matter.* This speaks well for the scarcity of the bacteriæ and other ferment-producing organisms in the desert air. The water has, after concentration, a slight acid reaction.

In one hundred thousand parts of water are contained—

Sodium chloride	170.8
Sodium sulphate	63.1
Calcium sulphate	21.2
Magnesium sulphate	8.5
Organic matter	19.0
Potassium	traces.
Lithium	traces.
Phosphoric acid	traces.
Silicic acid	traces.
	282.6

BITTER SPRING FROM MINERAL PARK, ARIZONA.

There are a number of springs in the vicinity of the little mining settlement of Mineral Park, situated in the Cerbat range in Northern Arizona, which have a disagreeable, bitter taste, making the water unfit for drinking or culinary purposes, and compelling the population of the town to procure potable water from a cañon 4 miles distant. Where the water of these bitter springs passes across rocks, it coats them gradually with a thin film of oxide of iron and manganese. It is neutral to test-papers, and on evaporation much gypsum is deposited.

It contains in one hundred thousand parts—

Calcium sulphate	118.5
Magnesium sulphate	65.3
Magnesium chloride	5.4
Sodium sulphate	trace.
Manganese sulphate	trace.
Iron sulphate	trace.
	189.2

No trace of potassium or lithium was found.

Another bitter spring, but not so strongly charged with salts, exists on the trail from Callville to Saint Thomas, Southern Nevada.

THE GYPSUM SPRING, NORTHWESTERN ARIZONA.

This spring is situated in a big dry wash that leads from the Detrital Valley to the Colorado River, which forms there the boundary-line between Arizona and Nevada. White soft crusts, for some distance from the spring, indicate that the water is charged considerably with mineral constituents. It has a faint odor of sulphureted hydrogen, and a strong saline and disagreeable taste. The geological formation, in which the spring is situated, consists of a red triassic sandstone, and conglomerate with gypsum and salt deposits.

To one hundred thousand parts of water are contained parts—

	I.	II.
Sodium chloride	397.8	12.23
Sodium sulphate	51.6	
Magnesium sulphate	172.8	23.73
Calcium sulphate	130.1	7.22
Calcium carbonate	12.0	74.80
Potassium chloride	trace.	13.44
Magnesium carbonate	trace.	9.62
	764.3	141.4

I have added here under II the composition of a spring from Kerami, in the Lybian Desert, of which a specimen was collected by the traveler G. Rholfs, and analyzed by T. Hessert, (Ann. de Chim., 1875.)

---

* Thousands of sheep die annually while driven through the desert into Arizona. Their decay is due not to bacteriæ of the desert, but to those in their bodies.

AP. J J—13

CONCLUSION AND REMARKS.

On subjecting the surface-conditions of California to a critical examination, one cannot fail to be struck forcibly by the great number of thermal springs; indeed, there are few countries in the world with such a large number upon an equal area. It is true, Montana and Idaho, with the famous geyser-regions, first explored by Lieut. W. Doane—then Colorado and New Mexico—contain also a considerable number, and still above them Nevada; but these Territories have more or less similar features as California in regard to the extensive volcanic formations. There is no doubt that a close connection exists between the latter and the thermal springs; both are due to the fact that the earth-crust covering the molten interior is thinner than elsewhere. Waters, after penetrating this crust, become heated, charged with salts, and are driven through other fissures to the surface by the power of the generated steam.

In connection with this relatively thin crust stands the great number of earthquakes felt annually in California. Alexander von Humboldt says, in the description of his travels through the equatorial regions of South America, that "the hypothesis of the relation between the volcanic formations and the existence of thermal springs seems not to be well founded;" simply because he encountered springs of nearly boiling temperature issuing from *Azoic rocks* at Mariara, on the Orinoco. He certainly would have formed a different opinion had he traveled through New Mexico, Nevada, and California, where gigantic peaks and wide-spread flows of *volcanic* material form most prominent features. It may safely be assumed it is rather the exception that thermal springs issue within very large distances from volcanic formations.

In Southern California thermal springs are found at Santa Barbara, San Juan Capistrano, San Diego, Yuma, San Bernardino, Benton, in Coahuila Valley, Long Valley, Death Valley, Panamint Valley, Salinas Valley, in the Coso Range, at Kernville, and Caliente. A number exist also in Northern California, among them several far-famed geysers.

Strange to say, most of these thermal waters contain comparatively small quantities of mineral constituents; indeed, less than many of New Mexico. Generally they contain carbonate of soda, a notable exception from those of San Bernardino; this salt is accompanied by sulphate of soda and chloride of sodium, in varying proportion. Carbonate of lime, with some carbonate of magnesia, form, in most of the cases, an additional mixture, while silicic acid never is wanting entirely; its quantity, however, is not considerable. As a general rule, the amount of potassium salts falls far behind that of sodium salts, which is probably due to the fact that the former are retained by the strata traversed. It is well understood how quickly potassium salts are absorbed by the soil in such a manner that water cannot wash them out.

One of the most remarkable thermal springs is that discovered by Lieut. Rogers Birnie, jr., heretofore mentioned. It has an intensely sour taste, due to free sulphuric acid, and, besides, it contains the sulphate of potassa, soda, lime, magnesia, alumina, and oxide of iron. The presence of small quantities of nitric acid and ammonia may be due to the formation of ammonium nitrite from water and nitrogen of the air, and to the immediately following combination of the ammonia with the sulphuric acid; thus the redecomposition of ammonium nitrite, taking place so readily under ordinary circumstances, being prevented.

With regard to the cold mineral springs, that of the Little Yosemite deserves especial mention, being the only soda spring with a notable quantity of iron. Next to this one, the soda spring of Encino Rancho, and the sulphur spring of San Fernando Mountain, are of medicinal value.

The brackish and bitter springs generally contain sulphate of magnesia, sulphate of soda, sulphate of lime, and chloride of sodium or magnesium; these waters, although mineral springs in a wider sense, are repulsive to taste, and men and animals avoid to drink of them, while springs of a character like that of the Little Yosemite are eagerly drunk, and preferred even by animals to ordinary water. The Mohave Desert, like many other deserts, abounds in such bad waters, the horror of the tired, thirsty traveler. Bitter springs and brackish waters were encountered in the mountains 20 miles north of Callville, in Southern Nevada, at Miningtown Mineral Park, in Northwestern Arizona, at Dos Palms, and thence 5 miles west from there, in San Diego County, on the saline flats of the Mohave, and, above all, in the Virgin River.

A fact of no little interest is the occurrence of a number of lakes in Eastern California that contain as main constituent carbonate of soda. These lakes, and a number of alkaline and saline flats, are situated east of the Sierra Nevada, and west of the great parallel ranges known as the White Mountains, Inyo and Argus ranges, which are filled with splendid fissure-veins. The alkaline lakes are Owens Lake, Mono Lake, Black Lake, and the slough of Bishop Creek.

In addition to this chapter, the notes on mineral and thermal springs visited by Mr. Douglas A. Joy, who was attached as geologist to the Division No. 2, in command of Lieut. W. Whipple, may find a suitable place. To Mr. Joy I am indebted for the following communication:

The first spring met with was near that at Encino Rancho. A pipe has been sunk in the rock, and the water rising in it fills an octagonal stone tank of about 10 feet diameter and 12 feet deep. It overflows at the top and runs into an artificial pond. I do not know at what depth the pipe was sunk, but undoubtedly deep enough to pierce the Tertiary (Pliocene?) limestone, dipping here from the mountains toward the plain.

The next mineral spring met with was in a cañon leading from Montan's ranch to Cespe Creek Cañon. It is 8 miles below the ranch and about 4 miles from the junction of the two cañons. The water issues through four or five fissues in the granite, one of them about 20 feet above the bottom of the cañon. Steam rises in thick clouds from the rocks, and has a peculiar odor. The rocks over which the water first flows are stained deep black, while those farther down in the stream have a bright red iron color. I do not know how to account for the black color unless it be a deposit of peroxide of manganese. The yield of water is about thirty gallons per minute, and its temperature was found 195° F.

In the Ojai Valley, about 6 miles from its junction with the Santa Clara Valley, there are a number of natural petroleum springs. A well sunk yielded two gallons per minute.

About 7 miles below Kernville, following the river, there is a hot spring, claimed to be a sulphur spring, but evidently is not, as it failed to blacken a bright silver coin left in the water for over half an hour. The temperature was 127° F., whilst the temperature of a spring not ten yards distant was only 70° F. The flow is about ten gallons per minute, the formation of the vicinity granite. A bath-house has been erected, and invalids suffering from rheumatism and other diseases come here for cure.

At a place called Agua Caliente, which is about 30 miles from the town of Caliente, there is a warm sulphur spring. The water bubbles up through the soil in numerous places. The largest spring has been dug out and is used by the Indians for washing and bathing. A bright silver coin is blackened almost immediately when dropped into this water. The temperature was 80° F., and the flow not more than two gallons per minute.

Another mineral spring was met with near the head of Walker's Basin. It issues from granite, and has a temperature of 100° F.; the flow is about three gallons per minute. A ranchman who lived near by built a little bath-house at the spring.

APPENDIX H 4.

RFPORT ON THE GEOLOGY OF THE MOUNTAIN RANGES FROM LA VETA PASS TO THE HEAD OF THE PECOS, BY A. R. CONKLING.

NEW YORK CITY, *April* 25, 1876.

SIR: Beginning at the La Veta Pass and proceeding south, the mountain ranges will be described in the following order: Spanish Peaks, Culebra range, Cimarron range, Taos range, Mora range, Santa Fé range, and Las Vegas range.

The Spanish Peaks form a minor range about 10 miles long. The peaks are two pyramid-shaped mountains and consist of pinkish trachyte. Perpendicular walls of trachytic rock diverge from the Spanish Peaks, extending into the plain for more than a mile in some cases. The walls are in general about 100 feet high. The top of the walls is flat and the jointed structure is well shown in them. In places these walls have fissures—breaks in the form of right angles, thus presenting the appearance of trap rocks. These rocky walls are dikes of trachyte upheaved through huge fissures in the earth's crust after the greater part of the Spanish Peaks had been formed. About 3 miles north of west Spanish Peak is a curious butte of basalt, having the form of a tower with a rounded top. The butte is about 250 feet high and stands alone in the midst of a plain. I propose the name La Torre for it. The south fork of the Cucharas River flows between the Spanish Peaks and the Culebra range. The river has cut its way through a steep wall of gray sandstone running north and south, but forming a break in the wall large enough to admit the passage of a wagon-road which is much used by the settlers. This gap is called the "shut in."

At Willis ranch, on the north side of the Cucharas, a ledge of fine-grained drab limestone outcrops, but I was unable to define the limits of it.

Following up the stream to the divide and a little beyond to the headwaters of the Purgatoire, the country is covered with a series of anticlinal ridges of white and yellowish sandstone with vertical joints, which rest on granite. The Spanish Peaks may be regarded as an outlier of the Culebra range, which will now be described.

The Culebra range extends from Trinchera Pass on the north to Costilla Creek on the south. The predominating rock is grayish granite. Hornblende porphyry occurs at various points. A series of low foot-hills of basalt bound the Culebra range on the west side, forming a portion of the plateau through which the Rio Grande runs.

No sedimentary rocks were observed in the Culebra range. Ore deposits are found at but one locality in this range, viz, one mile east of Culebra Peak. Mr. E. D. Bright, of Trinidad, Colo., who visited this locality informs me that there is a vein of quartz-bearing silver 7 feet wide running through hornblendic granite. The ore assays $75 a ton. Another locality 4 miles east of this point has been discovered where the ore yields 52 per cent. of copper and 6 ounces of silver to the ton. The granite forming the summit of the Culebra Peak contains a large amount of feldspar. The rock is traversed by numerous joints and fissures. It has also undergone much disintegration, and a large number of detached fragments is the result. Jasper is found in small quantities on top of Culebra Peak. There are three peaks south of Culebra that probably consist of granite, but I did not visit them. The rock in these peaks is colored red by oxidation of the iron. The range then trends southwest, and low ridges of sandstone appear for 15 miles south to Costilla Peak. The ridges slant gradually in going south. They are formed of a fine-grained yellowish sandstone that sometimes passes into a coarse conglomerate. No fossils were found in this rock. The sandstone shows mud-cracks and rill-marks, thus proving that it was formed in shallow water. The lofty peaks were islands in the primeval sea while the sandstone was formed. There are also on the eastern side of the Culebra range perpendicular walls of ferruginous sandstone. One of these walls at Beaver pond is of a brilliant red color, reminding one of the rock in the Garden of the Gods.

The Cimarron range contains a greater variety of rocks than the Culebra range. Beginning in the north with Costilla Peak, which slopes very abruptly on the northern side, we have diorite containing much olivine. Massive quartz is found on the top of Costilla Peak, and small bowlders of granite occur on its slopes. On the western side of the [Cimarron [range the granite assumes the columnar form, reminding one of the words of the poet:

> " The wild rocks shaped as they had turrets been,
> In mockery of man's art."

On the east side of Comanche Creek a variety of pinkish trachyte occurs in the forms of curved, long prismatic columns, resembling the basaltic columns in the island of Staffa.

Directly east of Costilla Peak is a short ridge of light-gray quartzite running parallel to the Cimarron range. The ridge is about 400 feet high. It is traversed by many fissures, and the rock is very much weathered. A small stream has cut its way through the ridge, thus forming a gap. On the west side of the ridge a bed of limonite outcrops. The rock is hard and fine grained, but breaks easily into thin fragments upon being struck with the hammer.

Proceeding south from Costilla Peak the rock is for the most part granite. Both red and gray varieties of granite occur, as well as coarse-grained granulite. On the western side of the Cimarron range feldspar-porphyry occurs, having a dark-gray matrix with white crystals. At the head of Moreno Creek the granite is poor in mica. Elizabeth Baldy Peak, the highest mountain in the Cimarron range except Costilla, is composed of fine-grained gray granite. A mass of dark-gray mica-schist outcrops on the western side of this mountain. Just north of Elizabethtown a mass of gray feldspar-porphyry outcrops. The town itself rests on granite, while blue limestone outcrops but a mile below it. In this blue limestone two species of inoceramus were found, which Dr. White informs me belong to the Cretaceous age. The occurrence of fossiliferous limestone in the Moreno Valley, between the Cimarron and Taos ranges, is in all respects singular. I was unable to define the limits of the limestone. Six miles below Elizabethtown a vein of bluish granite outcrops at the head of Cimarron River. From this point the range consists of reddish granite and granulite as far as Uraca Peak on the south. In the vicinity of Uraca Peak igneous rocks, such as trachyte and vesicular basalt, occur. Bluish hornblende schist is found on the south side of Uraca Mountain. Gray trachyte occurs along the banks of Uraca Creek, and south of this are messes of basalt extending beyond the foot-hills into the plain. The cavities in the basalt are sometimes filled with white calcite.

The western side of the Cimarron range is much steeper than the eastern side. A series of broad foot-hills, composed of sandstone, horizontally stratified, runs along the eastern side of the range. Much erosion has taken place in these foot-hills. They are covered with a net-work of cañons. In the Van Bremmer Park there is a detached mass of yellow sandstone, about 100 feet high, standing alone in the plain and at least a mile from the nearest foot-hill. This was the most striking example of erosion noticed. Considerable lignite and a few veins of coal are found in the foot-hills east of the Cimarron range.

The Taos range consist chiefly of granite and feldspar-porphyry, but many other rocks occur also. Taos Peak, the highest point in the range, (13,143 feet high,) is composed of gray granite and syenite, capped by mica-slate. This is the only locality of mica-slate between La Veta Pass and the Santa Fé range.

A fine section of the Taos range is seen in passing through the cañon of the Colorado Creek. In entering the cañon at the Placita de San Antonio and traveling eastward the following series of rocks were observed : Syenite, trachyte, syenite, feldspar-porphyry, quartz-porphyry, granulite, trachyte-porphyry, granite. On account of the rapid march through the cañon, I cannot give the exact limits of each zone of rock. In places the banks of the Colorado Creek contain auriferous drift, but not enough gold has been found thus far to pay for working. Just east of the head of the Colorado Creek, feldspar-porphyry occurs that has been much decomposed. The rock is soft enough at the surface to allow a trail to be made with but little difficulty, which is much used in crossing the range on the way to Elizabethtown.

Metalliferous deposits occur in the Colorado Cañon. The following particulars were furnished me by Mr. Hess :

ORE-DEPOSITS IN THE TAOS RANGE.—THE GOLCONDA MINE IN THE COLORADO CREEK CAÑON.

This mine was discovered by W. C. Hess in 1873, but no work was done until the spring of 1875. The lode runs northeast and southwest, and occurs in the main range. The country rock is granite. The lode varies from 6 to 10 feet in thickness. But two men were at work at the time of my visit. The chief ores found are those of copper and lead in the form of sulphides. The individual minerals occurring at this mine are described elsewhere in the list of minerals. The mine is owned by W. C. Hess, Charles Vernou, and Louis Noes. White men work for $2.50 a day and Mexicans for $1. There is plenty of wood and water in the vicinity. The Colorado is a swift-running stream, about 12 feet wide, and flows within a hundred yards of the lower part of the lode. A trace of gold has been found in the quartz along the banks of the Colorado Creek. There are outcroppings of ore in three places—on the lower part of the ridge near the creek, near the summit of the ridge, about half a mile south of the creek, and on the side of a gulch near the second outcrop. Game is abundant. Deer, mountain-sheep, and grouse are found. The cost of freight from Pueblo is 2 cents a pound. The owners of the mine are desirous of securing three claims of 300 feet on either side and 1,500 feet in length, making in all 600 by 4,500 feet. The owners contemplated building a blast-furnace at the time of my visit, in order to smelt the ore at the mine. The recorder of the Golconda Mine is the county clerk at Taos, N. Mex.

In passing through the Flechao Cañon, which separates the Taos and Mora ranges, yellow sandstone is seen, containing fossil leaves similar to those in the foot-hills east of the Cimarron range. The sandstone dips gently to the east. Passing over the divide, blue limestone outcrops on the side of the wagon-road running along the north side of Fernandez Creek. A zone of limestone runs north and south that contains many crinoids and brachiopods. Among the brachiopods are: *Productus semireticulatus, Productus costatus, Productus prattanianus, Spirifer rockymontanus, Spirifer (Martinia) lineatus.* According to Dr. C. A. White, these fossils belong to the carboniferous. This zone of limestone outcrops about 6 miles east of the town of Taos. Just west of the limestone yellow sandstone occurs again, beyond which is the alluvium forming the fertile plain of Taos.

The Mora range is chiefly composed of sandstone. No igneous rocks were observed within the limits of the range, excepting granite. The ridge-line of the Mora range is quite level, there being no prominent peaks as in the other ranges. No fossils were found in the parts of the Mora range examined, with the exception of a fine specimen of a fossil fern in the sandstone forming the eastern slope of Mora Mountain. This fern has not yet been determined. A few masses of eruptive gray granite occur in the eastern side of the Mora range. It is possible that the belt of carboniferous limestone observed just at the southern extremity of the Taos range runs through the Mora range. If so, it must be on the western side of it, as I failed to discover limestone on the east side.

Although the Santa Fé range was surveyed by the main party, to which I was attached, my examination of it was confined to the southern extremity. This was owing to a side trip taken to the " bad lands," in the northwestern part of New Mexico, which prevented me from accompanying the main party in their exploration of this range. From the descriptions of the topographer and previous explorers, it may be said that the predominating rock in the Santa Fé range is granite. I think the entire range consists of archæan rocks, excepting a narrow strip of blue limestone extending along the western side of the range near the base of it. This limestone outcrops within half a mile of Santa Fé, on the east of the town. The strata dip westward at an angle of about 25°, and hence underlie the town. The limestone contains well-known invertebrate fossils, such as *Productus* and *Spirifer*, similar to that found near Taos. I think there can be no doubt that the rock is of carboniferous age, and that a belt of this limestone extends northward as far as the Sangre de Cristo Pass, in Colorado. Mr. Justice, of Santa Fé, who has studied this limestone, informs me that it extends north for at least 100 miles. Although I traced the rock in very few localities, owing to the particular direction the party took in exploring the country, I am still of the opinion that the limestone at Santa Fé is identical in age with that at Taos and Trinchera Pass, on the north side of Culebra range.

The Las Vegas range was examined in the southern portions only. In crossing the range, en route from Santa Fé to Las Vegas, the valleys of the Rio Pecos, Rio La Cuera, and Rio Vaca were crossed. Each of these rivers affords a fine section of bluish limestone, containing many crinoids and brachiopods, similar to the rock at Santa Fé. I think the limestone forming the southern part of the Las Vegas range is identical in age with that of Santa Fé. The central and northern portions of the Las Vegas range contain archæan rocks, but as Dr. Oscar Loew has described this region in the annual report for 1875, reference may be made to this report for a detailed account of the geology of the range.

This chapter on the mountain-ranges may be concluded by a few general remarks on the individual ranges. The only ranges containing ore-deposits of any importance are the Taos and Cimarron ranges. In the Cimarron range both placer and vein mining are carried on, but in the Taos range there is vein-mining only.

The predominating rocks entering into the composition of the various ranges may be stated as follows: The Spanish Peaks are trachyte; the Culebra range is granite; the Cimarron range is granite and granulite; the Taos range is granite and syenite; the Mora range is sandstone; the Santa Fé range is granite; the Las Vegas range is granite and limestone.

The Culebra range contains the highest peaks; Culebra Peak, the culminating point, being 14,040 feet above the sea-level.

Cimarron range is the longest of all the ranges, being 50 miles long.

Respectfully submitted.

A. R. CONKLING.

Lieut. GEO. M. WHEELER,
Corps of Engineers, in charge.

APPENDIX H 5.

REPORT UPON THE OPERATIONS OF A SPECIAL NATURAL-HISTORY PARTY AND MAIN FIELD-PARTY NO. 1, CALIFORNIA SECTION, FIELD-SEASON OF 1875, BEING THE RESULTS OF OBSERVATIONS UPON THE ECONOMIC BOTANY AND AGRICULTURE OF PORTIONS OF SOUTHERN CALIFORNIA, BY DR. J. T. ROTHROCK, ACTING ASSISTANT SURGEON, UNITED STATES ARMY.

UNITED STATES ENGINEER OFFICE,
GEOGRAPHICAL SURVEY WEST OF THE 100TH MERIDIAN,
Washington, D. C., January 19, 1876.

SIR: In compliance with your instructions of January 12, 1876, I have the honor to submit the following report of my observations in connection with the operations of the special natural history party and main field party No. 1, of the California section, field season of 1875.

The work of the season may fairly be considered as commencing at the island of Santa Cruz, off Santa Barbara coast. This island, lying south of Santa Barbara and probably distant from it about 30 miles, is nearly 17 miles long, and at its widest portion about 6 miles across. It is much narrowed near its middle by bays making in from the northern and southern coasts, and from one point to the other a wagon-road or trail exists. The island is almost wholly given up to sheep-raising. It is estimated that in the spring of 1875 there were not less than 60,000 herds of them on the island. In June 15,000 were killed for the hide and tallow alone; the offal being carted down to the shore and cast into the water, attracting immense numbers of fish to the spot.

The island is rugged in the extreme; one point is said to attain to a height of 2,500 feet above sea-level. It is from one end to the other little else than a succession of rocky hills with intervening gulches rather than valleys. In a few places agriculture to a limited extent is carried on; only enough is raised to meet the needs of the population, if indeed it does this. Here and there a level surface intervenes between the base of the hills and the ocean. What its capacity for agricultural purposes may be I am not able to say, as there is not enough water for irrigation, and as, in my opinion, the mists from the ocean would be altogether too precarious to depend upon. I am, however, bound to state that in one or two places pepper-trees had been planted and were growing vigorously without care, and that some little grain is cultivated near Prisoner's Harbor.

It appears that at the time Cabrillo made his voyage along this coast, (1542,) these islands were timbered clear to the water's edge, and we now have abundant signs of forests that have disappeared at the sea-level, where their stumps and roots still remain *in situ.* At present the indigenous forest-growth is limited to the highest summits of the island. A dense under-growth does in many places descend lower, but it never obtains to the dignity of a forest. It is simply a thicket. Among this, however, is found one of the most beautiful and striking shrubs of the coast—*Dendromecon rigidum;* its beautiful yellow flowers shining conspicuously among a foliage that wore always a delicate glaucus bloom. It was the one redeeming feature of the vegetation.

On the grounds most visited by the herds of sheep, all vegetation, save sage-brush, *cactees* and the *erodium* or storksbill, had been entirely swept away. The grass had gone completely, and such plants of the island flora as sheep would eat, it was with difficulty that I could get even a decent botanical specimen of. In fact, pasture had become so thin that the sheep at the time of my visit mere wandering in very small bands that they might the more readily find food. Even the sage-brush was disappearing, as year after year the sheep had eaten away its leaves and younger shoots, until there was not left sufficient of the more green, succulent tissues to elaborate the sap.

It is impossible to conceive a more dreary waste than was here produced as the result of over-pasturage. The question may come up further on as to the reciprocal relations existing between vegetation and rain-fall. It would seem more than probable that ever since the discovery of the continent this and the adjacent islands had a more abundant supply of water than at present. Tradition as well as historic documents prove that in no distant past they supported a population that must have reached into the thousands. Indeed the burial-grounds, that are so numerous and so rich in articles of archæological interest, are often at points at which there is no water nearer than 3 or 4 miles, and there is abundant evidence that near the burial-places they had their permanent homes. What must have been the population that could cover, within a few centuries, an acre, to the depth of 10 or 20 feet, with the ordinary clam, muscle, and haliotus of the coast which were simply the refuse of their feasts. Yet, standing on one such shell-heap, I was able to count over twenty others within easy sight. This presupposes an immense population, and that, again, water in abundance at a point where none now exists. What has been the cause of this desiccation I am not able to say. The hypothesis has been advanced, that it is due to a greater elevation of the land. I have no evidence of this, as a fact having taken place so late as would be required; besides, it would imply also that the central portions of the island alone were changed, leaving the shore-line as it was, for the mounds and burial-places of those who formerly had water are found near their habitations, on the shore-line of to-day, a supposition which, though not impossible, is yet improbable. Supposing that a report on the results of excavations will be given, I omit any statement of them here.

It may not be out of place to call attention to the protecting influence the large sea-weeds have on shore-lines and on the harbors. Indeed to this alone, more than to anything else, is due the safety of the anchorage at Santa Barbara. It is a matter of regret that the authorities are willing to allow out-going and in-coming vessels, steamers particularly, to plough through and destroy, as they do, this the greatest protection to the harbor. I have stood on the hills to the north and west and seen the heavy swell come in from the ocean, watched it become less and less as it penetrated deeper and deeper into the "kelp" until, emerging on the shore side, its force was spent and its size gone. Instances are not wanting to show how great this protective power is. The better way it would seem would be to have certain channels through which steamers might go out and in.

From Santa Cruz Island we started to Los Angeles and temporarily joined the main party. While here I embraced the opportunity afforded by a letter of introduction from Lieutenant Wheeler to General Stoneman to visit the ranch of the latter. My time was exceedingly limited and another visit was contemplated. This I was, however, unable to make. I obtained the following facts relative to the productions of the region from General Stoneman, and they are, therefore, the result of a large and intelligent observation. In the neighborhood of Los Angeles from 40 to 60 bushels of corn (shelled) to the acre is about a fair estimate. Oats may be regarded as indigenous, and in early times the most fabulous crops of wild oats were known to grow on the soil as a volunteer crop. Frequently it was so high that it could be tied on the back of a horse. The wild oats was then the pasture of the country, and on it the thousands of "bronchos" lived without further attention from the owners. Of the oats produced under cultivation 32 pounds per bushel is regarded as the average weight. General Stoneman said that wheat could hardly be regarded as a reliable crop; it would fail, probably, four times out of five. I am led to think this is a mere local peculiarity, as certainly within 50 miles I saw abundant evidence as to the possibility of raising fine crops of this most important of all the cereals.

It would be next to impossible to overrate the number and size of the pumpkins and squashes the soil of Southern California produces. Let the reader imagine the longest field he thinks possible and he will probably fall short of truth by 50 per cent. There are thousands of persons who have from the car-windows on the Central Pacific Railroad seen the ground along the line of that road actually covered with them and of fabulous size, who will approve my statement and my failure to give figures to the incredulous. Apples are a sure crop and the trees bear in six years; peaches and plums in four years and cherries in eight. California pears are too well known to bear more than a mere allusion here. The trees as a rule are apt to be overloaded and to break down under the superabundance of the delicious fruit they carry; hence, it is a rule with the most careful of the pear-producers to remove nearly or quite one-half the fruit to protect the tree.

The Mission grape does best, being most prolific in pounds to the acre, and yielding most wine. It is to be remembered, however, that being the longest in the country, it has this advantage as yet over the other varieties. I am, however, bound here to state, on my own responsibility, that on this fruit there are likely to be differences of opinion; some, with General Stoneman, believing the Mission grape to be the most profitable, and others the Malaga to be likely to pay best for raisins; still, differences of opinion on this subject probably indicate that both merit attention.

Coming now to the question of the growth of subtropical fruits in Southern California, there is one fact that seems hardly to have merited the attention it deserves; i. e., that this capacity for growing in immediate association the vegetable products of both temperate and subtropical climates, both attaining not a usual perfection, but as a rule quite an unusual one, must of itself mark something peculiar and unique in the combination of soil and climate of the region. Nowhere else do I know of such an illustration of superabundant productiveness. It would be hard to convince one that the adobe soil one sees so devoid of vegetable life during the dry season can ever be anything else. Let him, however, see that same country after the winter-rains have awakened to new life the germs that have been scattered over the surface; no transformation of fairy scene can be more wonderful, and it is this which constitutes the proverbial glory of a California spring.

Returning from this digression, however, we will take as the type of the subtropical fruit the orange. General Stoneman estimates 70 as a fair number of trees per acre; this a rather larger number than some others plant; but as it is the result of the large experience of a gentleman well known for the reliability of his judgment, it is safe to assert that it is not an overestimate. Orange-trees grown from the seed yield a good quality of fruit, and will bear crops for the market in from seven to ten years, and in from twelve to fifteen years from the planting of the seed it is safe, with favorable seasons and a steady market, to expect from an acre devoted to orange-trees an income of $1,000 at least. I make this estimate low, to be within the limits of truth. It is not improbable that a larger yield might be anticipated. I may here add, that the longevity of orange-trees is remarkable. I have heard it asserted that trees one hundred years old produce well in other parts of the world.

Lemons, though not so extensively cultivated as oranges, do well, and promise in future to receive greater attention. They, as well as the lime, bear in six years.

Olives may be considered a regular crop. The tree is hardy, and requires less care than the orange, and it will probably pay from its sixth year on, the fruit yielding from 20 to 25 gallons of oil per tree when in its prime. I learn from various sources that it pays best to turn the olive into oil. I have never eaten finer pickled olives than those grown and cured by Mr. G. C. Welch, at Los Pueblos, near Santa Barbara.

English walnuts can hardly be considered a sub-tropical fruit. The tree is a native of the Orient, usually being assigned to Persia. It is probable that it may be considered as indigenous also in the Caucasian region, and as growing almost spontaneously as far to the south and east as India, and reaching, under cultivation, as far north as England, hence the name English walnut. It also grows in Algeria, and is cultivated successfully in Chili. Probably none of the nuts are more deservedly popular, nor do I know of another allowing so wide a range in cultivation. It is of slower growth than the trees above mentioned, and one may hardly look for anything like marketable returns before twelve years from date of planting. It becomes a beautiful tree, as ornamental as useful, is hardy, requiring little care, and has few of the enemies so destructive to some other fruits. In some parts of Southern California, when from 25 to 30 walnut-trees are planted to the acre, almond-trees, as of quicker growth and productiveness, are temporarily planted between the rows of walnuts. Citrons do well, bearing in four or five years; though one I saw, on the estate of General Stoneman, set out as a cut, bore flowers and young fruit in sixteen months.

Pomegranates require little care; have few enemies; are not injured by the frosts there; allow a large number of trees per acre, and the fruit can be kept six months. With all these qualities, it is not strange that the tree should be considered a success in the region. I may add they do well even as far inland and to the north as the old Tejon ranch.

Figs may fairly be considered a success. They begin to ripen in June, and mature crop after crop until October. The first crop, however, has too much astringency of taste to fairly represent the average product.

Almonds bear early, and for several years freely. The tree is perfectly hardy and little molested by enemies.

On General Stoneman's ranch I measured a rose-tree that was 45 inches in circumference. Near by it stood a catalpa-shoot with an astonishing history. It was the result of one year's growth, was 18 feet high and 14½ inches in circumference, and on the top bore a crown of flowers. Here may be a proper place to allude to the giant grape-vine of Santa Barbara, which is 14 inches in diameter, and at 7 feet above the ground is divided into branches large as a man's thigh; the branches covering an area of 3,600 square feet, and its annual product is from two to three tons of grapes.

I owe much of the above information to General Stoneman, to whom I would here gratefully acknowledge my indebtedness. Before leaving the vicinity of Los Angeles I would state that the rock out-crops are mostly argillaceous, shaly sandstones, with dark-colored shales interstratified. In some places these shales and sandstones, are capped by horizontal deposits of coarse conglomerate, made up of pebbles of granite, quartz, and hornblende rock, and not firmly consolidated. This is evidently a very modern formation, though elevated at least 100 to 125 feet above the plain.

There are also numerous asphaltum springs in the vicinity, the product of which has been turned to commercial account, and promises to be more largely used in the future.

It may be best to allude here to the "Bee Ranches" of Southern California, which of late have been so productive. I am without facts to prove my belief, but there is enough evidence to make it probable that the bees derive much of the material to make the honey from plants of the buckwheat family named *Eriogonum*. Several species of the group bloom profusely in the honey-producing regions, and its analogy to the buckwheat further serves to confirm the view.

June 19, we left camp with the natural-history division for Santa Barbara. Leaving Los Angeles and going northwest, we passed many fine farms before reaching the Santa Susana range. Like all this wonderfully productive California soil, it looked unpromising, and but for the abundant evidence we had of its fertility we would have passed by as worthless. The water used for irrigating purposes is, of course, that derived from the San Francisco slope. Proximity to the ocean, however, renders much less water necessary here than in the inland valleys. About 40 miles northwest of Los Angeles we came to El Conejo Ranch. This has 49,000 acres. Wheat yields 16 bushels to the acre, (as I was informed by a resident.) Hitherto sheep-raising has been the principal interest of the ranch, and of this we had the most indubitable evidence in the appearance of the land, everywhere pastured off the very surface. How long it will take California to regain the rank pasturage the State once had is a question. Already it is overstocked, and herders are seeking feed in Arizona. It would seem as though regions like the latter, having so much land particularly adapted to the sheep-raising interest, should be sufficient ground to devote to it without devastating such portions of the country as are capable of better things.

June 22, we crossed the western end of the Santa Clara Valley, and found the farmers engaged in harvesting their barley. Much of it they simply headed, allowing the straw to remain. Large fields of good corn were seen. It was just in tassel, and gave abundant promise of a heavy crop. It is hardly overreaching the truth to say that on that day we saw thousands of acres actually overrun with wild mustard, which attained a height often of 8 or 10 feet, realizing the oriental idea that it should become large enough for the birds of the air to lodge in its branches. In some places, indeed, it might well be doubted as to whether it was a mustard or barley field we were passing, both of which were luxuriant enough; but the idea still suggested itself how much larger would either be without the other. What more than anything else surprised me in the day's march was that no attention was paid to fruit-culture. I find recorded in my notes that not a single fruit-tree was seen that day. There was no apparent reason for this. The farmers through this region (along the coast from Los Angeles to Santa Barbara) do not irrigate more than they can avoid, for the reason, as they state it, it brings the alkali to the surface. This I found to be a prevalent objection along an entire line of march during the past summer. I was much struck by the almost perpetual succession of flower-crops that the same species of plant would produce the season through. At no place was this so apparent as near Santa Barbara, which we reached on June 24.

Here my time was again thoroughly taken up by archæological work, and I was unable to devote to botany and collection of agricultural data the time I wished to, and my notes of observation lay principally along the coast by the main road from the town as far up as Los Pueblos. From the coast nothing of the surpassing richness of this strip of land is seen; the whole shore-line looks barren and univiting; once landed, however, the semi-tropical beauty of the town and its surroundings bursts upon the beholder. Pepper-trees and acacias, with their light feathery foliage, contrasting beautifully with the more stately eucalyptus and its leathery leaves. And in the gardens flowers bloom in such profusion as to utterly bewilder one unaccustomed to see such an unlimited floral wealth. Immediately out of the town, on the road above alluded to, fine groves of oaks make their appearance. They have a height, on an average, of 40 feet. With widely-spreading limbs and perfect symmetry, they are as nearly the ideal forms as it is possible to imagine. To add to their beauty, the surface is devoid of undergrowth, so commonly associated with them on the eastern continental slope.

In the main, the productions of Santa Barbara are those of Los Angeles, and it is hardly worth while to re-enumerate them here. Though it is probable wheat does bet-

ter at the former than at the latter place. or at least is a more sure crop, the farms
we passed looked well. especially those of the brothers Moore, Mr. Cooper, and Colonel Hollister. The last two mentioned have been going largely into cultivation of
almonds, olives, walnuts, and eucalyptus. The question as to whether these will
flourish in that region as in Los Angeles is affirmatively settled. Notwithstanding
all that has been said as to the vastness of the market and the impossibility of flood-
ing it (East and West) with these semi-tropical productions, I am by no means certain
that this may not occur, especially as we see each year the area devoted to such productions so rapidly increasing. Just now (January, 1876) oranges are a drug in the
city markets, and this, be it remembered, is before California has to any considerable
extent been adding her stock. The event in itself is a small one, yet it is not entirely
devoid of significance.

The wonderful growth of the Australian blue gum in Southern California, and its
direct economic value, promise for it an important mission on the Pacific coast. The
wants of a vigorous civilization are rapidly using up the forest of California, and it
must be remembered that the ratio of destruction will probably never be less than it
is now. Hence it is almost impossible to overestimate the prospective value to California of her growing blue gums. Indeed, we of the East must not lose sight of the
fact that if, as has been calculated, in less than twenty years our own apparently
exhaustless pine forests will have disappeared, we may have to ask aid in timber from
the eucalyptus groves of California.

I quote at second hand from Ferd. Von Mueller the following statement relative to
the tree in its native land: "This tree is of extremely rapid growth, and attains a
height of 400 feet, furnishing a first-class wood ; ship-builders get keels of this timber
120 feet long; besides this, they use it extensively for planking and many other
parts of the ship, and it is considered to be generally superior to American rock-
elm. A test of strength has been made between some blue gum, English oak, and
Indian teak. The blue gum carried 14 pounds more weight than the oak, and 17
pounds 4 ounces more than teak, upon the square inch. Blue-gum wood, besides for
ship-building, is very extensively used by carpenters for all kinds of outdoor work,
also for fence-rails, railway-sleepers, (lasting about nine years,) for shafts, spokes of
drays, and a variety of other purposes." * Concerning the value of the blue gum as
an antiperiodic in the various forms of remittent and intermittent fever, I think it is
yet premature to; offer a decided opinion. That it is of no value I could not venture
to affirm. In my hands, though freely administered, it has never given satisfaction.
That may be due to the circumstances that I have had to encounter, the severest and
most persistent types of the disease. I do not think it in any sense a substitute for
quinine, and I think the profession will yet settle down to this belief; and yet the
favorable reports we have of its action must have some foundation in fact. Besides
its alleged antiperiodic properties there are others to which the physicians will yield
a more ready assent. I do not consider the question settled as to the presence of the
trees in a malarial region acting as a preventive of " chills and fevers " and allied diseases, though I am bound to admit that the balance of evidence is in favor of their
healthful influence. Such problems involve too many elements to be so speedily settled.
And more than once in the history of medicine have such ideas risen, been accepted for a
few years, and then forgotten. Hence the need for not only caution in generaliza-
tion, but for a most careful, critical scrutiny of facts. This may lead to important
results.

As worthy of attention in this connection, we may allude to the jarrah, or mahogany
tree, (*Eucalyptus marginata.*) That it would thrive in California is beyond all reasonable
doubt. Its chief merit being its strong, close grain and its great durability, or rather
indestructibility. It is proof against all the marine enemies of ordinary wood, and
hence much prized in sea-going vessels. For railroad ties it lasts a long time.

The red gum (*Eucalyptus rostrata*) is also of great economic value, on account of its
durability for underground uses; lasting, when properly selected, over a dozen years
for railway-ties. It has also a good reputation in ship-building.

It is a peculiarity of these Australian gums that they are soft when first cut, but become very hard when dried.

While on this question, it may be well to allude here to the idea of Mr. Stearns, that
several of the acacias from Australia could also be readily cultivated on some of the
drier, treeless portions of the State. Knowing the habits of the acacias generally, including the smaller ones of Arizona, I have but little hesitancy in commending his
suggestion as one peculiarly worthy of a fair test.

It is impossible to overestimate the importance of these rapidly-growing gums to the
Californian agriculturist. With little trouble or expense to himself, he may in a very
few years have around his home a shady, health-giving grove, that will, besides

---

* I am indebted for this quotation and for other information to a short but valuable
paper by Mr. Robert E. C. Sterns on *The Economic Value of certain Australian Forest
Trees and their Cultivation in California.*

beautifying his residence, furnish him with all needed timber and fuel. In special cases, it can be made to furnish a protection to his crops and his stock from violent periodical winds. As outside the realm of exact science, but still within the sphere of legitimate speculation, it may not be amiss to inquire, too, whether enough land in some limited areas might not be gained to cultivation by planting largely the trees we have above named. It may be accepted as an axiom that the tendency of decreasing the waste, sandy areas of evaporation is, other things being equal, to increase the areas capable of cultivation. We do not know that forests actually increase the annual rain-fall, but we do know that they enable us to gain greater permanent advantage from that which is precipitated.

In Southern California, the necessity for irrigation is the rule. However, where a region lies within reach of such banks of fog as at some seasons drift in from the ocean, (and notably so at Santa Barbara,) the necessity for irrigation is much diminished.

While at Santa Barbara, we examined the asphaltum-deposit on the property of Mr. J. Wallace Moore. The point at which we saw it was between 6 and 7 miles west of the town, where there is a fine exposure of it, depending on the encroachment of the sea upon the shore bluff undermining and allowing a toppling down of the overhanging earth and asphaltum. Rising from beneath the level of the ocean to a little above it are the bituminous shales, much disturbed and inclined at a high angle.

Over them is a Post-pliocene deposit with a varying thickness of from 60 to 90 feet. The asphaltum appears to be diffused through the underlying bituminous shales, and only found in a pure condition after expressed, as it were, from the slates by pressure, and probably heat. It then rises through the crevices or less compact portions of overlying soil. In this passage it is mixed with sand and gravel, and thereby made more hard. In this condition we find it lying in masses on the shore. It appears to be entirely similar to the product of the reputed oil-wells near Ojai, which, when first leaving the ground, is soft enough to flow down a moderate incline, but which soon becomes oxidized.* Though the probability of these bituminous exudations ever furnishing enough of oil to pay for working in competition with the wells of Pennsylvania is small in the extreme, yet the asphaltum is a commercial product of no small importance. Its use for roofing houses and making pavements is well known. On the route from Santa Barbara over to the island of Santa Cruz, the Steamer Hassler (to whose gentlemanly officers we are so much indebted) passed through an "oil spring," where something like oil could be distinctly seen floating on the surface. I am unable to obtain a specimen of the substance, but it is probably intimately connected with the shore-deposit. The list of uses to which the aborigines put this asphaltum is a long one.

We found it everywhere present among the archæological treasures we exhumed at Santa Barbara. They made their rush baskets water-tight with a covering of it; pitched their canoes; made ornaments; used it as a glue to mend their broken household pottery; and, for anything I know to the contrary, employed it as a paint for the face in times of mourning, as some tribes still do. It was their panacea for everything broken that required pasting, for everything pervious that must be made impervious to water.

North of the Santa Barbara, and running nearly east and west toward Point Concepcion, are the Santa Inez Mountains. Their trend corresponds exactly with the coast-line. What their influence may be in limiting the fog-bank (to which I have already alluded) to the belt of arable land between themselves and the shore, I am not prepared to say. Notwithstanding the high temperature to which the summits attain, I can hardly help thinking they act the part of condensers or limitations to the sweep of the fog, thus making its influence more positively beneficial over a smaller area. Hence the slight necessity for irrigation at Santa Barbara and in its vicinity. In places they are covered with a dense growth of scrub-oaks, *Adenostoma*, *Arctostaphylos*, and *Ceanothus*, constituting the densest and most soul-trying of chaparral.

The red-wood does not appear to grow south of Point Concepcion, and the Monterey cedar does not reach so far south. Along the streams which put down from the mountains to the ocean cotton-woods and button-woods are found, forming a narrow line of shade bordering the more open ground. Oaks of two or three different species form the mass of the trees from Santa Barbara back to the mountains. It may not be inappropriate to glance a moment *en passant* at the climate of the coast-line from Santa Barbara to San Diego. As our observations were simply those taken *in transitu*, I am here obliged to profit by the labors of others. I am indebted to the Santa Barbara Press for the following table of temperature at that place:

	1870-'71.	1871-'72.
Coldest day	42°	44°
Warmest day	92°	74°

* Passing down the line of flow from the Asphaltum spring, we found a living rattlesnake. It had attempted to cross, but had become fixed in the pasty mass. This may serve as a homely illustration of the consistency of the asphaltum at 100 yards from the point of exit from the ground. *Crotalus* was allowed to remain.

	1870-'71.	1871-'72.
Mean of spring	60°	60°
Mean of summer	69°	67°
Mean of autumn	65°	62°
Mean of winter	53°	53.5°
Yearly mean	60.2°	60.6°

In a series of observations extending over three years and a half we find, from the same authority, that on the coldest day the mercury stood at 42°, and on the warmest it ran up to 90°. On two occasions (due to extensive fires in the neighboring mountains) it reached 100°. This last extreme must, of course, be eliminated from any table of normal temperatures of the place.

"Comparing the hygrometer and thermometer for one year, from September, 1872, to September, 1873, we find the average difference between them for one month to be: September, 4°; October, 3.5°; November, 5.6°; December, 3.75°; January, 3°; February, 3.25°; March, 3.50°; April, 4.20°; May, 5.5°; June, 6.02°; July, 4.25°; August, 5.75°. The average difference for the year is 4.09°."

From Truman's Semi-Tropical California I obtain the elements for the following table of temperature at Los Angeles:

Month.	Sunrise.	9 a. m.	3 p. m.	9 p. m.	Monthly mean.
	°	°	°	°	°
January	40	55	64	50	52.25
February	41	56	64	48	52.25
March	40	60	69	54	55.75
April	53	66	73	57	62.25
May	56	65	71	60	63.00
June	61	70	77	64	68.00
July	66	74	80	67	74.25
August	65	75	81	69	72.50
September	61	75	85	67	72.00
October	59	74	79	62	68.50
November	49	67	69	57	60.50
December	47	57	62	51	54.25

The above observations were made by Mr. Broderick for the year 1871.

Figures from San Diego are very meager All I actually know of the climate is, its January temperature is 52°, and its July 72°. The annual rain-fall at Los Angeles is 18 inches; at Santa Barbara 15 inches; at San Diego 10 inches.

I will also extract the following from Semi-Tropical California:

"The deaths for each one thousand inhabitants in several of the leading cities of the United States are presented in the following table, and the comparison cannot fail to be suggestive:

Saint Louis	21	New York	29
San Francisco	21	New Orleans	37
Boston	24	Los Angeles	13
Chicago	24	San Diego	13
Philadelphia	25	Santa Barbara	13
Baltimore	27		

This table does not inform us whether the statistics include the host of consumptives who visit the three places last named or not. If it does these they are still, in spite of their favorable showing, charged with a percentage of mortality that is not inherent in the climate.

July 13 we broke up camp on Moore's Island and started for Fort Tejon. I would take this opportunity to state our profound obligations to the proprietors of the land on which we had been seeking for Indian antiquities, the brothers Moore; also to Mr. Joseph Park and to the Rev. Stephen Bower, the latter of whom was with us most of the time, and to whose active assistance, along with that of Mr. Park, we are so largely indebted.

According to our orders we were to have reached Fort Tejon by crossing the mountains over the trail directly from Santa Barbara. After a fair trial this was found utterly impracticable, on account of our mules being so greatly overloaded. We were then driven to take the trail via Cassitas Pass to the Ojai, and thence to the Santa Clara Valley, a little below Camulas ranch; hence left the coast at Rincon, 16 miles south of Santa Barbara. We camped for the night at a little cañon on the southern side of Cassitas Pass. The forage was poor, owing both to the presence of sheep and to the unusually dry season. There were two small but well-cultivated ranches in the cañon. The corn, though there was not much of it, looked extremely well, and the proprietor of one of the farms asserted that whatever would grow at Los Angeles could also be raised on his farm, the altitude of which my aneroid indicated as 540 feet above the sea. The potatoes that we dug on his farm certainly were above the average, both in yield and in quality. I am prepared to believe his statement that he would get 150 bushels per acre of them. Corn yielded 75 bushels to the acre. The barley, which had just been cut, averaged 65 bushels to the acre. The gentleman boasted on the apples he raised there. Barley was sowed in January or February; corn was planted May 1, and potatoes from December to June, depending somewhat on whether they were intended for early or late use. We found the oaks, pines, and everlasting chapparal the same as at Santa Barbara.

July 17 we camped on a tributary of the Ojai and found, for the first time this season, some grass that had escaped the sheep. It stood about us over a foot high, but was so dry and tasteless that our mules would not touch it. The capacity of the soil for yielding paying-crops of the cereals was good. Indeed the environment was the most pleasing we had seen. A real lively brook, such as would have been no discredit to the Sierras, flowed over its rocky bed, worn out through the rich oak-covered plain. We found beautiful clumps of *Eriogonum, Adenostoma*, and, on a spray-covered rock, a handsome *Lobelia*.

Along the road from here to the dividing-line separating the Santa Clara Valley we found well tilled farms and abounding signs of a comfortable community. The divide was 1,200 feet above the sea-level, and on it we found a soil that yielded good returns of wheat, corn, and melons, with how much more besides I know not. A vigorous grove of *Eucalyptus* stood by the road-side, proving that the altitude did not interfere with its growth. We camped on the Sespo Creek at an altitude 1,025 feet.

July 20 we entered Santa Clara Valley, and as we did so passed by a large flourishing mill which was evidently doing a good business. Well-tilled farms became more common, and there seemed to be still more room and water sufficient for a much larger population. The ground reaching down from the hills was a sage-brush covered slope, while on the flats bordering the river we found a greensward, much of which was made up of sour grass, that a mule will eat rather than starve. Earlier in the season it appears there is forage of a better character to be had in the valley. The water is alkaline, but less markedly so in the river. Northeast of the valley the whitish bituminous shales are visible on the hills, and the decomposition of them contributes largely to the surface of the valley.

Camulas ranch is quite noted for the quantity of wine it produces. We found the red wine good, cheap, and with a large percentage of alcohol—quite conducive to early slumber and headache next morning. In addition to vine-growing the other interests of the region are not neglected. Altitude, 750 feet.

Following the valley up we entered by one of its arms the San Francisquito Cañon, and then the pass of the same name, crossing the summit at an altitude something over 3,500 feet. Lake Elizabeth lay from this summit almost below us. The reading of my aneroid indicated 3,170 feet as its elevation above tide-water. The lake at which we camped is one of a series of three. It is about a mile long, and shallow throughout; water trickling down to it from the hills around indicates that it is simply the result of surface-drainage. The lower lakes of the series are deeper. It is said that the waters in the lake are rising each year. I can give no positive data, however, on this point. Good crops of corn, potatoes, and barley are raised here. The whole country has been overrun by sheep, until there was not a vestige of pasture left. We were on the western edge of the desert at Lake Elizabeth.

From this place we crossed the edge of the desert to La Lievre ranch, a distance of about 25 miles. The scenery was mountainous in the extreme; here and there along the road we encountered a large yucca, that reminded us more of some similar scenes in Arizona than anything else we had seen on the trip. At this season (July) there was a bare sprinkling of the driest grass; a possible reminder that during the rainy season it would be better. Bands of sheep and a few cattle manage to eke out a slender living. Water was not passed once after leaving Mud Spring. There is usually water at one other place, but it was gone at the time of our visit. The Lievre Ranch House, however, redeems the region, for here we found quite a little stream. The water was good, but it was all evaporated or absorbed by the thirsty earth before it fairly reached the plain below. Along the edge of the desert, rocks of the Tertiary period along with some of volcanic origin are found. The former are much fractured, and are tossed in all directions. The volcanic rocks in places look almost as though stratified; they are very hard, and filled with cavities in which specimens of chalcedony are found.

From the Lievre ranch house to Mrs. Gorman's the ground is of much the same character as that just passed over; perhaps more irregular in surface, with an occasional basin in which water collects during the rainy season. Alongside the road standing water was found in several places, and on one hillside a bog indicated that a good spring could be developed by a little digging.

The Gorman ranch is the most productive in the region, at least the largest crops are raised there. Whether this is due to better work, (as is probable,) or to a better natural locality, I am not sure. At present it suppl'es most of the forage used in the immediate country, furnishes the transient custom with feed for the animals, and I am told has beside a remainder for shipment to Bakersfield and Caliente.

Thence our road lay past the Castac Lake, (then a dry alkali-covered basin,) and through a valley, becoming each mile more attractive, until we reached old Fort Tejon.

It would be difficult to imagine a more fit site for a military post in this region; in a valley supplying plenty of ground fit for cultivating all the ordinary vegetables needed to maintain the health and promote the comfort of the troops stationed there; with abundance of forage for the animals in the valley and on the hills adjacent to the post; with good, cool, (62° Fahrenheit,) clear water bubbling up from several springs at the bases of the hills; with delicious shade from the ample expanse of oak foliage, and above all, in this torrid region, a constant breeze passing to and fro through the funnel-like valley at the hours that otherwise would be most unendurable.

I measured one oak tree on the now desolate parade-ground that was probably 60 feet high, and had a diameter of 8 feet 2 inches at 5 feet above the ground, with three branches each as large as a good-sized tree, carrying the shade out on all sides full 40 feet from the trunk. My aneroid barometer read 3,150 feet above the sea. The hillsides around were, besides the oaks already alluded to, covered with a dense growth of scrub-oak, California buckeye, and a hard shrub which, for want of a more intelligible name, I will allude to as *Cexcocarpus parrifolius.* These combined form the impenetrable thicket about the post. There had been sheep everywhere to leave behind them a waste almost destitute of grass or any green herb, yet from the dead wild oats on the tops of the hills I could readily see how abundant the pasturage had once been. The water is heavily charged with carbonate of lime, which for a time induces looseness of the bowels in those using it, but ordinarily in a time becomes as healthy as it is palatable. Where the water percolates through the soil it forms considerable deposits of calcareous tufa. All the cooking-utensils in which this water was boiled speedily became coated with carbonate of lime.

From Fort Tejon the natural history party started for Cuddy's ranch, situated about 6 miles east of Mount Piños, and at an altitude of 5,150 feet above the sea. We found on the way up that the piñon pine began to be common at 4,200 feet above the sea, and extended to nearly or quite 6,000 feet, this being the most characteristic tree at these altitudes. Incidentally I will allude to the fact that while the piñon nuts of New Mexico are round and with an average diameter of three-eighths of an inch, those of California are three-fourths of an inch long and have a diameter of one-fourth of an inch; the leaves and cones, however, of the trees being to ordinary observation much alike.

East and west of Mr. Cuddy's are hills the culminating points of which are from 7,500 feet to 8,500 feet above the sea. They are made up of granite and metamorphic rocks to the summit. During the course of ages an immense quantity of detritus has been washed down from the summits, and at an altitude of about 5,000 feet above the sea has accumulated to form the flats now covered with verdure, and known as cienegas or meadows. Looking first at the rough aspect of the surrounding hills and then at these cienegas, the latter are indeed oases. I have already alluded to the fertility of the Gorman ranch. That of Mr. Cuddy though not over 2,000 feet higher, is, from his testimony, utterly unfit to raise any of the cereals except rye, which he says does fairly. He does not grow any of the garden vegetables, and hence depends entirely upon his purchases for his vegetables, the chief interest being stock-raising. He has fine bands of horses and cattle roaming at will over the hills within a dozen miles of his home.

On the part of myself and my associates there I wish to make public acknowledgment of all the assistance and kindness we received from that whole-souled gentleman.

Just here I will allude to the fact stated by Mr. Robert Prado (some 20 miles distant from Mr. Cuddy's ranch, and on the extreme headwaters of the Lockwood Creek) that June frosts killed about all the vegetables he attempted to raise. The altitude of his ranch was the same as that of Mr. Cuddy, and also about the same as the Motor ranch, where I saw a very sickly-looking field of wheat that appeared to confirm all I had heard as to the impossibility of raising any cereals amid the mountains at that altitude.

July 30.—Ascended Cuddy's Peak, southeast of Cuddy's. It was by the aneroid 7,750 feet high, and covered with a growth of bull and yellow pines to the top, along with which were growing *Eriogonum flavum* and *Artemisia tridentata,* which (the latter) is there the commonest sage-brush. Mount Piños was found also, by an aneroid reading, to be 8,500 feet above the sea. This is also known as Saw-mill Mountain. The principal timber is bull and yellow pine, which Mr. Magill is now rapidly working up into lumber for the wants of the adjoining region. He sells it at the mills at $20 per hundred feet, the yellow pine making by far the better lumber. A thermometer placed in the spring under the saw-mill read 52° F., which may be taken as not far from the mean temperature of the earth at that point. The valuable timber does not appear to grow at an altitude much lower than 6,000 feet on the side of Mount Piños. I found a peculiar-looking dwarf-oak 20 feet high at an altitude of 7,000 feet. In addition to the pines already named I found near Mr. Prado's, on the headwaters of Lockwood Creek, some sugar-pines. They closely resembled, in their straight trunks and soft wood with long hanging cones, the white pine of the East.

August 13.—Bakersfield. This town enjoys a most unenviable notoriety for the fever and ague. Out of 17 patients in the county hospital I was informed by the attending physician 15 are victims to malaria, which here gives its own distinctive character to the pneumonia, diarrhœa, and dysentery. The cause of this prevailing disease is not far to seek. The town is situated in a low, level country, now much of it for the first time being opened up to cultivation. Much of the soil is so rich with decaying vegetable matter that it is simply a vast compost pile. During high water in June Kern River overflows its banks, and as the water recedes during the following months the vegetable matter thus left on the ground and that turned up by the plow decay and poison the atmosphere with disease-germs. Adding to this already sufficient cause, we have irrigating ditches along the streets, stagnant water in the streets, and kitchen-offal in back yards. The only·wonder is that there is not more disease. Much of this, however, is rather the course of events in any new town. We may confidently look for a great improvement as the citizens bring their surroundings to their own ideas. They are energetic and know what is to be done. The place is healthier now than Illinois was twenty years ago. Among the great sanitary measures needed, none are more important than one or two artesian wells, thus furnishing a substitute for the surface-water now used. It is extremely probable that water can readily be had in this way here.

The plants poisoning stock in this region are attracting much attention. The damage done by them to stock-raisers is immense. In one instance several hundred head of fine sheep were poisoned outright. The destruction among horses, though not so great in numbers, has probably been of an equal value in money. It is remarkable that in a family of plants the *Leguminosa,* which has until of late been regarded as possessed of so few hurtful plants, that all at once we should hear of the great damage being done over wide areas and in each location mainly by different plants of this order. In Colorado and New Mexico by *Oxytropis lamberti;* in Arizona by *Hosackia purshiana,* (so reported;) and in California by two or three species of *astragalus.* The animal becomes as addicted to its use as ever an opium-eater to his drug. It becomes a habit which we may appropriately call the *loco* habit; ("*loco*" is the Spanish for "fool.") The animal becomes demented—at one time hopelessly stupid, at another timid, and again actually dangerous. He grows progressively weaker and thinner until death comes to his relief. Sheep are more frequently poisoned by the woolly milk-weed of the region.

August 24.—We camped in the Tejon ranch. Water was so scarce that it was brought for some distance for the house, and the creek upon which a large band of Indians formerly depended for irrigating water was well-nigh dry. Along its rocky bed we saw perfect thickets of wild grape-vine, but no fruit this year. The ranch-garden was a semi-tropical wilderness; figs, grapes, pomegranates, peaches, pears, and melons, with what more I do not know, all thrive and bear profusely. We could see what the garden had been. Squirrels visit it by droves and are fast destroying the choicest trees and vines. The thermometer on August 23 stood at 110° F. in the shade at the house.

Passing by an interesting visit to the newly-opened ranch of Mr. Souther, south of Bakersfield, for the present I will simply allude to the artesian well that he now has in successful operation. It is the first one in the county, (Kern;) water was struck in quantity at a depth of 250 feet, and according to Mr. Souther's estimate the well delivers 35 gallons a minute; this I think is an underestimate. The water is good and has a temperature of 65° F., as nearly as I can guess, (my thermometer was accidentally broken the day before.) Mr. Conner informed me that Cotton ranch, near Bakersfield, was a success so far as growth of good cotton was concerned, and that the yield was fair.

August 27 found us at Walker's Basin, at an altitude of about 8,500 feet above the sea. The hills around are well covered with a growth of oaks and pines, among which (the latter) we for the first time found the Digger nut-pine. The tree may be known at once by its enormous cones, with sharp, hooked tips to the scales, over three-fourths of an inch long, and by its long leaves, which do not appear to be nearly so numerous as in the other pines of the region. Thus the tree presents a remarkably open appearance. The oaks were fairly full of acorns, and the "hog-herders" had driven their bands into the basin to take advantage of the crop. Barley grows well here, as do potatoes also, and the hardier garden vegetables, though the squirrels had destroyed almost all the fruit-trees.

Leaving the basin we crossed the divide to Havilah by wagon-road. The oak disappeared, and the forest on the hills was now made up exclusively of coniferous trees, and cottonwoods along the edges of the water-courses. After crossing another divide northward we were fairly in the plain through which the Kern River flows. There was not, except at a few spots, water enough for irrigation, and hence agriculture was limited. It is probable that during certain months the region would support a goodly number of sheep and cattle. The hot springs (though I am not aware of their having any marked medical properties) will doubtless become important as a health resort, on account of the high temperature of the water and the proximity to interesting mountain regions. Still, even this is in the future.

Weldon is, according to my aneroid, 2,900 feet above the sea-level. Soil is sandy and alkaline, though when cultivated capable of producing good crops of corn, wheat, barley, watermelons, squashes, tomatoes, and cucumbers. Potatoes do not do so well. The most common grass on the flats just above the level of the river is the so-called sour-grass, so hard and so acid that no stock will touch it except in extreme want. It was abundantly covered with an exudation on which the sourness depends. The higher ground had a vegetation composed of *Senecio, Bigeloria,* and *Baccharis,* with some sage-brush intermingled. A belt of cottonwood trees skirted the river, largely supplying the fuel for the region.

September 4.—We were at La Motte's ranch; altitude was 6,700 feet. Of course this precluded, if not all hope of raising crops, at least all attempts at it. We had our first frost of the season here. The meadow (of which this is the first of a series extending quite up to the base of Mount Whitney) was fenced in, thus protecting the grass from the immense bands of sheep passing. The grass sedges and rushes making up the mass of pasturage was over a foot high—eagerly eaten by our animals as grass, but rather reluctantly taken as hay. The hills were covered with piñon pine, the altitude of this tree here nearly coinciding with that observed at Cuddy's ranch. Here, through the kindness of Mr. Kennedy, I was put in possession of some *chia,* an article well known to the Mexicans and Indians, who use it as a food on their long trips, and also mix it with water to render it (water) more palatable and refreshing, and to do away with the necessity of drinking so much. Chia is the seed of *Salvia columbariæ.* It is roasted, then ground, looking much like flaxseed meal. It is mixed with water to which a little sugar has been added, and then the whole stirred for a few minutes until a thick, mucilaginous mass is developed. The taste, which grows upon one, is at first rather pleasant. I anticipate for this an important remedial use. In inflammation of the stomach and bowels, in dysentery, and even in hemorrhoids, it has proven a remarkably soothing mixture for internal administration, combining food and demulcent medicine in one article—just one of the desiderata of the physician to-day. Besides this, I have had experience enough with it to satisfy myself of its great value as a poultice. Measures have been taken to secure enough of the seed to test its virtues more thoroughly in medical treatment.

From La Motte's we gradually ascended through meadow after meadow (all closely pastured off) until we crossed (twice) the South Fork of Kern River and camped at the foot of Mount Olanche. Here we remained some days, and I succeeded in obtaining about 100 species of plants that were in bloom. The camp was 8,200 feet above the sea, and the whole character of the timber had changed. The piñon pines were below us, and we were among the *Pinus contorta,* (twisted pines,) *Pinus ponderosa,* and spruces. I could see that they extended well up toward the summit of the Olanche. Sage-brush covered the valley on which we camped. Higher up I found many beautiful composite and a splendid lupine in full bloom. The timber is of considerable value, but so remote from market that it is not likely to be utilized for many years. At 10,500 feet I found a cedar 4 feet thick growing vigorously. It was not more than 40 feet high, however. At 10,000 feet I observed a peculiar-looking pine, which I take to be *Pinus deflexa,* Torr. Having only insufficient specimens and no authentic herbarium specimen, I am unable to speak with certainty on this point.

From this camp, until we reached the base of Mount Whitney, there was no noteworthy change in the flora. There, however, we encountered, in addition to the coniferous trees already alluded to, *Pinus albicaulis* and *Pinus Breweri,* (what, at least, I take to be the latter species provisionally so named.) Observation on a peak south of Mount Whitney gave 12,100 feet as the height of timber-line. The dead and dying trees at what is now timber-line were a puzzle to us. What can have caused this I do not know. A halt of two days was made at a point midway between Mount Whitney and the Soda Springs, on the North Fork of Kern River, to await a portion of the party. Here Mr. Klett made the ascent of Meadow Mountain. I am indebted to Mr. Klett for the observation on timber-line. On the north side of the mountain it is 11,200 feet, and on the southern side 12,000 feet above the sea. Soda Spring, on the North Fork, was our next camp, and here we remained several days. The forest-growth at once assumed nobler proportions. One fallen *Pinus ponderosa* was measured, as it lay upon the ground, and found to be 160 feet high. It was 6 feet through at the base, and was straight as an arrow. *Librocedrus decurrens* was also quite common. Both of these trees must yet be made available as lumber. The size and quality of such trees will make them sought for at almost any cost ere long. We found here good pasturage for our animals. It was, however, almost entirely sedges and rushes, with but little genuine grass among it. Higher up on the mountains, where the sheep had not been, a still finer quality was found. The water is remarkable for the quantity of iron it contains. There is also enough carbonic acid escaping to make it pleasantly pungent : besides, it is decidedly alkaline in taste; it has not (that I could discover) any laxative properties. That the water will be proven to have an alterative effect I have no doubt; and that the iron will confer upon it a tonic effect I am equally sure. With such a combination of properties and situated in a perfect center of mountain scenery, with a fine climate and good fishing superadded, it will be among the marvels of the world if this spot does not in time become a most popular resort. We saw here the first oak we had seen since leaving Walker's Basin. It was a mere shrub, never exceeding 10 feet high. The mountain-sides had a thick chaparral of *Ceanothus* and *Castanopsis.*

Back of the Soda Springs, at an elevation of 10,000 feet, I came upon a deep wash in the side of the mountain. It was over a mile long and literally strewn with tree-trunks that had been washed (probably by melting snow or a water-spout) from the steep hill-sides. Below, this gulch terminated in a flat bog. In this bog were found layer upon layer of logs that at one time or another had been swept into it. Over all and among all was the wet mud—the mass being transformed into, may be, a peat-bog or, may be, a mere marsh. Still, the case was so striking and so like some of the events of former geological times that I could not avoid making a note of it.

Going south from Soda Springs we crossed on the old Visalia trail a western fork of the North Fork of Kern River, which was half as large as the main stream. It ran through a valley deeply scored and furrowed by glacial action. South of this, crossing a divide, we came upon the headwaters of Tule River. Dense forests of spruce covered the higher ridges, and in the valleys between were open meadows that, but for the sheep, would have been well covered with long grass.

Sequoia gigantea, the giant of our New World forests, grows, at a mean altitude of something over 6,000 feet, quite abundantly on these tributaries, as well as on those of the North Fork of Kern River to the east. They are fully as large as the more noted ones so frequently the wonder of tourists in the Calaveras grove. Here they are seen along the hill-sides in association with the other usual coniferous growth. It is rare to find one of the larger specimens unhurt at the base by fire, or with a top that is not in some way injured. I was struck with the apparent scarcity of young trees of this species. It is not an unusual thing to find these trees with a diameter of thirty or thirty-five feet, and attaining a height of two hundred and fifty feet. We found the "sugar-pine" (*Pinus Lambertiana*) on one occasion growing on the head of-Tule River, at a little over 7,000 feet of altitude. This, to my mind, is the most graceful of all our western conifers, as is its relative, the white pine, the most comely of our eastern pines.

It is usually straight as an arrow, with a clean trunk towering to 200 or more feet high, and a diameter of from 5 to 10 feet. From the ends of the branches the cones, a foot or more long, hang straight down in beautiful clusters of two or three. Its range in altitude is from 3,000 to 7,000 feet, attaining probably its finest proportions at from 4,000 to 6,000 feet. The resin, especially in trees that have been injured by fire, is markedly sweet, hence the name of "sugar-pine." The timber is of great value, being soft, light, and free from knots.

After crossing into the valley of Deer Creek there was another change in the arborescent vegetation. The pines were no longer the characteristic feature of the landscape. Oaks supplanted them, and among these *Quercus lobata,* resembling the white oak of the East, was most common. It is remarkable for the great length of its acorns, and furnishes good fuel. Though by no means equal to our eastern white oak as an article of lumber, it is not without its uses. In the valley we found fine crops of corn, wheat, and barley. Potatoes were especially fine, being large, sound, and boiling into a delicious "mealiness." I measured one potato-vine and found it to be eleven feet long.

Linus Valley we found to be in its flora much the same as that of Deer Creek. It was larger, more thickly settled, and supported large herds, but none of the cereals are raised there. Even barley, I am told, is brought in from Kern River, and there is not a thrashing-machine in the valley. This appears unaccountable. There was a grist-mill there at one time, and the soil is capable of raising almost anything.

From Linus Valley we crossed to the valley of Kern River, passing on our way a saw-mill that supplies from the well-timbered slope all the lumber needed in and about Kernville.

From the last-named point to Caliente our route lay through the region already gone over. It was noticed that though we had had but a few showers the hill-sides near Caliente were becoming quite green, as those of Walker's Basin, fourteen miles away, were indicating the approach of winter.

I have the honor to be, very respectfully, your obedient servant,

J. T. ROTHROCK,
Acting Assistant Surgeon, U. S. A.

Lieut. GEO. M. WHEELER,
Corps of Engineers, in charge.

AP. JJ—14

APPENDIX H 6.

REPORT ON THE PHYSICAL AND AGRICULTURAL FEATURES OF SOUTHERN CALIFORNIA, AND ESPECIALLY OF THE MOHAVE DESERT.

UNITED STATES ENGINEER OFFICE,
GEOGRAPHICAL SURVEYS WEST OF THE 100TH MERIDIAN.
Washington, D. C., April 7, 1876.

DEAR SIR: I have the honor to submit herewith a report upon the physical and agricultural features of Southern California. It treats of the coast counties, of the island of Santa Cruz, and of the Mohave Desert, with the valleys, oases, and soils.

Respectfully, your obedient servant,

OSCAR LOEW.

Lieut. GEO. M. WHEELER,
Corps of Engineers, in charge.

California is pre-eminently the land of great contrasts. From the living glaciers of Mount Shasta down to the coast of the grand Pacific exists every variety of climate. There the snow-capped peaks, here regions where snow-fall is unknown. From the bald peaks beyond the growth of trees are passed, in descending the mountain-slopes, dense forests of pine, with beautiful, charming scenery; next below, in altitudes from 5,000 to 6,400 feet, follows a belt of trees of 30 feet and more in diameter, of from 280 to 320 feet in height, and reaching an age of three to four thousand years; then succeed the fine agricultural valleys of the Sacramento, of San José, San Joaquin, Napa, Sonoma, and Tulare, and lastly the sea-coast region.

Ascending the gradually-rising western slopes of the Sierra and glancing at the other side of the picture, what a contrast is offered by nature to view! Beyond the precipitous eastern slopes a country spreads destitute in the extreme, and instead of the game regions of the Sierra is seen a region in which the jack-rabbit is nearly if not quite the only game. From here toward the south, passing the eastern slopes of the Sierra, the Owens Lake and the Panamint Valley to the Coahuila Valley, one finds for a distance of 300 miles a dry country, where springs are scarce, and in some cases of saline and bitter taste, where the hill-chains exhibit the plain rock, and the valleys are made up of coarse sand with scanty vegetation, or in not few instances of naked clay, resembling a barn-floor.

Now from the Coahuila Valley to the westward, crossing the San Bernardino and San Jacinto Mountains, the flourishing oases of San Bernardino and Los Angeles are entered. Here are grown—with artificial irrigation—the orange, the almond, the fig, the castor-bean, mulberry, olive, lemon, the pepper-tree, the palm, and the grape-vine. Three millions of oranges are annually produced at Los Angeles, while the wine-production of that county is estimated at 1,000,000 gallons. As there is an overproduction of wine, considerable quantities of grapes are exported as raisins, or serve for manufacture of syrup and grape-sugar. Again, what a strange contrast those tropical gardens present with the barren hills of the vicinity of Los Angeles! Nature here takes her vacation in summer-time, the vegetation sleeps, the grass dries up and disappears, and only when assisted by art she develops her tropical powers. In winter, however—if a warm season with frequent showers can thus be denominated—the hills and valleys of the coast counties become covered with rapidly-developing verdure.

The peculiarity of the California climate exercises considerable influence on agricultural pursuits, and demands a mode of farming different from that of Europe and the east of America. Plowing is not possible during the summer, dryness baking the earth, and the farmer has to wait until the rainy season has set in and softened the soil. The prolonged dryness of the summer-season is not favorable to the cultivation of Indian corn, (without irrigation,) hence such cereals as soonest mature form the principal object. Wheat stands foremost in the California farming, then follows barley. Of less importance are corn, rye, oats, and buckwheat. Tobacco, potatoes, sweet-potatoes, tomatoes, sugar-cane, hemp, flax-seed, mustard, hops, castor-beans, cotton, and fruit-trees are also cultivated to a moderate extent.

The middle and northern counties are the principal wheat-producing regions, while the southern ones, depending exclusively on irrigation, raise comparatively more corn and less wheat.

Another peculiarity produced by the climate is seen in connection with the production of grasses, whose propagation depends solely upon the annual germination of seed, and not, as elsewhere, principally upon the spreading of the roots.

In California the roots dry up during the prolonged rainless season, and thus lose their vital power. Here meadows similar in character to those of the East are unknown, whose turf is an overproduction of living grass-roots. The grass crumbles, stem and seed fall to the ground; and while the uninformed eye of the stranger looks in vain for the grass, cattle and sheep still find subsistence on the apparently barren soil. But as the production of hay is naturally limited, a substitute was introduced, namely, the perennial alfalfa or the Chilean clover, (*Medicago sativa*,) which allows repeated cutting in summer. Its roots penetrate much deeper into the soil than those of the ordinary grasses, and hence survive a drought much easier; it is in this respect, of course, much preferable in counties with dry climates. However, nobody will deny that alfalfa is inferior to our eastern clover, the proportion of woody fiber being much larger. Also wheat and barley-straw, (unthrashed,) called "hay" in California, is used as a substitute for the natural grasses. While a great production must readily be conceded, especially to Middle and Northern California, the drawbacks cannot be underrated at the same time, of which, for instance, the great number and destructive habits of the gopher (*Spermophilus Beecheyi*) is not a small one, undermining the field and gnawing off the roots of the cereals.

A considerable share of wheat and barley is raised in volunteer* crops, but it is a poor method, and only the lazy farmer's way. As irrigation—especially in the southern half of the State—is an imperative necessity, consequently good farming country is not cheap, and all the lands which are supposed to be suitable for production of grain have passed out of the control of the Government, and large tracts are held by a few private persons. However, there is prospect that these lands will finally be subdivided, and thus the existence of a denser population made possible.

Near Santa Barbara and Los Angeles an acre of good land is worth $150 to $200; in the vicinity of San Bernardino not less than $100, and this in face of the fact that sometimes in dry years the crops are a failure, the creeks for irrigation running dry. Artesian wells, it is true, are sunk easily at San Bernardino; there are over 100 with a depth rarely exceeding 140 feet, but they are not practicable on the higher country 10 miles from town. The fine oasis of San Bernardino is one of the few spots of San Bernardino County that are valuable, the larger part of it belonging to the Mohave Desert.

As might naturally be expected, the settlements in Southern California are not large, facilities for irrigation being limited.

Another noticeable fact is the absence of forest on the Coast ranges, which to an altitude of 3,000 feet are mainly covered with brush. Neither exist forests of any material extent upon the islands of the Santa Barbara Channel; upon one of them—the island of Santa Cruz—Dr. Rothrock, Mr. Henshaw, and myself passed nearly a week.†

THE ISLAND OF SANTA CRUZ.

This island covers 73.4 square miles and is of mountainous character; the highest peak is Devil's Peak, with an altitude of 2,700 feet. Anacapa and Santa Rosa Islands are seen in immediate vicinity. While the islands San Clemente, San Nicolas, San Miguel, Anacapa, and Santa Barbara belong to the United States, Santa Cruz, Santa Rosa, and Santa Catalina are private property. All the islands of the Santa Barbara Channel were discovered in 1542, by Juan Rodriguez Cabrillo, who was sent out by the Spanish governor of Mexico on exploring voyages.

The names first given to these islands were repeatedly changed, making it difficult to follow the later Spanish writers. Cabrillo died while on one of the islands—probably Santa Cruz—and was buried there. Besides the islands mentioned, Cabrillo pretended to have discovered six others southwest of San Clemente, which, however, never were seen by subsequent explorers. Whether those six islands really existed and sank afterward below the waves of the ocean, or whether they were an ambitious invention, cannot be decided. The reporter of Cabrillo's voyage says: "El 18 Febero, 1543, conviento N. E. corriéron al SO (de la isla San Salvador) en busca de otras islas que habia; viéron seis, unas grandes, otras pequeñas, y sin tocar en ellas siguiéron al SO." To this a Spanish historian‡ of the last century makes the following remark: "No se puede saber quales sean estas islas, que vió Cabrillo, pues al rumbo SO de la isla San Salvador ó San Clemente no las hay ni se tiene noticias de ellas." (On the 18th of February, with northeast wind, they sailed in southwesterly direction from the island San Salvador in search for other islands, which indeed existed; they saw six, one large and others small, and without touching them continued to the southwest.)

Remark of the historian: "The islands Cabrillo saw are unknown; for there are none in existence southwest of the island San Salvador or San Clemente, nor do we have any notice of such."

The islands are chiefly of volcanic origin, while at Santa Cruz the heights consist of trachyte and trachydolerite; the valleys are covered with Quaternary deposits of clays, sands, and conglomerate. Where the latter form the coast-rock it succumbs

---

* The local expression for spontaneous.

† Thanks are due to Lieut. Commander H. C. Taylor, of the Coast-Survey steamer Hassler, who kindly conveyed us from the coast to the island and return.

‡ Relacion de la viaje por las galetas.

gradually to the power of the waves. There exist unmistakable signs of the island having been subjected to repeated changes of level.

The vegetation of Santa Cruz is scanty; grass is gradually disappearing before the clean sweep made by the sheep-herds, whilst cactus, especially *Opuntia*, is spreading more and more. The mountain-sides are here and there covered by small patches of pine, while the low brush attempts in vain to conceal the barren rock, prominent in all the hills. It is said that formerly the entire island was covered with forests; whether true or not, one fact was observed by us, namely, that roots of pine trees exist in the loose soil of the western shore, where no trees in wide circumference occur at present.

Like the vegetable, the animal life is restricted to comparatively few species. Wild hogs are said to roam in great numbers, the progeny of those introduced by Sebastian Viscaino, (1606.) Wild cattle are also said to occur. The most interesting animal, however, is a fox, (*Vulpes littoralis*, Baird,) not larger than a house-cat, and found solely on the three islands of Santa Cruz, Santa Rosa, and San Miguel. It agrees in almost every particular, except in size, with the twice-as-large *Vulpes Virginianus* of the mainland. The principal food of this small fox consists of grasshoppers, as Mr. Henshaw ascertained by dissection. There can be hardly a doubt that this animal was gradually produced by the diminishing of the size of the common fox, which searched on that island in vain for a more substantial food than grasshoppers, nothing else being attainable; and thus furnishing an argument in support of the theory of transmutation of species by isolation, recently advanced by Moritz Wagner.[*] Truly there are sea-birds whose eggs would be welcomed by this fox, but these birds are too cunning not to pick out the most secure spot for their young. The sheep existing on the island, introduced but very recently, are large enough to resist attack.

However, there exist a number of other species of animal life isolated upon these islands, as pointed out to me by Mr. Dall, engaged as naturalist for a number of years upon the Pacific coast. On the Santa Barbara Island is found a species of snail, (*Binneya notabilis*,) not found upon any of the other islands, (not on Santa Catalina, as stated in one publication.) Another snail (*Helix facta*) is peculiar to three islands, viz, Santa Barbara, San Nicolas, and Santa Catalina. *Helix Gabbii* is only encountered on San Clemente and Santa Catalina, while *Helix Tryoni* is limited to Santa Barbara and San Nicolas Islands. None of these snails occur on the mainland.

While these facts can only be satisfactorily explained by a gradual transmutation of species by isolation upon the islands, where the struggle for existence was diminished and the enemies were few, we have a case of a somewhat different kind in a small green crab, *chlorodius*, inhabiting shallow sea-banks near the coast, which, on several points west of the town of Santa Barbara, has migrated into the swampy flats, separated now by deposits of sand from the ocean, and accustomed itself to live out of the water the most of the time, under which circumstances some changes have begun, especially noticeable with the shears and legs, which have attained a stronger development in consequence of their employment in digging holes, which art the animal has acquired.

Forty years ago Santa Cruz came into possession of several Californians who established a sheep-ranch on it, and brought the last remnant of the aborigines across to the mainland, where they scattered and disappeared among the multitude. The graves and kitchen middings left behind on the island were the subject of examination by Paul Shoemaker, employed by the Smithsonian Institution, while some of those upon the mainland were discovered by Dr. Yarrow, Dr. Rothrock, and Mr. Henshaw, of your survey, and the rich treasures conveyed to Washington.

During the stay on the island the mornings were foggy, with southwest winds, while the northwesterly breeze of the afternoons brought clear weather. The peculiar changes in the height of the tides are a subject of great interest, one of the two high tides being much the higher in twenty-four hours. A full account of this peculiarity is found in the report of the Superintendent of the Coast Survey, 1853. The "southern swell" is another fact important for navigation in those regions.

THE MOHAVE DESERT.

In former times the whole country between the Missouri River and California was termed the "Great American Desert." As population spread toward the West, seeking new homesteads, it was discovered that immense areas of that so-called desert were fine arable land, notwithstanding the absence of trees. What immense tracts of Nebraska, Dakota, and Eastern Colorado have since been reclaimed! The opinion gained ground of the non-existence of a desert upon this continent; the assertions that there exist true deserts were and are still ridiculed in books and newspapers. But such is the fact: there are tracts in Nevada, California, and Arizona that rival the Great Sahara in every particular, except in the number of square miles. The naked

---

[*] Mr. Ridgeway has described recently eight new species of birds from the island of Guadaloupe, 200 miles southwest of San Diego, which may be cited as another argument in favor of this theory.

hills, the barren, stony plains, the sand-storms, the cloud-bursts, the intense heat, the daily great contrasts in temperature, the dryness of the air, the singular forms of the scanty vegetation, the bitter and salty waters, the saline efflorescences, the mirages, the dried-up lakes, all prove the existence of a true desert.

Three subdivisions of the former so called "Great American Desert" may be made:

First. The *treeless, grass-covered plains* (llanos) of Nebraska, Dakota, Western Kansas, Eastern Colorado.

Second. Semi-deserts. Trees are absent and grass is scanty below altitudes of 4,500 feet, but low brush is plentiful, as atriplex, (grease-wood,) artemisia, (sage-brush,) aplopappus, ephedra, yucca, cactus. In this category belong Northwestern Texas, (*llanos estacados*, or staked plains,) Western Indian Territory, and those portions of New Mexico, Nevada, Utah, and Wyoming that have less than 4,500 feet altitude. As these latter Territories are traversed by many mountain-ranges and plateaus of considerable altitude, (from 5,000 to 9,000 feet,) extensive oases are formed, with fine grass, many streams, and dense pine forests, permitting numerous settlements. The traveler cannot sufficiently admire the great contrast between the beautiful vegetation of the forest-covered mountains and the arid plains at their feet. The fact is that the small amount of rain visiting those Territories is principally deposited upon the mountains, which, being cooler than the valley, nearer to the clouds, and attracting them by gravitation, receive the precipitates. Furthermore, the relative humidity of the mountain-air is higher, the average temperature being lower; hence the splendid development of vegetation finds a simple explanation.

Third. True deserts. The rain-fall is so limited that the vegetation sinks to a minimum, and disappears in some instances altogether. To this class belong the Painted Desert in Northeastern Arizona and Southern Utah, the Gila Desert in Southwestern Arizona, and the Mohave Desert, comprising Southeastern California and the southwestern corner of Nevada. Lower California, although a portion of Mexico, belongs geographically to the Mohave Desert. Said three deserts would form one coherent great tract if the plateau of Central Arizona, of an average altitude of about 7,000 feet, and extending from the Sierra Blanca, in Southeastern Arizona, to the Cerbat range in the northwestern portion, and comprising the Pinaleño and Pinal Mountains, the Mogollon Mesa, the San Francisco Forest, and Bill Williams's Mountain, did not intervene, forming a part of the Colorado Plateau, and representing a splendid great oasis, with immense forests, luxuriant meadows, and numerous springs and mountain-streams. It is difficult for the imagination to conceive any greater contrast than that existing between the Mogollon Mesa and Death Valley.

The Mohave Desert is by no means a vast plain with loose sand, or a system of sand-dunes and rocks, but forms an almost uninterrupted series of hill-chains, valleys, and mountain-ranges. The latter increase in height the farther north they are situated, the Inyo and Panamint ranges reaching 9,000 to 10,000 feet altitude, while the Providence, Opal, Payute, and Dead Mountains, of the central portion of the desert, reach 5,000 to 6,000, and the Chocolate, Chucavalla, and Riverside Mountains, of the southern portion, average hardly 4,000 feet.

In regard to the intervening valleys, a marked distinction between deep and shallow basins becomes manifest. Death, Salinas, Owen's, Panamint, Ivanpah, and Coahuila Valleys are examples of the former, representing deep depressions, while the more common form is that of a plain, gradually ascending toward the mountain-slopes.

Next to the mountains and valleys the oases come into consideration. There is, however, only one of great importance, the valley of the Colorado River, after which follow the valleys of the Mohave and Virgen Rivers. Besides these, smaller oases, covering a few acres and caused by a spring, occur, although in limited number. Los Toros, Dos Palmas, Hundred Palms, Whitewater, and Agua Caliente, all in the Coahuila Valley; Bitter Springs, 20 miles north of Callville; Indian Wells, Granite Wells, Coyotehole, Little Lake, may be mentioned. Next to these, the higher mountain-ranges may be said to change into oasis-like regions above an altitude of 5,000 feet, verdure becoming plentiful, springs are met, and game (mountain sheep and deer) easily find existence.

The Opal Mountains, with Trance Spring, and Ivanpah, the Payute, and Providence Mountains, with the mining district New York, the Cerbat range, with Mountain and Quail Springs, deserve here especial mention.

The altitude in which the oasis region begins increases toward the north, and the Inyo range, although reaching nearly 10,000 feet in height, gives an example of an unparalleled scarcity of water, comparing in its vegetation unfavorably with the mountains farther south. In these northern portions of the Mohave Desert are situated the Salinas and the dreaded Death Valley, the latter 130 miles long, 15 to 40 miles wide, and in part 200 feet below the sea-level. The Telescope and Panamint ranges separate it from the Panamint Valley, a hot and deep basin, with a soil of barren clay in the central portion and of gravel in the peripheral parts.

It is a singular and favorable coincidence that in both Death and Panamint Valleys, where no springs are fed by rains, hot waters issue from the bowels of the earth, thus supplying what meteorological conditions refuse. The neighboring Owens Valley, bordered by the Sierra Nevada on the west, by the Inyo range on the east, represents an oasis, with the two settlements Lone Pine and Independence. It is traversed by the Owens River, one of the few streams originating upon the eastern and precipitous slopes of the Sierra Nevada.

Some distance south of Panamint and Death Valley the Ivanpah Valley is situated. This is a deep depression between the Payute and Opal range, from which it receives the drainage. It exhibits in an excellent measure the elutriating influence of the water, for, from the coarsest gravel along the base of the mountains to the fine clay in the middle of the valley, a full series of transitions is met with, affording a large belt of good soil for agricultural pursuits, if water for irrigation could be obtained by artesian wells. A fair test in this direction should be made, as the mountains on either side are of considerable height, and receive in summer-time a moderate amount of aqueous precipitates.

East of the Ivanpah Valley, between the Payute Range and Dead Mountain, extends a shallow, plain-like valley of much higher altitude (3,500 feet) than the former, (2,700,) with gravelly surface.

The next great basin east from there is the valley of the Lower Virgen, with an average altitude of 1,300 feet, and groves of mesquite trees fringing the river margins. The river is but small; its waters contain a reddish mud in suspension, and sulphates of soda, lime, and magnesia in solution, to such extent that the taste forbids its use; hence the absence of settlements. For further particulars the reader is referred to the chapter on the " analysis of mineral springs and saline waters."

South of this valley, in the northeastern corner of Arizona, extends the great " Detrital Valley," or the " Forty-mile Desert," from the Colorado River to the foot of the Cerbat range, bordered on the west by the Blue Ridge Mountains, and covering an area of about 1,200 square miles. Bleached bones of horses and cattle, graves of perished persons, tell of the dangers of that stretch; dangers much greater in proceeding toward the river than in the opposite direction when passing toward the far mountains; in the latter case higher and cooler regions are before the weary traveler; in the former, he is descending gradually to the hot and deep Colorado Valley, the fatigue and thirst thus increasing enormously, and death overcoming in not few instances those who have approached to within a few miles of the longed-for river. The greater portion of the " Forty-mile Desert" is of gravelly surface, strewn over with bowlders. The scanty vegetation is formed by *Larrea Mexicana*, *Yucca brevifolia*, *Echinocactus cylindraceus*, and *Atriplex*. Along the base of the Cerbat range, however, in altitudes of 3,600 feet, good bunch-grass makes its appearance, and a few hundred feet higher up the excellent mesquite grass, (*Bouteloua sesleria*,) *Holocantha acacia. Juniperus* contribute to improve the floral aspect.

The two important springs on the northern slope of the range are Mountain Springs, issuing from basaltic rocks, and 10 miles farther west, Quail Springs, from granite. At the latter a family recently settled, supporting themselves by stock-raising and by selling the water of their improved spring to travelers and teamsters. Twelve miles southwest of Quail Springs, in the foot-hills, is situated the now-abandoned mining settlement Chloride, consisting of about thirty adobe buildings. It was founded in 1870, and inhabited only for five years, when the yield of the mines diminished. Four miles southwest from this place is the now-flourishing mining settlement Mineral Park, a designation that certainly is not derived from a park-like condition of the surroundings.

The most eastern portion of the Mohave Desert comprises the northern part of Los Angeles County and the southern part of Kern County, and is traversed by a branch of the Southern Pacific Railroad. For great distances the only relieving sight in the sand-waste are isolated specimens of *Yucca brevifolia*, a plant that, singularly enough, has caused the name " Palm Plain." What a sweet deception for the unprejudiced who looks here for palms!

The largest valley in the southern portion of the Mohave Desert is the Coahuila Valley, being 90 miles long, 10 to 30 miles wide, and in part below the level of the sea. Beginning with the San Gorgonio Pass, it separates the San Bernardino from the Jacinto Mountains, and widens out toward the dry-lake bottom in its southern portion, where it is joined by the so-called " Colorado Desert," the low, hilly, barren stretch connecting this valley with the mouth of the Colorado River. Whilst the mountains near San Bernardino are clad with extensive pine forests and the lower Jacinto Mountains with a good deal of piñon, the Coahuila Valley presents a most desolating sight; its northern portions consist of a system of sand-dunes and sand-hills, whose formations are caused by the wind heaping the sand around every object higher than the soil—a bush, for instance, as the stunted acacia—while its southern portions of bare clay are covered with patches of saline efflorescences.

Fortunately there are a number of oases to relieve the monotony; the large ones, Toros and Martinez's, covering several square miles. At Toros, situated at the foot of the steep eastern slopes of the Jacinto Mountains, water is reached at a depth of 3 feet; grass is plentiful, and cattle roaming between the little groves of mesquite produce a most refreshing sight after the long march through the deep sand. At Martinez corn is raised with success, while our surprise reaches its maximum by the sight of palm trees (*Brahea*) in the little oasis Dos Palmas. I think it not improbable that artesian wells ould be bored with success, and large tracts of the barren sand converted into productive fields, north of the clay flat above mentioned. Between the Coahuila Valley and the Lower Colorado extends the broad divide of Chukavalla, with an elevation of 2,500 feet. Nowhere in the desert have I seen the dry-washes covered with such well-developed vegetation as here, and it appears almost like a contradiction of nature when we see them side by side with the stony plain, whose scanty vegetation mainly consists of isolated specimens of *Larrea Mexicana, Fouquieria splendens,* and *Opuntia ramosissima.* Looking at the spendid acacia trees of the dry-washes, one easily forms the opinion that water must be struck at a moderate depth; but, alas, attempts have failed to confirm this supposition, and in one case (at Mule well) water was found at a depth of not less than 80 feet. A large spring issues at the western foot of Chukavalla Peak, and here an individual has squatted who ekes out a livelihood by keeping the spring in good condition and selling the water to the travelers. The grass found in the neighborhood of the peak is gathered and sold to teamsters and emigrants. This man communicated a fact worth recording. He states that there lives in the grass a peculiar insect called *campo mocho* by the Mexicans, (the insufficient description given points toward a *Phasmida,*) which being of a green color, and therefore not being distinguished by domestic animals, and eaten, produces death in less than twelve hours. As no occasion was offered to collect it, attention of traveling naturalists is here called to the fact for investigation.

From Chukavalla Peak northward to the Mohave River extend a series of low mountain ranges and shallow valleys, almost destitute of vegetation and water. It is a dangerous undertaking to cross this stretch in summer-time, and several parties who attempted to pass from San Bernardino in a direct easterly line to the Colorado have never been heard from. Two years ago an enterprising man, intending to establish a short road to Arizona from that town, succeeded in crossing with a wagon provided with sufficient water for himself and animals.

BOTTOM-LANDS OF THE LOWER COLORADO.

The Colorado River may justly be termed one of the most remarkable rivers of the world. Where else exists a stream that cuts its way from 3,000 to 6,000 feet deep through solid rock? Where else such a vast system of gigantic cañons, impressing the beholder with admiration and awe?

Great as appears the scientific interest attaching itself to this river, the value for agricultural purposes is small compared with the immense stretch of country it traverses. After its formation by the Green and Grand Rivers, it enters the far-famed Great Cañon, 280 miles in length; then, after a stretch of 48 miles, the Bowlder Cañon, 20 miles long and 500 to 2,000 feet deep. Leaving this cañon at the ruins of the Mormon town, (Callville,) it traverses but for a short distance an open country and enters the Black Cañon, about of the same dimensions as the former. The next cañons are those of the Needles, 90 miles farther south; one at Mount Whipple; and another between Ehrenberg and Fort Yuma.

From the mouth of the Great Cañon down to Fort Mohave the sole bottom-land is Cottonwood Island and vicinity, the banks of the river being generally very high and the country rocky. This island, 6 miles long and about one-quarter mile wide, is the abode of a number of Payute families. A small farming settlement could find a flourishing existence here, where probably over several hundred acres of good alluvial soil are covered with mesquite, cottonwood, and grass. Farther down the river is situated the now-abandoned settlement Hardyville, 5 miles north of Fort Mohave, and here the bottom-lands gradually begin to expand and form a most valuable oasis. The river branches out here and there, forms lagoons, and spreads its waters through the soil. The fine corn-fields of the Indians and the groves of mesquite bestow an aspect on the valley which is the more pleasing in the contrast it affords to the destitution of the adjoining country. The wide valley, whose present inhabitants are three white settlers and several hundred Mohave Indians, commences at Fort Mohave, extends 25 miles to the southward, and terminates at the Needles, a portion of the Mohave range, through which the river has washed a cañon. South of that range exist several other large valleys, well suited for subtropical farming, and reaching in places a width of 10 to 15 miles. The shifting of the river-bed, it is true, is a great drawback—the settlement and military camp (La Paz) had to be abandoned on that account; but engineering skill could probably remedy this. A portion of that region is at present an Indian reservation; but the extent of the bottom-lands are great enough, especially south of the Riverside Mountains, to support thousands of settlers. The climate, although very hot during four months, (June to September,) presents no serious drawback, as the inhabitants of the small town of Ehrenberg can testify. The soil is rich in the chemical combinations required for fertility, and only in small patches rather impervious, containing too large a proportion of clay. In some places, also, it contains small quantities of chloride of sodium and sulphate of lime. A specimen of a somewhat clayey soil, 6 miles east of Ehrenberg, was analyzed, with the following result:

Mechanical condition :

Silt	52.30
Clay	34.20
Hydroscopic water	3.65
Chemically-bound water	8.91
	99.06

Chemical constituents :

Potassa	0.283
Chloride of sodium	2.047
Soda	0.182
Carbonate of lime	9.264
Sulphate of lime	1.321
Phosphoric acid	0.151
Oxide of iron }	5.160
Alumina }	
Insoluble in hydrochloric acid	77.2

(left margin bracket: Insoluble in hydrochloric acid)

The river itself carries the fertilizer for the soil, a reddish mud, (hence the name Colorado,) in some respects resembling that of the Rio Grande in New Mexico, (see Part VI, Vol. III, Geographical Surveys west of the One-hundredth Meridian,) and that of the Nile in Egypt; its quantity varies from 0.1 to 0.5 per cent. of the water. After standing several hours the clear water can be drawn from the sediment, and if the former be now evaporated, leaves 0.14 gram solid residue of a weak alkaline reaction, due to a small quantity of carbonate of soda. The larger part of this residue consists of sulphate and chloride of sodium; also traces of lime, magnesia, potassa, and phosphoric acid are present.

This amount of mineral matter is too small to impart any taste to the water, which may be drunk directly from the river, though discolored with mud held in suspension. The same is not true of water from the wells sunk in the adjacent bottom-lands, which generally yield brackish water.

In the following table A the composition is given of the Colorado mud collected at Cottonwood Island, and compared with the mud of the Rio Grande and Nile:

TABLE A.

Constituents.	Colorado.	Rio Grande.	Nile.
Hygroscopic water	3.27	1.890	....
Chemically bound water	1.14	3.122	....
Soluble in hydrochloric acid:			
Potassa	0.103	0.284	0.166
Soda, with trace of lithia	0.074	0.064	0.022
Lime	....	1.479	1.725
Carbonate of lime	12.50	5.190	....
Magnesia	0.69	0.090	0.046
Oxide of iron	....	3.640	} 8.804
Alumina	2.26	1.308	
Phosphoric acid	0.146	0.092	0.143
Sulphuric acid	trace	trace	trace.
Oxide of manganese	trace		
Insoluble in hydrochloric acid	78.1	82.55	....

None of these sediments contain more than traces of organic matter, another proof that the fertilizing properties mainly depend upon the inorganic material. It will be seen that the Colorado mud contains less potassa than the others, while the amount of phosphoric acid is larger. The proportion of carbonate of lime is unusually large in the Colorado silt, which is certainly a favorable circumstance. Its presence is due to the immense limestone beds through which the river flows.

Of the few streams of Eastern California, the Mohave River is doubtless one of importance. Rising in the San Bernardino Mountains, it takes an easterly course and thus forms a natural highway across the desert to Arizona. Between the sources and its final sink it disappears five times in the sandy bed and re-appears again, remaining in some places for ten or more miles under ground. In an agricultural point of view the value of this stream is but small, and only the upper third, where a number of ranches at present exist, is worthy consideration.

The bottom-lands are not of great width, but the soil is one of the best, as revealed by the analysis of a specimen taken in the vicinity of Point of Rocks.

Mechanical condition :

Gravel	23.70
Sand	28.00
Silt	34.80
Clay	8.60
Hygroscopic moisture	0.53
Chemically bound water	1.41
	97.04

Chemical constituents :

Soluble in hydro-chloric acid.

Potassa	0.214
Soda	0.093
Magnesia	0.042
Carbonate of lime	1.914
Oxide of iron	2.200
Alumina	
Phosphoric acid	0.247
Sulphuric acid	0.007
Chlorine	slight trace.
Insoluble in hydrochloric acid (granitic)	92.6

ON THE NATURE OF THE SOILS IN THE DESERT.

If upon entering desert tracts one observes an unusually coarse, pebbly surface, and on the other hand a smooth and hard one, like that of a pavement, the inquiry is naturally made how these two extremes happen to touch each other here, while the normal state of the soil appears to be wanting. One needs not to be long in the desert to learn the cause of this. It lies in the action of the sweeping winds upon the dry surface not protected by vegetation. For example, should rain cease to fall in a fertile country, the first consequence would be the disappearance of vegetation, the next the drying up of the soil; whereupon the action of the sweeping winds becomes manifest by carrying away from the surface of the good soil the fine silt as well as the clay particles, which not only contain the principal amount of the fertilizing combinations, but also are the means of retaining the hygroscopic moisture in the soil. Finally, the once fertile fields will present the appearance of a plain of coarse pebbles, underneath which, however, the original soil has preserved its composition. (See below, subsoil of Chukawalla Plain, B.) On the other hand, soils with over 50 per cent. clay, bake by the increasing dryness, harden, and crack, resist the sweeping power of the winds, and remain intact. Such soils, however, although generally rich in the principal fertilizing agencies, are wanting in porosity and are not suited for agriculture. (See below, D.) Only in cases where the sweeping winds are shut out by mountains, or where moisture is spread by a river through bottom-lands, do the soils preserve their original surface-conditions.

In the following the mechanical conditions of several desert soils are given:

	A.	B.	C.	D.
Pebbles	50.2	46.01	11.00	None.
Sand	32.1	19.03	62.01	22.00
Silt	10.4	28.05	22.02	30.01
Clay	4.3	2.01	4.04	40.08
Hygroscopic moisture	1.34	2.65	0.49	3.01
Chemically-bound water	0.50	0.31	0.98	3.84
Organic matters	None.	None.	None.	None.

A.—Soil of the "Forty-mile Desert" near the slopes of the Cerbat Range, taken one foot deep; the surface is a mass of pebbles, while the subsoil shows nothing abnormal.

B.—Soil of Chukawalla Plain; surface exclusively consisting of pebbles and bowlders. The subsoil, however, exhibits no abnormal mechanical conditions, as the above figures show.

C.—Fine silt soil of Ivanpah Valley, protected against the sweeping winds by the Payute Range. The very surface of a portion of said valley consists of fine soil.

D.—Soil of a "dry lake" in the vicinity of Grapevine Ranch, (5 miles north of Mohave River valley,) perfectly bare and baked by the dryness; its chief mass being clay.

The above figures show that a certain, though small, proportion of hygroscopic moisture was still retained, notwithstanding the great dryness of these regions. This explains why some vegetation can succeed in finding existence. Another fact is the absence of organic matter, (humus,) which cannot surprise us, if we consider that the latter is formed by decaying vegetation. In the desert, however, no true decay takes place; the small amount of vegetation present dries up after it dies without forming humus.

That a soil which from its barrenness would impress the visitor with an idea of inferiority, still may be well provided with the fertilizing combinations, may be seen from the analysis of the subsoil of Chukawalla Plain, (east of Coahuila Valley :)

Potassa	0.281
Soda	0.172
Carbonate of lime	2.043
Magnesia	0.178
Oxide of manganese	trace.
Alumina, Oxide of iron }	5.400
Phosphoric acid	0.166
Sulphuric acid	0.0024
Chlorine	traces.
Insoluble in hydrochloric acid, (chiefly granitic)	91.8

Here the phosphoric acid and potassa are present in quantities expected in good fertile soils.

APPENDIX H 7.

REPORT ON THE GEOGRAPHICAL DISTRIBUTION OF VEGETATION IN THE MOHAVE DESERT.

UNITED STATES ENGINEER OFFICE,
GEOGRAPHICAL SURVEYS WEST OF THE 100TH MERIDIAN,
Washington, D. C., February 14, 1876.

DEAR SIR: I have the honor to submit herewith a report upon the distribution of vegetation in the Mohave Desert. Inasmuch as conditions have been observed, rarely occurring elsewhere, the report may prove of some special interest.

Thanks are due to Dr. Vasey, botanist of the Agricultural Department, who kindly determined a large number of the species collected.

Very respectfully, your obedient servant,

OSCAR LOEW.

Lieut. GEO. M. WHEELER,
    Corps of Engineers, in charge.

The zones of vegetation appear to be more sharply defined in a dry country than in one blessed with sufficient rain. With regard to the Mohave Desert, not only the lower and the upper lines of these zones are found very marked, but also the change of flora with the latitude.

A decrease in latitude is connected in the Mohave Desert by no means with an increase in altitude, as is the rule in most climates; and with some plants just the reverse can here be observed. Thus you find the piñon (pinus edulis) upon the San Bernardino Mountains in an altitude of 5,000 feet and latitude 34° 10', while in latitude 36° this tree is not seen below 8,000 feet altitude. Quite an analogous case was observed by parties of your expedition in 1874. The lower line of the piñon is found in Southern Colorado to be of 200 feet higher altitude than in Southern New Mexico. The distribution of vegetation in the dry regions is determined more by the relative humidity of the air than by the mean temperature or atmospheric pressure.

The southern portion of the Mohave Desert has more rain than the northern, and this is the main reason that the ranges of the latter are distinguished by a great degree of barrenness even in considerable heights. If a line were drawn from the Gulf of Mexico to west-northwest, one easily conceives how the more northern portions of the Mohave Desert, comprising Death Valley, Panamint Valley, Funeral and Telescope Ranges, and a great portion of Inyo County, California, fall beyond the reach of the moist monsoon, (see Chapter on the Meteorological Conditions of the Mohave Desert;) hence

summer rains seldom fall in that part. If, on the other hand, is taken into consideration the Sierra Nevada, with nearly 15,000 feet altitude, bordering the northern portions of the Mohave Desert on the western side, we can again understand why the
northwest winds prevailing in winter time arrive deprived of their moisture in the
desert, and hence the scarcity of winter rains.

That some plants depend, as the main condition of their existence, upon very dry
climates, is proved by the *Larrea Mexicana*, a plant that grows in considerable number
in many parts of the Mohave Desert, but not in a single specimen in the coast counties; indeed, standing on San Gorgonio Pass, that forms a point of separation between
the desert and the coast region, you may see many specimens of this plant on looking
toward the *east*, but not a single one between this point and the coast to the *westward.*

Most striking, also, is the sudden disappearance of the Yucca-tree (*Yucca brevifolia*,
Eng.*) as soon as the natural boundaries between the desert and the coast counties are
crossed. Yet, while this plant is confined to respectively narrow zones of limited altitudes, the *Larrea Mexicana* ranges from about sea-level up to 4,400 feet, and is so widely
spread that it may be assumed that the botanical boundaries of the Mohave Desert
are defined by the range of *Larrea Mexicana*.

At every mountain-chain crossed, the lower and upper lines of the zones of the most
striking vegetable forms were determined by a pocket aneroid.† These lines, of course,
shift somewhat—sometimes several hundred feet—with the direction of the mountain-
slopes, extent of the plateau, or proximity of deep and hot valleys. The following
table will show approximately the average zones of the more conspicuous plants.

Distribution of plants in the Mohave Desert.

Name of species.	Altitude of zone.		Latitude.	Remarks.
	Lower line.	Upper line.		
Yucca brevifolia	3,400	4,500	34° to 36° 20′	On the western slopes of the Pay-ute Range, lower line = 4,200.
Opuntia ramosissima	2,000	5,000	32° 20′ to 35° 40′	In southern regions, lower line = 800.
Opuntia arborescens	3,600	5,000	Widely spread	*Opuntia Strigii, Cereus Berlandieri, Yucca baccata,* and *Ephedra anti-syphilitica* occupy the same region.
Cereus giganteus		3,500	Not north of 33° 40′.	Prefers lime soils and rocky slopes.
Echinocactus cylindraceus		2,300	33° 30′ to 35° 40′	Very isolated.
Mammillaria barbata	600	3,500	33° to 36°	
Fouquieria splendens	Sea-level†	2,500	Not north of 34°	*Cercidium floridum, Parkinsonia microphylla, Asclepias subulata,* and *Acacia Wrightii* occupy the same regions.
Dalea spinosa		800	Not north of 35° 10′.	In southern regions, upper line = 2,000.
Acacia Ramieriana	3,900	4,500	34° to 36° 20′	*Holocantha Emoryi* occupies the same regions.
Juniperus occidentalis	3,900	5,400	Ranges between wide latitudes.	Upper line about 8,000 in the north-ern part of Mohave Desert.

While it is seen from the foregoing table how well certain species are fixed between
lines of altitudes, other species, as for instance, *Pectis angustifolia, Datura meteloides,
Cucurbita Californica*, range between very wide limits; they were observed in moist
spots in altitudes of 600 as well as 5,000 feet, their existence depending upon an increased moisture in the soil.

The character of the vegetation of the Mohave Desert resembles in many respects
that of Southern Arizona, especially of the Gila Valley.

While the changes of the vegetation with latitude and altitude are well marked,
those depending upon the different conditions of the ground are still more striking;
the clay soil, the sandy dry wash, the coarse pebbly plain, the salt flat, the loose sand-
hills, the rocky slopes, each nourish a different flora, however poor. In the following
list, a classification in this direction is attempted:

The chief representatives of the Colorado Valley from the mouth of the Grand Cañon
of the Colorado to the southern spurs of the Riverside Mountains, are: *Atriplex
hymenilyttra; Atriplex polycarpa; Palafoxia linearis; Psathirotes ramosissima; Physalis*

---

* Described but few years ago by Engelmann, (see Clarence King's Report on the Explorations of the Fortieth Parallel, vol. 3.) I have never seen this most singular plant
in Texas, New Mexico, or Arizona.

† These observations comprise the San Bernardino, Opal, Providence, Cerbat, and Colorado Ranges.

lobata; Lippia cuneifolia; Aster spinosus; Sesbania macrocarpa; Abronia Amaranthus; Pluchea Sarcostemma; Algarobia glandulosa, Strombocarpa pubescens, (the two species of Mesquit-trees;) *Salix longifolia, Tessaria borealis, Baccharis cœrulescens, Baccharis salicina,* (form dense brush along the river-margins ;) *Malvastrum marubioides; Datura meteloides; Cucurbita Californica; Xanthium strumarium; Suœda diffusa; Panicum crusgalli; Chloris alba; Allionia incarnata; Populus monilifera; Lygodesmia Garrya.*

Flora of the Coahuila (Cabezon) Valley: *Baccharis salicina; Baccharis Emoryi; Aplopappus caricifolius; Dicoria canescens; Algarobia glandulosa; Halostachys occidentalis; Petalonix Thurberi; Linocyris caricifolia; Atriplex lentiformis;* a palm species, (*Brahea ?*).

Flora of the Dry Washes: *Cercidium floridum; Parkinsonia microphylla; Chilopsis linearis; Krameria parvifolia; Acacia Wrightii; Asclepias subulata; Dalea spinosa.*

Flora of dry sand-hills: *Coldenia Palmeri; Pleuraphis Jamesii; Eriogonum inflatum; Tricuspis pulchella,* (sand-grass ;) *Heptis Emoryi, Mammillaria barbata; Aristida,* (galleta-grass ;) *Chlorisanthe rigida; Boutelona,* (sand-grass;) *Psathirotes annua.*

Flora of the coarse pebbly plain and rocky slopes: *Larrea Mexicana; Yucca brevifolia; Opuntia ramosissima; Echinocactus cylindraceus;* * *Yucca bacata; Atriplex canescens; Opuntia arborescens; Fouquieria splendens.*

Flora of clayey soil charged with sodium salts: *Halostachys occidentalis; Salicornia; Bryzopyrum spicatum.*

Specific forms of the coast and adjacent regions: *Frankenia grandifolia; Œnothera viridescens; Styphonia serrata; Photinia arbutifolia; Artemisia Californica; Zauschneria Californica; Mesembryanthemum crystallinum; Abronia umbellata; Rhus aromatica; Isomeris arborea; Eriogonum fasciculatum; Arctostaphilos tomentosa; Cerasus salicifolia.*

In reviewing the specific floral characters of the desert, a series of most singular forms arrests the attention. Indeed, what surprising forms are the *Fouquieria splendens,* the *Dalea spinosa, Krameria parvifolia, Yucca brevifolia, Cereus giganteus!* Truly, who once has seen these striking characters will forever keep their aspects in memory. They bear testimony of their struggle for existence, a struggle, not with other species, but with a scorching and merciless climate, in consequence of which the main organs supporting the evaporation of moisture—namely, the leaves—were gradually diminished in size, (*Algarobia, Strombocarpa,*) or disappeared entirely, (*Cactus,*) or became rudimentary and falling off soon after their development, (*Dalea, Cercidium,*) or became covered with a thick stratum of fiber, (*Yucca,*) or so charged with resinous matters as to make the evaporations of small particles of water quite impossible, (*Larrea.*) But, as the leaves are the principal organs of assimilation, the first condition of life and growth of plants, a substitute was necessary where they disappeared ; and the trunk, branches, and thorns assumed the vital functions. To this end, the bark and surface of such trees are provided with a green layer of chlorophyl. What a singular impression make *Dalea* and *Cercidium*—green trees without leaves!

APPENDIX H 8.

REPORT ON THE ORNITHOLOGY OF THE PORTIONS OF CALIFORNIA VISITED DURING THE FIELD-SEASON OF 1875 BY H. W. HENSHAW.

UNITED STATES ENGINEER OFFICE,
GEOGRAPHICAL SURVEYS WEST OF THE 100TH MERIDIAN,
Washington, June 10, 1876.

SIR : I have the honor to transmit the following report upon the ornithology of the portions of California visited by me during the field-season of 1875.

As a field for ornithological research, the region as a whole was by no means a new one, parts of California having been traversed by several governmental parties, and more or less extensive collections of birds made by the naturalists of the several surveys. The more southern portions of the State, however, those visited by us, were believed to be possessed of much interest, special importance attaching to the region lying about Mount Whitney as being less known, and hence likely to present features of value when studied with a view to its avian fauna.

While these expectations were not wholly borne out, the results obtained are believed to be possessed of considerable value in their bearing on the distribution of the many species that came under observation. Quite a number of birds were found at points considerably farther south than had previously been chronicled, and the range of others extended westward. Such instances are mentioned in detail under the re-

---

* A specimen of this singular Cactus, nestling amid the barren rocks, with but very little soil accumulated between them, is shown in the accompanying figure.

ERRATA. (Appendix H 8.)

Page 225, thirty-third line from top, for "Vuepes" read "Vulpes."
Page 227, first line from top, for "itt" read "Nutt."
Page 228, seventh line from bottom, for "Chamæodae" read "Chamæidae."
Page 228, sixth line from bottom, for "Chamoea" read "Chamæa."
Page 232, tenth line from top, for "cucogaster" read "leucogaster."
Page 235, fifth line from top, for "lunifrous" read "lunifrons."
Page 237, eighth line from top, for "Miadestes" read "Myiadestes."
Page 237, twenty-fifth line from top, for "Swains" read "Swainsons."
Page 240, twenty-fifth line from top, for "P. savannas" read "P. savanna."
Page 241, seventeenth line from top, for "alandinus" read "alaudinus."
Page 243, thirteenth line from bottom, for "navadensis" read "nevadensis."
Page 245, twenty-fifth line from top, for "Peucœa" read "Peucæa."
Page 247, fifteenth line from bottom, for "mesolencus" read "mesoleucus."
Page 248, fifth line from top, for "Zorotrichias" read "Zonotrichias."
Page 251, eleventh line from top, for "Baxtr" read "Bartr."
Page 254, eleventh line from bottom, for "Contopis" read "Contopus."
Page 254, seventeenth line from bottom, for "Cantopus" read "Contopus."
Page 255, fifteenth line from top, for "flaviventri" read "flaviventris."
Page 257, fifteenth line from top, for "costoe" read "costæ."
Page 258, ninth, seventeenth, twenty-third, and twenty-sixth lines from bottom, for "soolaris" read "scalaris."
Page 262, twenty-third line from top, for "cuniculari" read "cunicularia."
Page 263, fourth line from bottom, for "hawesii" read "beecheyi."
Page 264, eleventh line from bottom, for "Cucurus" read "leucurus."
Page 265, tenth line from top, for "Cathardidæ" read "Cathartidæ."
Page 266, twenty-fourth line from bottom, for "Laphortyx" read "Lophortyx."
Page 270, seventeenth line from top, for "piso" read "precisely."
Page 270, ninth line from bottom, for "melanacephalus" read "melanocephalus."
Page 271, fifteenth line from top, for "Phalarophidæ" read "Phalaropodidæ."
Page 272, eighteenth line from top, for "Heteoscelus" read "Heteroscelus."
Page 276, seventh line from top, for "violacens" read "violaceus."
Page 278, twelfth line from top, for "columbia" read "columba."

[This appendix passed through the press during the absence in the field of Mr. Henshaw.]

spective species. Not a small portion of the country traversed by the Survey was found to be remarkably destitute of birds, not only as regards the number of species, but also the number of individuals seen was small. In some sections, as on the dry and arid plains, the nature of the country itself furnishes the cause of this; but elsewhere, I am inclined to attribute much of this paucity of bird life to the presence of sheep and to the effect they have had on the vegetation, for over a very large region in Southern California these animals exist in such numbers as to fairly render their pasture-grounds little else save howling deserts, attractive neither to themselves nor to any other living creature. Not only is this true of the lower, more accessible, portions, but it was found to be the case in the lofty mountains of the Sierra, where thousands of sheep are driven in summer, their combined numbers resulting almost in the obliteration of every green thing within their reach. Not only are destroyed the plants and flowers upon which depend the presence of insects, which furnish to many birds a large proportion of their food; but this is accompanied by the destruction of much of the undergrowth, so essential to the mode of life of many of the smaller species. In no other way could I account for the fewness of birds in districts that seemed possessed of all the natural requisites to attract them in great numbers, but where the painful desolation brought about from this cause was accompanied by a marked scarcity of feathered life.

The field-work began June 1, at which time Doctors Rothrock, Loew, and myself visited the island of Santa Cruz, the most inland of the group of islands lying in the Santa Barbara Channel. The two weeks spent on the island were mostly occupied in making general collections in natural history. The surface of the island is extremely rough, and broken up in every direction by rocky ridges, which render all travel exceedingly difficult; and in the little time that was spent in collecting the land-birds it is not probable that by any means all of the species were found. All that came under notice occur on the mainland, and differ in no respect. Many species of sea-birds resort to these islands for the purpose of reproduction; fewer, however, to Santa Cruz than to the others.

Two species, *Uria columbe* and *Fratercula cirrhata*, were found to breed here; this fact being of interest as indicating a range at this season much farther south than suspected before. Among other objects of value obtained here was a series of specimens of the little "Island Fox," (*Vuepes littoralis*,) an animal but little known to naturalists, and of great rarity in collections. They inhabit the islands in very great numbers, and are found, as far as known, nowhere else. A quite extensive collection of fish and mollusks was also made.

In connection with our work on the island, it is a pleasant duty to mention the courtesy and many favors our party received from the officers of the Hassler, who one and all interested themselves in the object of our visit, and contributed much to its success. Indeed, it was through the kindness of Capt. H. C. Taylor that we were enabled to visit Santa Cruz, which otherwise would have been most difficult.

Upon joining the main party at Los Angeles, June 15, the original plan for the season's work was changed so as to admit a small natural-history party in charge of Dr. Rothrock to return to Santa Barbara, and there meeting Dr. Yarrow to prosecute our work in connection with archæological researches in this neighborhood. The locality was found to be extremely rich in Indian mounds used as burial-places, and, as a result of labors here, a large collection of Indian remains and implements was exhumed.

The collections in zoology made here were also quite large, including not only a large number of birds, but also many insects, fish, reptiles, &c., for many of which we are indebted to the zeal and interest displayed by Mr. C. J. Shremaker.

Leaving Santa Barbara July 13, we proceeded to old Fort Tejon, there joining the main party. About this point rather more than a month was occupied, two short trips being made to the neighboring mountains, where several rare and interesting birds were found.

September 4 the party set out for the Mount Whitney country, where the time up to the middle of October was taken up. The mountains of this region are, many of them, well wooded, mostly with pine and tamerack, while the streams, as usual in these high altitudes, were more or less densely fringed with deciduous vegetation. The avian fauna was found at this time to be quite limited in the aggregate of species, and, as a rule, not numerously represented in individuals. The absence of the Warbler tribe (*Sylvicolidæ*) was especially noticeable. The only ones of this family seen here were the *Helminthophaga celata*, *Dendroica audubonii*, *D occidentalis*, and *Myiodioctes pusillus*, all of them being comparatively rare.

Returning from Mount Whitney, the remaining interval up to October 15 was spent near Kernville and at Walker's Basin, when the field-work ended.

The season's collection of birds amounted to 700 specimens, representing 127 species. In addition, a considerable number were observed in greater or less numbers, and find mention in the report. A list of the specimens, with the localities where collected, follow each species; and, in the case of those less known, careful measurements are given.

A synonymical list follows such of the species as were not thus treated in our previous report.

In most instances, the classification followed is that given by Baird, Brewer, and Ridgway in their recent work on North American Birds. In the Waders and Water-birds, that adopted by Dr. Coues is taken.

TURDIDÆ.—THRUSHES.

1. *Turdus migratorius*, L.—Robin.

Nowhere in the region south of San Francisco does the Robin appear to be a common bird, and, indeed, it was rarely seen by us till after September, when they were found here and there in the mountain-valleys, not in large flocks but leading rather a hermit life, and subsisting much upon berries.

2. *Turdus nævius*, Gm.—Varied Thrush; Oregon Robin.

Turdus nævius, Newb., P. R. Rep., vi, 1857, 81.—Bd., B. N. A., 1858, 219.—Heerm, P. R. R. Rep., x, 1859, pt. vi, 45.—Bd., Zool. Ives's Exped., 1860, 5 (Colorado Valley).—Xantus, Proc. Phila. Acad. Nat. Sci., 1859, 190 (Ft. Tejon, Cal.)—Coop. & Suckl., P. R. R. Rep., vol. xii, pt. ii, 1860, 172.—Coues, Proc. Phila. Acad. Nat. Sci., 1866, 88 (Colorado Valley).—Coop., Am. Nat., iii, 1869, 31 (Montana).—Coop., B. Col., 1, 1870, 10.—Coues, Key N. A. B., 1872, 72.—B., B., & R., N. A. B., i, 1874, 29.

In California, this Thrush is found only in the character of a fall and winter visitor, returning with the spring to congenial haunts in the far north, there to pass the season of reproduction. It is usually common about San Francisco in winter, and not a few suffer at the hands of the gunners, and are brought into the markets and sold for the table. Though finding its way to the south of this point, it is in diminishing numbers, and in the foot-hills and low mountains near Caliente it was far from numerous during the last of October and November. They kept in small flocks, and were exceedingly shy and suspicious. In habits, they seem to correspond pretty closely to the Robin, as does their food, which, as with that species, consists largely, in the fall, of berries of various roots.

No.	Sex.	Locality.	Date.	Collector.
642	♂ ad.	Walker's Basin, California......................	Nov. 5	H. W. Henshaw.
643	♀ ad.	......do	Nov. 5	H. W. Henshaw.

3. *Turdus pallasi*, Cab., var. *nanus*, Aud.—Dwarf Thrush Hermit.

Of this little Thrush none were seen previous to the very last of September. After this time, every little willow-thicket along the mountain-streams contained one or more, the migration being at its height from about the 5th to the 15th of this month. I cannot but think that both Drs. Heermann and Cooper had in mind some other species, probably *ustulatus*, when they spoke of the *Turdus nanus* as breeding about and to the south of San Francisco. In his description of the spotted eggs of this species, Dr. Cooper unquestionably had in mind those of the *T. ustulatus*, the eggs of the *T. nanus* being perfectly plain, and it seems most likely that this error of identification was carried still further, and all of his statements as to breeding habits and summer habitat be referable to the *T. ustulatus*. In the interior, in the same latitude, the Dwarf Thrush occurs only as a migrant. I am inclined to believe that the breeding of this bird so far south as California, even in exceptional instances, has yet to be substantiated. It certainly does not as a rule occur by any means so far south in summer.

No.	Sex.	Locality.	Date.	Collector.	Wing.	Tail.	Bill.	Tarsus.
509	♂	Near Mount Whitney, Cal....	Sept. 29	H. W. Henshaw...	3.58	2.72	0.47	1.15
514	♂	......do	Oct. 3	....do	3.67	3.15	0.54	1.10
515	♀	......do	Oct. 3	....do	3.53	2.87	0.48	1.10
516	♀	......do	Oct. 3	....do	3.32	2.65	0.48	1.12
517	♂	......do	Oct. 3	....do	3.33	2.80	0.51	1.06
535	♂	......do	Oct. 9	....do	3.31	2.85	0.50	1.12
536	♂	......do	Oct. 9	....do	3.58	2.57	0.48	1.12
542	♂	......do	Oct. 10	....do	3.42	2.65	0.54	1.06
543	♂	......do	Oct. 10	....do	3.50	2.81	0.53	1.14
550	♂	......do	Oct. 10	....do	3.47	2.92	0.52	1.16

4. *Turdus swainsoni*, Cab., var. *ustulatus*, itt.—Oregon Thrush.

Turdus ustulatus, Bd., B. N. A., 1858, 215.—Coop. & Suckl., P. R. R. Rep., vol. xii, pt. ii, 1860, 171.—
Coop., B. Cal., i, 1870, 5.—Lawr., Proc. Bost. Soc. Nat. Hist., June, 1871 (Tres Marias).
Turdus swainsoni var. *ustulatus*, Coues, Key N. A. B., 1872, 73.—B., B., & R., N. A. B., 1874, 16.—
Nelson, Proc. Bost. Soc. Nat. Hist., vol. xvii, 1875, 355 (California).

This race of the more eastern and northern Swainsoni Thrush is found in summer throughout California, where it breeds, resorting to the valleys and lowlands generally, rather than to the mountainous districts. It was in full song about San Francisco the last of May, and the species was probably at this time nesting. Its habits and very nature appear to be different from its nearest ally, the Swainsoni Thrush. Unlike that bird, instead of finding a congenial home only in the solitude of the remote northern wilds, it is perfectly content to live a near neighbor to, and a companion of, man, and dwells as contentedly as the Robin in the gardens and orchards on the outskirts of the towns. Its song I frequently heard coming from the midst of the shrubbery that environs the houses. It is exceedingly like the well-known strains of the Wilson's Thrush, though seeming to lack something of the depth of tone and wildness which gives that song its chief charm. It is, too, rather shorter. At Santa Barbara, I found the young fully fledged by the last of June.

No.	Sex.	Locality.	Date.	Collector.	Wing.	Tail.	Bill.	Tarsus.
66	♂ ad.	Santa Barbara, Cal........	June 25, 1875	H. W. Henshaw.	3.87	3.30	0.57	1.21
70	♀ ad.	......do	June 26, 1875	....do	3.73	3.17	0.57	1.17
71	♂ jun.	......do	June 26, 1875	....do				
147	♂ ad.	......do	June 29, 1875	....do	3.73	3.05	0.52	1.14
148	♀ ad.	......do	June 30, 1875	....do	3.72	3.05	0.57	1.12

5. *Mimus polyglottus*, L.—Mocking-bird.

According to Dr. Cooper, the Mocking-bird is said to occur in California as far north as Monterey. Along our route from Los Angeles to Santa Barbara, it was seen on a few occasions only, chiefly on the dry plains, where the prickly pears and other cacti grew in abundance. The Sage-thrasher, (*Oreoscoptes montanus,*) according to Dr. Heermann, is not rare about San Diego. It probably intrudes only into the extreme southern portion of the State.

6. *Harporynchus redivivus*, Cabanis.—California Sickle-bill Thrush.

Harporynchus redivivus, Bd., B. N. A., 1858, 349.—Xantus, Proc. Phila. Acad. Nat. Sci., 1859,
191.—Coop., B., Cal. i, 1870, 15.—Coues, Key, N. A. B., 1872, 75.—B., B., & R., N. A. B., i,
1874, 45.

This Thrush was found in various localities throughout Southern California, where it is a constant resident. Though preferring the lowlands, we occasionally saw these birds in the dense chaparral that clothes the bases of many of the low mountains. Like the others of this singular genus, it is eminently terrestrial in its habits, its stout, s ong claws, aided by its heavy bill, being well adapted for scratching among the leaves and *débris* for all sorts of insect life. It is shy and timid, and covets the seclusion of the hedges and thickets at all times. When alarmed, its wings serve to carry it for a short distance, till it has gained some covert, when its active feet enable it to keep out of sight by dodging here and there till its safety is assured.

No.	Sex.	Locality.	Date.	Collector.	Wing.	Tail.	Bill.	Tarsus.
45	Ad.	Santa Barbara, Cal	June 25	H. W. Henshaw.....	3.65	5.62	1.47	1.52
46	Ad.	......do	June 25	....do	3.95	5.45	1.39	1.57
95	Jun.	......do	June 27	....do				
203	♂ jun.	......do	July 6	....do				
241	♀ jun.	......do	July 10	....do				
563	♀	......do	Nov. 9	....do	4.12	5.65	1.63	1.53
741	♂ ad.	......do	June 14	....do	4.00	5.58	1.55	1.48

CINCLIDÆ.—WATER-OUZELS.

7. *Cinclus mexicanus*, Sw.—*Water-ouzel.*

Throughout the mountains of the West, it needs only the presence of a stream of water, whirling and foaming over its rocky bed through cañon and pass, to surely attract this little nondescript to take up its abode on its banks. It occurs on many of the streams of the Sierras, numbers occupying the same reach of stream when food is abundant.

No.	Sex.	Locality.	Date.	Collector.
474	♂	Near Mount Whitney, Cal	Sept. 15	H. W. Henshaw.
475	♂	......do	Sept. 15	Do.

SAXICOLIDÆ.—STONE-CHATS.

8. *Sialia mexicana*, Sw.—Western Bluebird.

This species replaces in California the common Red-breasted Bluebird of the East, and is very common.

No.	Sex.	Locality.	Date.	Collector.
266	♀ jun.	Tejon Mountains, Cal	Aug. 2	H. W. Henshaw.
273	♂ jun.	......do	Aug. 2	Do.
303	♂ ad.	Fort Tejon, Cal	Aug. 7	Do.
304	♀ jun.	......do	Aug. 7	Do.
308	♂ jun.	......do	Aug. 7	Do.
553	♂ ad.	Near Mount Whitney, Cal	Oct. 11	Do.
600	♂ ad.	......do	Oct. 23	Do.
659	♂ ad.	......do	Nov. 5	Do.
755	Jun.	......do	July —	Do.

9. *Sialia arctica*, Sw.—Rocky Mountain Bluebird.

Apparently much rarer than the preceding species; indeed, I am not positive that I detected its presence at all, though a flock of Bluebirds seen in the high Sierras late in October were supposed to be of this species. This seems the more probable, as Dr. Cooper speaks of finding it numerous about Lake Tahoe and the summits of the Sierras in September.

SYLVIIDÆ.—SYLVIAS.

10. *Regulus calendula*, (L.)—Ruby-crowned Kinglet.

This species is very abundant during the migrations, and may perhaps be yet found breeding in the high mountains of Southern California, as it probably does in the more northern half of the State.

No.	Sex.	Locality.	Date.	Collector.
506	♂	Near Mount Whitney, Cal	Sept. 26	H. W. Henshaw.
519	♂ jun.	......do	Oct. 3	Do.

11. *Polioptila cærulea*, (L.)—Blue-gray Gnatcatcher.

The neighborhood of Fort Tejon was the only locality where this Gnatcatcher was seen. It was here particularly numerous, the bushes along the sides of the cañons being for some reason or other especially favored by their numbers. Neither here nor elsewhere was the closely-allied species *P. melanura* detected.

No.	Sex.	Locality.	Date.	Collector.
257	♀ jun.	Fort Tejon, Cal	July 27	H. W. Henshaw.
324	♂ jun.	......do	Aug. 8	Do.
325	♂ jun.	......do	Aug. 8	Do.
323	♂ jun.	......do	Aug. 8	Do.
326	♀ jun.	......do	Aug. 8	Do.
327	♀ jun.	......do	Aug. 8	Do.

CHAMÆODÆ.—GROUND WRENS.

12. *Chamoea fasciata*, Gamb.—The Ground Wren.

Chamœa fasciata, Bd., B. N. A., 1858, 370.—Xantus, Proc. Phila. Acad. Nat. Sci., 1859, 191.—Coop. B., Cal., i, 1870, 39.—Coues, Key N. A. B., 1872, 79.—B., B, & R., N. A. B., i, 1874, 84.— Nelson, Proc. Bost. Soc. Nat. Hist., vol. xvii, 1875, 356 (California).

The Ground Wren appears to inhabit Southern California at large, and was detected by us at several widely-separated points both in the Coast range and the Sierras. Its habits are a queer compound, and, though often suggestive of the Titmice, with which, too, its colors are somewhat correspondent, they yet resemble still more closely the Wrens, while the bird has characteristics borrowed from neither of its prototypes, but all its own.

I first saw the species in July, in a tangled growth of vines and bushes, close to the seashore near Santa Barbara. So careful, however, were they to keep themselves close within the friendly shelter of matted undergrowth that, though I made out from their voluble sputterings that a whole family was there congregated, I was unable to push a very close acquaintance. Subsequently I found another group in a small cluster of willows that fringed a mountain-rivulet near Fort Tejon. A few faint, querulous, sputterings from the center of the clump first attracted my attention, and, sitting down, I awaited patiently till I could catch a glimpse of their authors. After a few moments further silence on my part they began to approach nearer and nearer, till, ere long, I saw one little brown bunch of feathers balancing itself on the upright stem of a willow and peering cautiously about, all the while communing with itself and its fellows in quaint undertones. They appear to be fond of each others' society, and socially inclined toward other birds of very different habits, for I never saw or heard one without soon learning of the presence of others hard by, while, late in the fall, I often found several adding their quota to the flocks of Sparrows and Snowbirds in their journeyings through the chaparral thickets on the mountain-sides.

They spend most of their time seeking food about the roots of bushes, and especially apt were they to be found in willow-clumps along the stream. Enough of their time is passed upon and near the ground to make the name of Ground Wren an appropriate one.

No.	Sex.	Locality.	Date.	Collector.	Wing.	Tail.	Bill.	Tarsus.
380	♀ ad.	Tejon Mountains, Cal.....	Aug. 17	H. W. Henshaw.	1.27	3.22	0.40	1.00
399	♂ ad.	......do............	Aug. 19	....do............	2.43	3.98	0.42	1.04
400	♀ ad.	......do............	Aug. 19	....do............	2.40	3.60	0.40	1.03
688	♂	Walker's Basin, Cal......	Nov. 10	....do............	2.40	3.40	0.43	1.00
704	♂ ad.	......do............	Nov. 10	....do............	2.45	3.73	0.43	1.00
705	♀	......do............	Nov. 11	....do............	2.35	3.58	0.43	0.88

PARIDÆ.—TITMICE.

13. *Lophophanes inornatus*, (Gamb).—Gray-tufted Titmouse.

This species appears to be a resident throughout Southern California, and is numerous here, as indeed almost everywhere in the Far West.

No.	Sex.	Locality.	Date.	Collector.
146	♂ jun.	Santa Barbara, Cal	June 29	H. W. Henshaw.
249	♂	Ojai Creek, Cal	July 17	Do.
264	♀	Fort Tejon, Cal	July 27	Do.
322	♀	......do.....................	Aug. 7	Do.
353	♂ jun.	......do.....................	Aug. 10	Do.
393	♂ ad.	......do.....................	Aug. 17	Do.
658	♀	Walker's Basin, Cal	Nov. 5	Do.

14. *Parus montanus*, Gamb.—Mountain Chickadee.

This appears to be the commonest representative of its tribe in Southern California, inhabiting chiefly the coniferous regions, and rarely descending to the low country.

The *Parus occidentalis* appears not to occur in the southern portion of the State; none at least were detected by us, nor do I find it quoted from this region. Its proper habitat is the Columbia River region and to the northward.

No.	Sex.	Locality.	Date.	Collector.
489	♀	Near Mount Whitney, Cal	Sept. 19	H. W. Henshaw.
490	♂ jun.	......do....................	Sept. 19	Do.
522	♂	North Fork Kern River, Cal....................	Oct. 7	Do.
523	♀	......do....................	Oct. 7	Do.
524	♂	......do....................	Oct. 7	Do.
525	♂	......do....................	Oct. 7	Do.
550	♀	Near Mount Whitney, Cal....................	Oct. 10	Do.

AP. J J—15

15. *Psaltriparus minimus*, (Towns).—Least Titmouse.

Parus minimus, Townsend, Jour. A. N. Sci. Phila., vii, 11, 1837, 190.
Psoltriparus minimus, Bd., B. N. A., 1858, 397.—Coop. & Suckl., P. R. R. Rep., vol. xii, pt. 11
　1860, 189.—Coop., B. Cal., i, 1870, 48.—Coues, Key N. A. B., 1872, 82.—B., B., & R., N. A. B.
　1874, 109.—Nelson, Proc. Bost. Soc. Nat. Hist., vol. xvii, 356 (California).
Psaltria minima, Heermann, P. R. R. Rep., xvi, 38.

This Titmouse, in external appearance so much like the allied form var. *plumbeus* from Arizona and the Southern Rocky Mountains, is its exact counterpart in habits and notes. Like that bird, it shuns the coniferous trees for which most of the family are so partial, and is found in the shrubbery and chaparral of the open country, particularly on the edges of cañons and along the broken, rocky ridges. In large flocks of so many individuals that the bushes seem sometimes fairly laden with the tiny busybodies, they move rapidly over the country, launching themselves in short flights from clump to clump, their notes telling of their whereabouts and serving to keep the flock well together. The sight of a wounded or dead comrade is sufficient to put the whole company in a flutter of commotion, and as they flock in to inspect their unfortunate associate their cries are redoubled, while they descend to the ground and vainly endeavor to ascertain the cause of the trouble and to be of assistance.

No.	Sex.	Locality.	Date.	Collector.	Wing.	Tail.	Bill.	Tarsus.
64	♂	Santa Barbara, Cal.	June 25	H. W. Henshaw.	1.80	2.16	0.28	0.63
112	♂ jun.	......do	June 27	....do	1.98	2.14	0.29	0.62
113	♂	......do	June 27	....do	1.75	2.05	0.29	0.63
114	♂ jun.	......do	June 27	....do	1.85	2.12	0.29	0.63
115	♂ ad.	......do	June 27	....do	1.85	2.17	0.30	0.63
305	♀	Fort Tejon, Cal	Aug. 7	....do	1.77	2.07	0.37	0.68
306	♀	......do	Aug. 7	....do	1.77	2.05	0.29	0.65
607	♂ ad.	......do	Aug. 7	....do				
758	ad.	......do	Aug. 7	....do				
669	♂	Walker's Basin, Cal	Nov. 10	....do	1.92	2.18	0.30	0.60
675	♀	......do	Nov. 10	....do	1.95	2.16	0.26	0.59
676	♂	......do	Nov. 10	....do	1.98	2.18	0.29	0.63
677	♂	......do	Nov. 10	....do	1.93	2.30	0.27	0.60
678	♀	......do	Nov. 10	....do	1.88	2.17	0.27	0.63
679	♂	......do	Nov. 10	....do	1.93	2.25	0.30	0.63
680	♂	......do	Nov. 10	....do	1.90	2.15	0.30	0.63
681	♂	......do	Nov. 10	....do	1.98	2.16	0.26	0.63
682	♂	......do	Nov. 10	....do	1.88	2.25	0.28	0.70
684	♂	......do	Nov. 10	....do	1.90	2.20	0.28	0.64
686	♂	......do	Nov. 10	....do	1.90	2.20	0.28	0.67
687	♀	......do	Nov. 10	....do	1.95	2.30	0.30	0.63
685	♂	......do	Nov. 10	....do	1.97	2.28	0.30	0.66
756	....	......do	Nov. 10	....do	1.92	2.15	0.26	0.62
757	....	......do	Nov. 10	....do	1.93	2.28	0.28	9.63

SITTIDÆ.—NUTHATCHES.

16. *Sitta carolinensis*, Gm., var. *aculeata*, Cass.—Slender-billed Nuthatch.

❦ This species was found numerously in the pine region of both the Coast and Sierra ranges. I am inclined to think it is a resident in the mountains well down to the southern border of the State, as is the case in Arizona.

No.	Sex.	Locality.	Date.	Collector.
554	♂	Near Mount Whitney, Cal.	Oct. 10	H. W. Henshaw.
555	♀	......do ...	Oct. 10	....do ....

17. *Sitta canadensis*, L.—Red-bellied Nuthatch.

This Nuthatch is possessed of a range considerably more northerly than any of the others of the family. Its occurence, therefore, in the southern sierras is to be looked upon perhaps as rather unusual, and possibly it may be only found here as migrant and in winter. I found it breeding, however, in Southern Colorado, where it was not rare, which would render the supposition of it remaining in the high mountains of Southern California more probable. It appeared to be not uncommon in the pine region near Mount Whitney in October.

No.	Sex.	Locality.	Date.	Collector.
598	♀ jun.	Near Mount Whitney, Cal.	Oct. 7	H. W. Henshaw.
38	♀ jun.	......do ...	Oct. 9	....do ....

18. *Sitta pygmœa,* Vig.—California Nuthatch.

This is by far the most abundant of the three species seen in California, and was common everywhere where the presence of pines affords them the hunting-grounds they most affect.

CERTHIIDÆ.—CREEPERS.

19. *Certhia familiaris,* L., var. *americana,* Bon.—Brown Creeper.

The Creeper breeds in the mountains of Southern California, where I took a young bird in the first plumage near Fort Tejon, August 2. It is, however, not common till late in the fall, when their numbers are increased by the arrival of migrants from more northern breeding-grounds.

No.	Sex.	Locality.	Date.	Collector.
759	♂	Tejon Mountains, Cal	Aug. 2	H. W. Henshaw.

TROGLODYTIDÆ.—WRENS.

20. *Campylorynchus brunneicapillus,* Lafr.—Cactus Wren.

Only in a few localities was this species met with, though its absence in Southern California as high as latitude 35° or 36° may be attributed chiefly to the lack of cactus plains, the cacti being almost a necessity in the domestic economy of the bird, both because these plants furnish it with its favorite hunting-grounds, and because it is in their branches that they love to place their nests. Up to the latitude indicated the species may be looked for with confidence whenever is found a district well supplied with these plants. One or two individuals were shot a few miles northeast of Kernville, but with plumage in such a state of moult that they were not considered worth preserving.

21. *Salpinctes obsoletus,* (Say).—Rock Wren.

The Rock Wren is perhaps not as abundant throughout Southern California as in many portions of the central region, yet it is found here and there in varying numbers, inhabiting the rocky, sterile, waste lands, which few other species care to share with it. It was noted also on the island of Santa Cruz.

No.	Sex.	Locality.	Date.	Collector.
633	♀	Near Sunday Peak, Cal	Oct. 25	H. W. Henshaw.
703	♀	Walker's Basin, Cal...................................	Nov. 10	....do ....

22. *Catherpes mexicanus,* (Swains.), var. *conspersus,* Ridgw.—White-throated Rock Wren.

Probably the latitude of San Francisco forms about the northward limit of this species, thus coinciding with its known extension in the interior. It was detected by our parties as far north as the neighborhood of Mount Whitney, where it was tolerably numerous, being only seen among the broken masses of rocks that lie at the bases of the perpendicular cliffs or along their faces. It was detected, too, at various points in the Coast range, so that its diffusion over Southern California may be said to be general.

Of all its tribe, save perhaps the Winter Wren, this species is the most liable to be overlooked, where, too, it may be tolerably common. To a preference for the wild solitudes of the mountains it adds a shy, suspicious nature, which prompts it to hide away from observation and all chance of danger whenever anything of a suspicious character is observed.

No.	Sex.	Locality.	Date.	Collector.	Wing.	Tail.	Bill.	Tarsus.
706	♂	Walker's Basin, Cal...........	Nov. 10	H. W. Henshaw.	2.30	2.25	0.85	0.71

23. *Troglodytes bewickii* (Aud.), var. *spilurus*, (Vigors).—Western Mocking-bird.

Troglodytes spilurus, Vig., Zool. Beechey's Voyage, 1839, 18, pl. 4, f. 1 (California).
Thryothorus spilurus, Coop., B. Cal., i, 1870, 69.
Troglodytes bewickii, Newb., P. R. R. Rep., vi, 1857, 80.—Herm., ibid., x, 1859, pt. vi, 40.—Coop. & Sackl., ibid., vol. xii, pt. ii, 1860, 189.
Thryothorus bewickii, var. *spilurus*. Bd., Rev. N. A. B , 1964, 126.—Coues, Key N. A. B., 1872, 36.—B., B., & R., N. A. B., 1874, 147.—Nelson, Proc. Boston Soc. Nat. Hist., vol. xvii, 357 (California).

In one or another of its three varieties, this bird is represented quite across the United States. The Bewick's Wren in the east, and its white-bellied variety (var. *cucogaster*) in the middle region, are both quite southern in their habitats, much more so than the extreme western form (var. *spilurus*), which, according to Dr. Cooper, winters in the mild regions as far to the north as Puget Sound.

Throughout the southern half of California it is a common resident during the summer, preferring to inhabit the more elevated regions, and descending thence to the lowlands to pass the winter.

It is a bird of the rather open districts, at least as compared with some others of the family, and, when its breeding duties have been fulfilled, wanders a great deal over the country at large. It is apt to be found in company with the restless flocks of Sparrows and Snowbirds, their general habits of keeping in bushy localities being sufficiently like its own to admit of this companionship.

No.	Sex.	Locality.	Date.	Collector.	Wing.	Tail.	Bill.	Tarsus.
385	♀ jun.	Tejon Mountains, Cal	Aug. 17	H. W. Henshaw .	1.90	2.05	0.54	0.73
414	♀	......do	Aug. 19	....do	1.94	2.18	0.54	0.71
595	♂	Kernville, Cal.	Aug. 23	....do	2.08	2.20	0.55	0.75
672	♂	Walker's Basin, Cal.	Nov. 9	....do	1.97	2.12	0.58	0.65
763	♀ jun.	Tejon Mountains, Cal	Aug. 2	....do	2.05	2.10	0.58	0.69
764	jun....	....	Aug. —	....do	2.12	2.28	0.56	0.73

24. *Troglodytes aëdon*, Vieill., var. *parkmanni*, Aud.—Parkman's Wren.

The most numerous of its tribe in California, inhabiting the wooded sections everywhere.

No.	Sex.	Locality.	Date.	Collector.
80	♂ jun.	Santa Barbara, Cal............................	June 26	H. W. Henshaw.
314	♂ jun.	Fort Tejon, Cal	Aug. 7	....do ....
463	♂	Near Mount Whitney, Cal	Sept. 10	....do ....
762	jun	California		....

25. *Cistothorus palustris*, Wils., var. *paludicola*, Bd.

This Marsh Wren is abundant in Southern California, especially in fall. Though possessed of much the same palustrini habits as in the east, the bird is not nearly so particular here, but will be found to make the most of the circumstances. As tule swamps and bogs grown up to rushes do not abound, the Wrens often take up their residence on the running streams, where covert is so scanty that their habits necessarily undergo considerable change.

No.	Sex.	Locality.	Date.	Collector.
414	♂	Fort Tejon, Cal	Aug. 19	H. W. Henshaw.

MOTACILLIDÆ.—WAGTAILS.

26. *Anthus ludovicianus*, (Gm.).—Titlark.

The Titlark occurs in California, at least in the southern portion, only as a late fall and winter visitant. It is then distributed over the State at large, moving in small parties here and there, its movement depending solely upon the food-supply. This it gleans from the stubble-fields, from the sandy shores of the rivers, and from the grassy plains.

No.	Sex.	Locality.	Date.	Collector.
584	♂	Near Kernville, Cal...............................	Oct. 20	W. H. Henshaw.
585	♂	......do	Oct. 20	Do.

SYLVICOLIDÆ.—WARBLERS.

27. *Helminthophaga celata,* Say.—Orange-crowned Warbler.

This Warbler is a common species in summer in Southern California, and indeed possesses in the West an almost unrestricted range, reaching on the coast from Cape Saint Lucas to the Yukon in Alaska, and being distributed throughout the interior. In Colorado, and the interior generally, it is in summer a bird of the mountains, reaching sometimes above the timber, and making its home in the scanty alpine growth of bushes about the lofty summits.

It is found, too, on the mountains of California (var. *lutescens,* Ridgway), but not at all exclusively. It was the only Warbler I found on the island of Santa Cruz, where it was quite numerous, and breeding in early June. The surface of this island, broken up and diversified by rocky ridges, is covered with a growth of chaparral, often very dense, and forms just the locality which this Warbler delights in. A female which I shot June 10 contained an egg which would have been ready for depositing in a few days.

No.	Sex.	Locality.	Date.	Collector.
17	♂ ad.	Santa Cruz Island, Cal.	June 10	H. W. Henshaw.
18	♀ ad.	......do	June 10	Do.
277	♀	Fort Tejon, Cal.	Aug. 2	Do.
508	♂	Near Mount Whitney, Cal.	Sept. 26	Do.

28. *Dendroica æstiva,* (Gm.).—Yellow Warbler.

A common species about Los Angeles in June. It breeds and is quite numerous through the northern half of the State, being confined entirely to the low districts.

No.	Sex.	Locality.	Date.	Collector.
352	♂ jun.	Fort Tejon, Cal	Aug. 10	H. W. Henshaw.

29. *Dendroica audaboni,* (Towns.).—Audubon's Warbler.

This Warbler does not appear to remain in the mountains of California during the summer, as it does in Colorado and Arizona, but repairs farther north to rear its young. It may yet remain to be detected in the high forests in the northern half of the State.

In the neighborhood of Mount Whitney it was common in September, being then on its way south. The common Yellow-rump (*D. coronata*) has not yet been found in California, though found at the Straits of Ituca in April by Dr. Cooper.

No.	Sex.	Locality.	Date.	Collector.
530	♂	Head Tule River, California.	Oct. 7	H. W. Henshaw.
539	♀	Near Mount Whitney, Cal.	Oct. 9	Do.
557	♀	Near Thunder Mountain, Cal.	Oct. 10	Do.

30. *Dendroica nigrescens,* (Towns.).—Black-throated Gray Warbler.

I found this species common in the mountains, near Fort Tejon, in early August, and think they find here in the pine region their summer haunts. After leaving the Coast range, the species was not seen again, not even in the pineries of the high mountains near Mount Whitney.

No.	Sex.	Locality.	Date.	Collector.
765	♀ pin.	Tejon Mountains, Cal	Aug. 2	H. W. Henshaw.
766	jun.	......do	Aug. 2	Do.
291	♂ jun.	......do	Aug. 3	Do.
292	♂ jun.	......do	Aug. 5	Do.
238	♂ jun.	......do	Aug. 5	Do.
377	♀ ad.	......do	Aug. 3	Do.
411	♀	......do	Aug. 17	Do.
412	♀	......do	Aug. 19	Do.

31. *Dendroica occidentalis*, (Towns.).—Western Warbler.

Concerning the occurrence of this Warbler in California, we have no very extended information. Dr. Cooper cites the capture of a single specimen at Petaluma, and considers the species a very rare one.

A single individual, taken near the head of Tule River in October, was the only one I saw. It probably then uses the Rocky Mountains as a highway in its spring and fall journeyings to and from higher latitudes to breed. It was quite common at Mount Graham, Arizona, in September of 1874, there affecting exclusively the spruce and fir woods.

No.	Sex.	Locality.	Date.	Collector.
537	♂	Head Tule River, Cal	Oct. 9	H. W. Henshaw.

32. *Geothlypis trichas*, (L.).—Maryland Yellowthroat.

Apparently not very common, though distributed pretty evenly over the southern portion of the State. Notes and habits as at the East. The Macgillivray's Warbler (*G. macgillivrayi*) was not detected by us, from which I infer its general rarity in the southern portion of the State. It, however, occurs here, as it is given from several localities by Dr. Cooper; also noted at Nevada City by Mr. Nelson.

No.	Sex.	Locality.	Date.	Collector.
431	♂ jun.	Walker's Basin, Cal	Aug. 28	H. W. Henshaw.
432	♂ jun.	.......do	Aug. 28	Do.

33. *Icteria virens*, (L.), var. *longicauda*, (Lawr.).—Long-tailed Chat.

The Chat is wide-spread over Southern California, where, however, we nowhere found it abundant. It inhabits the undergrowth and thickets of the streams, from the friendly shelter of which it rarely ventures forth. It is one of the noisiest of our small birds, and one cannot long remain in the vicinity of a spot inhabited by a pair without being made aware of the fact by their noisy outpourings.

No.	Sex.	Locality.	Date.	Collector.
48	♂ ad.	Santa Barbara, Cal..	June 25	H. W. Henshaw.
82	♂ ad.	.......do	June 26	Do.
293	♂ jun.	Tejon Mountains, Cal	Aug. 3	Do.

34. *Myiodioctes pusillus*, (Wils.), var. *pileolatus*, (Ridgw.).—Western Blackcap.

While at Los Angeles, the middle of June, I found this little bird not uncommon in the swampy thickets, just the places, in fact, most frequented by it during the migrations. They were in full song, and their short, rather faint ditties were heard as they swept in short flights about the extremities of the branches, snapping up their flying food.

The late date at which they were noted seems to preclude the possibility of their being mere migrants, though this departure from their usual habits is strange enough, when this low altitude be compared with the high mountains they resort to in the interior region.

About the middle of August they became common, moving southward from the breeding-grounds in the far north. The bulk of these, however, are the true *M. pusillus*.

No.	Sex.	Locality.	Date.	Collector.
336	♂ jun.	Fort Tejon, Cal	Aug. 9	H. W. Henshaw.
511	♀ ad.	Near Mount Whitney, Cal	Sept. 26	Do.

HIRUNDINIDÆ.—SWALLOWS.

35. *Progne subis*, (L..).—Purple Martin.

Of apparently not so general distribution in Southern California as throughout the territory of the interior region, occuring, however, in colonies here and there.

36. *Petrochelidon lunifrous*, (Say).—Cliff Swallow.

Occurs over the country at large, being perhaps fully as abundant along the seacoast as in the interior.

37. *Hirundo horreorum*, Barton.—Barn Swallow.

This Swallow is far less numerous than the preceding, though on the coast, at least, it is not rare. On the island of Santa Cruz a few pairs were seen, and still clinging to their primitive mode of living. Their nests were built either in caverns or in the sheltered depressions on the faces of the rocky cliffs.

38. *Tachycineta thalassina*, (Sw.).—Violet-green Swallow.

Along the coast this Swallow is very numerous, resorting, as noted by Dr. Cooper almost exclusively to the oak-groves, where, in the natural knot-holes and the deserted homes of Woodpeckers, it builds its nest. It was abundant in September in the high meadows near the base of Mount Whitney, though whether the species is limited to the Coast-range region in summer, and only occurs in the sierras as a migrant, I am unable to say.

No.	Sex.	Locality.	Date.	Collector.
118	♂ ad.	Santa Barbara, Cal..	June 28	H. W. Henshaw.
139	♂ ad.	......do	June 29	Do.
140	♀ ad.	......do	June 29	Do.
465	♂	Near Mount Whitney, Cal	Sept. 10	Do.
754	♂ ad.	Tejon Mountains, Cal......................................	Aug. 2	Do.

39. *Stelgidopteryx serripennis*, (Aud.).—Rough-winged Swallow.

Occurs commonly through Southern California, its distribution being regulated only by the presence or absence of suitable localities. Frequents chiefly the banks of the rivers.

VIREONIDÆ.—VIREOS.

40. *Vireo gilvus*, (Vieill.), var. *swainsoni*, Bd.—Western Warbling Vireo.

Occurs commonly in California, inhabiting the deciduous trees of the low districts, and extending upward on the timbered mountains to at least 10,000 feet.

No.	Sex.	Locality.	Date.	Collector.
130	♂ ad.	Santa Barbara, Cal..	June 28	H. W. Henshaw.
289	♀ ad.	Tejon Mountains, Cal.....................................	Aug. 3	Do.

41. *Vireo solitarius*, (Wils.).—Solitary Vireo.

Southern California does not appear to be included in the range of this species, except in so far as it occurs there during the migrations. Further north, on the Columbia River, they are, according to Dr. Cooper, common in summer.

No.	Sex.	Locality.	Date.	Collector.
376	♀	Tejon Mountains, Cal..	Aug. 17	H. W. Henshaw.

42. *Vireo solitarius*, (Wils.), var. *cassini*.—Cassin's Vireo.

In the mountains, near Fort Tejon, the locality from which the first specimen was obtained, I took a single individual in August. This was the only one seen, and I am inclined to think that with the preceding species the Cassin's Vireo retires in summer to more northern breeding-grounds.

No.	Sex.	Locality.	Date.	Collector.
376	♀	Tejon Mountains, Cal.............................	Aug. 17	H. W. Henshaw.

43. Vireo solitarius, (Wils.), var. plumbeus, Cones.—Western Solitary Vireo.

I procured a single specimen of this Vireo in the mountains near Fort Tejon, August 1. It is in much-worn plumage, and probably had bred in this locality. The species is, however, one belonging more particularly to the Southern Rocky Mountains.

No.	Sex.	Locality.	Date.	Collector.
767	♀ ad.	Tejon Mountains, Cal.............................	Aug. 1	H. W. Henshaw.

44. Vireo pusillus, Cones.—Least Vireo.

The Least Vireo was the most abundant of its tribe about Los Angeles in June, and their notes, remarkable only for their oddity and quaintness, were constantly heard issuing from the thickets, often several males singing at a time. The bird seems to be the counterpart of the eastern *Vireo belli*. It is never seen in the open, and very rarely in the taller, trees, but keeps within the shelter of the shrubbery, either along a stream or in the swamps. It is very active and restless, and, numerous as they were, I found it very difficult to get even a glimpse of them, as they flitted about, now just over the ground, now in the tops of the young trees, that grew so thickly as to limit my view to the space of a few yards.

As far north as Santa Barbara and Fort Tejon they were quite numerous, and their range will very probably be found to reach as far north as San Francisco.

No.	Sex.	Locality.	Date.	Collector.	Wing.	Tail.	Bill.	Tarsus.
24	♂ ad.	Los Angeles, Cal..............	June 17	H. W. Henshaw.	2.12	2.25	0.41	0.72
258	♂ jun.	Santa Barbara, Cal......	July 7	....do	2.08	2.22	0.38	0.77
312	♀ jun.	Fort Tejon, Cal............	Aug. 7	....do	2.07	2.05	0.42	0.74
335		.......do	Aug. 9	....do	2.20	2.14	0.40	0.75

45. Vireo huttoni, Cassin.—Hutton's Vireo.

Vireo huttoni, Cassin, Pr. A. N. Sc.Phila. v, Feb., 1851, 150.—Bd., B. N. A., 1858, 339.—Coop., B. Cal., i, 1870, 121.—Cones, Key N. A. B., 1872, 123.—B., B., & R., N. A. B., i, 1874, 387.

This species is one of the least known of all our Vireos; nor is this owing entirely to its rarity, for at Santa Barbara, in June, it was quite common, and according to Dr. Cooper this is true in other parts of California, it wintering plentifully as high as latitude 38°. It breeds, I am inclined to believe, through the whole of Southern California.

In habits it is arboreal, as much so, judging from those I saw, as the Warbling Vireo. It frequented the oaks exclusively, and was at this season entirely silent, so that, though I watched them for the express purpose of listening to their notes, I heard not a single strain. This was probably due to the fact that their broods were just out and required their full attention.

All their movements were marked with a quiet deliberation as they silently moved about the ends of the branches, searching them with the utmost care for food.

No.	Sex.	Locality.	Date.	Collector.	Wing.	Tail.	Bill.	Tarsus.
769	♀ ad.	Santa Barbara, Cal.............	June 14	H. W. Henshaw.	2.45	2.15	0.43	0.76
58	♂ ad.	.......do	June 25	....do	2.43	2.09	0.42	0.75
109	♂ jun.	.......do	June 27	....do	2.35	2.10	0.39	0.73
110	♀ ad.	.......do	June 27	....do	2.38	2.04	0.43	0.73
111	♂ ad.	.......do	June 27	....do	2.29	2.07	0.42	0.77
135	♂ jun.	.......do	June 29	....do	2.35	2.04	0.42	0.77
152	♂ jun.	.......do	June 29	....do	2.40	2.12	0.43	0.77
245	♀ ad.	.......do	July 17	....do	2.45	2.15	0.40	0.72

AMPELIDÆ.—CHATTERERS.

46. *Phænopepla nitens*, (Sw.).—Black Flycatcher.

I saw this species on but few occasions. They are, however, not rare in the southern half of the State. They inhabit the bushy cañons, and are found much about the oaks, upon which they find the berries of the mistletoe. In fall these and other kinds of berries form their chief sustenance, varied with insects which they capture on the wing. They are among the shyest of the small birds.

47. *Miadestes townsendi*, (Aud.).—Townsend's Solitaire.

This species probably resorts to the high mountains, as in the interior region, to pass the summer. I saw none till in September; when in the Sierras, they appeared here and there noiselessly pursuing their avocations. Though usually a bird of very unsocial disposition, the abundance of food at any special locality, as berries, attracts them in numbers, when they seem inclined to live more or less in company, and in late fall are apt to be seen in parties of four or five individuals. They never, however, flock, in the strict meaning of the word.

No.	Sex.	Locality.	Date.	Collector.	Wing.	Tail.	Bill.	Tarsus.
464	♂	Near Mount Whitney, Cal.....	Sept. 26	H. W. Henshaw.	4. 45	4. 24	0. 52	0. 79
498	♀	......do	Sept. 26	....do	4. 49	4. 09	0. 50	0. 82
498 A	♀	......do	Sept. 26	....do	4. 59	4. 15	0. 47	0. 83
499	♂	......do	Sept. 26	....do	4. 58	4. 12	0. 48	0. 80
500	♂	......do	Sept. 26	....do	4. 57	4. 18	0. 49	0. 82
501	♀	......do	Sept. 26	....do	4. 55	4. 16	0. 48	0. 82
502	♂	......do	Sept. 26	....do	4. 53	4. 15	0. 47	0. 82

LANIIDÆ.—SHRIKES.

48. *Collurio ludovicianus*, (L.), var. *excubitoroides*.—Swain's White-rumped Shrike.

This Shrike is numerous in California, where its habits of life throughout appear not different from its usual mode of existence elsewhere. As noticed by Professor Baird, in his Review, there is observable in the birds from the west coast an appreciable difference from those of the interior, which latter represent what may be called the normal type of coloration of the var. *excubitoroides*. In our specimens from California the ash above is darker, the hoariness of the forehead of less extent, the white of scapulars more restricted. All the specimens, however, taken on the mainland have the white rump clearly defined. In this connection, two young birds in nesting plumage taken on Santa Cruz Island are especially noteworthy. These appear in all respects to be typical *ludovicianus*. In the depth of the plumbeus shade above and along the sides, in the lack of any hoariness on the forehead, and, above all, in the absence of any whiteness of the rump, this being like the back, they exactly resemble young birds from Florida.

No.	Sex.	Locality.	Date.	Collector.
87	♀	Santa Barbara, Cal..	June 26	H. W. Henshaw
205	♀ jun.	......do	July 6	Do.
467	♂ jun.	Near Mount Whitney, Cal		Do.
750	....	......do		Do.
625	♀	Kernville, Cal..	Oct. 27	Do.
		Ludovicianus.		
15	♂	Santa Cruz Island, Cal..	June 10	Do.
16	♂	......do	June 10	Do.

TANAGRIDÆ.—TANAGERS.

49. *Pyranga ludoviciana*, (Wils.).—Louisiana Tanager.

In one of the small cañons issuing from the mountains near Santa Barbara I found several of these Tanagers in July, at which time they were feeding their young. Elsewhere in Southern California they were most unaccountably rare, and, all told, I do not think I saw over a dozen during the entire summer. Probably the bulk of their numbers pass farther north to breed.

50. *Carpodacus purpureus*, (Gm.).—Purple Finches.

This species appears to be at least not a common one in Southern California, a single specimen being all obtained or seen by us. Dr. Cooper speaks of finding them on the summits of the Coast range, toward Santa Cruz, in May, where they had nests.

No.	Sex.	Locality.	Date.	Collector.	Wing.	Tail.	Bill.	Tarsus.
563	♀	Near Mount Whitney, Cal..	Oct. 10	H. W. Henshaw...	3.00	2.34	0.45	0.72

51. *Carpodacus frontalis*, (Say).—House Finch.

In Southern California, in summer this Finch is perhaps the most numerous of any of the small birds. Their diffusion is very general, the mountains alone being unvisited by them. On the island of Santa Cruz, their numbers are as great as on the mainland. They are always found in greatest numbers in the vicinity of houses, where there are scarcely any bounds to their familiarity. On the uninhabited portions of the island a few had taken up their abode, resorting to little niches in the face of cliffs to place their nests. Their disposition toward each other is sociable in the extreme, and wherever found they will be seen to have established themselves into communities, often of many individuals, while the air is fairly filled with their songs, which continue from morning to night.

No.	Sex.	Locality.	Date.	Collector.
72	♀ ad.	Santa Barbara, Cal	June 26	
73	♂ ad.	......do	June 26	
74	♂ jun.	......do	June 26	
129	♂ ad.	......do	June 28	
150	♂ ad.	......do	June 29	
151	♂ ad.	......do	June 29	
159	♂ ad.	......do	June 29	
160	♂ ad.	......do	July 1	
161	♂ ad.	......de	July 1	
248	♀ ad.	Walker's Basin, Cal	July 1	
342	♀ jun.	Fort Tejon, Cal	July 17	
770	♂ ad.	Santa Barbara, Cal	Aug. 9	

52. *Chrysomitris tristis*, (Linn.).—Goldfinch.

This is an abundant species throughout Southern California, avoiding only the high mountainous districts. It was particularly numerous at Los Angeles, and as early as the middle of June was breeding plentifully. Their eggs at this time were in most cases far advanced toward hatching, though in one instance fresh eggs were found, and in another the nest had been just begun.

In a dense willow-thicket within an area of a few yards no fewer than seven of their nests were counted. They were all placed quite low, the highest about 12 feet from the ground, and, save in being less compactly woven, resembled the usual style of structure in the East. The eggs are of an unspotted greenish-white color.

No.	Sex.	Locality.	Date.	Collector.
26	♀ ad.	Los Angeles, Cal	June 17	H. W. Henshaw.

53. *Chrysomitris psaltria*, (Say.).—Arkansas Finch.

Of the three species inhabiting Southern California, this Goldfinch appears to be the most widely spread, as perhaps also the most numerous. It was seen at many different localities, and, like the other two, inhabits the valleys. The reeds of grasses and weeds appear to form the chief part of its food.

No.	Sex.	.	Locality.	Date.	Collector.
247	♀ jun.		Ojai Creek, Cal	July 17	H. W. Henshaw.
294	♂ jun.		Fort Tejon, Cal....................................	Aug. 3	Do.
338	♂ ad		......do	Aug. 9	Do.
338	♂ ad.		......do	Aug. 9	Do.
339	♂ jun.		......do	Aug. 9	Do.
340	J un.		......do	Aug. 9	Do.
341	♀ jun.		......do	Aug. 9	Do.
344	♀ jun.		......do	Aug. 9	Do.
426	♂ jun.		Walker's Basin, Cal	Aug. 27	Do.
427	♀ ad.		Fort Tejon, Cal	Aug. 28	Do.
430	♂ ad.		Walker's Basin, Cal	Aug. 28	Do.
433	♀ jun.		......do	Aug. 28	Do.
434	♀ jun.		Fort Tejon, Cal	Aug. 28	Do.
435	♀ jun.		......do	Aug. 28	Do.

54. *Chrysomitris lawrencii*, (Cassin).—Lawrence's Goldfinch.

Carduelis lawrencii, Cassin, Proc. A. N. Sc., v, Oct., 1859, 105, pl. v (California).—Heermann, P. R. R. Rep., x, 1859, vi, 50, (California).
Chrysomitris lawrencii, Bd., B. N. A., 1858, 424.—Xantus, Proc. Phila. Acad. Nat. Sci., 1859, 191.— (Ft. Tejon, Cal.).—Coop., B. Cal., i, 1870, 121.—Coues, Key N. A. B.. 1872, 132.—B., B., & R., N. A. B., 1874, 478.

This Goldfinch appears to be more particularly a Californian species, and I've do not find it reported from outside the State, except from Camp Whipple, Arizona. Its distribution here seems confined to a comparatively narrow area coastwise, from the southern border to the most northern portions of the State, where Heermann gives it as very abundant throughout the mining-regions. It thus may, and probably does, extend somewhat into Oregon, though it has not been detected at Camp Harney by Lieutenant Bendrie.

Near Santa Barbara, which was the only place where I met with the bird, it was a numerously represented species, though even there my observations respecting it were confined to a single locality, the neighborhood of some springs of fresh water to which the birds resorted in great numbers all through the day to slake their thirst. They certainly did not breed in the immediate locality, and I was at a loss to imagine the particular attraction the spot had for them.

No.	Sex.	Locality.	Date.	Collector.	Wing.	Tail.	Bill.	Tarsus.
53	♂ ad.	Santa Barbara, Cal..........	June 25	H. W. Henshaw ..	2.53	2.03	0.32	0.50
56	♂ ad.	......do	June 25	....do	2.80	2.24	0.33	0.54
57	♂ ad.	......do	June 25	....do	2.70	2.08	0.35	0.54
75	♂ ad.	......do	June 26	....do	2.79	2.18	0.33	0.53
97	♂ ad.	......do	June 27	....do	2.62	2.16	0.33	0.52
163	♂ ad.	......do	July 1	....do	2.70	2.12	0.32	0.52
164	♂ ad.	......do	July 1	....do	2.73	2.12	0.35	0.52
50	♀ ad.	......do	June 25	....do	2.56	1.97	0.33	0.52
51	♀ ad.	......do	June 25	....do	2.50	2.03	0.34	0.53
165	♀ ad.	......do	July 1	....do	2.54	2.05	0.33	0.47
49	♂ jun.	......do	June 25	....do	2.64			
52	♂ jun.	......do	June 25	....do				
54	♀ jun.	......do	June 25	....do				
55	♀ jun.	......do	June 25	....do				
76	♀ jun.	......do	June 26	....do				
62	♂ jun.	......do	June 25	....do				
98	♀ jun.	......do	June 27	....do				
99	♀ jun.	......do	June 27	....do				
100	♂ jun.	......do	June 27	....do				
176	♂ jun.	......do	July 2	....do				
177	♂ jun.	......do	July 2	....do				

55. *Chrysomitris pinus*, (Wils.).—Pine Finch.

In winter, the Pine Finch overspreads California, probably visiting all portions. It apparently does not, as in the same latitude in the interior, resort to the high mountains in summer, but all retire to the far north. In the interior, about Kernville and elsewhere, it was present the last of October in small flocks, finding in the weedpatches an abundance of food.

No.	Sex.	Locality.	Date.	Collector.
641	♀ jun.	Walker's Basin, Cal	Nov. 5	H. W. Henshaw.

56. *Passerculus savanna*, (Wils.), var. *alaudinus*, Bp.—Western Savanna Sparrow.

We have no positive proof of the occurrence of this variety in California in summer, and all the evidence I could obtain seems to point to the opposite conclusion.

During the fall migrations it makes its appearance from the North, and then occurs over the State at large.

I found it early in September on the streams high up in the mountains, near Mount Whitney, while in November it was exceedingly numerous about Oakland, across the bay from San Francisco, frequenting the plowed lands, gardens, and grassy fields everywhere, almost to the shore. Whether it ever is found in the salt-meadows along the shore I do not know, but believe it never does occur in such places, even during the migrations.

No.	Sex.	Locality.	Date.	Collector.
443	♀	Walker's Basin, Cal.	Sept. 6	H. W. Henshaw.
510	♀	North Fork Kern River, Cal.	Sept. 29	Do.
587	♀	Near Kernville, Cal.	Oct. 20	Do.
588	♂	......do	Oct. 20	Do.
594	♂	......do	Oct. 23	Do.
713	♂	Walker's Basin, Cal.	Nov. 11	Do.

57. *Passerculus savanna*, (Wils.), var. *anthinus*.—Fitlark Sparrow.

Passerculus anthinus, Bonaparte, Comptes Rendus, xxvii, Dec., 1853, 919 (Russian America).—Bd., B. N. A., 1858, 445.—Coop, B. Cal., i, 1870, 183.
Passerculus savanna var. *anthinus*, Coues, Key N. A. B., 1872, 136.—B., B., & R., N. A. B., i, 1874, 539.

This sparrow, so far as known, is confined to California,* where it inhabits exclusively the coast, being found in the salt-meadows and beds of rushes. Its habits seem to resemble very closely those of the eastern Savanna Sparrow (*P. savannus*) as seen under similar circumstances. They lie close hidden in the grass, rise with extreme reluctance, and fly with apparent difficulty to a short distance, alighting usually on the tops of the mattocks of grass, or upon the mazing reeds, there to reconnoiter for a moment ere taking refuge among the roots.

No.	Sex.	Locality.	Date.	Collector.	Wing.	Tail.	Bill.	Tarsus.
101	♀ jun.	Santa Barbara, Cal	June 27	H. W. Henshaw.	2.57	2.00	0.43	0.80
103	♀ ad.	......do	June 27	....do	2.40	1.90	0.46	0.74
104	♀ ad.	......do	June 27	....do	2.45	1.80	0.46	0.77
105	♂ jun.	......do	June 27	....do	2.65	2.15	0.43	0.79
106	♀ jun.	......do	June 27	....do	2.54	2.00	0.47	0.79
112	♂ ad.	......do	June 27	....do	2.64	2.03	0.47	0.83
119	♀ ad.	......do	June 28	....do	2.40	......	0.40	0.82
120	♂ ad.	......do	June 28	... do]	2.56	2.00	0.47	0.89
121	♀ ad.	......do	June 28	....do	2.50	2.00	0.50	0.81
122	♂ ad.	......do	June 28	....do	2.72	2.06	0.47	0.82
156	♀ ad.	......do	July 1	....do	2.43	1.90	0.42	0.77
157	♀ ad.	......do	July 1	....do	2.46	2.00	0.46	0.77
237	♂ ad.	......do	July 8	....do	2.49	2.06	0.46	0.78

58. *Pooecetes gramineus*, (Gm.), var. *confinis*, Bd.—Grass Finch.

I did not meet with the Grass Finch in summer in Southern California, and believe with Dr. Cooper that if it breeds within the State, it is only in the more northern parts. Like the Lark Sparrow, it is more a bird of the dry interior regions, being found, however, over California during the migrations.

No.	Sex.	Locality.	Date.	Collector.
466	♀	Near Mount Whitney, Cal	Sept. 12	H. W. Henshaw.
630	♀	Near Kernville, Cal.	Oct. 25	Do.

59. *Coturniculus passerinus*, (Wils.), var. *perpallidus*.—Western Yellow-winged Sparrow.

This Sparrow is chiefly a bird of the interior region, where it is rather southerly in its habitat. It has not hitherto been known certainly to occur on the Pacific coast.

* The locality of Bonaparte's type-specimen was probably transferred with that of *P. alaudinus*. (See B., B., & R., N. A. B.)

At Santa Barbara, directly on the coast, I. found the species breeding, and took the young in nesting plumage the last of June. Elsewhere I did not see it, though, as it is a bird of very unobtrusive habits, it may have easily been overlooked; hence its diffusion over Southern California is by no means improbable.

No.	Sex.	Locality.	Date.	Collector.
142	Jun.	Santa Barbara, Cal	June 29	H. W. Henshaw.
162	♀ ad.	......do	July 1	Do.

60. *Chondestes grammaca*, (Say).—Lark Finch.

As might be expected from its almost universal dispersion over the West, the Lark Finch is found in California, wintering, according to Dr. Cooper, in the southern part of the State. That this is somewhat out of their usual range is shown by the fact of their general scarcity as compared with the great number to be seen in the interior sections.

61. *Zonotrichia leucophrys*, (Forst.).—White-crowned Sparrow.

This species is not known to breed in Alaska, or, indeed, within the Pacific coast region. I found it in the high sierras in September, in company with the succeeding variety, forming, however, but a very small proportion of the vast flocks of those birds. In common with some other species, as the *Passerculus alaudinus*, large numbers of this Sparrow in pursuing their migration southward in fall, instead of following a direct course, radiate out of the line, and are thus found far to the west and east of the region inhabited by them in summer.

A more notable instance of this irregular mode of migrating is seen in the *Junco oregonus*. This, though a species belonging to the western province, is found in fall and winter diffused over all the interior province, and to the edge of the eastern. In these cases it seems to be merely a question of the abundance of food which determines their path. In a great measure, independent of climatic conditions, the birds wander almost at will, wherever they find their wants most easily satisfied.

No.	Sex.	Locality.	Date.	Collector.
470	♂ ad.	Mount Whitney, Cal	Sept. 12	H. W. Henshaw.
471	♂ ad.	......do	Sept. 12	Do.
491	♀ ad.	......do	Sept. 19	Do.
495	♀ ad.	......do	Sept. 21	Do.

62. *Zonotrichia leucophrys*, (Forst.), var. *intermedia*, Ridgw.

By the middle of September this bird had become very common in the sierras, and at an altitude of 12,000 feet was seen in large flocks feeding among the low willows and alpine shrubbery that fringe the little streams of this elevation. As we descended thence into lower regions, it grew still more abundant, till in the low valleys they numbered thousands. About San Francisco, the middle of November, they were seen in throngs in the shrubbery of the gardens, and they doubtless spend the winter here. This variety does not breed in California.

No.	Sex.	Locality.	Date.	Collector.
478	♂ jun.	Near Mount Whitney, Cal	Sept. 18	H. W. Henshaw.
479	♂ ad.	......do	Sept. 18	Do.
480	♀ jun.	......do	Sept. 19	Do.
481	♂ jun.	......do	Sept. 19	Do.
482	♂ ad.	......do	Sept. 19	Do.
483	♀ ad.	......do	Sept. 19	Do.
484	♀ ad.	......do	Sept. 19	Do.
485	♀ ad.	......do	Sept. 19	Do.
486	♀ jun.	......do	Sept. 21	Do.
492	♂ ad.	......do	Sept. 21	Do.
494	♀ ad.	......do	Sept. 21	Do.
503	♂ ad.	......do	Sept. 26	Do.
549	♀ jun.	Near Kernville, Cal.	Sept. 10	Do.
593	♂ jun.	......do	Sept. 23	Do.
575	♂ jun.	......do	Oct. 16	Do.
585 A	♂ ad.	......do	Oct. 98	Do.
585	♂ ad.	......do	Oct. 19	Do.
586	♀ jun.	......do	Oct. 98	Do.
680	♂ ad.	Walker's Basin, Cal	Nov. 5	Do.
591	♀ jun.	Near Kernville, Cal.	Oct. 23	Do.
592	♂ ad.	......do	Oct. 23	Do.

63. *Zonotrichia gambeli,* (Forst.).—Gambel's Finch.

The true Gambel's Finch is confined to the Pacific province, where it breeds as far to the north as Kodiak. I was unable to detect its presence in the mountains about Fort Tejon in summer, and I am inclined to believe that it does not breed in the sierras south at least of the latitude of San Francisco.

Among a large number of the preceding birds, shot near Mount Whitney in September, were only two of this variety. About San Francisco, too, where in November the other variety was so numerous, I succeeded in finding but a single pair, an adult and a young bird. They are thus probably quite local in their habitat, and resident to a great extent, though in their wanderings for food they extend some distance farther south in the fall and winter than their regular habitat.

No.	Sex.	Locality.	Date.	Collector.
485	♀ ad.	Near Mount Whitney, Cal	Sept. 19	H. W. Henshaw.
486	♀ jun.	......do	Sept. 19	Do.

64. *Zonotrichia coronata,* Pallas.—Golden-crowned Sparrow.

Emberiza coronata, Pallas Zoog. Rosso-Asiat., ii, 1811, 44.
Zonotrichia coronata, Bd., B. N. A., 1858, 461.—Xantus, Proc. Phila. Acad. Nat. Sci., 1859, 191.—Coop. & Suckl., P. R. R. Rep., vol. 12, pt. ii, 1860, 201.—Coop., B. Cal., i, 1870, 197.—B., B., & R., N. A. B., ii, 1874, 573.—Nelson, Proc. Bost. Soc. Nat. Hist., vol. xvii, 359 (California).
Zonotrichia aurocapilla, Newb., P. R. R. Rep., vi, 1857, 88.

In its fall migration, this Sparrow appears to follow pretty exclusively the mountain-ranges, where it is found from their bases up to an altitude of about 6,000 or 7,000 feet, thus avoiding the higher summits and not descending into the valleys. It is a brush-loving species, and inhabits the thickest chaparral of oak-scrub or "blue brush," sometimes in flocks of its own kind, oftener in company with the other *Zonotrichia* and the *Pipilos*. Its habits differ in no noteworthy respect from those of its congeners. Its food, which in the fall consists almost entirely of the seeds of grasses and weeds, is obtained from the ground, the various species mingling together in perfect amity as they conduct their search. By the 10th of November most of the young birds had passed north, those remaining being for the most part in the adult plumage. Later in the fall and in winter their diffusion becomes more general. In company with the var. *intermedia,* they were seen in the hedge-rows and weed-patches about Oakland, where they spend the winter. Probably more or less remain in the mountains of Northern California during the summer. Heerman, as quoted, mentions finding a nest of this species near Sacramento.

No.	Sex.	Locality.	Date.	Collector.	Wing.	Tail.	Bill.	Tarsus.
564	♀ ad.	Mountains near Kernville, Cal.	Oct. 16	H. W. Henshaw.	3.30	3.37	0.47	0.95
562	♂ ad.	......do	Oct. 16	....do	3.28	3.45	0.50	0.88
668	♂ ad.	Walker's Basin, Cal.	Nov. 9	....do	3.20	3.48	0.50	0.95
609	♂ ad.	Mountains near Kernville, Cal.	Oct. 25	....do	3.32	3.63	0.46	0.92
611	♀	......do	Oct. 25	....do	3.08	3.17	0.47	0.93
561	♂ jun.	......do	Oct. 10	....do	3.26	3.40	0.45	0.97
565	♂ jun.	......do	Oct. 16	....do	3.22	3.37	0.50	1.00
566	♀ jun.	......do	Oct. 16	....do	3.15	3.25	0.45	0.92
565	♂ jun.	......do	Oct. 10	....do	2.93	3.05	0.48	0.90
603	♀ jun.	......do	Oct. 25	....do	3.07	3.95	0.49	0.93
604	♀ jun.	......do	Oct. 25	....do	3.14	3.33	0.48	0.95
567	♀ jun.	......do	Oct. 16	....do				
605	♂ jun.	......do	Oct. 25	....do				
607	♂ jun.	......do	Oct. 25	....do				
608	♂ jun.	......do	Oct. 25	....do				
715	♀ jun.	......do	Nov. 11	....do				
540	♀ jun.	Near Whitney, Cal.	Oct. 10	....do				
610	♂ jun.	Mountains near Kernville, Cal.	Oct. 25	....do				
618	♀ jun.	......do	Oct. 25	....do				
614	♂ jun.	......do	Oct. 25	....do				
612	♀ jun.	......do	Oct. 25	....do				
613	♂ jun.	......do	Oct. 25	....do				
615	♂ jun.	......do	Oct. 25	....do				
616	♀ jun.	......do	Oct. 25	....do				
617	♂ jun.	......do	Oct. 16	....do				
569	♂ jun.	......do	Oct. 16	....do				
619	♂ jun.	......do	Oct. 25	....do				
695	♀ jun.	Walker's Basin, Cal.	Nov. 11	....do				
696	♂ jun.	......do	Nov. 10	....do				
737	Jun.	Mountains near Kernville, Cal.	Oct. 25	....do				

65. *Junco oregonus*, (Towns.).—Oregon Snowbird.

This Snowbird is probably a summer resident in the high mountains throughout California.

As late as August 19 I obtained the young fully fledged, though still retaining their nest plumage, in the mountains near Fort Tejon, where the species was very abundant. In September the number in the State is increased by the arrival of immense flocks from the north, when they overspread the whole country, remaining till the following spring.

No.	Sex.	Locality.	Date.	Collector.
274	♂ jun.	Mountains near Fort Tejon, Cal.............................	Aug. 2	H W. Henshaw.
275	♀ ad.	......do..	Aug. 2	Do.
407	♀ jun.	......do..	Aug. 19	Do.
452	♂ ad.	Mountains near Mount Whitney, Cal........................	Sept. 10	Do.
504	♂ ad.	......do..	Sept. 26	Do.
544	♀	......do..	Oct. 10	Do.
547	♀ ad.	......do..	Oct. 10	Do.
548	♀ ad.	......do..	Oct. 10	Do.
556	♀ ad.	......do..	Oct. 10	Do.
650	♂ ad.	......do..	Nov. 5	Do.

66. *Poospiza belli*, (Cassin).—Bell's Sparrow.

Emberiza belli, Cassin, Pr. A. N. Sc. Phila., v, Oct., 1850, 104, pl. iv, 41 (San Diego, Cal.).
Poospiza belli, Bd., B. N. A., 1858, 470.—Kennerly, P. R. R. Rep., x, 1859, 29.—Heerman, ibid., 46.—
    Coop., B. Cal., i, 1810, 204.

The Bell's Finch appears to be confined to the southern half of California, where it is a resident species. It inhabits to some extent the chaparral on the mountain-sides, but is more particularly a bird of the sage-brush plains, no spot being too desolate to suit the taste of this Sparrow. In the mountains near Fort Tejon, it breeds abundantly at an elevation of 5,000 or 6,000 feet. At this date, August 4, the young were just moulting and about to don the adult feathering.

No.	Sex.	Locality.	Date.	Collector.	Wing.	Tail.	Bill.	Tarsus.
295	♀ jun.	Mountains near Fort Tejon, Cal	Aug. 4	H. W. Henshaw.	2.72	2.96	0.44	0.80
296	♂ jun.	......do..................	Aug. 4	....do..........	2.74	2.88	0.38	0.81
297	♂ jun.	......do..................	Aug. 4	....do..........	2.82	2.87	0.38	0.82
369	♂ jun.	......do..................	Aug. 17	....do..........	2.75	2.91	0.37	0.80
378	♂ jun.	......do..................	Aug. 7	....do..........	2.80	2.89	0.39	0.83
390	♂ ad.	......do..................	Aug. 17	....do..........	2.75	2.88	0.37	0.81
391	♀	......do..................	Aug. 18	....do..........	2.65	2.82	0.39	0.77
392	♀ ad.	......do..................	Aug. 18	....do..........	2.63	2.75	0.35	0.74
749	....	......do..................	....	....do..........	2.79	2.95	0.43	0.79

67. *Poospiza belli*, (Cassin), var. *navadensis*, Ridgw.—Artemisia Sparrow.

This well-marked variety of the Bell's Finch is found throughout the middle region' being limited in its westward extension by the Sierra Nevada, on the eastern slope of which it was found by Mr. Ridgway.

Though, in the strict meaning of the word, not a migratory species, these Sparrows do yet wander in the fall and winter to very considerable distances. As it is of a hardy nature, these journeyings are undertaken more in quest of food than through the exigencies of climate; though, doubtless, both causes are, to some extent, operative.

It is hence the less surprising that this species should cross the range and be found in the winter on the ground occupied in summer by the other variety alone. At Kernville, I took a single individual, October 28, and saw others.

No.	Sex.	Locality.	Date.	Collector.	Wing.	Tail.	Bill.	Tarsus.
636	♂ jun.	Near Kernville, Cal.........	Oct. 28	H. W. Henshaw...	3.13	3.04	0.43	0.86

68. *Spizella socialis,* (Wils.), var. *arizonæ,* Cones.—Western Chipping Sparrow.

According to Dr. Cooper, the Chipping Sparrow is an abundant bird in Northern California, and according to our observations it is pretty well diffused too in the Southern half of the State. The young and old were seen in great numbers in the mountains in the vicinity of Fort Tejon in early August. This species was also seen in June on the island of Santa Cruz.

No.	Sex.	Locality.	Date.	Collector.
141	♂ jun.	Santa Barbara, Cal...	June 29	H. W. Henshaw.
269	♂ jun.	Mountains near Fort Tejon, Cal	Aug. 2	Do.
270	♀ jun.	......do	Aug. 2	Do.
271	♀ ad.	......do	Aug. 2	Do.
487	♀ ad.	Mount Whitney, Cal...	Sept. 19	Do.

69. *Spizella breweri,* Cass.—Brewer's Sparrow.

Of the distinctness of this species from the *S. pallida* I am well assured, believing that the differences seen in the plumage, which are perfectly appreciable and always constant, the different character of songs and habits, and the totally different habitats of the two are points of distinction too great to be reconciled on the assumption of a mere varietal difference. No intergradation between the two has ever been attempted to be proven, their sameness specifically having apparently been taken for granted on the strength of the superficial resemblance of the two birds.

The mountainous country adjoining Fort Tejon was the only locality where this sparrow was found by our parties. It was here rather numerous in August, and I am inclined to believe that those seen here were summer residents. They perhaps winter in the extreme southern portion of the State.

No.	Sex.	Locality.	Date.	Collector.	Wing.	Tail.	Bill.	Tarsus.
413	♂ ad.	Mountains near Fort Tejon, Cal	Aug. 19	H. W. Henshaw.	2.44	2.73	0.33	0.67
476	♀	......do	Aug. 14	....do	2.40	......	0.35	0.68
772	Jun.	......do	Aug. 14	....do	2.43	2.63	0.34	0.67

70. *Melospiza melodia,* (Wils.), var. *heermanni,* Bd.—Heerman's Song Sparrow

This Song Sparrow is found all over the southern half of California, and like its allies is, wherever found, an abundant species.

They like best the vicinity of water, and will always be found in the thickets of the small streams, preferring, however, not to follow these upward as they course down from the high mountains, but keep pretty exclusively in the low altitudes. Precisely like their relative in the East, they are always to be seen in the cultivated fields of the farmer, and build even in the hedgerows that surround the houses in the outskirts of the cities. In short, the bird is almost an exact reflection of the Eastern Song Sparrow. Their songs, however, while in general style similar to that bird, are very readily distinguished. Their tones are deeper, the songs longer, and of a much more varied character than the monotonous ditties of the eastern *Melospiza.*

They were quite numerous about San Francisco in November, and I presume they are permanent residents of the same locality throughout the year.

On the borders of Kern Lake, these Sparrows were found in the swamps of Tulle Rushes, their only companions being the Rails and Marsh Wrens.

This was the only *Melospiza* seen by us in the south of the State, and is *par excellence* the Californian Song Sparrow.

No.	Sex.	Locality.	Date.	Collector.	Wing.	Tail.	Bill.	Tarsus.
12	♂ ad.	Santa Cruz Island, Cal......	June 10	H. W. Henshaw.	2.45	2.62	0.47	0.87
13	♀ ad.	......do......	June 10	....do............	2.48	2.72	0.46	0.85
14	♂ ad.	......do......	June 10	... do	2.37	2.52	0.48	0.83
68	♂	Santa Barbara, Cal.........	June 26	....do............	2.53	2.70	0.48	0.90
423	♂	Walker's Basin, Cal.........	Aug. 27	....do............	2.43	2.67	0.47	0.84
693	♂ ad.	......do......	Nov. 10	....do............	2.65	2.97	0.47	0.92
711	♂	......do......	Nov. 11	....do............	2.56	2.77	0.43	0.88
712	♂	......do......	Nov. 11	....do............	2.53	2.67	0.43	0.80
59	♀ jun.	Santa Barbara, Cal.........	June 25	....do............				
59	♀ jun.	......do......	June 25	....do............				
60	♀ jun.	......do......	June 25	....do............				
61	♂ jun.	......do......	June 25	....do............				
69	♀ jun.	......do......	June 26	... do				
77	♀ jun.	......do......	June 26	... do				
107	♂ jun.	......do......	June 27	....do............				
145	♂ jun.	......do......	June 29	....do............				
155	♂ jun.	......do......	June 29	....do............				
357	♂ jun.	Kern Lake, Cal	Aug. 15	....do............				
358	♂ jun.	......do......	Aug. 15	....do............				
359	♀ jun.	......do......	Aug. 15	....do............				
359 A	♀ jun.	......do......	Aug. 15	....do............				
361	♂ jun.	......do......	Aug. 15	....do............				
362	♀ jun.	......do......	Aug. 15	....do............				
365	♂ jun.	......do......	Aug. 15	....do............				

71. *Peucæa ruficeps*, (Cass.).—Red-capped Finch.

I notice this species here merely to call attention to the negative evidence afforded by its entire absence from our collections of the past season. The original specimen came from California, and the species has since been obtained by one or two collectors only, principally by Heerman, who found it abundant near the Calaveras River. Dr. Cooper refers to the species as inhabiting the Catalina Island, where he saw a few. Though I searched carefully for this bird in localities exactly similar to those which were always inhabited by the closely-allied variety (var. *boucardi*) in Arizona, I did not succeed in finding a single individual.

It is certainly not a widely-distributed species, and is probably quite rare. I have recently been informed that Mr. Allen, of Marin County, just north of San Francisco, has found this bird breeding in his locality.

72. *Passerella townsendi*, (Aud.).—Townsend's Sparrow.

Early in October the mountains in the vicinity of Mount Whitney began to be thronged with these birds, strangers from the far north, and now the chaparral and thickets on the steep mountain-sides, as well as the bushy ravines, were crowded by their numbers. From their abundance as far south as Caliente, I should suppose they spread over quite the entire southern part of the State. Lower than an elevation of 5,000 feet I did not find them. In spring they retire to more northern parts to breed, none being known to remain in the State.

No.	Sex.	Locality.	Date.	Collector.	Wing.	Tail.	Bill.	Tarsus.	Depth of bill.
596	♀	Mount Whitney, Cal....	Oct. 7	H.W. Henshaw.	3.12	3.05	0.48	0.97	0.38
527	......	Kern River, Cal.......	Oct. 7	....do	3.17	3.27	0.50	0.90	0.34
534	♂	Mount Whitney, Cal....	Oct. 9	....do	3.33	3.22	0.50	0.93	0.36
541	♂	......do	Oct. 10	....do	3.23	3.12	0.50	0.97	0.36
562	♂	......do	Oct. 11	....do :...........	3.19	3.10	0.48	0.93	0.36
644	♀	Walker's Basin, Cal.....	Nov. 5	....do	3.12	2.95	0.48	0.98	0.37
645	♂	......do	Nov. 5	....do	3.33	3.27	0.50	0.95	0.36
646	♀	......do ....`....	Nov. 5	....do	3.26	3.30	0.52	0.95	0.37
665	♀	......do	Nov. 9	....do	3.08	2.98	0.46	0.88	0.34
666	♂	......do	Nov. 9	....do	3.25	3.30	0.50	0.93	0.38
690	♂	......do	Nov. 10	....do	3.10	3.15	0.48	0.93	0.37
691	♂	......do	Nov. 10	....do	3.15	3.16	0.45	0.92	0.33
707	♀	......do	Nov. 11	....do	3.28	3.26	0.51	0.90	0.36
708	♀	......do	Nov. 11	....do	3.30	3.24	0.48	0.93	0.34
709	♂	......do	Nov. 11	....do	3.38	3.29	0.51	1.02	0.39
667		......do	Nov. 9	....do	3.18	3.12	0.48	0.97	0.35

73. Passerella schistacea, Baird, var. megarynchus, Baird.—Thick-billed Sparrow.

Passerella schistacea, Baird, B. N. A., 1858, 490 (only in part) (Fort Tejon).
Passerella megarynchus, Coop., B. Cal., i. 1870, 231 (Fort Tejon and northward).
Passerella townsendi var. *schistacea*, Couee, Key N. A. B., 1872, 352 (includes this form).
Passerella townsendi var. *megarynchus*, B., B., & R., N. A. B., ii, 1874, 57, pl. 22, f. 10.

Of the four species or varieties of *Passerella*, the present bird is the most remarkable of all. In coloration it approaches, most closely to the form of the northern middle region, *P. schistacea*, from which indeed it differs but little, if color alone be taken as a test. It has the same slate-gray, perhaps slightly darker, as the prevailing tint, contrasted on the wings and upper coverts with brownish rufous. It has associated with an unusual development of the hind claw, an increased size of bill, paralleled perhaps in no other case. This is so thick as to appear actually deformed. In the large series of the preceding bird collected there is no approach to this form in the size of these parts, while the type of coloration peculiar to either is always perfectly tangible and well preserved. Besides being actually larger, the relative proportions of wing and tail are very different. In the present bird, as in the *schistacea*, the tail is very much longer than the wing. In *townsendi* the tail is usually the shorter, sometimes, however, equaling the wing. I have, therefore, thought best to consider *P. schistacea* as distinct from either *townsendi* or *iliacus*, assigning to it as a local variety *megarynchus*, which agrees with it in color and proportions. The relationship of the other two is probably similarly intimate.

The Thick-billed Sparrow appears to be quite confined to California, where it is an exclusive inhabitant of the mountains, chiefly in the middle and southern parts.

Mr. Ridgway found it abundant at Carson City, on the eastern slope of the Sierras, which is the northermost locality recorded. In the mountains about Fort Tejon it was numerous enough in the month of August, but from its habits it was difficult to become very familiar with it, or even to procure specimens. Besides being of a naturally timid disposition it was only found in the chaparral, which was here composed chiefly of oak scrub; I did not find them lower down than about 5,000 feet.

When found feeding upon the ground on the outskirts of the thickets, they threw themselves with a peculiar loud sharp chirp into the undergrowth, and usually resisted all attempts to dislodge them by keeping in the thickest parts low down among the roots, and only flying, when absolutely compelled, to the next hiding-place.

No.	Sex.	Locality.	Date.	Collector.	Wing.	Tail.	Bill.	Tarsus.	Depth of bill.
371	♂ ad.	Mountains near Tejon, Cal.	Aug. 17	H. W. Henshaw	3.42	3.96	0.63	0.98	0.52
404	♂ ad.	....do	Aug. 19	....do	3.35	3.83	0.64	0.98	0.52
405	♀ jun.	....do	Aug. 19	....do					
290	♂ jun.	....do	Aug. 3	....do					

74. Guiraca melanocephalus, (Swains.).—Black-headed Grosbeak.

This species, common throughout the middle region, is no less so on the Pacific slope. It occurs in all portions of California. During the summer it is rather partial to mountainous retreats, where it is found as often in the pine region as elsewhere; but it also graces the lower regions, and is found in the low valleys coming often about the houses.

Its song is the most interesting part of its history, and in its melody this species is excelled by very few others.

No.	Sex.	Locality.	Date.	Collector.
96	♂ ad.	Santa Barbara, Cal...	June 27	H. W. Henshaw.
25	♂ jun.	Los Angeles, Cal..	June 29	Do.
133	♂ ad.	Santa Barbara, Cal...	June 17	Do.

75. Guiraca cœrulea, (Linn.).—Blue Grosbeak.

Though quite southern in its distribution, this Grosbeak appears to reach much farther north on the Pacific coast than in the interior, and Dr. Newberry has reported it from the extreme northern part of the State. We met with it at several places in Southern California, where it is pretty well diffused. It is never, I believe, found in the mountains, but inhabits the warm, sheltered valleys.

No.	Sex.	Locality.	Date.	Collector.
98	♀ ad.	Los Angeles, Cal..	June 17	H. W. Henshaw.
309	♀ jun.	Fort Tejon, Cal...	Aug. 7	Do.
310	♀ jun.	.....do	Aug. 7	Do.
332	♀ jun.	....do	Aug. 9	Do.
333	♀ jun.	....do	Aug. 9	Do.
425	♂ jun.	Walker's Basin	Aug. 27	Do.

76. *Cyanospiza amœna*, (Say).—Lazuli Finch.

This Finch, so much like the Indigo-bird in voice and habits, entirely replaces that species in the far west. Its organization seems to unfit it for a residence in high latitudes, and it also shuns the bracing air of the mountains, not occurring, according to Mr. Trippe, higher than 8,000 feet, an altitude at which I have never seen it. It is found in great abundance in the sheltered valleys, living for the most part along the streams, but at any rate the locality chosen must be more or less grown up to brush and bushes, among which it places its nest and spends the greater part of the time.

77. *Pipilo maculatus*, (Swains.), var. *megalonyx*, Bd.—Long-spurred Towhee.

This *Pipilo* is spread in great numbers over the southern half of California, ranging from the shrubbery of the lowlands well up on the mountains. On Santa Cruz Island it was one of the most numerously represented species; indeed, the surface of this island, broken and cut up in every direction by ridges and corresponding ravines, and everywhere covered with chaparral, forms just the abode suited to the habits of this bird. Accordingly, I think I never saw in a limited area such numbers of these birds, their mewing calls sounding in all directions. They are probably resident in Southern California, where, too, their numbers in fall are still further swelled by additions from more inclement regions farther north.

No.	Sex.	Locality.	Date.	Collector.
256	♂ jun.	Fort Tejon, Cal	July 27	H. W. Henshaw.
267	♀ ad.	Tejon Mountains, Cal..	Aug. —	Do.
234	♂	......do	Aug. 17	Do.
370	♂ ad.	......do	Aug. 17	Do.
375	♂	......do	Aug. 17	Do.
397	♂ jun.	......do	Aug. 18	Do.
398	♂ jun.	......do	Aug. 19	Do.
406	♂	......do	—	Do.
647	♀	Walker's Basin, Cal..	Nov. 5	Do.
692	♀	......do	Nov. 10	Do.
743	♂ jun.	Tejon Mountains, Cal..	Aug. 2	Do.
744	Jan.	......do	Aug. —	Do.

78. *Pipilo fuscus*, Swains., var. *crissalis*, Ridgw.—Brown Finch.

This Finch was found by our parties in great abundance from San Francisco southward. The appellation Cañon Finch is not a very happy one, since it would suggest a preference for the rocky cañons, an inference by no means borne out by the habits of the bird. It is indeed an inhabitant of the mountains, being there, however, partial to the open thickets on the slopes, rather than to the recesses of the ravines. Moreover, it is found in much greater numbers in the level country and low valleys. In essential particulars it is a true *Pipilo*, having many of the habits common to the birds of this family, but especially resembles the var. *mesolencus* from the southern interior region, its mode of life being indeed almost identical with that of this bird, except in so far as it has been modified to suit the somewhat different nature of the region it inhabits. It is never found far from cover, though venturing into the open oftener and to a greater distance than is the case with the shyer, more retiring, black *Pipilos*. Its whole nature seems to be more reliant, and in some places I have seen them venturing to the very door of the houses, and hopping with the utmost freedom about the yards, picking up crumbs, in company, perhaps, with their smaller friends, the Snowbirds. Their flight is better sustained and less "jerky" than most of the family, and is not so very unlike that of the Sickle-billed Thrush but that, when taken in connection with its large size, colors, and its long tail, it may often mislead one as it goes flirting through the foliage. When one comes upon them suddenly they throw themselves into the nearest clump with all haste, but should a convenient tree be at hand they will quickly be seen among the branches, where mounting to some convenient perch they sit and watch the cause of all the trouble, the various individuals meanwhile responding to each others' calls by constant sharp chirps. For birds of this group they are more than usually gregarious. Through the summer each family maintains a close connection. In fall, their wanderings begin, and then they come together in large companies, the numbers being still further augmented by the addition of other species, as the Snow-birds and *Zonotrichias*, the whole forming a merry and united flock. Two broods are reared in a season. At Santa Barbara in June the young were very numerous, while I took the young still in nesting-plumage as late as August 10. I found the species in November about San Francisco, and they doubtless winter here.

No.	Sex.	Locality.	Date.	Collector.	Wing.	Tail.	Bill.	Tarsus.
84	♂ jun.	Santa Barbara, Cal.............	June 26	H. W. Henshaw.	......	......	......	......
85	♂ jun.	......do	June 26	....do	......	......	......	......
86	♀ jun.	......do	June 26	....do	......	......	......	......
128	♂ jun.	......do	June 22	....do	......	......	......	......
351	♂ jun.	......do	Aug. 10	....do	......	......	......	......
47	♂ ad.	......do	June 25	....do	3.74	4.37	0.60	1.04
422	♂	Walker's Basin, Cal.........	Aug. 27	....do	3.84	4.68	0.63	1.05
570	♂	Near Kernville, Cal.........	Oct. 16	....do	4.05	4.95	0.67	1.17
571	♀	......do	Oct. 16	....do	3.55	4.55	0.58	1.04
572	♀	......do	Oct. 16	....do	3.90	4.89	0.63	1.12
634	♂	......do	Oct. 28	....do	3.97	4.83	0.55	1.10
648	♀	Walker's Basin, Cal	Nov. 5	....do	3.67	4.57	0.58	1.07
649	♂	......do	Nov. 5	....do	3.68	4.67	0.58	1.08
710	♀	......do	Nov. 11	....do	3.78	4.52	0.59	1.05

79. *Pipilo chlorurus*, (Towns.).—Green-tailed Finch.

The present bird appears to be by no means as common in Southern California as throughout the interior country. It is pre-eminently a mountain-loving species, and in California I did not find it lower than 5,000 feet. At this elevation it was breeding in the mountains near Fort Tejon; the young, perhaps of a second brood, being taken August 1. It inhabits the tangled brakes and thickets nearly always close to the streams.

No.	Sex.	Locality.	Date.	Collector.
771	♂ jun.	Tejon Mountains, Cal......	Aug. 1	H. W. Henshaw.
408	♂	......do	Aug. 19	Do.
409	♂ jun.	......do	Aug. 19	Do.
444	♀	Near Mount Whitney, Cal......................	Sept. 1	Do.

ALAUDIDÆ.—LARKS.

80. *Eremophila alpestris*, (Forst.), var. *chrysolæma*, Wagl.—Southern Horned Larks.

The small bright-colored race of the Horned Lark is a common summer resident along the coast of Southern California, and is found, too, at this season, according to Dr. Cooper, as far north as Puget Sound. In certain parts of the island of Santa Cruz, it was very numerous in June, as well, too, as along the adjoining shore of the mainland at Santa Barbara. The immense flocks of these birds that gather together in the fall are well known, but I was surprised to find to what extent this sociable feeling was carried during the breeding-season. Both on the mainland and on the island they were seen all through June in scattered flocks of both sexes, though nearly all, perhaps all, were at this time nesting. Both sexes incubate, and it appeared to be the habit of the birds when off duty to repair together in small flocks, and thus to wander in search of food. At this season they do not resort much to the sandy beaches, but keep on the upland, where among the herbage they find more easily, and in greater abundance, the insects and seeds which they are fond of. Their time of breeding must be quite irregular, as I found a fully-fledged young one June 1, though after this I took two nests, with fresh eggs, and the greater number, I am persuaded, still had eggs.

The nests were but rude attempts, being nothing more than a small pile of dried grasses, sufficiently hollowed to admit the reception of the eggs. One is deserving of notice as being placed within the cavity of an "abalone" shell, one of a large heap, lying half overgrown with herbage. The whole cavity of the shell was filled by the material, and the eggs looked prettily enough as they lay contrasted with the shiny, pearly shells clustered about them. The eggs have a grayish-white background, spotted quite uniformly with fleckings of reddish brown. In one set of these, the background is almost obscured by the markings, which are aggregated together in blotches.

Two sets measured 0.86 × 1.81 — 0.85 × 0.63 — 0.88 × 0.63; 0.80 × 0.63 — 0.80 × 0.63 — 0.77 × 0.62.

ICTERIDÆ.—ORIOLES.

81. *Agelæus phœniceus,* (Linn.), var. *gubernator,* (Wagl.).—Red-shouldered Blackbird.

Psarocolius gubernator, Wagl., Isis, 1832, 281.
Agelaius gubernator, Woodh., Sitgr. Exp., 1853, 89 (California).—Newb., P. R. R. Rep., vi, 1857, 86 (California).—Bd., B. N. A., 1858, 529.—Coop., B., Cal., i, 1870, 263.—Bendire, Proc. Bost. Soc. Nat. Hist., vol. xviii, 1815, 158.
Agelaius phœniceus var. *gubernator,* Coues, Key N. A. B., 1872, 156.—Bd., B., & R., N. A. B., ii, 1874, 163.

Though in perfectly adult plumage easily distinguishable from *phœniceus,* this bird is very closely allied to that species, but may perhaps properly be set apart from it as its western varietal form. It occurs throughout California, being, however, according to Dr. Cooper, chiefly a bird of the warm interior. I saw these birds in but few instances, and had no opportunity to observe their habits, which, however, according to other observers, are quite identical with those of the Eastern Red-wing.

So far as I am aware, no specimens of *A. phœniceus* of unquestioned identity have ever been taken in California, and I am led to believe that this bird does not occur there at all. The immature stages of *A. gubernator* are so much like the corresponding conditions of *phœniceus* that they may readily be mistaken, the one for the other, and in this way *A. phœniceus* has erroneously been ascribed to California.

No.	Sex.	Locality.	Date.	Collector.
31	♂ ad.	Los Angeles, Cal...	June 18	H. W. Henshaw.
366	♀ ad.	Fort Tejon, Cal	Aug. 16	....do ....

82. *Agelæus tricolor,* Nutt.

Icterus tricolor, Nutt. Man., i, 2d ed., 1840, 186.
Agelæus tricolor, Newb., P. R. R. Rep., vi, 1857, 86.—Bd., B. N. A., 1858, 530.—Xantus, Proc. Phila. Acad. Nat. Sci., 1859, 192 (Fort Tejon, Cal.).—Coop., B. Cal., i, 1870, 265.—Bd., B., & R., N. A. B., ii, 1874, 165.
Agelæus phœniceus var. *tricolor,* Coues, Key N. A. B., 1872, 156.

The isolation of this form from its allies seems to be warrantable in view of the tangible differences that distinguish them in all stages, especially when taken in connection with the different habits and notes which most observers have remarked.

The species is quite strictly confined to California, possibly reaching on the north into Southern Oregon. In the southern portion of the State it is resident both on the sea-coast and in the interior. I found the species breeding in but one locality, in Santa Clara Valley, June 21. Noticing large numbers of Blackbirds flying across the road and into an adjoining pasture, I followed their flight till I found myself before a patch of nettles and briers that must have covered three or four acres. The place was not at all swampy, but was a dry pasture, differing in this respect entirely from the breeding-places selected by the Red-wings in the East. I noticed that each bird as it darted down into the clump bore in its bill a large object, which I subsequently found to be grass-hoppers. The cause of their journeyings was then explained. They had found some spot where these insects were very numerous, and back and forth they poured all day long, bringing in their bills all they could carry for their hungry young. The nettles grew so dense and high, some attaining to 12 feet, that I found it almost impossible to force my way into their midst, nor did I succeed in penetrating beyond a few yards. I speak within bounds when I say that two hundred pairs had here congregated to rear their young, and the odor arising from some portions was almost as strong as from the Cormorant rookeries. The nests were there by hundreds, nearly every bush holding several. They were, however, mostly old, showing that the place had served for a breeding-resort for probably many years. A few of the nests were this season's and contained young, none that I saw having eggs, though could I have extended my search some would doubtless have been found. The nests were rather slight, flimsy structures, but in general resembled those made by the Red-wing of the East, and were fastened on the bushes in the same way. My presence among them created a great disturbance, and the trees were soon covered with the parent birds, one and all resenting this intrusion on their old-time possessions in no gentle tones. A few days later I came across an immense flock of young birds in the streaked nesting-plumage. Able to take care of themselves, they had gathered thus together, and fairly covered several small trees by the roadside. In all the number there was not a single adult bird. Later still, July 6, a similar flock was found at Santa Barbara, the young having parted from the old birds and made an independent party. Possibly this early separation is due to the fact that the parents, having seen their charges fairly fledged and able to shift for themselves, shook them off and busied themselves with preparations for a second brood.

Heermann notes another very similar breeding-ground in the north of California;

so that it may be the regular habit of the species to thus gather together into rookeries. In fall, the white bordering the Red-wing patch changes to pale buff, being then precisely as seen in southern examples of *A. phœniceus*. The red, however, is of a totally different hue, being many shades darker. The black is of a brilliant metallic luster, very much as in the *Scolecophagus cyanocephalus*, never dull, as in *phœniceus*. The female and young are readily distinguishable from that species.

The Yellow-headed Blackbird is, according to Dr. Cooper, a common resident of the warm valleys of the interior of the State. The species was, however, not met with by us, owing to the fact that scarcely a locality was visited which would meet the necessities of their mode of life.

No.	Sex.	Locality.	Date.	Collector.	Wing.	Tail.	Bill.	Tarsus.
33	♂ ad.	Santa Clara Valley, Cal	June 22	H. W. Henshaw	4.83	2.72	0.92	1.19
34	♂ ad.	......do	June 22	...do	4.70	3.66	0.92	1.23
35	♂ ad.	......do	June 22	...do	4.90	3.16	0.75	1.03
37	♂ ad.	......do	June 22	...do	4.93	3.79	0.92	1.17
363	♂ ad.	......do	Aug. 16	...do	4.58	3.76	0.86	1.22
364	♂ ad.	......do	Aug. 16	...do	4.80	3.77	0.88	1.22
636	♂ ad.	......do	Oct. 25	...do	4.79	3.74	0.90	1.13
745	♂ jun.	......do	Aug. —	...do	4.58	3.62	0.88	1.23
189	♂ jun.	Santa Barbara, Cal	July 6	...do				
190	♀ jun.	......do	July 6	...do				
191	♂ jun.	......do	July 6	...do				
192	♂ jun.	......do	July 6	...do				
193	♂ jun.	......do	July 6	...do				
194	♂ jun.	......do	July 6	...do				
195	♀ jun.	......do	July 6	...do				
196	♀ jun.	......do	July 6	...de				
197	♀ jun.	......do	July 6	...do				
198	♀ jun.	......do	July 6	...do				
199	♂ jun.	......do	July 6	...do				
200	♂ jun.	......do	July 6	...do				
201	♀ jun.	......do	July 6	...do				

83. *Sturnella magna*, (Linn.), var. *neglecta*, Aud.—Western Meadow-lark.

The distribution of the Lark in California agrees with its general dispersion over the West. It is found in the fertile valleys and on the plains, even when the latter are dry. In summer it is more restricted to the meadowy lands where there is herbage sufficient to serve as a cover for its nest. I was much surprised to see a pair upon the island of Santa Cruz, this being about as unpromising a locality for birds of their habits as could well be imagined. The single couple had taken up their abode in a small garden, the green things in which were nourished by a small rivulet of water, and their nest had probably been made in a patch of grain, which they frequented much of the time.

No.	Sex.	Locality.	Date.	Collector.
21	♂ ad.	Santa Cruz Island, Cal	June 10	H. W. Henshaw
172	♂ ad.	Santa Barbara, Cal	July 1	......do
173	♀ jun.	......do	July 1	......do
490	♂	Walker's Basin, Cal	Aug. 27	......do
502	♂	Near Mount Whitney, Cal	Sept. 26	......do
582	♂ ad.	Near Kernville, Cal	Oct. 21	......do
583	♂ ad.	......do	Oct. 20	......do
628	♂ ad.	......do	Oct. 27	......do
629	♀	......do	Oct. 25	......do

84. *Icterus bullocki*, (Swains.).—Bullock's Oriole.

This Oriole occupies in the West the same place so conspicuously filled by the well-known Baltimore in the East. It comes freely into the precincts of village and city, suspending its nest from the swaying limbs of the shade-trees. It was very common about Los Angeles in June, and probably had young at that time.

No.	Sex.	Locality.	Date.	Collector.
260	♀ jun.	Fort Tejon, Cal	Aug. 27	H. W. Henshaw

85. *Scolecophagus cyanocephalus*, (Wagl.).—Brewer's Blackbird.

A very abundant species throughout the State and a constant resident. In summer they prefer the neighborhood of the streams to the marshes proper, though found in the latter in company with the Tri-colored Blackbirds, there existing between the two species an unusually close intimacy. Many of these birds were breeding in company with a large colony of the *A. tricolor* before mentioned.

No.	Sex.	Locality.	Date.	Collector.
390	♀	Fort Tejon, Cal	Aug. 8	H. W. Henshaw...
627	♀ ad.	Near Kernville, Cal	Oct. 27	......do ....

CORVIDÆ.—CROWS.

86. *Corvus corax*, Linn., var. *carnivorus*, Baxtr.—American Raven.

The Raven is an abundant resident in California, and is found without much reference to locality. Its omnivorous tastes and its great usefulness as a scavenger are well known. I saw Ravens occasionally on Santa Cruz Island, and, on inquiry, learned that they were no favorites with the sheep-raisers here, on account of their habit of occasionally destroying the lambs. Captain Forney informed me that he had been an eye-witness to the destruction of a lamb by one of these birds, the attack being made first upon the eyes, which were torn out. This habit of the Raven, he states, was well known to the shepherds.

87. *Corvus caurinus*, (Bd.).—Western Fish-Crow.

Corvus caurinus, Bd., B. N. A., 1858, 569.—Coop. & Suckl., P. R. R. Rep., vol. xii, pt. ii, 1860, 211.—Dall & Bann., Tr. Chic. Acad., i, 1869, 286.—Coop., B. Cal., i, 1870, 285.—B., B., & R., N. A. B., ii, 1874, 248.—Bendire, Proc. Bost. Soc. Nat. Hist., vol. xviii, 1876, 159 (Camp Harney, Oreg.).
Corvus americanus var. *caurinus*, Coues, Key N. A. B., 1872, 163.
Corvus ossifragus, Newb., P. R. R. Rep., vi, 1857, 83.

In the uncertainty respecting the relations of this bird, I am disposed to keep it apart from the *Corvus americanus*, with which it has been associated as a variety by some writers, till its relationship be established on a firmer basis than at present. It appears to be mainly distinguishable from its smaller size and certain apparent differences of habits. I regret I can add so little to our knowledge of the subject. On the road from Los Angeles to Santa Barbara, these Crows were seen on several occasions, always in large flocks, and at a distance from the coast of from 5 to 15 miles. In fact, in Southern California, the species does not appear to be specially all maritime in its habits, if, indeed, it is so to more than a moderate extent. In its northern home, however, on Puget Sound and elsewhere, it is essentially a bird of the coast, living there upon shell-fish and the refuse cast up by the waves.

In my own brief experience in California I saw nothing in their manner incompatible with the normal habits of the Common Crow. In this respect, however, it is not different from the Fish-Crow (*ossifragus*) of the Gulf States, which, save in its maritime proclivities, presents little to distinguish its habits from those of the Common Crow; yet the Fish Crow in Florida is found very often miles away from the coast, while not infrequently I have there seen the *Corvus americanus* associated with it in its excursions along shore. The truth seems to be that with birds possessing the omnivorous tastes of the Crows, it is the quantity and ease with which food is obtained that directs their choice more than anything else. Hence, about Puget Sound and this region generally, as in the warm waters of Florida, where mollusks and crustaceans exist in greatest abundance, the habit of resorting to the shores for the chief part of their living has become a fixed one, while elsewhere they find it easier to obtain their food from the interior.

The notes of *caurinus*, as I heard them in California, were different from those of the *Corvus americanus*, and I should say they resemble very closely those of the true Fish Crow. Certainly, no one hearing their hoarse calls could for a moment mistake them for the Common Crow. Like the Fish Crow, the *C. caurinus* keeps very much in flocks, and it is said to even build in communities.

No.	Sex.	Locality.	Date.	Collector.
186	♂ jun.	Santa Barbara, Cal	July 5	H. W. Henshaw.

83. *Picicorvus columbianus*, (Wils.).—Clarke's Nutcracker.

During the month of September, this curious bird was met with in great numbers, and, according to its usual habit, in large flocks in the high sierras, where it kept entirely among the yellow pines. These were hanging full of seeds, and to extract these from the cones was their chief, indeed their only, occupation. Their loud, shrill cries went echoing through the deep woods, as they flew about in noisy bands intent only on cramming their stomachs. The seeds are obtained with much ease and dexterity, as the birds hang back downward, clinging to the ends of the branches or to the cones themselves. A seed fairly extracted, it is taken to a horizontal limb of some size, and there the covering shelled off by a few sharp blows with their heavy bills, when it is quickly disposed of.

The *Gymnokitta cyanocephala* was not noted in any part of the region traversed by the Survey. Though recorded from California, it does not appear to be a common bird on the Pacific slope, and may perhaps be wanting in the more southern parts of the State.

No.	Sex.	Locality. .	Date.	Collector.
734	♂	Tejon Mountains, Cal..	Aug. 2	H. W. Henshaw.

89. *Pica melanoleuca*, (Linn.), var. *nuttalli*.—Yellow-billed Magpie.

Pica nuttalli, Aud., Orn. Biog., iv, 1838, 450, pl. 362.—Woodh., Sitgr., Exp. Zuñi & Col. Riv., 1854, 77.—Newb., P. R. R. Rep., vi, 1857, 84.—Bd., B. N. A., 1858, 578.—Heerm., P. R. R Rep., x, 1859, pt. vi, 54.—Coop., B. Cal., i, 1870, 295.
Pica melanoleuca var. *nuttalli*, Coues, Key N. A. B., 1872, 164.
Pica caudata var. *nuttalli*, B., B., & R., N. A. B., ii, 1874, 270.

This is the form prevailing in all the region west of the Sierras. They inhabit the valleys, being rather partial to a rough broken surface, interspersed with groves of oaks. I saw many of their nests placed in these. Like their relative from the interior, anything edible suits their appetite, though, like them, flesh is preferred to almost everything else. They are thus, with the Ravens, very useful as scavengers, and, having found the body of a dead animal, never leave the vicinity till the bones and skin alone remain. In the Sierras proper we did not meet with these birds, but in various parts near the sea-coast they were very numerous.

No.	Sex.	Locality.	Date.	Collector.	Wing.	Tail.	Bill.	Tarsus.
32	♂ jun.	Santa Clara Valley, Cal ...	June 22	H. W. Henshaw .	7. 50	9. 50	1. 27	1. 38
36	♂ jun.	......do	June 22	....do	7. 35	8. 15	1. 35	1. 82
134	♀ jun.	Santa Barbara, Cal........	June 29	....do	7. 15	8. 65	1. 21	1. 79
182	♀ jun.	......do	July 5	....do	7. 30	9. 50	1. 32	1. 85
183	♀ jun.	......do	July 5	....do	7. 15	8. 30	1. 16	1. 76
184	♂ jun.	......do	July 5	....do	....	....	....	....
185	♂ jun.	......do	July 5	....do	....	....	....	....

90. *Cyanura stelleri*, (Gm.), var. *frontalis*, Ridgw.—Steller's Jay; Blue-fronted Jay.

Cyanocitta stelleri, Newb., P. R. R. Rep., vi, 1857, 85.—Bd., B. N. A., 1858, 581 (includes var. *frontalis*.)—Xantus, Proc. Phila. Acad. Nat. Sci., 1859, 192.—Coop. & Suckl., P. R. R. Rep., vol. xii, pt. ii, 1860, 215.—Coop., B. Cal., i, 1870, 298 (includes var. *frontalis*).—Coues, Key N. A. B., 1872, 165 (var. *frontalis* also).—Nelson, Proc. Bost. Soc. Nat. Hist., vol. xvii, 360 (California).—Bendire, Proc. Boston Soc. Nat. Hist., vol. xviii, 1875, 160 (Camp Harney, Oregon; probably var. *frontalis*).
Cyanura stelleri var. *frontalis*, B., B., & R., N. A. B., ii, 1874, 279.

This Jay is a common inhabitant of the mountains throughout California, rarely being seen in summer below 5,000 feet, and extending from about that point to the very limit of the timber-line. During the breeding-season, they separate into pairs, and are then very silent and retiring. After the broods are out and well on the wing, they begin their roving, independent life, and their enforced silence gives way to their more usual frame of mind, when noisy outpourings herald their presence in every direction. The bird is a true resident of the pine-woods, and from the pines is had no small part of its subsistence.

No.	Sex.	Locality.	Date.	Collector.	Wing.	Tail.	Bill.	Tarsus.
282	Jun.	Tejon Mountains, Cal	Aug. 3	H. W. Henshaw...				
421	♂ jun.	Walker's Basin, Cal.........	Aug. 27	....do				
469	♂ jun.	Near Mount Whitney, Cal...	Sept. 12	...do				
461	♀	......do............	Sept. 10	...do				
598	♀ ad.	......do............	Oct. 16	...do	5.65	5.25	1.12	1.63
599	♀ jun.	......do............	Oct. 23	...do	5.66	5.71	1.28	1.63
620	♂ ad.	......do............		...do	5.58	5.35	1.26	1.43
621	♂ jun.	Near Kernville, Cal		...do	5.75	5.62	1.33	1.65
					8.62	5.30	1.36	1.63

91. Cyanocitta floridana, (Bartr.), var. californica, (Vigors,).—Californian Ground Jay.

Garrulus californicus, Vigors, Zool Beechey's Voy., 1839, 21, pt. v.
Cyanocitta californica, Newb., P. R. R. Rep., vi, 1857, 85.—Bd., B. N. A., 1858, 584.—Xantus, Proc. Phila. Acad. Nat. Sci., 1859, 192; ibid., 1859, 305 (Cape St. Lucas).—Coop., B. Cal., i, 1870, 302.
Aphelocoma floridana var. *californica*, Coues, Key N. A. B., 1872, 166.—Nelson, Proc. Bost. Soc. Nat. Hist., xvii, 1875, 360 (California).
Cyanocitta floridana var. *californica*, B., B., & R., N. A. B., 1874, ii, 291.

So far as habits are concerned, this bird is simply the Florida Jay transferred from the scrub of that peninsula to the chaparral of California. Its vertical range is exactly complementary to that of the Steller's Jay. It is found from well down in the valleys to a height on the mountains of about 5,000 feet, farther up than which it begins to be rare, while here the other, a true Mountain Jay, begins to put in an appearance.

Its mode of life offers little that is distinctive. When disturbed, it clings to the thickets for protection, and if much alarmed makes off at its best speed under their cover. When slightly startled, its curiosity compels it to linger, and in the unfrequented parts of California, and especially upon the island of Santa Cruz, the report of a gun was not sufficient to excite its fears. In this unsophisticated nature, it is rather peculiar; both the Woodhouse's and Florida Jays possessing their full share of the wariness characteristic of the family.

No.	Sex.	Locality.	Date.	Collector.
19	♀ ad.	Santa Cruz Island, Cal...............	June 10	H. W. Henshaw.
20	♂ ad.	......do	June 11	Do.
731	♀ ad.	......do	June 2	Do.
331	♀ ad.	Fort Tejon, Cal............	Aug. 9	Do.
429	♀ ad.	Walker's Basin, Cal............	Aug. 28	Do.
656	♂	......do	Nov. 5	Do.
657	♀	......do	Nov. 11	Do.
664	♂	Near Kernville, Cal............	Nov. 9	Do.
697	♂	......do	Nov. 10	Do.
698	♀	......do	Nov. 10	Do.
699	♀	......do	Nov. 11	Do.
732	......	......do	Nov. 11	Do.
733	......	......do	Nov. 11	Do.

TYRANNIDÆ.—FLYCATCHERS.

92. Tyrannus verticalis, Say.—Arkansas Flycatcher.

This Flycatcher extends from Kansas, and even farther eastward (Iowa), across the central plains, and so on to the Pacific. South it is found well into Arizona, and north into British Columbia. Over much of this region it is abundant, and it is absent only from the high mountain-ranges. In the southern half of California, it is quite numerous, perhaps as many so as anywhere in its wide habitat. Its habits are the same everywhere. The *Tyrannus vociferans*, I find recorded in my note-book as occurring about Los Angeles in June, but I did not secure any specimens, nor was it seen elsewhere.

According to Dr. Cooper, it is quite common in Southern California, and winters about Los Angeles.

No.	Sex.	Locality.	Date.	Collector.
35	♂ ad.	Los Angeles, Cal............	June 17	H. W. Henshaw.
28	♀ jun.	Santa Barbara, Cal............	June 26	Do.
91	♂ jun.	......do	June 26	Do.
149	♂ jun.	......do	June 29	Do.
268	♂ jun.	Tejon Mountains, Cal............	Aug. 2	Do.

93. *Myiarchus cinerascens*, Lawr.—Ash-throated Flycatcher.

Generally distributed over the southern portion of the State and common, avoiding the heavy timber and the mountains. Habits very similar to those of *crinitus*. The young were fully fledged by the middle of July.

No.	Sex.	Locality.	Date.	Collector.
81	♀ ad.	Santa Barbara, Cal	June 26	H. W. Henshaw.
246	♀ jun.	Ojai Creek, Cal	July 17	Do.
261	♂ jun.	Fort Tejon, Cal	July 26	Do.
262	♂ jun.	......do	July 26	Do.
341	♀ jun.	......do	Aug. 9	Do.
279	♀ jun.	......do	Aug. 2	Do.

94. *Sayornis nigricans*, (Swains.).—Black Flycatcher.

This Flycatcher is quite numerous in summer in California, especially in the southern portion, where its habits and method of nidification recall those of the eastern Phœbe.

No.	Sex.	Locality.	Date.	Collector.
43	♂ ad.	Santa Barbara, Cal	June 24	H. W. Henshaw.
143	♀ ad.	......do	June 29	Do.
263	♂ jun.	Fort Tejon, Cal	July 27	Do.

95. *Sayornis sayus*, (Bon.).—Say's Flycatcher.

The proclivity of this species for a rather northerly habitat is seen in California, where it is not found at all in summer in the southern portion, though possibly it occurs in the more northern parts. In the fall, it makes its appearance south of San Francisco late in September, and remains through the winter.

No.	Sex.	Locality.	Date.	Collector.
583	♀	Near Kernville, Cal	Oct. 20	

96. *Contopus borealis*, (Swains.).—Olive-sided Flycatcher.

Coincident with its diffusion over North America generally, this Flycatcher is found on the Pacific coast, though in Southern California at least it appears to be not so numerous as in the middle region. It is pretty closely confined to the mountains.

No.	Sex.	Locality.	Date.	Collector.
753	♂ ad.	Mountains near Fort Tejon, Cal	Aug. 5	H. W. Henshaw.

97. *Contopis virens*, (Linn.), var. *richardsoni*, Swains.—Short-legged Pewee.

In California, as elsewhere throughout the far west, this Pewee is by far the most numerous of the Flycatchers. It is found in every piece of woodland, though in summer the greater number retire to the depths of the mountains.

No.	Sex.	Locality.	Date.	Collector.
144	♀ ad.	Santa Barbara, Cal	June 29	H. W. Henshaw.
321	♀ ad.	Fort Tejon, Cal	Aug. 7	Do.
354	♂ jun.	Tejon Mountains, Cal	Aug. 10	Do.

98. *Empidonax trailii*, (Aud.), var. *pusillus*, Swains.—Little Flycatcher.

Dr. Cooper was certainly in error in considering the *E. trailii* as identical with the form found in California. This is the same as that occurring in the middle region, known as var. *pusillus*. Specimens from the two regions are indistinguishable. The habits of the bird in the two regions are quite identical. The bird is an abundant one in Southern California, and especially so in the swampy thickets about Los Angeles.

A nearly-completed nest of this bird, which I found June 17, was placed in a crotch of a swinging grape-vine, and its structure was unusually neat and firm for a Flycatcher. A curious departure from the usual method of nidification has just come to my notice. Mr. Allen, of ——, discovered the nest of this species, with eggs, built in the hollow of a tree. The parent was secured, and its identity thus firmly established.

No.	Sex.	Locality.	Date.	Collector.
27	♂ ad.	Los Angeles, Cal	June 17	H. W. Henshaw.
63	♂ ad.	Santa Barbara, Cal	June 25	Do.
79	♀ ad.	......do	June 26	Do.
310	♂ jun.	Fort Tejon, Cal	Aug. —	Do.
364	♂ ad.	Tejon Mountains, Cal	Aug. 17	Do.

99. *Empidonax flaviventri*, Bd., var. *difficilis*, Bd.—Western Yellow-bellied Flycatcher.

A not uncommon summer resident in Southern California. They spend the summer from sea-level up to 7,000 feet, but are most numerous among the mountains.

No.	Sex.	Locality.	Date.	Collector.
78	♂ ad.	Santa Barbara, Cal	June 26	H. W. Henshaw.
154	♂ ad.	......do	June 29	Do.
410	♂ ad.	Tejon Mountains, Cal	Aug. 19	Do.
415	♀ ad.	......do	Aug. 19	Do.
424	♂ jun.	Near Kernville, Cal	Aug. 27	Do.

100. *Empidonax obscurus*, (Swains.).—Wright's Flycatcher.

I saw but a few of this species in the Sierras, near Mount Whitney, in September. One specimen obtained here was in such immature plumage that I think it had been reared in the neighborhood.

No.	Sex.	Locality.	Date.	Collector.
451	♀ jun.	Near Mount Whitney, Cal	Sept. 10	H. W. Henshaw.

101. *Empidonax hammondi*, (Xantus).—Hammond's Flycatcher.

I could find no evidence that this Flycatcher breeds in Southern California, though I am by no means positive that the deep mountains do not afford it a summer home. Dr. Cooper's account of its method of nidification refers with but little doubt to the var. *pusillus*.

After September, the species became a common one in the mountains. It remains till into October, but finally retires farther south.

No.	Sex.	Locality.	Date.	Collector.
450	♂	Near Mount Whitney, Cal	Sept. 10	H. W. Henshaw.
518	♂	......do	Oct. 3	Do.
551	♀	......do	Oct. 11	Do.

ALCEDINIDÆ.—KINGFISHER.

102. *Ceryle alcyon*, (Linn.).—Belted Kingfisher.

Present here in about the usual numbers. Every small stream which is stocked with fish is occupied by one or more of these birds.

CAPRIMULGIDÆ.—GOATSUCKERS.

103. *Chordeiles popetue*, (Vieill.), var. *henryi*, Cass.—Western Night-hawk.

This Hawk is extremely abundant throughout all of the middle region, but appears to be wanting in much of Southern California. We did not meet with the species at all; and, with Dr. Cooper, I am inclined to believe that it is wanting through the Coast range. It is spoken of as quite numerous in the Sacramento Valley in summer by Dr. Newberry, and not unlikely occurs in the interior and western portions of the State at this season.

104. *Antrostomus nuttalli,* (Aud.).—Poorwill.

On the summits of the mountains near Fort Tejon the Poorwills were remarkably numerous, keeping hidden during the day among the dense chaparral, where they crouched so close that I several times almost trod on them ere they took to wing.

No.	Sex.	Locality.	Date.	Collector.
751	♂ jun.	Tejon Mountains, Cal	Aug. 2	H. W. Henshaw.
752	♂ jun.	.......do	Aug. 2	Do.

105. *Chætura vauxii,* Townsend.—Oregon Swift.

A Swift was present in the Tejon Mountains in August, which I believe was this species.

TROCHILIDÆ.—HUMMING-BIRDS.

106. *Stellula calliope,* (Gould).—Calliope Humming-bird.

This species was most unaccountably rare in the mountains of Southern California, and I saw but a single individual in the Tejon Mountains, August 17. Even this may have been a migrant, and the species may not occur at all in summer in Southern California. It is very abundant in the Cascade Mountains in the northwest, where it breeds.

107. *Trochilus alexandri,* Bourcier & Mulsant.—Black-chinned Humming-bird.

This Hummer was not found by our parties very common in any portion of California. They are probably most numerous in the early part of the season, when flowers are most abundant.

No.	Sex.	Locality.	Date.	Collector.
403	♂ pin.	Tejon Mountains, Cal ....:......................	Aug. 19	H. W. Henshaw.

108. *Selasphorus rufus,* (Gmel.).—Rufous-backed Humming-bird.

This species is quite common in summer throughout California, and breeds apparently as commonly in the valleys as in the mountains. It occurs at this season all along the coast as far to the north at least as Sitka. A few probably remain during the winter, in the warm, sheltered valleys of the western part of the State, though the species, as a rule, retires farther south for winter-quarters. In comparing a series of these birds taken in California and to the northward with a full *suite* secured by the expedition in Arizona and New Mexico, I was struck with certain differences in coloration which appeared, and though these, after full consideration, appear of too slight and inconstant a nature to warrant the definition of a varietal form upon them, they are yet of sufficient interest as to be worthy of mention. Briefly, the differences resolve themselves into a somewhat deeper tone of coloration in individuals from the Pacific coast. The slight cinnamon of the interior type becomes, in some specimens, a deep rufous, and in all it is noticeably darker. In them is seen also a tendency to retain the metallic-green color on the dorsal surface, which is common to the females and young, and which, in the interior, is always replaced in the adult males with clear light cinnamon.

In no small proportion of what appear to be adult males from the Pacific coast the metallic green extends from the head entirely over the back and even over the upper tail-coverts, leaving only the tail rufous. Some males, also adult, are found, which have the back of an unmixed rufous, while many occur which exhibit both phases of coloration in varying measure—green mixed with rufous, rufous mixed with green.

As these different conditions may be found in the same locality in California, the impossibility of drawing a varietal line is here seen.

No.	Sex.	Locality.	Date.	Collector.
65	♀ ad.	Santa Barbara, Cal	June 25	H. W. Henshaw.
81	♂ jun.	......do	June 26	Do.
132	♀	......do	June 28	Do.
382	♀ ad.	Tejon Mountains, Cal	Aug. 17	Do.
389	♀	......do	Aug. 18	Do.
774	♂ jun.	......do	Aug. —	Do.
775	♂ jun.	......do	Aug. —	Do.

109. *Calypte anna*, (Lesson).—Anna Humming-bird.

During the summer we saw none of this Hummer in the low valleys, but found it reasonably numerous in the mountains, where it is likely most of them retire to breed. Dr. Cooper, however, found them breeding about San Francisco as early as March. They appear to winter there, as I found them quite numerous in the gardens late in November.

The *Calypte costoe*, according to Dr. Cooper, occurs as far north as San Francisco, where it is rare. None were detected by our parties. Its general distribution is southern, being very abundant in summer at Cape Saint Lucas.

No.	Sex.	Locality.	Date.	Collector.
259	♂ jun.	Fort Tejon, Cal	July 26	H. W. Henshaw.
272	♂ ad.	......do	Aug. 2	Do.
328	♀ ad.	......do	Aug. 8	Do.
381	♀ ad.	Tejon Mountains, Cal	Aug. 17	Do.
387	♂	......do	Aug. 18	Do.
396	♀	......do	Aug. 18	Do.
385	♀	......do	Aug. 18	Do.
401	♂	......do	Aug. 19	Do.
402	♀	......do	Aug. 19	Do.
776	....	California	....	Do.
777	jun.	......do	....	Do.
778	jun.	......do	....	Do.

CUCULIDÆ—CUCKOOS.

110. *Geococcyx californianus*, (Lesson).—Chaparral Cock.

The Ground Cuckoo is an abundant resident through Southern California. It is found in all sorts of localities, though the hill-sides, covered with a more or less dense growth of bushes, and interspersed here and there with rocks, are as well suited to its habits as any. Its food consists of all sorts of insects, of lizards, and the smaller reptiles generally; in fact, of all kinds of animal life that its speed, aided by its powerful bill, enable it to overtake and kill. In many parts of the State, it appears to have become familiarized, to a certain extent, with man, and to regard him with very little fear.

When running at full speed, the long tail is lowered till its end almost touches the ground, when the bird seems fairly to glide over the earth, so easy are its movements. When hurrying, the tail is made of considerable use to enable it to turn quickly, being thrown with a jerk from side to side, according to the direction to be taken. Having gained the cover of the bushes, its safety seems assured, and it usually pauses in the first cover and stands with head erect and listening ears, the tail vibrating with nervous haste, ready to recommence its flight at a moment's warning.

No.	Sex.	Locality.	Date.	Collector.
238	♂ ad.	Santa Barbara, Cal	July 8	H. W. Henshaw.
601	♀	Kernville, Cal	Oct. 25	Do.

PICIDÆ.—WOODPECKERS.

111. *Picus villosus*, (Linn.), var. *harrisi*, Aud.—Harris's Woodpecker.

The Harris's woodpecker is a more or less common summer resident of the mountains throughout Southern California, finding its home chiefly among the pine-forests.

No.	Sex.	Locality.	Date.	Collector.
317	♂ jun.	Fort Tejon, Cal	Aug. 8	H. W. Henshaw.
552	♂ ad.	Near Mount Whitney, Cal	Oct. 11	Do.

112. *Picus pubescens*, (Linn.), var. *gairdneri*, Aud.—Gairdner's Woodpecker.

The disproportion existing in the number of this bird in the interior region, as compared with the preceding species, is not observable in California, at least to anything like the same extent. In Northern California, Cooper appears to have found it not uncommon, and a similar experience was had by us the past season in the region south of San Francisco. In distribution it is not so boreal as the Harris's Woodpecker, and coincident with this difference we do not find it among the high mountains in California, save occasionally, but with the Nuttall's it resorts to the low districts, and frequents, to a great extent, the deciduous timber, especially the oaks.

No.	Sex.	Locality.	Date.	Collector.
126	♂ jun.	Santa Barbara, Cal	July 28	H. W. Henshaw.
138	♀ ad.	......do	June 29	Do.
204	♀ jun.	......do	July 6	Do.
574	♀ jun.	Near Kernville, Cal.	Oct. 16	Do.
689	♂ ad.	Walker's Basin, Cal	Nov. 10	Do.
736	♀	......do	....	

113. *Picus nuttalli*, Gambel.—Nuttall's Woodpecker.

Picus nuttalli, Gambel, Pr. A. N. Sc., i, 1843, 259 (Los Angeles, Cal.).-Woodh., Sitgr. Exp. Zuñi & Col. River, 1854 (California).—Newb., P. R. R. Rep., vi, 1857, 89.—Bd., B. N. A., 185?, 93.— Xantus, Proc. Phila. Acad. Nat. Sci, 1859, 190.—Coop., B. Cal., i, 1870, 378.—Bd., B., and R., N. A. B., ii, 1874, 521.
Picus scolaris var. *nutalli*, Coues, Key N. A. B., 1872, 193.—Nelson, Proc. Bost. Soc. Nat. Hist., vol. xvii, 1875, 362 (California).

From the *P. scolaris* of the southern interior region and Mexico this bird appears sufficiently distinct. Though in general the two resemble each other, the points of discrepancy are yet sufficiently tangible and are not found to intergrade. The relationship of the *P.* var. *lucasanus* of Cape Saint Lucas seems to be with *scolaris*, and is, I think, to be considered with that bird as distinct from *nuttalli*. Considerable differences exist, I think, in the habits of *scolaris* and *nuttalli*, though in birds like the Woodpeckers, where general family characteristics are to be seen in every species, it is not easy to emphasize these in such manner as to make them very apparent to others, though they may be evident enough in the field.

The notes, especially as I have heard them, differ totally in character. Those of *scolaris* are quite like the usual ones of the well-known *pubescens*. No such similarity can be traced in the *nuttalli*. The usual notes of this species consist of a series of loud, rattling notes, much prolonged, and can be compared with no other Woodpecker with which I am acquainted.

This Woodpecker is a bird particularly of the oak-groves, and ranges from the lower valleys up into the mountains to a height of at least 6,000 feet, where, near Fort Tejon, I found it fairly numerous among the pines; this being the only locality where I found it among the conifers. *P. scolaris*, on the other hand, inhabits the low, hot valleys of the interior, being most partial to the mesquite-thickets. It is never, I believe, at least in Arizona, found in the mountains nor among the pines, and rarely among the oaks, and though I have frequently seen it in places where it would easily have found the surroundings if so minded.

The Nuttall's Woodpecker is pretty strictly confined to California, barely reaching into Oregon on the north, and limited in range eastward by the western slope of the Sierras. It appears to be most numerous in the valleys of the Coast range, though I found it quite common at Fort Tejon, and in October secured specimens at Kernville.

No.	Sex.	Locality.	Date.	Collector.	Wing.	Tail.	Bill.	Tarsus.
137	♀ ad.	Santa Barbara, Cal........	June 29	H. W. Henshaw.	4.06	2.97	0.80	0.71
136	♂ ad.	......do	June 29	....do	4.07	3.05	0.86	0.75
744	♂ jun.	...... do	June 17	....do	4.03	2.79	0.80	0.75
255	♂ ad.	Fort Tejon, Cal	July 27	....do	4.25	......	0.88	0.82
318	♀ jun.	......do	Aug. 8	....do	4.13	2.92	0.78	0.73
349	♂ ad.	......do	Aug. 10	....do	4.06	3.97	0.88	0.73
350	♂ ad.	......do	Aug. 10	....do	4.22	3.20	0.86	0.73
394	♂ jun.	Tejon Mountains, Cal	Aug. 17	....do	....	....	....	....
395	♂ jun.	......do	Aug. 17	....do	....	....	....	....
396	♂ jun.	......do	Aug. 17	....do	....	....	....	....
573	♂ ad.	Near Kernville, Cal.......	Oct. 16	....do	4.18	2.98	0.82	0.74
589	♂ ad.	......do	Oct. 23	....do	4.18	3.05	0.80	0.73

114. *Picus albolarvatus*, Cassin.—White-headed Woodpecker.

Leuconerpes albolarvatus, Cassin, Pr. A. N. Sc., v, 1850, 106, California.
Picus albolarvatus, Bd., B. N. A., 1858, 96.—Coop. and Suckl., P. R. R. Rep., vol. xii, pt. ii, 1860, 160.—Coop., B. Cal., i, 1870, 382.—Cones, Key N. A. B., 1872, 192.—B., B., and R., N. A. B., ii, 1874, 521.—Nelson, Proc. Bost. Soc. Nat. Hist., vol. xvii, 362 (California).

This fine species was found by us tolerably numerous in the pine-woods of the mountains near Fort Tejon, and also in the Mount Whitney region, and I am inclined to think that it is a resident in the high mountains throughout Southern California. It appears to keep pretty much among the pines, and is thus a bird of the high altitudes. In habits it shows no peculiarities from those of the *Pici* generally, and its notes are in no wise peculiar.

No.	Sex.	Locality.	Date.	Collector.
373	♀ jun.	Tejon Mountains, Cal...	Aug. 17	H. W. Henshaw.
545	♀ ad.	Mount Whitney, Cal..	Oct. 10	Do.
546	♀ jun.	......do	Oct. 10	Do.
622	♀	......do	Oct. 25	Do.
623	♂ ad.	......do	Oct. 25	Do.
661	♂	Walker's Basin, Cal	Nov. 9	Do.
662	♂	......do	Nov. 9	Do.

115. *Sphyropicus varius*, (Linn.), var. *ruber*.—Red-breasted Woodpecker.

Picus ruber, Gm., Syst. Nat., i, 1788, 429.—Heerm., P. R. R. Rep., x, 1859, pt. vi, 57.
Sphyropicus ruber, Bd., B. N. A., 1858, 104.—Coop. and Suckl., P. R. R. Rep., vol. xii, pt. ii, 1860, 160.—Cous, K. N. A. B., 1872, 195.—Nelson Proc. Bost. Soc. Nat. Hist., vol. xvii, 362.
Sphyropicus ruber, Coop., B. Cal., i, 1870, 392.—Xantus, Proc. Phil. Acad. Nat. Sci., 1859, 190.
Sphyropicus varius var. *ruber*, Bd., B., and R., N. A. B., ii, 1874, 544.

In its typical dress this is purely a Pacific-slope form. It has been shown by Mr. Ridgway to grade gradually into the var. *nuchalis* of the interior, which in Eastern North America gives place to the *varius*, in which the red and black workings are at their minimum.

The Red-breasted Woodpecker is decidedly northern in its distribution, being found in greatest abundance in Oregon and Washington Territory. It breeds about as far south as Fort Tejon, as I took a young bird in the mountains in August, and saw several more. Later, in October, I took a pair near Kernville, though in this extreme southern limit of its range it is rare.

No.	Sex.	Locality.	Date.	Collector.	Wing.	Tail.	Bill.	Tarsus.
373	♀ jun.	Tejon Mountains, Cal...	Aug. 17	H. W. Henshaw ..	4.73	3.17	0.96	0.84
677	♂ ad.	Near Kernville, Cal.....	Oct. 30	....do	4.83	3.35	0.93	0.78
638	♂ ad.	......do	Oct. 28	....do	4.82	3.47	0.93	0.77

116. *Sphyropicus thyroideus*, (Cass.).—Black-breasted Woodpecker; Williamson's Woodpecker.

This Woodpecker was quite common in the heavy pine and red-wood forests in the Sierras, near Mount Whitney, in September, and they doubtless breed here. The males were in about equal numbers with the females, as I have always found them to be in each of the many and widely-separated localities where I have met with the species.

They are among the most silent of the tribe, not only in respect to their notes, but in their manner of procuring food, the most of this being obtained from the crevices of the bark rather than dug out with the noisy hammerings characteristic of many of the family. No other of the tribe is so constant a resident of the conifers as this. It appears to live in them exclusively, and if it ever descends into the lower regions and frequents the deciduous timber it must be only in the depths of winter.

No.	Sex.	Locality.	Date.	Collector.
488	♀ ad.	Near Mount Whitney, Cal..................................	Sept. 19	H. W. Henshaw.
560	♂ ad.	......do	Oct. 11	Do.

117. *Hylotomus pileatus*, (Linn.).—Pileated Woodpecker.

Picus pileatus, Linnæus, Syst. Nat., i, 1766, 173.
Drycopus pileatus, Woodh., Sitgreave's Exp. Zuñi & Col. Riv., 1854, 90 (Indian Territory, Texas. New Mexico.)
Hylotomus pileatus, Bd., B. N. A., 1858, 107.—Coop. & Suckl., P. R. R. Rep., vol. xii, pt ii, 1860, 161.—
Coop., B. Cal., i, 1870, 396.—Coues, Key N. A. B , 1872, 192.—Bd., B., & R., N. A. B.,ii, 1874, 550.—
Nelson, Proc. Bost. Soc. Nat. Hist , vol. xvii, 1875, 362 (California).—Bendire, Proc. Bost
Soc. Nat. Hist., vol. xviii, 1875, 160 (Camp Harney, Oregon).

This "Log Cock" is found in the Sierras as far south as latitude 37°, where I saw two individuals in October. It is not unlikely that the heavily-timbered districts may give this bird shelter throughout the extent of the Sierras. It was found near Nevada City by Mr. Nelson, but is more numerous farther north, becoming abundant, according to Dr. Cooper, near the Columbia.

118. *Melanerpes torquatus*, (Wils.).—Lewis's Woodpecker.

I did not see this species till reaching Fort Tejon, in August. It was here, and at other places in the Sierras, common in certain localities. In summer, it seems to prefer the pineries of the mountains, but in fall descends, and then inhabits the oak-groves in common with the next species, without, however, mingling with them.

In habits, the species is somewhat anomalous among its relatives. Like the Californian, it is rarely found alone, but associates in bands of many individuals, the gathering taking place as soon as the young are well on the wing. In the late fall, these companies appear to be pretty nearly stationary, not roving over the country at large, but remaining in some favorable spot where food is plenty. Here they may always be found either at play, chasing each other in and out the branches, or industriously hunting for insects. These are obtained with the expenditure of very little labor in digging, as they prefer to take them from the accessible crevices in the bark or even to capture them on the wing. Berries, too, when they can be had, form a part of their varied diet. Their peculiar manner of circling about the tree-tops in wavering circles is well known, and is one of the most noticeable characteristics of its appearance. They are endowed by nature with a shy, suspicious disposition, and always regard the appearance of man with distrust.

No.	Sex.	Locality.	Date.	Collector.
374	♂ jun.	Fort Tejon, Cal	Aug. 17	H. W. Henshaw.
418	♂ jun.	Walker's Basin, Cal..	Aug. 27	Do.
419	♂ ad.	......do	Aug. 27	Do.
717	♂ ad.	......do	Nov. 11	Do.

119. *Melanerpes formicivorus*, (Swains.).—Californian Woodpecker.

The habitat of this Woodpecker, in California as in Arizona, seems to be determined by the range of the oaks; the presence or absence of these trees, their abundance or scarcity, affording a pretty sure index of the numbers of this bird. In California, they are certainly the most abundant of the tribe, as they also are in Arizona in the sections they inhabit.

The social instinct is developed in them to a degree equaled in no other species, and they are almost never found other than in large communities, while as often as otherwise they take up their residence in the oaks that overspread the farmers' dwelling. Their most curious trait is seen in their habit, shared by no other Woodpecker, of storing up a supply of acorns in holes drilled for that purpose in the trunks of trees. a custom which seems to admit of no adequate explanation. They were most industriously at this work at Fort Tejon the last of August, and during the day this seemed to keep them busy pretty nearly all the time. Judging from their cries and earnest manner, as they bent themselves to the task of fitting the acorns into the holes, which had served the same purpose the last season, and perhaps many seasons before, the work must be an important one in their own estimation, whatever the object. With them, however, it is not by any means "all work and no play," but, on the contrary, the labor, if labor it be to them, is lightened by much gamboling and chasing each other in and out of the branches in circular sweeps, like boys playing at tag. Indeed, there is no reason why they should not make merry, for food is abundant and easily obtained, not only in the fall, when the acorns are thus laid away, but during all the winter, a fact which serves to make their economy appear all the more inexplicable and useless. The species is a resident one wherever found.

No.	Sex.	Locality.	Date.	Collector.
300	♀ ad.	Fort Tejon, Cal	Aug. 7	H. W. Henshaw.
301	♂ ad.	......do	Aug. 7	Do.
302	♀	......do	Aug. 7	Do.
316	♂ jun.	......do	Aug. 8	Do.
348	♂ jun.	......do	Aug. 10	Do.

120. *Colaptes mexicanus*, (Swains.).—Red-shafted Flicker.

This Flicker is found throughout Southern California, without reference to special locality, being common both in the mountains and in the low districts. Its habits agree essentially with those of the Common Flicker of the East.

No.	Sex.	Locality.	Date.	Collector.
187	♂ jun.	Santa Barbara, Cal	July 25	H. W. Henshaw.
319	♀ jun.	Fort Tejon, Cal	Aug. 8	Do.
624	♂ ad.	Kernville, Cal	Oct. 25	Do.

STRIGIDÆ.—OWLS.

121. *Strix flammea*, Linn., var. *americana*, (Aud.).—American Barn Owl.

Strix flammea, Linn., Syst. Nat., i, 1766, 133.
Strix americana, Aud., Syn., 1839, 25.
Strix pratincola, Newb., P. R. R. Rep., vi, 1857, 76.—Bd., B. N. A., 1858, 47.—Xantus, Proc. Phil. Acad. Nat. Sci., 1859, 190.—Heerm., P. R. R. Rep., x, 1859, pt. vi, 34.—Coop., B. Cal., i, 1870, 415.
Strix flammea var. *pratincola*, B., B., & R., N. A. B., iii 1874, 13.

The Barn Owl appears to be common throughout Southern California, and in some portions, as in the swamps near Los Angeles and again in the San Bernardino Valley, I found it in great numbers. This was in June, and they had gathered themselves into communities numbering, in one instance, at least twenty. They resorted in the day-time to the dense undergrowth of the swamps or the thickest foliage of the oaks, to doze away in quiet the hours of sunlight. Dusk fairly settling down, they may be seen silently issuing by twos and threes from their shady retreats in quest of food. It be-comes less numerous in the northern part of the State, though, according to Dr. Cooper, it is found to the Columbia River.

No.	Sex.	Locality.	Date.	Collector.	Wing.	Tail.	Bill.	Tarsus.
29	♂ ad.	Los Angeles, Cal	June 17	H. W. Henshaw.	13.23	5.50	1.33	2.82
773	♀	Santa Barbara, Cal	June 14	....do	13.75	6.00	1.35	2.85

122. *Bubo virginianus*, (Gmel.), var. *arcticus*.—Western Great Horned Owl.

This Owl is found throughout California, confining itself for shelter to the wooded districts. It is a solitary species, the pairs separating as soon as the young are out of the way. Except during the breeding-season, it hunts only by night, though its pow-ers of vision are excellent during the brightest hours of day.

No.	Sex.	Locality.	Date.	Collector.
330	♂	Fort Tejon, Cal	Aug. 8	H. W. Henshaw.

AP. JJ—17

123. *Scops asio,* (Linn.), var. *maccalli,* Cass.—Western Mottled Owl.

The little Screech Owl is a common resident of California, as it is indeed in all the wooded portions of the far west. Its habits, however, are so strictly nocturnal that its presence is easily overlooked.

I have never seen the var. *maccalli* in any but the gray plumage, nor can I ascertain that the red phase of coloration of this variety has been noted by others. The Pigmy Owl (*Glaucidium gnoma*) has been found by several observers to be quite numerous in the mountains of the State. Of the Flammulated Owlet (*Scops flammeola*), a single specimen was taken at Fort Crook. The Whitney's Owl (*Micrathene whitneyi*) occurs in the Colorado Valley, where the type-specimen was shot by Dr. Cooper. The two last may occur over much of the southern portion of the State, but their small size and nocturnal habits render them extremely liable to be overlooked.

No.	ex.	Locality.	Date.	Collector.
2-2	♂	Fort Tejon, Cal	July 26	H. W. Henshaw.

124. *Otus vulgaris,* (Linn.), var. *wilsonianus,* (Less.).—Long-eared Owl.

Like the Barn Owl, this species is prone to congregate together, and it is uncommon in the West for one to stumble upon one of these birds roosting in retirement without finding that the same thicket or grove shelters a number. Such was the case at Los Angeles, where the same swamps that gave protection to the Barn Owl also afforded a congenial retreat to this species, and while threading the tangled mazes I several times saw three or four start out from the same spot.

This owl is extremely averse to facing the sunlight, though when forced to do so its eyesight is pretty good.

125. *Speotyto cuniculari,* (Mol.), var. *hypugæa,* (Bon.).—Burrowing Owl.

Nowhere in the West does this Owl occur oftener or in greater numbers than in Southern California, and according to the observations of others it appears to be equally numerous in the northern part of the State.

The deserted holes of the destructive Ground Squirrel (*Spermophilus beecheyi*) furnish it with its usual abode. The birds are very often to be seen during the hours of sunlight sunning themselves at the mouths of the burrows. They are not, however, very active by day, except when disturbed in their meditations, when, with a few expostulatory notes, they fly off a few hundred yards to a neighboring hillock, whence they can keep a good lookout. Their sight under such circumstances is most excellent, and they have no difficulty when so minded in keeping themselves out of danger. Notwithstanding this, I have never seen them in pursuit of food during the day, and should say that this was obtained wholly after nightfall. In the uninhabited districts I have usually found them rather wary, but in the settled parts of California they are quite the reverse, and I have seen them sitting by the roadside paying no attention to the teams and passers-by. It is generally supposed that among other items of their fare are the young of the squirrels. This I have never confirmed, though presuming such to be the case. They are known to eat mice, lizards, and snakes.

FALCONIDÆ—FALCONS.

126. *Falco communis,* Gmel., var. *anatum,* Bon..—Duck Hawk.

This Hawk appears to be rather common in Southern California, being perhaps most so on the coast. It is numerous on the Santa Barbara Islands; also present around Kern Lake, where the water-fowl which reside here throughout the year furnish it with the most of its food.

127. *Falco columbarius,* Linn.—Pigeon Hawk.

At quite a number of localities in Southern California I noted Hawks which appeared to be of this well-known species. The following variety, however, is remarkably close to this species, and hence I may have confounded the two, and a portion of those supposed to belong here may have really been of the next variety, if that be really distinct.

The true Pigeon Hawk is, however, from the observations of others, well distributed over California.

128. *Falco columbarius,* Linn., var. *richardsoni,* Ridgw.—American Merlin.

It does not appear at all certain that this variety, established by Mr. Ridgway, will not be found to be merely a special plumage of the Pigeon Hawk. It was supposed to be confined to the interior region east of the Rocky Mountains. It is, however, found in Southern California, and I think not uncommonly, though I took but a single specimen. I frequently saw small Falcons, which I took to be Pigeon Hawks, but at such distances and under such circumstances that I did not succeed in procuring them.

No.	Sex.	Locality.	Date.	Collector.	Wing.	Tail.	Bill.	Tarsus.
437	♀ jun.	Walker's Basin, Cal.....	Aug. 28	H. W. Henshaw...	8.30	5.75	0.53	1.48

129. *Falco sparverius*, Linn.—Sparrow Hawk.

In California, as throughout the West generally, the Sparrow Hawk is very numerous. The dry hills along the coast near Santa Barbara were resorted to by great numbers of these birds in July, and in following the line of the telegraph one of them was to be seen perched on one of the poles at intervals of every few yards.

No.	Sex.	Locality.	Date.	Collector.
368	♀	Tejon Mountains, Cal...	Aug. 7	H. W. Henshaw.

130. *Pandion haliætus*, (Linn.), var. *carolinensis*, Gmel.—Fish Hawk.

Present throughout California, both in the interior where the streams are stocked with fish, and on the coast, but more particularly the latter.

131. *Circus cyaneus*, (Linn.), var. *hudsonius*, Linn.—Marsh Hawk.

The open country everywhere is visited by this Hawk, which is very numerous in California. Resident in the southern part.

No.	Sex.	Locality.	Date.	Collector.
596	♂ ad.	Walker's Basin, Cal	Oct. 23	H. W. Henshaw.
597	♀	......do	Oct. 23	Do.
673	♂ ad.	......do	Nov. 9	Do.

132. *Nisus fuscus*, Gmel.—Sharp-shinned Hawk.

A common resident throughout Southern California. This little Hawk is of a bold, dashing disposition, preying indiscriminately upon all the smaller kinds of birds as well as upon small mammals. In procuring these, it beats through the mazes of the woods, following the edges of the thickets, and passing through the leafy openings, and secures its victim, either by surprising and dropping suddenly down upon it, or else, having started it out, pursues it in open chase and clutches it while at full speed.

133. *Nisus cooperi*, (Bon.).—Cooper's Hawk.

The Cooper's Hawk seems to be about as numerous in Southern California as its smaller relative. In summer it is not often seen in the lower districts, but will then be found to have retired to the mountains, where it nests, choosing some lofty pine as the site of its domicile. In the fall, there appears to be a very decided migration from the north, and then the low country generally is occupied by this species, which winters in the southern half of the State.

No.	Sex.	Locality.	Date.	Collector.
383	♀ jun.	Tejon Mountains, Cal..	Aug. 17	H. W. Henshaw.
468	♂ jun.	Near Mount Whitney, Cal...................................	Sept. 12	Do.
640	♂ jun.	Walker's Basin, Cal..	Nov. 5	Do.

134. *Buteo swainsoni*, Bon.—Swainson's Hawk.

This Hawk appears to be pretty well distributed over the southern part of California, and is, in certain localities, very common. This was the case in the San Fernando Valley in July. Camping here one evening, our attention was directed to the great number of Gophers (*Spermophile haweaii*), which in large colonies inhabited some barren hills near the station. Toward dusk the place was visited by at least a dozen of these birds, which took up their positions on the little hillocks thrown up by the animals in front of their burrows, and awaited with patience the moment when a favorable opportunity should occur to snatch a supper. Elsewhere I have frequently seen them thus employed, and their persistence in destroying these pests should entitle them to due consideration at the hands of the farmer. Large numbers of insects, particularly grasshoppers, are destroyed by these birds, whose abilities as purveyors of food are thus of the lowest order.

No.	Sex.	Locality.	Date.	Collector.
740	♂ ad.	Los Angeles, Cal	June —	H. W. Henshaw.

135. *Buteo borealis*, (Gmel.), var. *calurus*, Cassin.—Western Red-tailed Hawk.

The present species is of almost universal distribution in the West, and, though most abundant in the mountains during summer, is by no means confined exclusively to them, but is found more or less commonly in the low country, according to the conveniences it finds for nidification. In California are seen the most extreme examples of the dark fuliginous style of coloration, which is known under the above varietal name. The lighter condition of plumage, which was known to earlier writers as *B. montanus*, is also found, though the proportion of these is not large, and probably it would not be easy to find in California an individual which was not appreciably darker than the usual type of this Hawk from the interior region. On the other hand, the extreme melanistic conditions, in which the rufous markings are only present in slight dashes here and there, and the prevailing color an extreme blackish brown, is also not common. Most individuals range between the two extremes, while no two are exactly alike.

In its wide range, the habits of this Hawk undergo but little change. It is everywhere the same heavy-winged, sluggish bird, its nature causing it to prey upon the very humblest kind of game, and even to eat carrion when this is handiest. In company with the Swainson's Buzzard, it may often be seen in the villages of the Gophers, and like that bird, is more prone to capture these animals by lying in wait for them than by seizing them from above after the manner of the true Hawks.

No.	Sex.	Locality.	Date.	Collector.
477	♂ ad.	Near Mount Whitney, Cal...................................	Sept. 18	
532	♂ ad.	......do	Oct. 7	
576	♂ ad.	Walker's Basin..	Oct. 16	
655	♂ ad.	......do	Nov. 5	

136. *Archibuteo ferrugineus*, (Licht.).—Californian Squirrel Hawk.

In my note-book I find reference made to some large Hawks which, in company with the Swainson's, I saw in a Gopher village in the San Fernando Valley in July, and which I believed to be of this species. It does not appear, however, to be at all common in Southern California in summer, but becomes numerous in fall, making its appearance either from the high mountains, or, as is more probable, from farther north. It is more active in its motions and more Falcon-like in its method of hunting than either of the preceding species. It is usually seen beating over the open country on vigorous wing, and keeping a few feet above the ground, ready on the instant to close with any unlucky mammal it may chance to surprise. As implied by its name, it is a determined enemy of the Ground Squirrels, and, with the other two species, must annually destroy an immense number of them. The *A. sanctijohannis* is, according to Dr. Cooper, a winter visitor to the State. I saw in possession of Mr. Gruber, of San Francisco, a fine specimen of this bird, representing the most extreme condition of melanism. It was shot, I believe, near San Francisco.

137. *Elanus Cucurus*, (Vieillot).—Black-shouldered Kite.

This species does not appear to occur in the southern parts of California, where none were met with by our parties. It is found about San Francisco in considerable numbers, and is there a resident.

138. *Haliaëtus leucocephalus*, (Linn.).—White-headed Eagle.

This Eagle is an abundant resident of California, particularly along the sea-coast. It is also not uncommon in the mountain districts. The islands in the Santa Barbara Channel are the resort of many pairs that remain during the year. The broken ledges on the faces of the cliffs, sometimes overhanging the ocean, afford favorite spots for their nests. They are said to annually destroy many of the lambs. I am informed by Lieutenant Carpenter that this Eagle at the mouth of the Columbia River is exceedingly numerous, and that here its habits of feeding upon carrion are as regular and fixed as those of the true Buzzards. Its chief dependence is on fish, more particularly Salmon, of which vast numbers are cast up by the waves. On one occasion, he found half a dozen of these birds feeding upon the flesh of a putrid ox. With this they had become so gorged as to be utterly unable to fly. One of them had so completely filled itself with the foul food that a large piece which it had partially swallowed it was utterly unable to force further down, and in this situation, unable to move, it was approached and knocked on the head with a wiping-stick.

In this region, they nest almost entirely on the tall pines.

CATHARDIDÆ.—AMERICAN VULTURES.

139.—*Pseudogryphus Californianus*, (Shaw).—California Vulture.

Our opportunities for an acquaintance with this Vulture were most brief and unsatisfactory, and were limited to seeing two or three individuals warring on the wing in the mountains. So far as I could learn, they descend rarely into the valleys during the summer months, and only then when attracted by the sight of some dead animal; their keen sight enabling them to detect the presence of food at very long distances. Dr. Taylor informed me that at Santa Barbara they were of quite common occurrence, remaining, however, most of the time in the neighboring mountains. I hear they breed, seeking the shelter of caves, in the most inaccessible situations.

It seems probable that the numbers of this huge bird have very much diminished during the last few years. So large and conspicuous an object could scarcely fail to attract the attention of any chance rover of the wilderness, yet its presence was almost undetected by our parties. As is well known, this bird is easily killed by strychnine, and as this poison has been in almost constant use for a term of years in the destruction of wild animals, it seems highly probable that great numbers of these birds have suffered a like fate from eating the carrion.

According to the observations of earlier naturalists, it was numerous throughout most of California, and extended its range on the north to the Columbia. Near Mount Whitney, in September and October, I frequently saw the carcasses of sheep which had lain for days, and in one instance the body of a huge Grizzly Bear, which had died from poison, was in the final stages of decomposition, yet in no case had any of these been visited by Vultures, a fact which seemed to argue their total absence from this region.

140. *Rhinogryphus aura*, (Linn.).—Red-headed Vulture.

This bird is far more numerous throughout Southern California than its larger relative. It is less of a mountain-loving species, and is, indeed, much less shy and more domestic in its habits, coming freely about the ranches and houses whenever carrion or refuse of any kind is to be had. I saw numbers of them on the islands off Santa Barbara, and think likely they nest there.

COLUMBIDÆ.—PIGEONS.

141. *Columba fasciata*, (Say).—Band-tailed Pigeon.

This Pigeon occurs abundantly in California, retiring to spend the season of reproduction in the mountains, where it resorts very much to the pineries to nest. It does not appear, however, for some cause or other, to be found in any numbers in summer in the more southern portions of the State, and was not seen by us till in the fall, when, in the course of the migrations, it makes its appearance in bands from the far north. In November I often saw them in flocks of from ten to one hundred, flying swiftly about from one oak-grove to another, for, at this season, acorns form their chief, indeed almost their sole, food. Their shyness now is very remarkable, and it is probably due to the fact that in their passage from the north they are compelled to run the gauntlet of hundreds of gunners, who in the neighborhood of cities and towns eagerly pursue them for the market.

142. *Zenaidura carolinensis*, (Linn.).—Carolina Dove.

The Turtle Dove is very numerous in Southern California, its limit northward on this coast being reached at about the forty-ninth parallel, as in other portions of the country.

The dry sandy deserts, which repel nearly all the feathered tribe, form favorite resorts for these Doves. Their powerful wings easily bear them out on the barren wastes, where, it might seem, they would find little to attract them, but where they secure a good supply of seeds from plants whose hardy natures enable them to withstand the drought. The very nature of this dry hard food renders frequent visits to water a necessity, and hence, in the vicinity of any of the rare pools that grace these saharas, the Turtle Dove may always be seen.

TETRAONIDÆ.—GROUSE.

143. *Canace obscurus*, (Say.)—Dusky Grouse.

This Grouse is an inhabitant of high latitudes, but finds in the Rocky and Sierra Nevada Mountains a climate and vegetation analogous to the far northern districts. In California, it is found in both the Coast and Sierra ranges as far south as latitude 35°, and probably even lower. It was present, though not very common, in the mountains near Fort Tejon, and was rather numerous in the region about Mount Whitney. Its presence depends much upon the conifers. It cares less for the pines, but the thick tangled forests of spruce, fir, and tamerack will rarely be entered without grouse sign being very soon apparent. In the Sierras, they are very fond of staying about the vicinity of the little grassy cienagas that are found scattered here and there on the flanks of the mountains, sometimes entirely environed with the coniferous trees.

Lieutenant Carpenter, who has enjoyed most excellent opportunities for observing this bird both in Oregon and the Rocky Mountains, thus speaks of their habits: " Late in the fall, the Dusky Grouse disappear entirely from the grounds frequented by them in summer. At this latter season, their range is much wider. They leave, to a great extent, the thick woods, and are found much in the open glades, where many kinds of berries, as the wild strawberry, afford them a varied and luscious fare. About November, however, they wholly disappear, and a person looking for game in the places where in summer there were an abundance of these birds, would now see no sign of their presence. The idea credited by some, that they have migrated to warmer climes, or that they are passing the long winter hidden away in a torpid state, is alike erroneous. In the Rocky Mountains, about the time of the first heavy snows, they betake themselves to the densest pine-woods, where they live entirely in the conifers. The buds of the pine and spruce now furnish them their only food, and upon these they subsist till the next spring, when the genial sun, with returning warmth, having released the streams and removed the snow, they again descend to mother-earth. In Oregon, too, even along the coast where no snow falls, this same habit obtains. They leave the ground entirely, resort to the pines, and their terrestrial mode of life does not begin till the next summer, when berries and small seeds afford a greater attraction than their usual piny fare. About the 1st of April, the males begin their booming notes, which may now be heard coming from all parts of the forest as the emulous birds begin their courtships. It is at this time that many are shot, the gunners now having a sure guide to their prey in the love-notes, which seem to proceed from the mid-air, as the birds give utterance to them when perched on the branch of some tall pine."

These notes, which are so characteristic of the species in Washington Territory and Oregon, do not appear to have been noticed by any observer in the Rocky Mountains, and Lieutenant Carpenter tells me that not only has he himself not heard this, but all his inquiries among hunters and trappers have failed to establish this habit as belonging to the bird in the various parts of the Rocky Mountains he has visited.

PERDICIDÆ.—QUAILS.

144. *Laphortyx Californicus*, (Shaw).—California Valley Quail.

Tetrao californicus, Shaw, Nat. Miss., pl. 345.
Callipepla californica, Newb., P. R. R. Rep., vi, 1857, 92.—Heerm., ibid., x, 1859, pt. vi. 60.
Lophortyx californica, Bd., B. N. A., 1858, 644.—Xantus, Proc. Phila. Acad. Nat. Sci., 1859, 192.—
 Bd., Xantus, ibid., 305 (Cape St. Lucas).—Coop. & Suckl., P. R. R. Rep., vol. xii, pt. 11, 1860,
 225.—Coop., B. Cal., i, 1870, 549.—Coues, Key N. A. B., 1872, 238.—Nelson, Proc. Bost. Soc.
 Nat. Hist., vol. xvii, 365 (California).

The Valley Quail, as its name implies, is an inhabitant of the lower districts in California, where it is found overspread over all the country to the west of the Sierra Nevada range. On the north it reaches to the Columbia River. The most extreme limit at which I found it was in the mountains near Fort Tejon, where I saw the species on several occasions at an altitude of 6,000 feet. At this height, I found the young. Here they meet the Mountain Quail, or rather the ranges of the two were found at this point to overlap each other; for the Mountain Quail was found somewhat lower than this. Such, however, is rarely the case, as the Valley Quail is a much less hardy bird than its mountain-loving relative, and courts the warmth of the pleasant valleys. As the number of its natural enemies, in the shape of wild animals and snakes, has been very much diminished through the agency of man, and its increase goes on almost without check, its numbers in some sections of the State are simply enormous. On the island of Santa Cruz, the attempt has been made to introduce them, but with only measurable success, and it is not likely that they will ever there become very numerous; for the number of foxes on this island would be sufficient to keep them in check, were every other condition favorable.

The average time for laying in Southern California seems to be along in April or early May, and by the last of June large numbers of the young are out and able even to fly short distances. The time, however, for nesting must be quite variable, or else the great disparity in the ages of the broods is due to the fact that the later nestlings are the product of a second clutch of eggs, the first having perhaps been destroyed. Thus, though I have seen many young able to fly in the month of June, I have found others of about the same size and age late in August. Two broods may occasionally be reared in a season.

As soon as the young are out, it is usual for several broods to unite together, and in this way it is not unusual to find in one company birds representing several progressive stages of plumage, and more or less advanced toward maturity. Within the limits of its range, this Quail affects almost all situations. Often during the day, the bands will be found in the vicinity of water, the nature of their food requiring much to soften and aid in its digestion. The bushy pastures, grain fields, and the foot-hills, all in turn invite attention, and are visited by the busy flocks that thus spend the greater part of the day in a constant search for food. Whether it is a constant habit with them to seek shelter during the hottest part of the day, I do not know; but I have often found the bevies about noon in the shade of the bushes that fringe the margin of some favorite spring, where they have come to slake their thirst and apparently pass the heated hours of day in shady seclusion. This I think is a fixed habit with them. In October and November, the young birds are full-grown, and as strong on the wing as their parents.

They now gather into very large bevies, or rather an assemblage of bevies, sometimes a hundred or more, though the average would be less than this. As a rule, their ways are not such as to endear them to the sportsman; for they are apt to be wary, and, unless under specially favorable circumstances, are not wont to lie closely. I have, however, flushed a large bevy contiguous to a bushy pasture where the scrub was about knee-deep, with cattle-paths through it, and have had glorious sport. The birds lay so close as to enable me to walk almost over them, when they got up by twos and threes, and went off in fine style. The sportsman may now and then stumble upon such chances, but they do not come often. A bevy once up, off they go, scattering but little unless badly scared, the main body keeping well together, and having flown a safe distance they drop, but not to hide and be flushed one after another at the leisure of the sportsman. The moment their feet touch firm ground, off they go like frightened deer, and if, as is often the case, they have been flushed near some rocky hill, they will pause not a moment till they have gained its steep sides, up which it would be worse than useless to follow. Should they, however, be put up hard by trees, they will dive in among the foliage and hide, and there standing perfectly motionless will sometimes permit one to approach to the foot of the tree they are lodged in ere taking wing.

They roost always in bushes or trees, and almost invariably in those which are hard by water, which they resort to in the early morning ere setting forth on the business of the day.

When anxious and disturbed, the members of the flock call to each other in querulous tones, the notes resembling the syllable *pit, pit*, constantly repeated; this, too, just as they are taking to wing. Besides this, the males have a loud call, which answers, when the band has been dispersed, to bring them together. This has been variously interpreted. It resembles perhaps as much as anything, when put into English, the words *come-right-here*, the last syllable lengthened and much emphasized.

No.	Sex.	Locality.	Date.	Collector.
83	♂ ad.	Santa Barbara, Cal.	June 26	H. W. Henshaw.
92	♂ ad.	.....do	June 26	Do.
93	♀ ad.	.....do	June 26	Do.
94	♀ ad.	.....do	June 27	Do.
116	♂ ad.	.....do	June 28	Do.
117	♀ ad.	.....do	June 28	Do.
253	♂ ad.	Fort Tejon, Cal.	July 26	Do.
254	♂	.....do	July 26	Do.
278	jun.	.....do	Aug. 2	Do.
578	♂ ad.	Near Kernville, Cal.	Oct. 19	Do.
579	♂ ad.	.....do	Oct. 19	Do.
580	♀	.....do	Oct. 19	Do.
580 A	♀	.....do	Oct. 19	Do.
581	♀	.....do	Oct. 20	Do.
581 A	♀	.....do	Oct. 19	Do.
587	♀	.....do	Oct. 20	Do.
588	♀	.....do	Oct. 23	Do.
589	♂ ad.	.....do	Oct. 23	Do.
653	♂ ad.	Walker's Basin, Cal.	Nov. 5	Do.
654	♂	.....do	Nov. 5	Do.
716	♂	.....do	Nov. 5	Do·
725	jun.	Fort Tejon, Cal.	July 26	Do.
726	jun.	.....do	July 26	Do.

145. *Oreortyx picta*, Douglas.—Mountain Quail.

Ortyx picta, Dougl., Trans. Linn. Sc., xvi, 1829, 143.
Callipepla picta, Newb., P. R. R. Rep., vi, 1857, 93.—Heerm , ibid, x, 1859, Birds, 61.
Oreortyx pictus, Bd., B. N. A., 1858, 642.—Xantus, Proc. Phila. Acad. Nat. Sci., 1859, 192.—Coop. &
  Suckl., P. R. R. Rep., vol. xii, pt. ii, 1860, 225.—Coop., B. Cal., i, 1870, 546.—Coues, Key N. A. B.,
  1872, 237.—B., B., & R., N. A. B., iii, 1874, 475, pl. 63, f. 5.—Nelson, Proc. Bost. Soc. Nat. Hist.,
  vol. xvii, 364 (California).

This, the most beautiful of all our game-birds, is limited in its distribution to California and Oregon, and, as its name well implies, is strictly a bird of the mountains. We found it in the mountains near Fort Tejon, and in the Sierras in a sufficient number of localities as to justify the belief that its distribution in Southern California is at least quite general, and dependent only upon the mountainous nature of the country. In summer, it is not found lower than 4,000 feet, and is not so common at this elevation as somewhat higher. Above 9,000 feet, it was not seen, and this is presumably about its limit. Its habitat is thus complementary to that of the Valley Quail, the higher and lower limits of either species occasionally overlapping each other. It seems nowhere to be an abundant species. As compared with the preceding, the bevies are very small, and I do not remember to have ever seen more than fifteen together, oftener less. It is a wild, timid bird, haunting the thick chaparral-thickets, and rarely coming into the opening. When a band is surprised, they are not easily forced on the wing, but will endeavor to find safety by running and taking refuge in the thickness and impenetrability of their favorite thickets. If forced, however, they rise vigorously and fly swiftly and well, and sometimes to a considerable distance, and then make good their escape by running. During the heat of midday, they will be found reposing under the thick shade of the chaparral, and there they remain till the cooler hours invite them to continue their quest for food. When the covey has been scattered, the males have a loud call, which consists of a series of notes clearly given, the whole recalling to mind the notes of the Golden Flicker. Besides this, both sexes have the more commonly heard piping-notes, which they emit just as they take to wing, and when they are agitated, or moved by fear.

No.	Sex.	Locality.	Date.	Collector.	Wing.	Tail.	Bill.	Tarsus.
283	♂ ad.	Mountains near Fort Tejon, Cal	Aug. 3	H. W. Henshaw	5.50	3.70	0.60	1.41
417	♂	......do	Aug. 19	....do	5.25	3.65	0.57	1.47
724	♀ ad.	......do	Aug. 1	....do	5.30	3.43	0.60	1.40
723		......do	Aug. 1	....do	5.40	3.97	0.62	1.37
780	♀ ad.	......do	Aug. 1	....do	5.25	3.45	0.57	1.45
779	♀ ad.	......do	Aug. 1	....do				
640	♂ ad.	Walker's Basin, Cal...........	Nov. 5	....do	5.43	3.77	0.58	1.48
700	♂	......do	Nov. 5	....do	5.43	3 53	0.58	1.47
701	ad.	......do	Nov. 5	....do				
702	♂ ad.	......do	Nov. 5	....do				

CHARADRIIDÆ.—PLOVERS.

146. *Ægialitis vocifera*, (Linn.).—Killdeer Plover.

By the side of every lake and along all the streams, as well as on the shores of California, this Plover is found in great abundance. It is only partially migratory, numbers remaining in Southern California till the ensuing spring sends them farther north.

The Mountain Plover (*A. montana*) occurs and is numerous in certain localities in Southern California, as on the plains about Los Angeles.

No.	Sex.	Locality.	Date.	Collector.
671	♀ jun.	Walker's Basin, Cal..	Nov. 9	H. W. Henshaw.

Charadonis cantianus, Lath., Birds, vol. viii, 328 —Heerm , P. R. R. Rep., x, 1859, pt. vi, 64.
Ægilitis (Leucopolius) nivosa, Cass., Bd., B. N. A., 1858, 696.—Coues, 1866, 274 (San Pedro, Cal.).
Ægilitus cantianus, Coues, Key, 1872, 245.

147. *Ægialitis cantianus*, Lath., var. *nivosa*, (Cass.).—Snowy Plover.

This species is an abundant one on the coast of California, though by no means a strictly coastwise bird. I did not see it in the interior, though Mr. Ridgway found the species at Great Salt Lake, where it was breeding. At Santa Barbara, it was numerous, frequenting here only the sandy shores, not following the creeks inland, and never visiting the marshes, though within a few yards of its breeding-ground. Its habits seemed exactly like those of the common Piping Plover, and their notes are very similar. Its food consists of all sorts of worms and marine crustacea which it finds close to the water's edge, following the retreating waves down and scurrying back as they come rolling in.

July 7, I found two broods of young which had left the nest but a few hours before. They were clothed in down, and were yet so weak as scarcely to be able to stand. Subsequently I found quite a number of nests containing eggs. The spot selected for a breeding-ground was a strip of bare white sand, a hundred yards, perhaps, from the ocean. The nest was simplicity itself. In all but one instance the eggs were deposited in a slight hollow scratched in the sand, without lining of any sort. In the exceptional case the owners must have been of an artistic turn of mind, for they had selected from along the shore little bits of the pearly nacre, the remnants of broken sea-shells, and upon a smooth lining of this material were placed their treasures. The effect of the richly-colored eggs as they lay on the cushion of shining nacre was very pleasing. So slight is the contrast between the eggs and the drifted sand about them that they would be difficult enough to find were it not for the tracks about the nest. As the mates came to relieve each other from setting or to bring each other food, they alighted near the nest, and thus for a little distance about each one was a series of tracks converging to a common center, which too surely betrayed their secret. Great was the alarm of the colony as soon as my presence was known, and, gathering into little knots, they nervously attended my steps, following at a distance with low sorrowful cries. The female, when she found her nest was really discovered, hesitated not to fly close by, and used all the arts which birds of this kind know so well how to employ on like occasions. With wings drooping and trailing on the sand, she would move in front till my attention was secured, when she would fall helplessly down, and burying her breast in the soft sand, present the very picture of utter helplessness, while the male with the neighboring pairs expressed his sympathy with loud cries. The full nest complement appears to be three, and in no instance did I find more. These are of a light clay-color, numerously marked with blotches and scratchy markings of black. In size and appearance they approach most closely to those of *A. melodus*, but may be easily distinguished by the different style of the spotting.

Examining a good series of the eggs of *melodus* in the Smithsonian, I find them to vary among themselves but little in the character of their markings. These take the form of small circular dots, very rarely becoming aggregated into blotches, and without penlike scratchings. Those of *nivosus* are more heavily marked with irregular blotches, while the scratchy marks are conspicuous. Three sets measure, respectively, 1.30 by 0.93; 1.27 by 0.92; 1.25 by 0.93; 1.29 by 0.93; 1.27 by 0.89; 1.24 by 0.95; 1.22 by 0.90.

No.	Sex.	Locality.	Date.	Collector.	Wing.	Tail.	Bill.	Tarsus.
178	♂ ad.	Santa Barbara, Cal	July 4	H. W. Henshaw.	4.23	1.93	0.60	0.97
179	♂ ad.	......do	July 4	....do	4.22	2.00	0.64	0.92
180	♂ ad.	......do	July 4	....do	4.20	1.98	0.62	0.98
181	♀ ad.	......do	July 4	....do	4.08	1.85	0.61	6.88
209	♂ ad.	......do	July 7	....do	4.02	1.98	0.64	0.96
210	♂ ad.	......do	July 4	....do	4.08	2.08	0.63	0.92
212	♂ ad.	......do	July 7	....do	3.90	1.92	0.59	0.92
214	♀ ad.	......do	July 7	....do	4.20	1.97	0.60	0.92
216	♀ ad.	......do	July 7	....do	4.13	1.96	0.63	0.96
217	♀ ad.	......do	July 7	....do	4.13	2.03	0.63	0.93
213	♂ ad.	......do	July 7	....do				
215	♂ ad.	......do	July 7	....do				
218	♂ ad.	......do	July 7	....do				
219	♀ ad.	......do	July 7	....do				
230	♂ ad.	......do	July 8	....do				
232	♀ ad.	......do	July 8	....do				
233	♀ ad.	......do	July 8	....do				
234	♀ ad.	......do	July 8	....do				
235	♀ ad.	......do	July 8	....do				
236	♂ ad.	......do	July 8	....do				
237	♂ ad.	......do	July 8	....do				
220	♀ jun.	......do	July 8	....do				
760	jun.	......do	July 8	....do				
761	jun.	......do	July 8	....do				

HÆMATOPIDÆ.—OYSTER-CATCHERS.

148.—*Hæmatopis niger*, Pallas.—Black Oyster-catcher.

Hæmatopis niger, Pallas,'Zoog. Rosso-Asiat., ii, 1811, 131.—Towns., Narr., 1839, 348.—Cass., Bd.
 B. N. A., 1854, 700.—Coop. & Suckl., P. R. R. Rep., xii, pt. ii, 1860, 233.—Cones, Key N. A. B.
 1872, 246.
Hæmatopis townsendii, Heerm., P. R. R. Rep., 1859, 65.

This curious bird is found in considerable numbers on the island of Santa Cruz, and, as I was informed, on the others of the group. They much of the time frequented the little islets which were separated from the main islands by narrow channels, probably finding on them breeding-grounds safe from the intrusion of all enemies. Their short, extremely stout legs and feet would seem to adapt them for a life among the rocks, and they probably do obtain much of their food among the kelp and sea-weed that covers the slippery rocks and shelters all sorts of crustaceans and mollusks. The long, strong, wedge-like bill is admirably adapted for the purpose of prying open the bivalve shells. On the island, however, they seemed to obtain a plenteous supply of food by a much easier and readier method, and did not resort to this mode at all. They fed much of the time on the sandy beaches piso, where the Sand-pipers, had there been any, would have resorted, and, like them, found all they wanted on the surface, where it was cast up by the waves. Their stout robust form would not seem to indicate much agility, and their movements were rather clumsy, as though they felt a little out of place. On the level beaches, they were quickest when they followed the retreating waves to the lowest point, whence they would have time to snatch a titbit and run back in season to avoid the on-coming surf. The birds were not at all shy, and would permit me to approach easily enough within 30 yards of them as they rambled along the beach, pausing now and then, and looking back as if not quite assured of my intentions.

Of all the feathered tribe that frequented the island, they were the noisiest, and their harsh vociferous cries could be heard all day long, coming from their island strongholds.

After some search, I succeeded in finding two nests: the first containing a single freshly laid egg was taken June 6, on the extreme point of a high cliff jutting over the sea; the second, a few days later, was found on one of the islets adverted to. The nests proper were rude enough affairs, being simply slight hollows made in the pebbly detritus, which in each case had been added to by bits of stone brought from elsewhere. In neither case was there any grass or other lining softer than the stones themselves. The two eggs in one case were slightly incubated, and probably were all that would have been laid. These are indistinguishable from those of the better known species *H. palliatus*. Their ground-color is a faint, grayish drab, profusely spattered with irregular blotches of black. They measure 2.27 by 1.59; 2.29 by 1.48; 2.18 by 1.52.

The Surf-bird (*Aphriza virgata*) was not found by us on the island of Santa Cruz, and I do not think it breeds on this group. Heermann mentions finding numbers on the Farallone Islands in June, and here it is likely it remains all summer. Mr. Gruber showed me a fine specimen which he obtained at Santa Barbara in spring. It seems to be a rather uncommon species on the Californian coast, and one whose habits are very little known.

No.	Sex.	Locality.	Date.	Collector.	Wing.	Tail.	Bill.	Tarsus.
8	♀ ad.	Santa Cruz Island, Cal........	June 4	H. W. Henshaw	9. 75	4. 73	3. 00	2. 03
727	♀ ad.	.......do	June 4	....do	10. 25	5. 03	3. 12	2. 12
728	♂ ad.	.......do	June 4	....do	9. 50	4. 70	2. 73	1. 87
729	♂ ad.	.......do	June 5	....do	10. 15	4. 83	2. 88	2. 05
730	♀ ad.	.......do	June 5	....do	10. 50	5. 02	2. 75	2. 04

149. *Strepsilas melanacephalus*, Vigors.—Black Turnstone.

This bird occurs on the islands, and all along the coast of California, during the spring and fall migrations. All pass to high northern latitudes to breed.

The *S. interpres* also occurs along the coast.

RECURVIROSTRIDÆ.—AVOCETS; STILTS.

150. *Recurvirostra americana*, Gmel.—American Avocet.

The Avocet occurs in California, though not, I think, in such extreme abundance as in many sections in the interior. On the island of Santa Cruz I saw several, and these had paired, and were probably breeding. As there were no ponds they were driven to a different mode of life from their usual one, and lived on the beaches, picking up sea-slugs and small crustaceans from the surface. They were present near Los Angeles in June, and apparently on their return journey from more northern parts had stopped in quite large flocks at Kern Lake, where it is possible some may remain all summer.

No.	Sex.	Locality.	Date.	Collector.
738	Ad.	Santa Cruz Island, Cal..	June 10	H. W. Henshaw.
739	Ad.	......do	June 10	Do.

151. *Himantopus nigricollis*, Vieill.—Black-necked Stilt.

The Stilt was present in large numbers on the borders of Kern Lake August 15. Usually when I have found these two birds together, the Avocets have far outnumbered the Stilts. Here the case was reversed, and the number of the latter species was largely in excess of any other bird found here.

No.	Sex.	Locality.	Date.	Collector.
355	♂ jun.	Kern Lake, Cal...	Aug. 15	H. W. Henshaw.

PHALAROPHIDÆ.—PHALAROPES.

152. *Lobipes Hyperboreus*, (L.).—Northern Phalarope.

This Phalarope is numerous along the coast in Washington Territory, and probably also in California during the migrations. I shot a single one, evidently a wanderer, on a small meadowy mountain-stream in the Sierra Nevada September 15, at an elevation of about 9,000 feet.

No.	Sex.	Locality.	Date.	Collector.
473	♂	Head of Kern River, Cal..........................	Sept. 15	H. W. Henshaw.

SCOLOPACIDÆ.—SNIPES.

153. *Gallinago Wilsoni*, (Temm.).—Wilson's Snipe.

This Snipe is very abundant in all localities suited to its wants during the migrations, and probably more or less winter in the southern part of the State.

154. *Macrorhamphus griseus*, (Gmel.).—Red-breasted Snipe.

During the migrations, occurring in large flocks along the coast, and also on the lakes and ponds in the interior of the State. Present in numbers at Kern Lake in August.

155. *Ereunetes pusillus*, (Linn.).—Semi-palmated Sandpiper.

I saw a small flock of these "Pups" on the sea-shore near Santa Barbara, in July. Shooting several, I found, upon dissecting, that they were barren birds, which accounted sufficiently for their presence here at this time. During the migrations it is abundant.

No.	Sex.	Locality.	Date.	Collector.
221	♀	Santa Barbara, Cal...	July 7	H. W. Henshaw.
222	♀	......do	July 7	Do.

156. *Calidris arenaria*.—Sanderling.

This bird occurs more or less numerously on the coast during the migrations. I took a single specimen on the Santa Cruz Island in June. Its journey northward had been interrupted by an injury, possibly a gunshot wound.

No.	Sex.	Locality.	Date.	Collector.
740	Ad.	Santa Cruz Island, Cal	June —	H. W. Henshaw.

157. *Limosa fedoa,* (Linn.).—**Great Marbled Godwit.**

The Godwit appears in large flocks on the Californian coast in spring and fall. June 16, a fine specimen was brought me by Mr. J. A. Hasson, who shot it on some salt-ponds near Los Angeles, where he stated he saw many others. The bird was in worn breeding-dress, and I am inclined to judge that many find here their summer-home. According to Dr. Cooper, it abounds at Shoalwater Bay, Washington Territory, though he thinks all pass north to breed. The Willet (*Totanus semipalmatus*) also occurs abundantly on the coast.

No.	Sex.	Locality.	Date.	Collector.
23	Ad.	Los Angeles, Cal..	June 16	H. W. Henshaw.

158. *Totanus melanoleucus,* (Gmel.).—**Greater Yellowlegs.**

Occurs numerously during the migrations both on the coast and in the interior. I am not aware that the Lesser Yellowlegs has actually been recorded from the Pacific coast. Its occurrence here is, however, extremely probable, as the range of the two species is almost exactly coincident.

159. *Tringoides macularius,* (Linn.).—**Spotted Sandpiper.**

An individual of this species seen now and then on the fresh-water streams of the interior.

160. *Heteoscelus incanus.* (Gm.).—**Wandering Tattler.**

The Wandering Tattler, as this bird is aptly named, possesses a very extreme distribution, being found on the islands of the Pacific generally, and from Russian America to Australia. It has been found in Washington Territory by Dr. Cooper, who, however, it was not common. Santa Cruz Island was the only place where I enjoyed the opportunity of seeing the bird, though it is found, as I learned from others, on the other islands also. Captain Forney, of the Coast Survey, who has paid considerable attention to the birds of these islands, presented me with a specimen, one of quite a number he secured on San Miguel, where I should judge the bird must occur in considerable numbers. They appear not to be a bird of the sandy shores at all, but resort exclusively to the rocks covered with sea-weed, where they follow the tide as it ebbs and flows, running back and forth and picking up the minute worms and marine animals, of which they find a great abundance. In motions, they simulate exactly the little Spotted Sandpiper, and have the same curious "tip-up" motion of the body, which they indulge in at moments of rest from feeding or when attentively looking about them. They fly, too, with the same deliberate wing-beats, the pinions being slightly decurved, the tips pointed downward. Their voices are, however, wholly different, and the notes are very loud and harsh when compared with the smooth whistle of the other species. I found them usually solitary and quite watchful and full of distrust, though I found myself once or twice within a few feet of one of them, and was allowed a most excellent chance to watch their motions. This was June, and the species was unquestionably paired and breeding, though I obtained no hint of their method of nidification.

No.	Sex.	Locality.	Date.	Collector.	Wing.	Tail.	Bill.	Tarsus.
6	♀ ad.	Santa Cruz Island, Cal	June 4	H. W. Henshaw...	6.70	3.13	1.54	1.37

161. *Numenius longirostris,* Wils.—**Long-billed Curlew.**

This Curlew is numerous during the migrations. It was present in large flocks on the borders of Kern Lake in August. The Esquimaux Curlew (*N. borealis*) is also said by Heermann to be common in the San Francisco market.

TANTALIDÆ.—IBISES.

162. *Ibis thalassinus,* Ridgw.—**Glossy Ibis.**

This Ibis is probably a summer resident in suitable localities throughout the interior of the State. It was a common bird at Kern Lake in August; flocks of considerable size being seen there.

ARDEIDÆ.—HERONS.

163. *Ardea herodias*, Linn.—Great Blue Heron.

Of common occurrence.

164. *Herodias egretta*, (Gmel.).—Great White Egret.

This Heron was seen on quite a number of different occasions in various parts of Southern California. It appears to be a rather common summer resident. The little Egret (*Garzetta candidissima*) is also said by Heermann to be numerous.

165. *Butorides virescens*, (Linn.).—Green Heron.

Common.

166. *Botaurus minor*, (Gmel.).—Bittern.

Quite numerous on the marshes throughout the State.

GRUIDÆ.—CRANES.

167. *Grus canadensis*, (Linn.).—Sand-hill Crane.

Of common occurrence in California.

RALLIDÆ.—RAILS.

168. *Rallus elegans*, Aud.—King Rail.

This Rail was common in certain marshy spots close to the sea at Santa Barbara. They retired during the day into the beds of tall rushes, which served to screen them from all enemies, as well as from the glaring sun. By July 1 the young were out and able to accompany their parents after food. They began to be active about sunset, heralding the approach of dusk by loud outcries. They were not altogether quiet during the day, and they are probably forced to forage more or less during the uncongenial hours of day to satisfy the hunger of their young.

169. *Rallus virginianus*, Linn.—Virginia Rail.

This is perhaps the most generally distributed of the family throughout the United States. It appears to be quite numerous in Southern California; as much so in certain localities as anywhere in the East. I found it abundant at Walker's Basin in November, and it probably winters throughout the southern half of the State. The Sora Rail (*Porzana carolina*) was not noticed by us, nor do I find it recorded from the west coast. The Black Rail (*P. jamaicensis*) appears to be fully as numerous in California as in any other part of its habitat. From information from Mr. Gruber I should judge it was rather common in the extensive tulle swamps in the State. It has also been found by this gentleman on the Farallone Islands. Its small size and skulking habits, combined with the inaccessibility of its swampy haunts, render the procuring of specimens exceedingly difficult.

170. *Fulica americana*.—Coot ; Mud-hen.

Very abundant on the fresh-water ponds throughout the State where they breed. The species is a resident one, though a migration in spring and fall occurs with perfect regularity.

ANATIDÆ.—GEESE AND DUCKS.

171. *Anser hyperboreus*, Pallas.—Snow-goose.

Great numbers of this Goose were seen on the prairies and in the stubble-fields south of San Francisco in November.

172. *Anser albifrons* var. *gambelii*, Hartlamb.—White-fronted Goose.

Immense numbers of this species winter in California, returning from their northern breeding-grounds in October and November.

173. *Branta canadensis*, (Linn.).—Canada Goose.

Very numerous in fall and winter.

174. *Branta canadensis*, (Linn.), var. *hutchinsii*, Rich.—Hutchins's Goose.

Vast numbers throng the State, both along the coast and on the interior prairies.

175. *Anas boschas*, Linn.—Mallard.

The Mallard is found in great abundance in fall and winter, while many doubtless remain to breed in the lakes. It is found on the mountain-streams to a height of even 9,000 feet. In fall, it is quite terrestrial in its mode of life, and gleans a rich harvest from the grain-fields.

176. *Dafila acuta*, Linn.—Pin-tail.

Numerous during the migration.

177. *Chaulelasmus streperus*, Linn.—Gadwall.

More or less numerous in the fall and winter. I saw numbers in the San Francisco market in November.

178. *Mareca americana*, (Gmel.) —American Widgeon.

Numbers seen in Walker's Basin in November. Abundant in winter.

179. *Querquedula carolinensis*, (Gmel.).—Green-winged Teal.

Abundant on the fresh-water courses throughout the State. Both the Blue-winged and the Red-breasted Teal occur in abundance in California; the latter as a summer resident leaving the State early in fall for farther north.

180. *Spatula clypeata*, (Linn.).—Shoveler.

Very numerous in the fall and winter.

181. *Fuligula marila*, (Linn.).—Greater Blackhead.

Abundant in fall and winter; chiefly along the coast.

182. *Fuligula ferina*, (Linn.), var. *americana*, Eyton.—Redhead.

Common in fall and winter.

183. *Fuligula vallisneria*, (Wils.).—Canvas-back Duck.

This Duck was present, though not numerous, in Walker's Basin in November. Dr. Newberry speaks of the species as being found in large numbers in the bays and rivers of the State in fall and winter.

184. *Bucephala clangula*, (Linn.).—Golden-eye.

An abundant species, visiting the State in fall and remaining through the winter. I was informed by Mr. Gruber that the Barrow's Golden-eye (*B. islandica*) was found occasionally in the San Francisco markets, where he had procured specimens. As on the east coast, it breeds quite far to the north, descending chiefly along the coast in winter.

185. *Bucephala albeola*, (Linn.).—Butter-ball.

Perhaps the most common and widely-distributed of the genus.

186. *Harelda glacialis*, (Leach).—South Southerly; Old Squaw.

Doubtless an abundant visitor to the sea-coast of the State in winter. A single specimen was shot at Santa Barbara in June. This bird, a female, had from some cause or other remained behind, and the plumage was so much worn and in such a faded condition as to be scarcely recognizable.

No.	Sex.	Locality.	Date.	Collector.
9	♀	Santa Barbara, Cal..	June 9	H. W. Henshaw.

187. *Œdemia perspicillata ; Melanetta velvetina ; Pelionetta perspicillata.*

The three species of Sea Coot occur abundantly all along the Californian coast in winter.

In passing down the coast in June from San Francisco to Santa Barbara, I saw large numbers of Coot all along the shore and in the little bays. These were probably the young and barren birds, which did not go north to breed. Of what species these were, or whether, as is probable, the three were not represented, I am not able to say. About the island of Santa Cruz, they were to be seen at this time by hundreds. A single one shot here proved to be the *Melanetta*; but I am reasonably sure that all three birds were present. Their fishy diet and coarse flesh render them, if edible at all, anything but palatable food, and hence they are scarcely ever disturbed. As a result, they have become very tame, and approach close to the wharves and vessels in the harbor of San Francisco with the utmost unconcern.

188. *Mergus serrator,* (Linn.).—Red-breasted Merganser.

Very numerous in fall and winter, both on the coast and inland. A single bird, worn and faded, was shot, June 7, at Santa Barbara. The *Mergus merganser* also occurs in large numbers.

189. *Mergus cucullatus,* (Linn.).—Hooded Merganser.

Appears in fall in large numbers as a migrant.

PELECANIDÆ.—PELICANS.

190. *Pelecanus trachyrynchus,* (Lath.).—White Pelican.

The most conspicuous of all the feathered tribe that we found assembled at Kern Lake were the White Pelicans, noticeable both from their great size and the extreme whiteness of their plumage. This was in August, and the birds had probably remained here all summer, breeding somewhere about the lake. During the hours of mid-day they appeared to give up fishing entirely, and, betaking themselves to some dry spot along the lake, they dozed away the unoccupied hours, standing motionless in long rows, with their heads drawn on their breasts, and appearing lost to all around them. They were not, however, so taken up with their own meditations as to be forgetful of safety, and roused themselves always in time to be up and away ere I could get fairly within gunshot. They breed very early. Captain Bendire found the eggs of this bird in Oregon as early as April 12, though they continued laying eggs till into May. They are present upon all the inland waters of any size in California, and less often and in fewer number are found upon the coast.

191. *Pelecanus fuscus,* (Linn.).—Brown Pelican.

Pelecanus fuscus, Newb., P. R. R. Rep., vi, 1857, 108.—Bd., B. N. A., 1858, 870.—Heermann, P. R. R. Rep., x, 1859, 72.—Coop & Suckl., ibid., 12, 1860, 966.—Coues, Key N. A. B., 1872, 300.

In contrast to the habits of its more showy white cousin, which resorts to the fresh waters of the interior, breeding and living there, the Brown Pelican is found exclusively on the sea-coast, resorting to the bays and shallow inlets where are found the small fry which constitute its food. The waters about San Francisco are particularly favored by this bird, and in a trip across the bay one may see hundreds of these huge, uncouth birds winnowing their way from one fishing-ground to another with slow, measured wing-beats, or diving with sure aim from mid-air on some luckless fish swimming near the surface. Undisturbed, they roam the bay at will, viewing the approach of steamer and vessel with utmost unconcern, and often, indeed, remaining on the water till almost run down by the approaching craft, when they lazily clear the water with heavy strokes and fly from almost under the bows. On account of their heavy bodies and the length of wings, they raise themselves with some little difficulty, and it requires a number of quick, vigorous strokes, delivered upon the surface of the water, ere they can get fairly on the wing. They progress easily and firmly, now flopping their broad wings till the desired momentum is obtained, now gliding without motion on outstretched pinions. When fishing, they keep a few feet, from 10 to 20, above the water, and when a fish is discovered they gather themselves for the effort by a few short strokes of the wing; then with head down descend, making the water foam with the violence of their plunge. At night-fall they retire from the bay to distant sleeping-grounds, probably, as noticed by Dr. Newberry, to the broad expanse of the ocean, and when going and coming they fly in lines, and just clear the surface of the water, falling and rising with the heaving waves.

About the island of Santa Cruz, these birds were uncommon, and I saw but few.

GRACULIDÆ.—CORMORANTS.

192. *Graculus dilophus.*—Gray Double-crested Cormorant.

This species nests on the Farallone Islands in great abundance, as also upon the Santa Barbara group. It is common along the coast, and is also found on the large bodies of water inland, as at Kern Lake, where it was numerous.

193. *Graculus penicillatus,* (Gray).—Brandt's Cormorant.

This is one of the Cormorants found upon the Farallone Islands in summer, and no doubt breeds also on the Santa Barbara Islands, though I was not able to satisfy myself perfectly of its presence on Santa Cruz in June. A specimen, however, taken on San Miguel was very kindly presented me by Captain Forney, who shot numbers of the same kind.

194. *Graculus violacens,* (Gray), var. *bairdii,* Cooper.

The Violet-green Cormorant of Oregon, Washington Territory, and to the northward, is represented on the Californian coast by a smaller bird, which appears to be its southern race. The difference is one chiefly of size, the discrepancy being in this particular considerable and out of the range of purely individual differentiation. The proportions, colors, &c., of the two appear to be identical.

This bird is very numerous all along the coast of Southern California, and probably reaches northward into Oregon. I saw many in San Francisco Bay in May, and on reaching the islands of the Santa Barbara Channel it was found congregated in great numbers. Most of the places they had selected as nesting-sites were inaccessible to me. At low tide I succeeded in entering one of the gloomy caverns, where a dozen pairs had established themselves. The nests were merely collections of weeds and sticks matted together and placed upon the shelves of rock sufficiently high to be out of danger from the tide. This was June 4, and they all contained young in the downy state. The old birds forsook the place in a mass, and flew wildly about the entrance, but without attempting to re-enter, though the young birds kept up a vociferous calling all the while. In flying about the island, the old birds passed within easy gunshot of the rocky points, and I could have procured all the specimens I desired had it not been for the strong surf which swept the shores and made their recovery very hazardous. They never ventured over the land. It is a constant habit with these birds, having spent the morning in fishing, and having appeased their hunger, to sit in groups on the cliffs which immediately overhang the water, and often in such numbers as to blacken the rocks. When disturbed, those nearest to the edge drop overboard, while those in the rear scramble forward in the most awkward way, and, having made the plunge, swim underneath the water till they have gained a safe distance.

The present species was immediately recognizable among its congeners by its small size. The white flank-tufts are, I think, a distinguishable feature of the breeding-period, and are soon lost. They were seen only in the males, and the size is extremely variable; being in some individuals scarcely discernible, while in others they were conspicuous at a long distance.

No.	Sex.	Locality.	Date.	Collector.	Wing.	Tail.	Bill.	Tarsus.
784	♂ ad.	Santa Cruz Island, Cal	June —	H. W. Henshaw...	9 75	7.00	1.85	1.90
785	♂ ad.	......do	June —	....do	9.75	7.00	1.78	1.85
783	♀ ad.	......do	June —	....do	10.10	7.50	1.97	2.10

LARIDÆ.—GULLS; TERNS.

195. *Larus argentatus,* (Brünn.,) var. *occidentalis,* Aud.—Pacific Herring Gull.

This Gull is very numerous in San Francisco Harbor, as it is indeed in all the bays and inlets of the coast, and its numbers are perhaps greater the year round than any other species. Free from molestation, they have become almost semi-domesticated, and fly about the wharves and over the vessels with an impunity only born of long immunity from danger. The rocky islets along the coast furnish them with safe and plentiful breeding-grounds. At Santa Cruz, thousands had congregated and were nesting in early June. In a few instances, they had attempted to nidificate on the mainland, but a few feathers and bits of egg-shells about the nests told in each case the fate of parent and eggs; their enemy was the foxes, whose numbers are scarcely without limit. Only one of the small adjoining islets was accessible to me. A few pairs had nested here. The nests were made of a good generous supply of sea-weed and like material, well matted together, the cavity being quite deep. The eggs are of a greenish olive, spattered profusely and irregularly with blackish markings.

No.	Sex.	Locality.	Date.	Collector.	Wing.	Tail.	Bill.	Tarsus
11	Ad.	Santa Cruz Island, Cal..	June 9	H. W. Henshaw...	16.75	7.00	2.29	2 5
22	♀ ad.	......do	June 11	....do	15.50	6.60	2.08	2.34
781	Ad.	......do	June 11	....do	16.10	6.00	2.34	2 4

196. *Larus (Blasipus) heermannii,* (Cass.)—Heermann's Gull.

A very large flock of these Gulls was seen pursuing their way along the shore near Santa Barbara, and two or three hours later I came upon them where they had settled upon a rocky point which jutted out into the water. Many were engaged fishing, hovering over the half-submerged kelp-covered rocks, the shallow water surrounding which evidently harbored the smaller kinds of fish. In three or four discharges of my gun I obtained a dozen specimens, when the whole mass flew wildly about as though fascinated by the sight of their dead comrades, and it was some time ere they left the place, which they did in a long straggling flock. The whole flock was composed of old males, and from their long and direct flight it seemed pretty evident that the journey was one between their breeding-grounds and this fishing-place, where was had an abundance of some food, perhaps peculiarly fitted for the young. The species is a common one all along the coast, and breeds upon many of the sea-islands.

Several other species of Gulls and at least one species of Jaeger were observed along the coast; but my acquaintance with them was altogether of too unsatisfactory a nature to warrant any mention of them here.

No.	Sex.	Locality.	Date.	Collector.	Wing.	Tail.	Bill.	Tarsus.
39	♂ ad.	Santa Barbara, Cal	June 24	H. W. Henshaw...	13.25	6.00	1.70	1.98
40	♂ ad.	......do	June 24	....do	13.90	6.20	1.83	1.98
41	♀ ad.	......do	June 24	....do	13.70	6.00	1.77	2.03
42	♂ ad.	......do	June 24	....do	13.25	5.75	1.75	1.90
44	♂ ad.	......do	June 24	....do	14.25	6.00	1.85	2.09

197. *Sterna regia,* (Gamb.)—Royal Tern.

This Tern probably reaches no farther to the north than the coast of California, where it appears to be of rather common occurrence. I saw it up as far as San Francisco. A specimen obtained on San Miguel was presented by Captain Forney, to whom I am under obligation for other similar kindnesses. Upon this and perhaps others of the Santa Barbara group, the species breeds. At least one other of the small Terns was seen about San Francisco, but its identity could not be established.

198. *Hydrochelidon lariformis,* (Linn.).—Black Tern.

Saw this species but once, in the interior, north of Los Angeles; according to other observers, the bird is numerous on the inland waters.

PROCELLARIIDÆ.—PETRELS.

199. *Cymochorea homochroa,* (Coves.)—Lesser Black Petrel.

Petrels are quite numerous along the coast of California. A specimen of the above species, so identified by Dr. Coves, was given me by Captain Forney, who found these birds breeding in great numbers on San Miguel. As usual, they were nesting in burrows. The relationship of this bird with the *C. melania* is somewhat obscure, and a good series to confirm this supposed distinctness is greatly to be desired.

No.	Sex.	Season.	Locality.	Collector.	Wing.	Tail.	Bill.	Tarsus.
.....	Ad.	Summer.	San Miguel Island, Cal........	Captain Forney.	5.35	8.35	0.60	0.88

COLYMBIDÆ.—LOONS.

200. *Colymbus torquatus,* (Brünn.)—Great Northern Diver.

Numerous on the California coast in fall. The *C. arcticus* var. *pacificus* is also known to be common in winter.

PODICIPIDÆ.—GREBES.

201. *Podiceps auritus,* (Linn.), var. *californicus,* Heerm.—Eared Grebe.

Coincident with its general dispersion in the West, this Grebe appears to be distributed over California. We only saw it in the fresh-water ponds, though it also occurs along the shore.

No.	Sex.	Locality.	Date.	Collector.
29?	♂ ad.	Near Fort Tejon, Cal	July 24	H. W. Henshaw.

278

202. *Podilymbus podiceps*, (Linn.).—Carolina Grebe.

Present in numbers on many of the fresh-water ponds of the interior; found also on the coast. The *P. cornutus* was not recognized by us, though it, too, is found numerously in fall.

ALCIDÆ.—AUKS.

203. *Fratercula cirrhata*, (Pallas).—Tufted Puffin.

This Puffin, though more commonly known as a resident of the far north, was ascertained by us to inhabit the islands of the Santa Barbara group in summer. It was not uncommon, and was nesting apparently in the crevices of the cliffs, from which I frequently saw it flying back and forth. Heermann likewise found it breeding in numbers on the Farallone Islands.

204. *Uria columbia*, (Pallas).—Western Guillemot.

The Santa Barbara Islands form, too, it is probable, about the southern limit for this species in summer; among them it is, however, numerous—breeding in the caves and hollows of the generally inaccessible cliffs. Early one morning, while out collecting, I noticed many of these birds frightened at the report of my gun, streaming out of a little ravine hemmed in by high rocky cliffs, and terminating at the upper end in a low narrow cave. The tide being at its lowest, I succeeded in gaining the entrance, and, crawling on my hands and knees for a short distance, I soon had the satisfaction of placing my hands on the eggs. Their housekeeping arrangements are of the simplest kind. No nest at all is prepared to receive the eggs; but these were deposited on the sandy floor of the cavern, and at its farther end, where it was so dark that I had to strike a match to see them at all. Other pairs had availed themselves of the nooks and fissures in the face of the wall, and laid their two eggs on the bare rock. I succeeded in finding a few only of the many eggs that must have been deposited here, as the shelves of the rocks were, in many instances, too high to be reached. The birds submitted to the pillage without a murmur, though not without solicitude, as their anxious manner as they swam back and forth at the entrance to the ravine, keeping, however, well out of gunshot, sufficiently evinced.

The eggs are a faint greenish white, spotted mostly at the larger end with irregular blotchings.

Very respectfully, your obedient servant,

H. W. HENSHAW.

Lieut. GEO. M. WHEELER,
   Corps of Engineers, in Charge.

APPENDIX H 9.

REPORT ON THE ORTHOPTERA COLLECTED BY THE UNITED STATES GEOGRAPHICAL SURVEYS WEST OF THE ONE HUNDREDTH MERIDIAN, UNDER THE DIRECTION OF LIEUTENANT GEORGE M. WHEELER, DURING THE SEASON OF 1875, BY SAMUEL H. SCUDDER.

CAMBRIDGE, MASS., *May* 29, 1876.

SIR: The explorations during the past season covered a region from which we have hitherto received very little material; on the eastern slope of the Rocky Mountains scarcely any *Orthoptera* have been collected south of Colorado, while the Orthopterous fauna of the entire Pacific coast has been but little studied. It is, then, hardly surprising that of the fifty species here enumerated, mostly forwarded to me in half a dozen small bottles, nearly one-half should prove new to science. It would have been more satisfactory if the descriptions given here could have been generally based on a richer material, and upon specimens which had never been subjected to an alcoholic bath; for not only are most of the colors obliterated by long immersion in alcohol, but the structural features are falsified by the unnatural prominence given to all angulations, and the deeper hollowing to sulcations, or even to flat surfaces. The nature of the explorations, however, renders it nearly impossible, at times quite impossible, to preserve and transport objects of natural history in any other way; and the rich proportion of novel forms which the moderate collections of a single year have exposed will in part make amends for the somewhat unsatisfactory nature of the material.

It should be added in this connection that all the specific descriptions which follow (and also all of the generic descriptions, which cover only the species mentioned in this report, excepting *Hadrotettix*) have been based upon specimens more or less imperfect from their preservation in alcohol, with the exception of *Steirozys melanopleura*, *Arphia teporata, Camnula atrox*, and *Œdipoda venusta;* of all of these I had in my own collection ordinary cabinet-specimens.

It had been my intention to present at this time some general observations upon the geographical distribution of North American *Orthoptera*, and also a revision of the North American *Pezotettigi* and *Œdipodidæ*, so richly represented in this collection. Other engagements prevent the present fulfilment of this intention, and I can only offer the student, in place of the latter, brief tables of the species obtained by the Survey during this single season, which will be found preceding the groups mentioned.

A few specimens in an immature condition have not been determined.

Respectfully submitted.

SAMUEL H. SCUDDER.

Lieut. GEO. M. WHEELER,
Corps of Engineers, in Charge.

GRYLLIDES.

1. *Gryllus abbreviatus* Serv.—Two males are referred to this species with some doubt; one is from the Colorado River, California, collected July 28 by W. Sommers, No. 867; the other from Santa Cruz Island, collected in June by Dr. O. Loew, No. 840.

2. *Gryllus neglectus* Scudd.—A number of specimens were taken on the plains of Northern New Mexico, eastern slope, October 14–31, and others in Southern Colorado June 11–20, by Lieut. W. L. Carpenter.

3. *Gryllodes lineatus*, nov. sp.—Head above blackish brown, conspicuously striped longitudinally with pale testaceous; a broad band follows the upper limits of the antennal sockets and the eye, and above the middle of the latter emits an oblique, slender shoot to the pronotum; midway between this and a testaceous median line is a slender, rather short, testaceous stripe, reaching neither extremity of the vertex; upper limit of clypeus brownish, rest of face and mouth-parts pale testaceous, the lower part of clypeus white; antennæ brownish testaceous; apical joint of maxillary palpi tipped with brown, obliquely docked on the apical third. Pronotum quadrate, the sides slightly rounded, black, with small, lateral, pale, testaceous spots, and a distinct median sulcation. Tegmina half as long as the abdomen, testaceous, the lateral field with four nervures, of which the lowest is forked from the base, the upper half of the field blackish brown; wings of the same length as the tegmina. Legs yellowish, the hind femora heavily marked with dark brown in crowded, oblique, abbreviated lines; hind tibiæ dusky beneath, the extreme tip of spines black; first hind tarsal joint half as long as the hind tibiæ, armed above on either side with a row of half a dozen rather stout recumbent spines. Abdomen black, the last superior segment and the cerci light brown. Length of body, 11.5^{mm}; of tegmina, 3.5^{mm}; of hind tibiæ, 4.6^{mm}; of cerci, 5.5^{mm}. One ♂ taken between Virgin River and Fort Mojave, Arizona, August, W. Sommers, No. 858.

This species is closely allied to *G. pusillus* (Burm.), but has short wings.

LOCUSTARIÆ.

4. *Stenopelmatus talpa* Burm.—Specimens in various stages of development were taken on Santa Cruz Island, in June, by Dr. O. Loew, No. 851; at Los Angeles, Cal., in June, by J. Brown, No. 808; at Los Angeles and Santa Barbara, in June, by C. J. Shoemaker, No. 798; at Santa Barbara, June 10, by Dr. O. Loew; and, in July, by H. W. Henshaw, No. 880; same place, July 1, H. W. Henshaw, No. 1005; and, in the Mojave Desert, in July, by Dr. O. Loew, No. 935.

5. *Udeopsylla robusta* (Hald.) Scudd.—One ♂, Northern New Mexico, August to September, Lieut. W. L. Carpenter.

6. *Ceuthophilus pallidus* Thom.—Two ♂, five ♀, Southern Colorado, June 11–20, Lieut. W. L. Carpenter; plains of Northern New Mexico, eastern slope, October 14–31, Lieut. W. L. Carpenter.

7. *Ceuthophilus denticulatus*, nov. sp.—Yellowish brown, the segments narrowly margined posteriorly with brown, the face and under surface of the body paler; frontal tubercle a scarcely perceptible swelling; eyes subpyriform, small, not very prominent. Fore coxæ unarmed; all the femora, as well as the front and middle tibiæ, minutely serrulate beneath; fore femora with but a single apical interior spine; fore tibiæ with three pairs of spines besides the apical ones; middle femora with three or four minute spines on the outer carina; middle tibiæ with three pairs of inferior spines besides the apical ones, and two distant spines on the posterior face, nearly equidistant from each other and either extremity; posterior femora without spines; posterior tibiæ rather coarsely serrulate throughout on either carina, and provided also with four pairs of long superior spines, besides those at the apex. Ovipositor a little longer than the

middle femora, the exterior valves turned abruptly upward and very finely pointed at the tip, the inner valves furnished with five curved aciculate teeth, growing larger apically, and there equaling the breadth of the ovipositor. Length of body, 15ᵐᵐ: hind tibiæ, 9.25ᵐᵐ; of ovipositor, 4.75ᵐᵐ.

1 ♀, Santa Barbara, Cal., June 10, Dr. O. Loew.

This species is closely allied to *C. californianus* Scudd., but differs decidedly from it in the length of the ovipositor, which is nevertheless shorter than in most of the other species, and from all species known to me in the extreme length of the teeth of the ovipositor. Thomas's *C. castaneus* is probably to be referred, as he suggests, to my *C. californianus*, as the measurement given of the hind femora of the latter should read .32 in instead of .22 in, a typographical error to which my attention was first called by his description. I may add that 'while the hind femora of *C. californianus* are not spined beneath, they are minutely serrulate throughout, as in the present species.

8. *Anabrus coloradus* Thom.—This species differs from *A. simplex* Hald., in its smaller size, shorter pronotum, in the character of the posterior lobe of the same, which is slightly broader, distinctly though but slightly carinate, with lateral carinæ, distinct though obtuse and slightly more docked posterior border; the lower border of the deflected lobes from front to back is broadly margined with pale yellow, made more conspicuous posteriorly by being broadly edged above with reddish brown. As in that species, the tegmina and wings are completely covered by the pronotum in both sexes. The ovipositor is gently and regularly curved, while in *A. simplex* it is straight, and is black, excepting next the base of the lower valves and next the extreme base of the upper valves, while in *A. simplex* it is black only next the apex; the male cerci are very sparsely and briefly pilose, rather stout, subcylindrical, dividing at about the middle into two fingers, an inner, short, cylindrical, thumb-like process, directed a little upward, less than twice as long as broad, and a gently incurved, tapering, pointed finger; the male cerci of *A. simplex*, on the other hand, are laterally compressed, and have both the appendages abbreviated. Length of body, ♂, 22ᵐᵐ; ♀, 20ᵐᵐ; of antennæ, ♂, 20ᵐᵐ (!); ♀, 25ᵐᵐ; of pronotum, ♂, 8ᵐ·³; ♀, 8.75ᵐᵐ; of hind tibiæ, ♂, 12.25ᵐᵐ; ♀ 13.25ᵐᵐ; of ovipositor, 18.5ᵐᵐ.

2 ♂, 4 ♀, Southern Colorado, June 11-20, Lieut. W. L. Carpenter; Taos Peak, Sangre de Cristo Mountains, New Mexico, at a height of 13,000 feet, (above timber,) Lieut. W. L. Carpenter.

This species is closely allied to *A. simplex* Hald. The species described and figured under this name by Herman (Verh. zool.-bot. Gesellschaft, Wien, xxiv, 209, tab. vi, figs. 76-86), though undoubtedly a true *Anabrus*, can certainly not be the *A. simplex* of Haldeman, but is an allied species, unknown to me in nature, in which the ovipositor and male cerci are different, and in which the tegmina (at least of the male) project beyond the pronotum. Apparently it is Herman's species which is figured under the name *A. simplex* by Glover (Ill. Ent. N. Amer. Orth., pl. ix, fig. 1).

It may also be remarked in this place that Thomas (Proc. Acad. Nat. Sci. Phil., 1870, 74 et seq.) has strangely separated *A. purpurascens* Uhl., from this genus, because the prosternum is not spined, whereas the genus was founded upon *A. simplex*, which has an amucronate prosternum! and although he subsequently explains it (Geol. Surv. Montana, 438 et seq.) from his never having seen *A. simplex* when his first paper was written, he still further confuses his readers by stating that *A. Haldemanii* Gir., has the prosternum distinctly spined, whereas it is as clearly amucronate as is the prosternum of *A. simplex*.

9. *Steiroxys melanopleura*, nov. sp.—The top of the head is tumid, well rounded; pronotal shield short but narrow, compressed, the deflected lobes falling off so that the lower part of the prothorax is three times as broad as the upper; three equal but slight longitudinal carinæ extend its entire length, the median broken near the middle by a ψ-shaped impression and crossed anteriorly at the end of the anterior sixth by a slight transverse sulcus, which cuts the lateral carinæ and passes on to the deflected lobes: from this sulcus forward the lateral carinæ diverge slightly; the deflected lobes are divided by a straight transverse impression into two unequal halves, and the deepest point of the lobes is just behind this impression; behind this the hinder edge of the lobe passes upward in a nearly straight, oblique direction, and the posterior margin of the lobe is very broadly rounded. Tegmina very short, broadly rounded, somewhat shorter than the pronotum, with very prominent veins. Fore tibiæ with four spines on the anterior face; outer face of middle tibiæ with five or six alternate spines in a double row; hind legs very long, the terminal interior tibial spine much longer than the first tarsal joint. Male cerci short, stout, cylindrical, suddenly recurved, and tapering, forming a short, horny, black hook; subgenital plate excised, with a slightly obtuse angle apically, the styles as long as the width of the extreme apex of the plate. Length of body, 26ᵐᵐ; of pronotum, 7.75ᵐᵐ; of tegmina, 5.5ᵐᵐ; of hind tibiæ, 27ᵐ·· .

1 ♂, Los Angeles or Santa Barbara, July, C. J. Shoemaker, No. 224.

I place this species in Herman's genus *Steiroxys*, although it differs from it in some structural features. Another species of the same group from California is in my collection.

PLAGIOSTIRA (πλάγιος, στείρα), nov. gen.

Head of moderate size, well rounded, more of the hinder part than usual concealed beneath the pronotum, which is slightly expanded to receive it; the projecting part of the vertex rather slight, narrow, tapering, deeply cleft transversely between the antennæ; eyes round, small, rather prominent. Pronotum short, equal, shallow, flat, and depressed above, the sides nearly vertical; dorsal and deflected lobes separated by very prominent, but obtusely-rounded, transversely wrinkled carinæ, subobsolete anteriorly; anterior margin straight, posterior margin broadly rounded, next the middle slightly excised; dorsal surface flat, depressed, with a very slight longitudinal impressed line, a transverse, slightly arcuate, broad furrow just behind the middle, and a similar but narrow furrow near the anterior margin, the latter extending nearly across the deflected lobe; between these two furrows the dorsum is slightly tumid; the deflected lobes are exceedingly shallow, deepest at about the middle, but scarcely one-third as deep as long, very obtusely angled, and a little rounded at this point, the lower margin a little oblique, and meeting the front margin sharply at a little more than a right angle, the posterior margin scarcely convex, nowhere excavated; prosternum unarmed. Tegmina squamiform, densely reticulated, not gibbous. Fore tibiæ with a single outer row of four spines on the anterior face; middle and hind tibiæ with a similar double row of nearly opposite spines on the outer surface of the former and on the inner surface of the latter; hind legs very slender, and not remarkably long. Mesosternum and metasternum, as well as all the coxæ, bluntly mucronate laterally. Abdomen bluntly ridged above, but not carinate. Supraanal plate acutely triangular with apically submucronate lateral swellings occupying most of the base. Ovipositor as long as the body, straight or nearly so, not serrate at tip; subgenital plate rather deeply and very angularly excised at apex, otherwise simple. The male unknown.

The shallow, flat, quadrate pronotum, carinate laterally only, the anterior portion protecting the head, together with the slender femora and sharply angulate supraanal plate, are striking features of the genus.

10. *Plagiostira albonotata*, nov. sp.—Very pale brownish yellow, the face paler, the whole marked conspicuously with chalky white; a transverse, faint, white stripe follows the lower edge of the cheeks; another, broader and more distinct, starts at the lower edge of the eye and passes backward along the lower margin of the pronotum nearly to the hinder edge; another runs back from the top of the eye to the first transverse sulcus of the pronotum; the middle of the deflected lobes of the pronotum is more or less clouded with it; dashes occur on prominent points on the pleura and outer face of the coxæ; the top of the head bears a pair of diverging longitudinal streaks, equidistant from each other and the eyes; the middle dorsal lobe of the pronotum has two small round spots at the middle of the sides, and two other smaller ones in front of and within them; the posterior lobe has a broad longitudinal dash in the middle of either lateral half; and the abdomen has a pair of dorso-pleural, subcontinuous streaks, and a pair of pleural, discontinuous, oblique streaks, the former with a minute round spot of the ground-color at the distal end of the streak of each segment; tegmina dark brown, the fine reticulations paler; wings shining black, nearly as large as the tegmina. The vertical process of the head is very slightly sulcate longitudinally; the ovipositor is scarcely curved downward, equal nearly to the tip, tapering to a very fine point. Length of body, 21.75^{mm}; of pronotum, 7.9^{mm}; of tegmina, 5^{mm}; of hind tibiæ (which are longer than the femora), 18.2^{mm}; of ovipositor, 23.5^{mm}.

1 ♀, Northern New Mexico, August, September, Lieut. W. L. Carpenter.

11. Another species of this group (*Decticidæ*) was brought from the same locality as the last, but in too immature a condition to determine.

12. A species of *Phyllophoridæ*, of a genus not yet recognized in the United States, was obtained at Los Angeles, Cal., by Wm. Sommers, No. 822; but it is in so poor a condition that it is better to await further material before description.

ACRYDII.

13. *Acridium vagum* Scudd.—2 ♀, Santa Barbara, June 10, Dr. O. Loew, No. 977; Mojave Desert, J. Thompson, No. 826.

14. *Caloptenus spretus* Uhl.—1 ♂, 1 ♀, Taos Peak, Sangre de Cristo Mountains, Northern New Mexico, at a height of 13,000 feet (above timber-line), in July, Lieut. W. L. Carpenter.

15. *Caloptenus minor* Scudd.—1 ♂, Southern Colorado, June 11-20, Lieut. W. L. Carpenter.

The species of *Pezotettix* in this collection fall into five very distinct groups. The first comprises those of Caloptenoid form and structure, like most of those that have been hitherto described. The second comprises but a single species, *P. marginatus*, remarkable for the simple pronotum, inclined face, and prosternal spine. A third comprises also but a single species, *P. vivax*, in which the head is much broader and higher than the rest of the body, the antennæ are long and rather stout, the first joint of the same unusually small. the prosternal thorn inclined backward, and the male abdomen scarcely upturned. The fourth consists of a pair of species, *P. jucundus* and *P. enigma*, of a remarkable appearance; they are clumsy-bodied, the front of the prothorax a little tumid, the deflected lobes having their front border nearly as oblique as their hind border, and the hind legs stout; while the antennae and the prosternal thorn are slight, and the joints of the female abdomen, especially near the tip, are very short and round, and the appendages short, so that the abdomen is almost bluntly rounded at the tip. The fifth group consists again of a single species, *P. pictus*, remarkable for its variegation and its cylindrical body, in which the prosternal spine is bent backward, the tegmina are more than usually lateral, the vertex of the head not at all prominent, and the hind legs very short. I am not aware that any of these groups, excepting the first, is represented in the Old World or upon both sides of the Rocky Mountains in America. These species may be distinguished by the following table:

1 (8) Metasternal lobes of ♂ distinctly though not widely separated.
  2 (3) Tegmina of opposite sides widely separate throughout; sulcus of anterior border of pronotum as distinct above as on the sides.............*pictus*.
  3 (2) Tegmina of opposite sides attingent, in the middle at least; sulcus of anterior border of pronotum distinct only on sides.
    4 (5) Median carina of pronotum equal throughout; hind femora longitudinally striped...*tellustris*.
    5 (4) Median carina of pronotum less distinct on the middle of the anterior than on the posterior lobe; hind femora angularly striped.
      6 (7) Disk of pronotum more or less depressed on either side of the median carina at the posterior sulcus........................*stupefactus*.
      7 (6) Disk of pronotum not distinctly depressed on either side of the median carina..............................*Marshallii*.
1 (8) Metasternal lobes of ♂ attingent.
  9 (12) Anterior lobe of pronotum a little gibbous above.
    10 (11) Pronotum scarcely angled posteriorly; tegmina scarcely longer than, or not so long as, the pronotum............................*jucundus*.
    11 (10) Pronotum distinctly though obtusely angled behind; tegmina distinctly longer than the pronotum........................*enigma*.
  12 (9) Anterior lobe of pronotum of the usual form.
    13 (14) Median carina of pronotum distinct throughout...........*marginatus*.
    14 (13) Median carina of pronotum distinct only on posterior lobe.
      15 (16) Tegmina and wings more than half as long as the abdomen..*plagosus*
      16 (15) Tegmina and wings not so long as the pronotum ....*virax*

16. *Pezotettix tellustris*, nov. sp.—Head light brown, mottled heavily above with dark reddish brown, which is absent from a median longitudinal line on the vertex, and a narrow stripe which runs backward from the summit of the eye; eyes pretty large and prominent, the front edge almost straight; vertex between the eyes as broad as the narrowest part of the frontal costa, and twice as broad as the basal joint of antennæ; fastigium scarcely depressed, expanded a little in front of the eyes; frontal costa sparsely punctate, flat, depressed, and broadened a little at the ocellus, below slightly broader than above; antennæ brownish yellow. Pronotum nearly flat above, broadening regularly backward, the posterior margin obtusely and roundly angulated; median carina slight, but equally distinct throughout; lateral carinæ obtuse, but marked by the abrupt descent of the deflected lobes and by the extension over them of the narrow pale stripe behind the eye; upper surface of anterior and front part of posterior lobe dark reddish brown, extending over the head to the eye; remainder of upper surface dark brownish, obscurely and sparsely punctate; deflected lobes of the color of the head, the upper third or more of the anterior lobe dark fuliginous; all the transverse sulcations, which are distinct, marked with black on the sides and the middle half of the dorsum; prosternal spine short, stout, blunt, subconical. Tegmina nearly half as long again as the pronotum, ovate-lanceolate, the tip roundly pointed; they are dark brown, with small quadrate, blackish, scattered spots, and many of the nervures yellowish, or brownish yellow, over short distances; wings about half as long as the tegmina. Hind legs yellow; the femora dark reddish brown, or blackish outside, excepting beneath; above with a basal, an apical, and two intermediate broad patches of the same; apex, excepting lower geniculate lobe, black; spines of tibiæ black; arolium subpyriform, nearly as long as the claws; abdomen yellow; all but the apical margins of the dorsal segments dark reddish brown, deepening into black. Length of body, 18.5^{mm}; of antennæ, 6.5^{mm}; of tegmina, 6^{mm}; of hind tibiæ, 8.5^{mm}.
1 ♀, Northern New Mexico, August and September, Lieut. W. L. Carpenter.
The narrow pale stripes diverging from the summit of the eye backward give this a very different appearance from the two following species.
17. *Pezotettix Marshallii* Thom.—Head light brown, mottled lightly on the face. heavily on the vertex, with brownish fuscous, becoming blackish in a small triangular median spot at the back of the head, and often in a rather broad arcuate band run-

ning from the middle of the vertex between the eyes along their upper edge, but separate from it; eyes moderately large and prominent, the front edge nearly straight in the female; vertex between the eyes broader than the middle of the frontal costa in the ♀, of about equal breadth in the ♂; fastigium broadly and shallowly channeled, most deeply in the ♂, with distinct, and nearly straight lateral carinæ, which melt into the edge of the frontal costa, the latter distantly punctate throughout, slightly narrowed, and scarcely convex at its upper extremity, broadening slightly at the ocellus, which is sunken, and followed beneath by a slight, short sulcation; antennæ uniform brownish yellow. Pronotum above with the posterior lobe nearly flat, the anterior faintly convex, the whole broadening slightly behind, especially in the female, the posterior border convex, with somewhat straight sides, so as to appear almost angulated; median carina very distinct, though slight on the posterior lobe, but faint on the anterior lobe, and scarcely elevated except at the extreme front; lateral carinæ distinct, except anteriorly, but obtuse, the deflected lobes descending vertically; both dorsal and deflected lobes of the color of the head, the posterior lobe a little clearer above and the lower part of the deflected lobes a little paler; upper third or more of the deflected lobes with a dark reddish brown, fuliginous, or blackish stripe extending from the posterior sulcus forward to the eye, narrower on the head, and there margined above with a yellowish line; transverse sulcations distinct, occasionally marked above with black; prosternal spine stout, conical, very blunt at apex. Tegmina less than half as long as the abdomen, subfusiform, the tip roundly pointed; they are dark brown, obscured, clouded, and dotted with black, most of the veins dark castaneous; wings a little shorter than the tegmina. Hind femora dull brownish yellow, the outer and upper faces with two submedian, rather broad, oblique, and angulate blackish fuscous stripes, besides a basal and an apical cloud of the same; hind tibiæ of a warmer yellow, the apical half of the spines black; arolium of same legs yellow, margined with blackish, broad obovate, as long as the breadth of the apex of the hind tibiæ (♂), or obpyriform, as long as the claw and as broad as the apex of the last tarsal joint (♀). Abdomen yellow, heavily mottled above with dark reddish brown, deepening laterally into black at the bases of the segments; last abdominal segment of male terminating in a blunted point; anal cerci of male as long as the width of the hind femoral genicular lobes, slender, straight, tapering on the basal half, the tip bluntly rounded. Length of body, ♂, 17ᵐᵐ; ♀, 25ᵐⁱⁿ; of antennæ, ♂, 6.5ᵐᵐ; ♀, 7.5ᵐᵐ; of tegmina, a, ♂, 5.5ⁱⁿⁿ; ♀, 7.75ᵐᵐ; b, ♂, 6.5ᵐᵐ; ♀, 9.7ᵐˡᵉ; of hind tibiæ, ♂, 7ᵐᵐ; ♀, 9ᵘⁿᵐ.

4 ♂, 5 ♀, 4 immature, Taos Peak, Sangre de Cristo Mountains, Northern New Mexico, at a height of 13,000 feet, Lieut. W. L. Carpenter; also, Northern New Mexico, August to September, and Southern Colorado, June 11-20, Lieut. W. L. Carpenter.

I have referred this species to *P. Marshallii* Thom., with some doubt, and therefore have described it anew. Specimens from Taos Peak and Southern Colorado seem to differ only in the length of the tegmina, those marked *a* in the table of measurements coming from Taos Peak and those marked *b* from Southern Colorado. A single specimen from Northern New Mexico (♀) agrees wholly in this respect with one from Taos Peak.

18. *Pezotettix stupefactus*, nov. sp.—Head light brown or yellowish brown, the upper half and sometimes the whole head mottled rather heavily, on the top of the head very heavily, with brownish fuscous, often becoming blackish in a median band on the top of the head, and, less distinctly, above the upper edges of the eyes, as in *P. Marshallii*; eyes, particularly of the male, large, the front border nearly straight in the female; vertex between the eyes broader than the middle of the frontal costa in the female, narrower than it in the male; fastigium distinctly channeled, most deeply in the male, with distinct and nearly straight lateral carinæ, which run into the outer edges of the frontal costa; the latter distinctly punctate near the edges like the whole of the face, nearly equal, but slightly narrower above, the surface flat, with a slight, short, narrow sulcation, in the upper part of which the ocellus is situated; antennæ brownish yellow, becoming dusky toward the tips. Pronotum above nearly flat, the anterior lobe with scarcely perceptible fullness, and on either side of the median carina, at the posterior sulcus, a slight oblique depression; the whole pronotum broadens a little and regularly in passing backward, the posterior margin obtusely and roundly angulated; median carina distinct though slight on the posterior lobe, inconspicuous, excepting in front on the anterior lobe, and in the female nearly obsolete; lateral carinæ distinct, though not prominent; surface of pronotum profusely punctate, almost rugulose behind. Pronotum brownish yellow, darkest on dorsum and profusely flecked with darker colors; upper third or half of deflected lobes with a brownish fuliginous belt extending from the last transverse sulcus to the eye, narrower at the extreme front of the pronotum and on the head; transverse sulcations distinct, only seldom, and then but slightly, marked with black; prosternal spine short, stout, bluntly conical, in the male thickened apically. Tegmina fully half as long as the abdomen, elongate, subfusiform, in the male almost linear, the tip roundly pointed, dark brown, more or less variegated with yellowish and blackish, the small spots showing a tendency to a longitudinal arrangement; most of the veins light; wings a little shorter than the tegmina. Hind femora light yellowish brown, with a pair of conspicuous sub-

median **V**-shaped dark brown or blackish bands externally, crossing the upper surface transversely; the extreme base and tip are marked with the same color; hind tibiæ yellow, the spines black to the base; arolium as in *P. Marshallii*. Abdomen yellowish beneath, mostly reddish brown above, deepening into black: last abdominal segment of male terminating with a quadrate, slightly and broadly notched margin; anal cerci of male short, very broad, nearly equal, strongly compressed, laminate, the tip broadly rounded, slightly incurved, so that the outer margin is broadly convex, the inner shallowly concave. Length of body, ♂, 17^{mm}; ♀, 20^{mm}; of antennæ, ♂, 7.5$_{mm}$; ♀, 7^{nm}; of tegmina, ♂, 7.7^{mm}; ♀, 8^{mm}; of hind tibiæ, ♂, (?); ♀, 9.5^{mm}.

1 ♂, 3 ♀, besides many immature, Taos Peak, Sangre de Cristo Mountains, Northern New Mexico, at a height of 13,000 feet, Lieut. W. L. Carpenter.

This species might easily be confounded with the preceding, but is distinct from it in the character of the tegmina, the dorsum of the pronotum, and the anal cerci of the male.

19. *Pezotettix plagosus*, nov. sp.—Brownish yellow marked with dark brown or brownish fuscous; especially noticeable is a medio-dorsal dark stripe, extending from the middle of the vertex between the eyes, where it is not half so broad as the vertex, to or nearly to the end of the pronotum, broadening as it goes, on the posterior half of the pronotum inclosing a median pale line and fading out at the extremity of the posterior lobe; and also a broad belt at the upper limit of the deflected lobes in front of the posterior sulcus extending forward to the eye and fading inferiorly. Vertex between the eyes slightly broader than the frontal costa; fastigium broadly and rather shallowly sulcate, the frontal costa equal, narrowly sulcate below the ocellus. Pronotum broadening slightly posteriorly, the posterior lobe punctate, the median carina distinct only on this lobe, and their slight lateral carinæ moderately abrupt, obtuse, the posterior border obtusely angulated, the angle rounded; prosternal spine very short, straight, stout, pyramidal, pointed. Tegmina not much shorter than the abdomen, obscure brown, mottled with many paler and darker spots (due to the color of the veins), mostly arranged longitudinally in the median field; the costal field is broadly swollen near the base, and beyond it the whole wing tapers nearly to the rounded tip; wings well formed, the veins of the apical half of the preanal field dusky or blackish. Hind femora with two median, angulate, moderately broad, brownish fuscous bands, the arc of the geniculation black; hind tibiæ pale dull glaucous, pale at the base, the spines black-tipped. Anal cerci of male broad at base, rapidly tapering on basal compressed conical half, very slender and nearly equal on the apical half, a little incurved at tip. Length of body, ♂, 18.5^{mm}; ♀, 21^{mm}; o antennæ, ♂, 8^{mm}; ♀, 7.5^{mm}; of pronotum, ♂, 4.75^{mm}; ♀, 5^{mm}; of tegmina, ♂, 11^{mm}; ♀, 11.2^{mm}; of hind tibiæ, ♂, 9^{mm}; ♀, 10.25^{mm}.

1 ♂, 1 ♀, Northern New Mexico, August to September, Lieut. W. L. Carpenter.

20. *Pezotettix marginatus*, nov. sp.—Dull pale olivaceous brown, slightly darker above, with a broad black stripe, occasionally obsolescent, extending from behind the eye, along the upper border of the deflected lobes of pronotum, to the posterior transverse sulcus; pleura sometimes marked with black and the abdomen with a lateral black band, sometimes continuous and equal, sometimes confined to small triangular spots on the segments of the anterior half; hind femora sometimes a little infuscated externally, the genicular lobes sometimes blackish, the hind tibiæ rather dark olivaceous, the apical half of the spines black; the summit of the head is sometimes marked with black in broad median and diverging supraorbital stripes. Face unusually oblique, forming, with the descending fastigium of the vertex, a little more than a right angle; fastigium rather deeply channeled in the male, slightly in the female; frontal costa equal, shallowly sulcate throughout. Pronotum rather long, the dorsum equal, with slightly sloping sides, distinct but rather slight and equal median carina and distinct though very obtuse lateral carinæ; hind border scarcely angulate; prosternal spine rather small, bluntly subconical, inclined a little backward. Tegmina a little longer than the pronotum, simple, but at the extreme tip a little pinched, and tapering to a blunt point; wings a little shorter. Hind legs rather slender, the femora compressed. Last abdominal segment of male terminating in a pyramidal point; anal cerci of same straight, rather stout, moderately long, noticeably but broadly constricted in the middle, the tip larger than the base, gibbous, the whole scarcely depressed, curving slightly downward beyond the middle. Length of body, ♂, 17^{mm}; ♀ (contracted), 14.5^{mm}; of antennæ, ♂, 7.5^{mm}; ♀, 6^{mm}; of pronotum, ♂, 4^{mm}; ♀, 4.6^{mm}; of tegmina, ♂, ♀, 6^{mm}; of hind tibiæ, ♂, 8.5^{mm}; ♀, 9.5^{mm}.

2 ♂, 4 ♀, Southern California, No. 921, H. W. Henshaw; Fort Tejon, California, July 26, No. 905, H. W. Henshaw.

21. *Pezotettix rivax*, nov. sp.—Head large, prominent, yellowish green, mottled with brown, which on the summit forms a very broad longitudinal stripe; vertex between the eyes as broad as the frontal costa, the fastigium slightly sulcate, the frontal costa equal, rather deeply sulcate below the ocellus; antennæ light brown, the basal joint unusually small. Pronotum small, equal, compressed, the dorsum flat, the whole so much smaller than the head as to give the insect a strangulated appearance, brownish

green, mottled with darker and lighter markings, the lateral carinæ with a yellowish stripe and the deflected lobes with a similar oblique stripe descending to the lower anterior angle; the posterior lobe is profusely punctate, the transverse sulci deeply impressed, the median carina obsolescent, the lateral carinæ wholly obtuse, the posterior margin very obtusely angulate; prosternal spine not very stout, cylindrical, very bluntly tipped, inclined rather strongly backward. Tegmina about as long as the pronotum, slender, short, lanceolate; wings rudimentary. Hind femora slender, yellow, tinged on upper half with brownish and obscurely, narrowly, and transversely bifasciate above with the same; hind tibiæ glaucous (?), the spines reddish, tipped with black; arolium extremely large. Abdomen yellowish, tinged above with greenish brown, the last segment of the male scarcely upturned, terminating in a blunt point; anal cerci of male depressed laminate, scarcely longer than the sides of the last dorsal segment, gently incurved, tapering on the basal half, scarcely enlarging beyond, the tip broadly rounded. Length of body, 18.5^{mm}; of antennæ, 9.5^{mm}; of pronotum, 4.25^{mm}; of tegmina, 4.15^{mm}; of hind tibiæ, 9^{nm}.

1 ♂, Plains of Northern New Mexico, eastern slope, October 14-31, Lieut. W. L. Carpenter.

22. *Pezotettix jucundus*, nov. sp.—Yellow, marked with brownish fuscous; top of the head behind the narrowest part of the vertex a little tumid, marked with an elongated, triangular, blackish fuscous dash, through the middle of which runs a yellow line, and by a supraorbital arcuate band of a similar color usually broken and terminating just below a narrow, short, yellow stripe behind the upper part of the eye; vertex between the eyes rather narrower than the frontal costa, the fastigium broadening more than usual in front of the eyes, and longitudinally, broadly sulcate throughout; frontal costa broad and nearly equal, broadest just above the ocellus, rather sparsely punctate, and at the ocellus very shallowly sulcate, often nearly imperceptible. Pronotum short and rather stout, the anterior and posterior halves of the deflected lobes nearly symmetrical; dorsum with equal sides, the posterior lobe scarcely more than half as long as the anterior, the former divided in the middle by a straight sulcus, extending only just beyond the lateral carinæ, and immediately behind this, one-third the distance only to the posterior sulcus, by the sinuate sulcus, which passes across the deflected lobes; the whole anterior lobe slightly gibbous, particularly in the female; the median carina, which is marked with dark brown and is distinct, though slight, on the posterior lobe, is here obsolete, represented only by the dark line, sometimes faintly impressed; the lateral carinæ are very obscure, converging anteriorly, and distinguished by a narrow, dull, yellow stripe, the rest of the dorsum and the upper part of the deflected lobes being obscurely marked with dusky brown, which, on the deflected lobes, is darkest in the sulci; a distinct longitudinal sulcus, more distinct for its deeper color, unites the two percurrent sulci of the deflected lobe in the middle; anterior border marked by a submarginal continuous sulcus, distinct only on the sides; posterior border very broadly rounded or subangulated; prosternal spine straight, rather slender, subconical, bluntly pointed. Tegmina subovate, slightly longer than the pronotum, the apex roundly angulated, the veins mostly of a light color, the middle field furnished with three or four small, quadrate, dark spots in a longitudinal row; wings rudimentary; pleura with an oblique bright-yellow stripe edged with black above the hind coxæ. Outer disk of hind femora marked within the carinæ by a large, apical, yellowish-brown spot, a very broad, angulate, transverse, median band of the same, and a similar basal band, sometimes obsolete or obsolescent, on the lower half; outer arc of genicular lobes black; tibiæ yellow, the apical half of the spines black; arolium either quadrate, rather narrow, longer than the claws (♂), or obpyriform, small, but little more than half as long as the claws (♀). Abdomen yellow, the sides chafed by the femora dark fuscous; the joints of the abdomen of the female are less compressed than usual, and also contracted, and, the appendages being short, it has a peculiarly compact appearance; joints of the male abdomen normal, the last joint upturned, the apex rounded and entire; anal cerci of male very broadly expanded at the base, tapering rapidly and regularly just beyond the middle, beyond less rapidly, forming a delicate, slender, but bluntly pointed tip, slightly hooked downward. Length of body, ♂, 15.5^{mm}; ♀, 17^{mm}; of antennae, ♂, 6.75^{mm}; ♀, 6^{mm}; of pronotum, ♂, 4.1^{mm}; ♀, 4.75^{mm}; of tegmina, ♂, 4^{mm}; ♀, 5.1^{mm}; of hind tibiæ, ♂, 9.25^{mm}; ♀, 10^{mm}.

This species is described from 11 ♂, 10 ♀, collected by Dr. Edw. Palmer at San Diego, Aguas Calientes and Tighes Station in Southern California. A male and female in not so good condition, taken by Dr. O. Loew, near Mojave River in July, No. 870, are larger than the average (♂, 19.5^{mm}; ♀, 18.5^{mm}), and have proportionally longer tegmina (♂, 6.5^{mm}; ♀, 6^{mm}). Especially is this the case with the male; agreeing, however, in all other respects, and having identical anal cerci, there can be little doubt that they belong to this species.

23. *Pezotettix enigma*, nov. sp.—Pale brownish yellow, marked with darker brown and fuscous. Head large, tumid, all the angles rounded, the summit darker, with a sometimes inconspicuous median blackish stripe, broadening from in front backward; vertex between the eyes narrower than (♂) or equal to (♀) the frontal costa; fastigium

very broadly and shallowly sulcate, most distinctly in the male; frontal costa broad
and equal, very faintly punctate, with a scarcely perceptible narrow sulcus below
the ocellus; antennæ slightly infuscated at the tip. Pronotum shaped as in the preceding species, but more distinctly tumid on the dorsum of the anterior lobe, the middle transverse sulcus nearly as close to the posterior sulcus as to the short one in front
of it, and the posterior lobe fully three-quarters the length of the anterior; posterior
margin angularly rounded ; median carina, as in the preceding species, marked in form
like all the transverse sulci; dorsum mottled with dark brown, the lateral carinæ
marked with a more or less distinct narrow yellow stripe; the anterior margin of the
deflected lobes clear yellow or pallid; prosternal spine straight, small, conical, bluntly
pointed. Tegmina rather broad, ovate, overlapping, the tip scarcely produced, fully
half as long as the abdomen, brownish fuscous, marked with yellow longitudinal veins,
and flecked, principally along the median area, but also elsewhere, with longitudinal
series of subquadrate blackish fuscous spots; wings a little shorter than the tegmina.
Hind femora stout and full, yellow, the outer face marked with alternate, narrow,
angulate, yellow and black stripes, often fainter in parts than in others, so as to show a
tendency to transverse bands arranged as in *P. jucundus;* outer arc of genicular lobes
broadly black ; hind tibiæ yellow, the apical half of the spines black ; arolium of either
sex as in the preceding species. Abdomen yellow, the upper portion infuscated, the
middle of the dorsum marked frequently with a series of approximate, subdorsal,
roundish, black spots, often inclosing white spots nearly as large as themselves, those
of opposite sides separated only by a slender yellow line; the abdomen of the two
sexes has the peculiarities of that of the preceding species, the last joint of the male
being also entire; the anal cerci of the male scarcely differ from those of that species,
the slender apex only being a little less suddenly contracted. Length of body, ♂,
2.5ᵐᵐ ; ♀, 24ᵐᵐ; of antennæ, ♂, 9.25ᵐᵐ; ♀, 7.5ᵐᵐ; of pronotum, ♂ 6ᵐᵐ; ♀, 6.9ᵐᵐ;
tegmina, ♂, 8.25ᵐᵐ; ♀, 10.75ᵐᵐ; of hind tibiæ, ♂, 12.5ᵐᵐ; ♀, 13.5ᵐᵐ.

25 ♂, 48 ♀, Santa Barbara, Cal., July 1, No. 1005, H. W. Henshaw ; Los Angeles and
Santa Barbara, July, No. 224, C. J. Shoemaker.

24. *Pezottetix pictus* Thom.—1 ♂, 1 ♀, plains of Northern New Mexico, eastern slope.
October 14–31, Lieut. W. L. Carpenter; Northern New Mexico, August, September.
Lieut. W. L. Carpenter. This highly interesting species should be referred to a distinct genus ; it is unique in structure and coloration among the species of this group
of Acridians.

25, *Hesperotettix viridus* (Thom.) Scudd.—1 ♂, Mojave Desert, California, No. 829, Dr.
O. Loew.

26. *Gomphocerus clepsydra* Scudd.—A single female of this species, originally described and hitherto only known from British America, was taken in Southern Colorado June 11–20, Lieut. W. L. Carpenter. 5 ♂ were also taken at the same place
and time, and in Northern New Mexico, August to September, by Lieut. W. L. Carpenter. As this sex has not been known, the following description is appended :
Head pale brownish yellow, excepting on the summit and sometimes on the cheeks,
where it is yellowish brown, more or less tinged with reddish ; a pair of moderately
broad, arcuate, blackish streaks, sometimes obsolete excepting in front, run from
the middle of the summit of the eye to the back of the head above the lateral
carinæ of the pronotum ; fastigium of vertex depressed, flat, separated from the
lateral foveolæ by a distinct ridge, that of one side meeting the other at a little
less than a right angle; lateral foveolæ distinct, depressed, forming a slightly
arcuate, oblong parallelogram, at least three times as long as broad, the inner
extremity rounded ; frontal costa a little narrowed above, otherwise nearly equal,
punctate, sulcate below and for a slight distance above the ocellus; antennæ more
than twice as long as the pronotum, testaceous, the club black, composed of five or six
joints, the middle ones but slightly larger, though much more depressed, than those of
the stalk. Pronotum dark yellowish brown, the lateral carinæ as distinct as the median carina, arcuate, twice as close together just in advance of the middle as at the
posterior extremity, pallid, edged exteriorly and especially in front with black, and
interiorly on the posterior lobe with the same ; lower portion of deflected lobe with an
anterior blackish triangle, the longest side facing upward and forward, followed behind by a slender yellowish stripe, directed a little downward and sometimes edged
above with black. Tegmina reaching the tip of the abdomen with the costal field
broad, pellucid, the nervules scalariform ; the remainder testaceous, with minute
faint fuliginous clouds in the more or less pellucid middle field. Hind legs long and
slender, the femora generally more or less marked longitudinally with black along
the upper exterior carina. Sides of the abdomen marked with black on the basal
half of each segment. Length of body, 18 ᵐᵐ; of antennæ, 9 ᵐᵐ; of pronotum, 4.1 ᵐᵐ;
of tegmina, 12.85 ᵐᵐ; of hind tibiæ, 12 ᵐᵐ.

This species seems to be allied to *G. clavatus* Thom., but differs from the description of that species in many important particulars.

27. *Gomphocerus navicula*, nov. sp.—Pale dull brownish yellow, the upper surface
of the head and pronotum darker ; summit of the head with a delicate, straight, black

line running from midway between the extreme summit of the eye and the medio-dorsal carina to the back of the head, midway between the dorsal and lateral carinæ, and edged within by an entirely similar yellowish or roseate line; both these lines scarcely taper anteriorly; there is a slight but distinct medio-dorsal carina extending from the back of the head through the fastigium, where it is more distinct, to its very tip; the lateral carinæ of the fastigium are equally distinct, and together form a U; lateral foveolæ wanting; frontal costa slightly narrowed above, otherwise nearly or quite equal, punctate, below the ocellus a little sulcate; antennæ less than twice as long as the pronotum, brown, the club duskier, made up of seven or eight joints, occupying fully a quarter of the antennæ, fusiform, and in the male distinct. Pronotum with the lateral carinæ as distinct as the median carina, a little arcuate, only a fifth nearer each other a little in advance of the middle, than at the posterior border, pale yellow, narrowly edged within throughout and without in the middle with black; sides of the deflected lobes with an arcuate, yellow, black-edged streak extending a little below the middle from the anterior sulcus to the hind border. Tegmina extending to the tip of the abdomen, the costal field (♂) somewhat expanded beyond the middle, with oblique subscalariform nervules, the remainder (♂♀) testaceous, with longitudinal fuscous, subconfluent streaks in the middle area; wings with the apical nervures slightly thickened and blackish. Hind femora moderately slender, the upper outer carina more or less edged beneath with blackish fuliginous; tibiæ yellow, with spine black on the apical half. Length of body, ♂, 14.75^{mm}; ♀, 18.5^{mm}; of antennæ, ♂, 5.5^{mm}; ♀, 5.4^{mm}; of pronotum, ♂, 3.25^{mm}; ♀, 3.8^{mm}; of tegmina, ♂, 10.5^{mm}; ♀, 12.5^{mm}; of hind tibæ, ♂, 8^{mm}; ♀, 9^{mm}.

2 ♂, 1 ♀, Southern Colorado, June 11-20, Lieut. W. L. Carpenter; Northern New Mexico, August to September, Lieut. W. L. Carpenter.

This species bears most resemblance, among species known to me, to the preceding, from which it is readily distinguished by the much shorter antennæ, the want of lateral foveolae on the vertex, and the lesser curvature of the lateral carinæ of the pronotum.

28. *Dociostaurus ornatus,* nov. sp.—Pale brown, above darker; an arcuate row of fuscous dots from the posterior extremity of the fastigial carinæ to the back of the head midway between the dorsal and lateral carinæ of prothorax, and a similar straight row from the middle of the posterior edge of the eye backward; fastigium rather deeply sulcate posteriorly with a low median carina, the lateral carinæ prominent, meeting at an acute angle in front; lateral foveolæ very distinct, pretty large, subquadrate, a little longer than broad, the lower edge horizontal; frontal costa much narrowed above, nearly equal below, shallowly but broadly and abruptly sulcate throughout, and punctuate; antennæ dark brown, becoming duskier toward the tip, the apical joints slightly enlarged. Pronotum slightly convex anteriorly, broadly angulated posteriorly, the lateral carinæ as distinct anteriorly as the median carina, thickened posteriorly, very strongly arcuate, so as to be fully twice as distant at the posterior border as a little in advance of the middle, edged outwardly with black on the anterior lobe and front part of posterior lobe, and interiorly more broadly on the front part of the posterior lobe; deflected lobes, marked with dark colors and especially with a subcentral, quadrate, blackish fuliginous spot. Tegmina longer than the body, the costal area subpellucid, the rest testaceous, with a longitudinal row of conspicuous fuliginous and blackish fuliginous spots in the median area. Hind femora sparsely dotted externally with blackish fuscous along the edges, above showing signs of clustering into spots, and on the upper surface a median spot crossing the upper carina and bordered distinctly with black on the outer side; there is also an oblique black basal streak, and an apical and subapical fuscous spot; hind tibiæ clear pale yellow, the spines black-tipped. Length of body, 12.5^{mm}; of antennæ, 5.3^{mm}; of pronotum, 2.6^{mm}; of tegmina, 11.8^{mm}; hind tibiæ, 7.1^{mm}.

1 ♂, Northern New Mexico, August to September, Lieut. W. L. Carpenter.

In North America, the *Œdipodidæ* are far more abundantly represented than any other group of Acridians. The present collection is rich in novel forms, and affords also the opportunity of continuing my efforts to reduce the known species to symmetrical relations. The following table, partly based upon that by Stål (Recens. Orthopt., I), has accordingly been prepared for the readier determination of the genera; it is intended, however, to apply only to the genera mentioned in this paper, and even only to such species of these genera as are catalogued. Doubtless, it may have a wider application but in its preparation no species were examined but those mentioned below.

1 (30) Opposite tegmina attingent when closed.
  2 (23) Median carina of pronotum, with a single submedian incision.[*]
    3 (12) Mesosternal lobes of female nearly or fully twice as distant as the
        metasternal lobes.
      4 (9) Median carina of pronotum much more elevated than the lateral
          carinæ.
        5 (6) Hind border of pronotum very obtusely rounded...... *Œdocara.*
        6 (5) Hind border of pronotum bent at a right angle or less.
          7 (8) Intercalary vein of tegmina running through the middle of
            the postradial area.......................... *Chimarocephala.*
          8 (7) Intercalary vein of tegmina approaching the radial apically,
                                                    Arphia.
      9 (4) Lateral carinæ of pronotum nearly or quite as elevated as the me-
          dian carina.
        10 (11) Axillary vein of tegmina free.............. *Stirapleura.*
        11 (10) Axillary vein joining the anal vein in the basal half of
            the tegmina... *Psoloessa.*
    12 (3) Mesosternal lobes of female scarcely or not at all more distant than the
        metasternal lobes.
      13 (16) Median carina of pronotum very inconspicuous; axillary vein of
          tegmina uniting with the anal without branching.
        14 (15) Tegmina with a close, irregular reticulation on the basal
            four-fifths... *Hadrotettix.*
        15 (14) Tegmina with a close, irregular reticulation on the basal
            fourth only.. *Anconia.*
      16 (13) Median carina of pronotum distinct, sometimes very prominent;
          axillary vein of tegmina free or branching before joining
          the anal vein.
        17 (18) Median carina of pronotum uniform throughout.... *Camnula.*
        18 (17) Median carina of pronotum irregular.
          19 (20) Pronotal carina nearly obsolete on posterior portion of
            anterior lobe... *Hippiscus.*
          20 (19) Pronotal carina crested on anterior lobe.
           21 (22) Dark band of wings extending nearly or quite to the
            base... *Dissosteira.*
           22 (21) Dark band of wings only as broad as, or but little
            broader than, the tegmina......................... *Œdipoda.*
  23 (2) Median carina of pronotum with a deep secondary incision. [See, how-
        ever, note under 2 (23).]
    24 (25) Summit of head conspicuously rugulose, or furnished with sharp
         transverse carinæ........................... *Trachyrachys.*
    25 (24) Summit of the head with the usual configuration.
      26 (27) Axillary vein joining the anal in the basal half of the tegmina,
                                   Psinidia.
      27 (26) Axillary vein terminating on the hind border of tegmina.
        28 (29.) Posterior lobe of pronotum scarcely longer than the anterior.
                                    Derotmema.
        29 (28.) Posterior lobe of pronotum nearly twice as long as the anterior
                                      Trimerotropis.
30 (1) Closed tegmina separated by more than their own width .... *Brachystola.*

29. *Chimarocephala viridifasciata* (De Geer).—3 ♂, 1 ♀, were taken in Southern Colorado,
June 11-20, and in Northern New Mexico in August and September by Lieut. W. L. Carpenter. In the cloudiness of the wings, they agree best with Texan specimens as described in my Entomological Notes (IV, 81). The generic name (χίμαρος, κεφαλή) is proposed for the species (*viridifasciata, brevipennis, cubensis, pacifica*) placed by me (*loc. cit.*)
under *Tragocephala;* the latter name, as M. Auguste Salló has pointed out to me, being
pre-occupied in *Coleoptera* (Dupont, 1834).
  30. *Psoloessa maculipennis* Scudd.—2 ♀, Southern Colorado, August to September,
Lieut. W. L. Carpenter.
  31 *Arphia teporata,* nov. sp.—This red-winged species is so nearly allied to *A. frigida*
Scudd., of the high north, that it need only be compared with it; the upper extremity
of the frontal costa of the head has no transverse carina setting off a pair of minute
frontal foveolæ; the fastigium of the vertex is very slightly narrower. The tegmina are

---

[*] The single species of *Hippiscus* mentioned below has the anterior lobe slightly impressed by a transverse sulcus, and in specimens dried after soaking in alcohol such an
impression may be accidentally intensified. In *Hadrotettix* and *Anconia* there is a slight
transverse sulcus near the middle of the anterior lobe, which in *Anconia*, and sometimes
in *Hadrotettix,* severs the carina, but so slightly that I have placed the genera in this
division.

slightly narrower and are flecked almost uniformly throughout with fuscous dots, smaller and a little less frequent on the apical third; entire inner area pale yellowish testaceous; wings with a transverse band, and base exactly as in *A. frigida*, but with the entire apex uniformly pellucid, obscured only by the blackish veins. · Hind tibiæ glaucous, with a very broad pale yellow annulation at the base, and a slight testaceous tinge at extreme tip; the apical half of spines black. Length of body, ♂,19ᵐᵐ; ♀, 23.2ᵐᵐ; of antennæ, ♂, 8ᵐᵐ; ♀.7.5ᵐᵐ; of pronotum, ♂, 4.5ᵐᵐ; ♀, 6.25ᵐᵐ; of tegmina, ♂, 19ᵐᵐ; ♀, 26.5ᵐᵐ; of hind tibiæ, ♂, 8.8ᵐᵐ; ♀, 12.5ᵐᵐ.

2♂, 2 ♀, Southern Colorado, June 11-20, and Northern New Mexico, August to September, Lieut. W. L. Carpenter. I have also received the species from Pecos River, Texas, Captain Pope.

32. *Camnula atrox* (*Œdipoda atrox* Scudd.; *Camnula tricarinata* Stål). A considerable number of specimens were taken in Southern California; at Santa Barbara, June, No. 885, C. J. Shoemaker; June 10, Dr. O. Loew; July 1, No. 1005, H. W. Henshaw; in the Mojave Desert, July, No. 935, Dr. O. Loew; and on Santa Cruz Island, in June, No. 967, H. W. Henshaw. I have also received it from other points in California, viz: from Tighes Station, and Julian, in the southern part of the State, Dr. E. Palmer; and from Santa Rosa Island, Central California, Nevada, and Vancouver's Island, Henry Edwards, esq. This material shows that the species varies greatly in the markings of the tegmina. The usual distribution of the fuscous spots seems to be the following: the middle area is filled with large, transverse, quadrate spots, separated by rather narrow interspaces; those on the basal half of the wings more or less confluent, particularly below. On the outer half, they become smaller and less conspicuous toward the tip, and are usually confined to a couple of patches, somewhat curtailed beneath in the third quarter of the wing and scattered dots beyond; besides these, there is, usually, a small, oblique, subquadrate, dark fuscous spot on the costal border, just beyond the highest point of the costal arch, and near the middle of the same border two or three short oblique streaks. Not infrequently, however, all these markings are much reduced, the quadrate spots become rounded, and the result may be simply a series of three or four subequal, round, fuscous spots in the median area, some scattered dots beyond them, and slight touches along the costal border; or there may be a couple of narrow, transverse streaks at and beyond the middle of the wing, made up of clustered dots, with one or two dots beyond, a small, quadrate, longitudinal spot in the middle of the basal half of the middle area, and a small spot at the costal arch.

33. *Hippiscus corallipes* (Hald.) Scudd.—1 ♀, Southern Colorado, June 11-20, Lieut. W. L. Carpenter.

ŒDOCARA, (οἰδέω, κάρα), nov. gen.

Allied to *Œdaleus*. Head large and tumid, the face vertical; vertex between the eyes twice as broad as the frontal costa; the anterior half of the fastigium suddenly contracted to about one-third its previous width, the lateral carinæ prominent; lateral foveolæ very distinct, with prominent walls, pretty large, triangular, pointed interiorly; frontal costa rather strongly contracted above, a little expanded at the ocellus, somewhat sulcate throughout and especially just below the ocellus; eyes rather small, not very prominent; antennæ about as long as the hind tibiæ, the joints of the basal half a little depressed. Pronotum small, greatly constricted in the middle, scarcely longer than the head, the anterior and posterior lobes of nearly equal length; the constriction is nearly confined to the anterior lobe, which is furnished posteriorly with two deeply-impressed transverse sulci, extending (deeply) a short distance into the deflected lobes, and extending up to, but not traversing, the median carina; this is of equal and slight elevation throughout, and the lateral carinæ are present only as a shoulder to the flat posterior lobe; the anterior lobe, on the other hand, is nearly tectiform, and its anterior border is full and rounded, expanding slightly upon the surface of the head; posterior border very obtusely and roundly angulated; pleura of metathorax carinate, especially below. Tegmina extending beyond the abdomen, the costal margin considerably expanded in the middle of its basal half, the intercalary vein rather inconspicuous, minutely tortuous, dividing the postradial field; axillary vein free. Hind femora moderately slender, with sharp, unarmed, superior carina; interior, apical, curved spines of hind tibiæ subequal.

34. *Œdocara strangulatum*, nov. sp.—Yellowish brown, the face and cheeks paler, and, like the top of the head, profusely mottled with small, darker fuscous spots; antennæ pale, the apical third blackish fuscous. Dorsum of pronotum pallid, the posterior lobe, excepting a broad, lateral, pallid stripe along the carinæ, reddish brown and punctate, a quadrate patch of the same color on the upper half of the deflected portion of the anterior lobe. Tegmina dead brown, the inner edge paler, the rest rather sparsely and almost uniformly flecked with small brownish spots; wings pellucid; most of the veins of the outer half of the front portion black. Hind femora with two oblique bars of reddish brown, crossing the upper half of the outer face and the upper face, the distal one also traversing, not obliquely, the inner face; a few dots of the same color fleck the lower outer carina; upper half of the outside and whole of the inside

of the genicular lobes black: tibiæ pale yellow, fuscous at extreme tip, the apical half of spines black. Length of body, 21.5^{mm}; of antennæ, 11^{mm}; of pronotum, 4.1^{mm}; of tegmina, 20^{mm}; of hind tibiæ, 11.75^{mm}.

1 ♀, 7 pupæ, Southern Colorado, June 11–20, Lieut. W. L. Carpenter.

STIRAPLEURA (στείρα, πλευρά), *nov. gen.*

Allied to the preceding. Head moderately large, the face vertical; vertex between the eyes rather broader than the lower extremity of the frontal costa; fastigium depressed, with very high and sharp bounding-walls, which are parallel through most of their course, incline slightly toward each other as they disappear posteriorly, and bending sharply in front meet at a right angle; lateral foveolæ rather large and distinct, with high walls, the posterior at right angles to the inferior and but little shorter, the other portion of the wall forming a sharply arcuate hypothenuse of the triangle; frontal costa strongly compressed above, expanding to near the ocellus, then parallel, and below the ocellus again expanding, throughout sulcate; eyes of medium size, not very prominent; antennæ (♀) slightly depressed, short, scarcely reaching the tip of the pronotum. Pronotum small, slightly constricted in the middle, the posterior slightly longer than the anterior lobe, the dorsum nearly flat, the median carina undivided on the anterior lobe, equal and slight throughout, the lateral carinæ similar but strongly arcuate; posterior margin bent at slightly more than a right angle, the angle rounded; pleura of metathorax with a distinct sharp carina on the outer face, extending from the edge of the coxæ close to the edge of the closed tegmina. Tegmina extending beyond the tip of the abdomen, the costal area slightly expanded at the end of the basal third, the intercalary and axillary veins as in *Œdocara;* wings rather ample. Hind femora moderately slender and short, scarcely reaching the tip of the abdomen, with superior carina unarmed.

35. *Stirapleura decussata*, nov. sp.—Wood-brown above, paler below; face and mouth parts tinged with yellow, the former flecked with reddish brown; antennæ yellowish brown; behind the eye a broad, dark band, expanding posteriorly, deepening into black above and edged with pallid yellow, extends to the pronotum. Lateral carinæ of pronotum a little paler than the disk, especially on the posterior lobe, where a distinct yellowish band follows its interior border, edged on either side by velvety black, followed by reddish brown; more or less of the velvety black follows the inferior edge of the carinæ anteriorly and the anterior and posterior borders of the deflected lobes; are distantly dotted with it; on the anterior section of the deflected lobe next to the lateral carinæ, the dark-brown postocellar band continues; just below it are some short longitudinal rugæ, and across the middle of the deflected lobe a second dark-brown band extends horizontally and a little arcuate, inclosing just behind the middle a small crescentic yellow spot. Tegmina dotted rather profusely, excepting at the extreme tip, with small, unequal, fuscous spots; wings pellucid, most of the veins in the apical half of the expanded wing black. Hind femora, with basal, median, and post-median dark-brown streaks on the upper half of the outer surface of the wing, growing more oblique apically, and connecting on the upper face with more distinct, triangular, transverse blotches, with darker edges, the inferior outer carina dotted with blackish fuscous; hind tibiæ yellow, a little infuscated at extreme tip, the spines black on the apical half. Length of body, 19^{mm}; of antennæ, 5.5^{mm}; of pronotum, 3.5^{mm}; of tegmina, 16^{mm}; of hind tibiæ, 9.2^{mm}.

1 ♀, Southern Colorado, June 11–20, Lieut. W. L. Carpenter.

36. *Phlibostroma parvum*, nov. sp.—Dull brown; the face infuscated; antennæ pale yellowish brown, a little infuscated at the extreme tip. Pronotum with the same dark markings as in *P. pictum* Scudd., the posterior margin with the angle a very little less rounded. Tegmina scarcely reaching the tip of the abdomen, pale cinereous, with four large, equidistant, rounded, triangular, fuscous spots, darkest on the edges, seated upon the ulnar veins, the middle ones larger than the outer; wings hyaline, the veins at the apex blackish. Hind legs as in *P. pictum*. Length of body, 14.5^{mm}; of antennæ. 9^{mm}; of pronotum, 3.5^{mm}; of tegmina, 9.5^{mm}; of hind tibiæ, 8.9^{mm}.

1 ♂, plains of Northern New Mexico, eastern slope, October 14–31, Lieut. W. L. Carpenter.

This species closely resembles the one formerly described under the name of *P. pictum*, but differs strikingly from it in the shortness of the tegmina and wing, which in *P. pictum* reach far beyond the tip of the abdomen; in both, though it is not mentioned in the description of either species, the tegmina have a longitudinal series of equidistant fuscous points just above the radial veins. The genus resembles more closely *Œdaleus* Fieb., than *Psinidia* Stål, to which I compared it, and, like several genera in its vicinity, but perhaps more than most of them, bears a striking resemblance to the *Stenobothri*.

37. *Œdipoda venusta* Stål.—2 ♂, 1 ♀, Santa Cruz Island, Cal., June, No. 853, Dr. O. Loew; Los Angeles and Santa Barbara, Cal., July, No. 224, C. J. Shoemaker.

To this species with little hesitation I refer one of the specimens obtained on Santa Cruz Island, although it differs in some points from another specimen from that island,

and from those received from the neighboring main (the above and San Diego, Dr. E. Palmer), particularly in the narrowness of the dark mesial band of the hind wings, which nowhere even appears to meet the hinder margin, and in the obscurity of the marking of the tegmina.

One specimen from Los Angeles has one antenna shorter than the other, and both much shorter than usual, although not broken; they were doubtless injured in early life and reproduced in an atrophied form; the apical joint of the longer antenna is half divided beyond the middle.

Stål's limitation of the genus *Œdipoda*, in his Recensio Orthopt., I, forces us to consider *Gryllus cœrulescens* Linn., as the type, and not, as stated by Thomas, *Œdipoda carolina* (Burm.). On this basis the genus is but feebly represented in the United States, and by species which differ considerably from the type. The nearest ally to *Œd. cœrulescens* is the species just recorded, in which the intercalary vein is much more closely approximated to the radials, the anal vein is connected with the axillary only by one of its branches, and the apical fourth of the tegmina is free from the intricate network of the middle of the same; the pronotal carina is also much more elevated anteriorly.

Next in position to this group, in their relation to the true *Œdipodæ*, are, in this country, *Œdipoda carolina* (Burm.) and *Œdipoda trifasciata* (Say), which are also included in the genus by Stål. The points wherein they differ, and upon which distinct genera should be based, will be given under the generic names *Dissosteira* and *Hadrotettix*, next following.

DISSOSTEIRA ($\delta\iota\sigma\sigma\acute{o}\varsigma$, $\sigma\tau\epsilon\hat{\iota}\rho\alpha$), *nov. gen.*

This genus, of which *Gryllus carolinus* Linn. is the type, differs from the true *Œdipoda*, as represented by *Gryllus cœrulescens* Linn., in the following points: The head is more prominent, the vertex being elevated and tumid; the antennæ of the male do not thicken before, and taper at, the tip; the front of the fastigium terminates, as in *Trimerotropis*, by an angulate depression, and not by a straight transverse ridge; the vertex is somewhat broader between the eyes, and the latter in the male are rounder. The enlargement of the pronotum is wholly confined to the posterior lobe; the median carina is greatly elevated, and that of the posterior lobe much arched; the metasternal lobes of the male are scarcely less distant than the mesosternal lobes. The tegmina are freer from the fine network of veins over a much larger part of the apex; all the veins are more prominent, and the anal vein is free from the axillary; the anal area of the wings is deeper, nearly or quite reaching the tip of the abdomen; the species are all insects of large size.

In very many of these points it will hardly fail to be noticed that this group approaches much more closely the American than the gerontogeic section of *Œdipoda*. *Œdipoda nebrascensis* Brun. and the following species also belong to this genus.

38. *Dissosteira longipennis* (*Œdipoda longipennis* Thom.).—1 ♂, plains of Northern New Mexico, eastern slope, October 14–31, Lieut. W. L. Carpenter.

HADROTETTIX ($\dot{\alpha}\delta\rho\acute{o}\varsigma$, $\tau\acute{\epsilon}\tau\tau\iota\xi$), *nov. gen.*

This group agrees with the typical *Œdipodæ* in the general structure of the tegmina, but differs in the comparative length of the anal area and the point of junction of the anal and axillary veins, and offers several other points of contrast; the whole body is stouter; the lateral foveolæ of the vertex are obsolete; the antennæ are uniform in size throughout in the male, and in both sexes are longer and much stouter. The hind lobe of the pronotum is slightly tumid, the median carina nearly obliterated (in which it differs strikingly from *Dissosteira*), and the lateral carinæ obtuse; the metasternal lobes of the males, instead of being only half as far apart as the mesosternal lobes, are very nearly as far apart, much as in the females in both genera; the inferior carina of the hind femora is also much broader.

39. *Hadrotettix trifasciatus* (*Gryllus trifasciatus* Say).—, 1 ♀, Northern New Mexico, August to September, Lieut. W. L. Carpenter.

TRACHYRHACHYS ($\tau\rho\alpha\chi\acute{\upsilon}\varsigma$, $\dot{\rho}\acute{\alpha}\chi\iota\varsigma$), *nov. gen.*

Allied to *Trilophidia* Stål. Head pretty large, broadening very slightly below; summit more or less rugose; the space between the eyes equal to the width of the eyes; the quadrate fastigium bounded by sharply-elevated carinæ, which run parallel to each other at the sides, but in front suddenly incline toward each other, but do not meet, leaving a deep sulcous between their separated tips; the fastigium is deeply depressed; lateral foveolæ rather large, triangular, deeply hollowed; frontal costa sulcate throughout, expanded slightly at extreme summit between the lateral foveolæ, just below it constricted, expanding again slightly at the ocellus, below which it is again, though very slightly, constricted, and then expands; eyes small, moderately prominent; antennæ slightly depressed, a little shorter (♀) than the hind tibiæ. Pronotum moderately small, the posterior lobe a little longer than the anterior, the disk rugose, the median carina moderately high, equal, compressed, on the anterior lobe severed behind the middle, and the portion between the two sulci accompanied by closely-approximated, nearly as elevated, more or less irregular, subdorsal carinæ, sometimes connected by a ridge with the median carina; posterior border rectangular;

anterior border slightly produced and angulate. Tegmina straight, the intercalary vein traversing the middle or near the middle of the postradial area, the axillary connected at its tip with the anal vein. Hind femora broad, rather short, the upper and lower carinæ elevated, the former suddenly decreasing near the middle of the apical half of the leg.

The callosities of the head and pronotum, and the structure of the hind femora, separate this genus from any other American group known to me. *T. coronata* Scudd., may be considered the type.

40. *Trachyrhachys aspera*, nov. sp.—Summit of the head between the middle of the eyes and backward furnished with many parallel approximate series of transverse rugæ, divided longitudinally by a pair of slight and inconspicuous subdorsal sulcations; fastigium with a very deep median transverse sulcation, in front of the middle of which is a minute tubercle; frontal costa deeply sunken between the lateral foveolæ: sulcus below this uninterrupted; whole head profusely punctulate. Subdorsal carinæ of the posterior portion of the anterior lobe of pronotum regularly crescentic, opening inward, posterior margin of the pronotum sharply angled. Light yellowish brown; summit of the head fuscous; a dark fuscous stripe crosses the eye from the lateral carinæ of the face to the back of the head, broad and directed downward in front of the eye, slender and horizontal behind it; the lower posterior corner of the cheeks are also dusky, and the middle of the face is more or less obscured with it; basal half of the antennæ of the general color, beyond deepening into dark fuscous. Pronotum more or less infuscated, especially on the posterior lobe and on the subdorsal carinæ of the disk, the anterior lobe next the ocellar stripe of the head, and the neighborhood of a short, rather broad, oblique dash of yellow on the lower posterior part of the deflected lobe. Tegmina just reaching the tip of the abdomen, flecked with fuscous, mostly collected into median, post-median, and costal spots on the anterior half of the tegmina, the latter in the middle of the basal half, the lower apical third subhyaline; wings hyaline (perhaps faint yellow on the basal half), with an arcuate, moderately broad, fuliginous belt, traversing the middle of the apical two-thirds, in the preanal area sending a broad, tapering shoot almost to the base, and accompanied by a few dusky fleckings at the apex, and a blackish fuliginous costal stigma, nearly half as long as the wing, from the middle of the wing outward. Hind femora with very obscure, broad, oblique, basal, and median brownish stripes on the outer face, and, on the upper face, basal, median, and post-median, darker, oblique, reversed blotches, edged with blackish; hind tibiæ yellowish, with a broad, apical, dark fuscous cloud; apical half of the spines black. Length of body, 23^{mm}; of antennæ, 8^{mm}; of pronotum, 4.5^{mm}; of tegmina, 16.7^{mm}; of hind tibiæ, 9.5^{mm}.

1 ♀, plains of Northern New Mexico, eastern slope, October 14–31, Lieut. W. L. Carpenter.

41. *Trachyrhachys coronata*, nov. sp.—Fastigium of the vertex bounded behind as well as in front by an oblique extension of the lateral carinæ, which fork at the posterior limit of their parallel course; behind this the summit of the head is furnished with moderately conspicuous, oblique rugæ parallel to the posterior bounding ridge of the fastigium, the anterior set broken into tubercles: middle of the fastigium with a transverse, bent carina, as high as, and parallel to, the anterior bounding ridge of the same; behind the middle of the lozenge-shaped space posterior to it is a tubercle; the carinæ of the frontal costa extend briefly into the front of the fastigium; in the middle of the expanded portion of the frontal costa, between the lateral foveolæ, is a slight longitudinal tubercle, and next the upper edge of the ocellus a transverse ridge breaking the sulcus. The subdorsal carinæ of the posterior portion of the anterior lobe of the pronotum are irregular in height and direction, but in general their highest point, with the slight elevations of the median carina on either side of the anterior sulcus, form a sort of quadrilateral; the rugosities of the disk of the pronotum are more elevated than in the preceding species, and more or less confluent, forming sharp, tortuous carinæ; posterior margin of the pronotum slightly sinuous on the sides, the angle rounded. Pale cinereous, the markings of the head, thorax, and hind femora much obliterated by the mode of preservation of the single individual before me, but apparently as in the preceding species, excepting that the yellow dash on the deflected lobe of the pronotum is near the center of the lobe, short, small, and horizontal, with a dusky quadrate cloud above it. Tegmina extending much beyond the tip of the abdomen, cinereous, profusely sprinkled with large, roundish, dark fuscous patches, edged with black ; in the costal area are five such equidistant spots, the innermost next the base a mere dot, the next united to the basal spot of the median area, the inner edge of the third and largest lying just beyond the costal angle; the median area has seven or eight such spots, equidistant, growing larger until close to the tip, the second and third from the base roundish, those beyond triangular or transverse and less distinct; the ulmar veins are yellowish, and below them is a basal cloud and a post-basal, rhomboid, pale fuscous spot; wings pale yellow at base, hyaline at tip, with a pretty broad, blackish fuliginous, arcuate band extending across the wing and curving next the border to the anal angle, its inner border crossing the middle of the wing; it sends a broad, tapering, rather abruptly terminating shoot half-way to the base in the preanal area, and just

beyond it on the costal margin is a whitish stigma extending nearly half-way to the tip of the wing; veins at apex black. Length of body, 24^{mm}; of antennæ, 9.25^{mm}; of pronotum, 4.6^{mm}; of tegmina, 22.5^{mm}; of hind femora, 10^{mm}.

1 ♀, Northern New Mexico, August to September, Lieut. W. L. Carpenter.

42. *Psinidia suloifrons*, nov. sp.—Very pale cinereous; upper half of the head, but especially the summit, sprinkled with blackish fuscous dots, and behind the eye a short, rather broad, longitudinal fuscous bar; summit of head much depressed between the eyes, and scarcely so broad as their width when seen from above; antennæ of male scarcely as long as the hind tibiæ, fuscous on the apical half. Median carina of pronotum not very elevated, nearly equal, the posterior portion of the anterior lobe very short, transversely corrugated; posterior lobe transversely rugose and punctate; the lateral carinæ with a narrow, blackish, fuscous stripe, in continuation of that on the head, the deflected lobe with an anterior, mesial, quadrate, blackish fuscous patch. Tegmina extending much beyond the tip of the abdomen, flecked with blackish dots at the extreme base, especially on the costal and anal fields, beyond with two large quadrate, blackish fuscous, costal patches, infringing a little on the median field, one in mid-wing, the other in the middle of the basal half; the apical half of the tegmina are nearly pellucid, especially in the middle field, with hoary veins; but it is flecked with a few dark fuscous dots, mostly clustered into minute equidistant spots, near, but not upon, the upper and lower borders; wings pellucid, the base suffused very faintly with lemon as far as the arcuate belt, which crosses the middle of the wing; the belt is moderately broad, reaches the lower border, but does not extend far toward the anal angle, and in the preanal area sends a rather slender tapering shoot nearly half-way to the base; all the veins and cross-veins of the wing are yellowish white, excepting at the extreme tip and next the costal margin, where they are black; the uppermost radial of the anal field is also black. Length of body, 18.5^{mm}; of antennæ, 9.75^{mm}; of pronotum, 3.8^{mm}; of tegmina, 21.5^{mm}; of hind tibiæ, 10^{mm}.

1 ♂, near Mojave River, Southern California, July, No. 870, Dr. O. Loew. In several particulars, but especially in the brevity of the antennæ, this is rather an aberrant member of the genus *Psinidia*.

DEROTMEMA, (δηρός, τμμία), nov. gen.

Closely allied to *Psinidia*. Head of moderate size, the face a little oblique, sharply ridged, the eyes large, globose, very prominent, farther apart above than twice the extreme width of the basal joint of antennæ; fastigium of vertex very deeply channeled, with exceedingly high, sharply-compressed, lateral carinæ, which, as soon as they have left the edge of the eyes, bend toward each other at an acute angle, closely approximate, and continue distinctly down the face as lateral raised edges of the frontal costa; summit of head with a slight median carina, which at the extreme front of the fastigium divides and strikes against the lateral carinæ; directly at this fork commences a deep sulcus, which unites uninterruptedly with that of the frontal costa; this costa, expanding a little at the ocellus, again contracts slightly, and then expands greatly; the lateral carinæ of the face equally prominent with, and parallel to, the borders of the frontal costa; antennæ very long, depressed cylindrical, with elongated joints. Pronotum much as in *Psinidia*, the median carina quite the same; surface of disk rugose, the posterior border rectangular. Tegmina reaching beyond the tip of the abdomen, straight, the costal shoulder rather prominent and angular; the intercalary vein is prominent, and runs along the middle of its area; the axillary vein is free, though occasionally united at its tip with the anal vein by a cross-nervure; meso- and metasternum about equally distant in the male. Hind femora extending beyond the tip of the abdomen, rather slender, the carinæ very moderate; arcuate, apical, inner spurs of hind tibiæ equal.

43. *Derotmema cupidineum*, nov. sp.—Cinereo-fuscous, darkest above; a slender, black band unites the middle of the eyes in front, directly above the base of the antennæ; a similar black or blackish stripe unites the middle of the eyes above, on the summit, traversing the fastigium; besides which, there is a more or less distinct, median, longitudinal, black stripe on the summit, and an arcuate black stripe back of the upper part of the eye; antennæ pale toward the base, blackish fuscous toward the tip, and, excepting at the tip, the alternate joints paler, giving them an annulate appearance. Posterior portion of anterior lobe of pronotum with subdorsal crescentic carinæ, much as in *Trachyrhachys aspera;* posterior edge of pronotum dotted with black; center of deflected lobes with a small, quadrate, black spot. Tegmina with the middle field nearly immaculate, the others with blackish fuscous spots linearly arranged, most abundant and most distinct on the basal two-thirds; wings lacteous at base (yellowish in life?) with a rather strongly arcuate, moderately broad, blackish fuliginous band, its inner edge crossing the middle of the wing, extending along the lower margin halfway to the anal angle, and in the preanal area sending a moderately broad, long, tapering shoot more than half-way to the base; the band is slightly obsolescent along the ulnar veins; the apex of the wing is pellucid, with blackish cross-veins, often edged with a fuliginous cloud, especially on either side of the ulnar vein apically. Hind femora with faint fuscous basal, median, and post-median blotches on the upper surface,

AP. JJ—18

and a few dusky dots along the inferior outer carina; hind tibiæ pale reddish, (?) the spines tipped with reddish fuscous. Length of body, 13^{mm}; of antennæ, 10.5^{mm}; of pronotum, 3^{mm}; of tegmina, 15.75^{mm}; of hind tibiæ, 8^{mm}.

2 ♂, Northern New Mexico, August to September, Lieut. W. L. Carpenter.

44. *Trimerotropis obscura*, nov. sp.—Pale brownish cinereous, the under surface and mouth-parts paler; head dotted profusely, pronotum less profusely, with blackish specks; an inconspicuous dusky stripe behind the middle of the eye; antennæ nearly as long as the hind tibiæ, annulate with pallid and fuscous. Deflected lobes of pronotum with a very small blackish or fuscous, central, longitudinal bar. Tegmina sprinkled profusely with blackish fuscous dots, partially collected into three, equidistant, indistinct, dusky clouds, the middle one in the middle of the wing; wings pale yellowish at base, beyond the middle with a very broad fusco-fuliginous band, in the preanal area sending a broad, tapering, rather bluntly-terminating shoot more than half-way toward the base; the inner margin of the band otherwise scarcely arcuate, at the lower margin scarcely extended toward the anal angle; apex of the wing faintly fuliginous, the veins blackish. Hind femora with exceedingly faint, dusky, broad, transverse, oblique bands on the outer face, made more distinct on the upper surface by a sprinkling of black dots, which also mark the inferior outer and inferior carinæ; hind tibiæ of the color of the femora, infuscated at tip, the spines black, excepting at base. Length of body, 27^{mm}; of antennæ, 12^{mm}; of pronotum, 6.5^{mm}; of tegmina, 29^{mm} (?); of hind tibiæ, 13^{mm}.

1 ♀, Northern New Mexico, August to September, Lieut. W. L. Carpenter.

This species in its coloration, and especially in the broad band of the wings, has much the aspect of an *Arphia*. The single specimen obtained is somewhat mutilated, but it is so distinct from any species known to me that I venture to describe it.

45. *Trimerotropis pseudofasciata*, nov. sp.—Brownish cinereous, more or less dotted with fuscous, the middle of the deflected lobes of the pronotum with a quadrate, blackish spot, followed beneath by yellowish; antennæ more or less distinctly annulate with fuscous. Tegmina extending far beyond the tip of the abdomen, with two transverse, blackish, fuscous bands, one median, the other in the middle of the basal half, tapering anteriorly, the inner edge of the inner one usually distinct, and always followed basally by a profuse sprinkling of fuscous dots; apical two-fifths similarly sprinkled, but with slightly larger quadrate spots, often irregularly clustered into spots, nearly or quite as conspicuous in the middle area as toward either border; wings very pale yellow at base, beyond pellucid, with a faint, slightly irregular, subarcuate, mesial band, made up altogether of the darkening (to blackish fuliginous) of the veins of this portion, and occasionally by a slight smokiness of the neighboring cells; the veins of the apex of the wing are again darkened, though not to so great an extent. Hind femora with three obscure, brownish, transverse, oblique belts on the outer face, sparsely sprinkled with blackish dots, becoming more distant, though still not very conspicuous, on the upper face; hind tibiæ yellow, fully one-half of the spines black apically. Length of body, ♂, 20^{mm}; ♀, 24^{mm}; of antennæ, ♂, 8^{mm}; ♀, 10^{mm}; of pronotum, ♂, 4^{mm}; ♀, 5.4^{mm}; of tegmina, ♂, 21.25^{mm}; ♀, 27^{mm}; of hind tibiæ, ♂, 9.25^{mm}; ♀, 12.2^{mm}.

1 ♂, Santa Cruz Island, No. 853, June, Dr. O. Loew. I have other specimens (♂, ♀) collected by Dr. Edw. Palmer, at San Diego, Southern California, in July.

[This species should not be confounded with a hitherto undescribed species from Tighes Station and Julian, Southern California, collected by Dr. Palmer, which I have marked in my collection as *Trimerotropis Juliana*. This latter species has the transverse bands of the wings formed of a rather faint fuliginous shade in the cells, than which the veins are scarcely darker; the contrast of colors upon the tegmina is greater, and the large dark spots are, if anything, broader next the costa, and certainly as distinct there as anywhere; the inner margin of the inner bar is nearly lost in the flecking of the base, and the spots of the apex of the tegmina are wholly or almost wholly confined to the upper and lower margins. The hind tibiæ have also a distinct, dark, basal annulus, and, finally, the median carina of the anterior lobe of the pronotum is not so elevated. In other respects, the two species can scarcely be distinguished; and one of the San Diego specimens of *T. pseudofasciata* approaches *T. Juliana* in having the cells in the banded area of the wings almost wholly fuliginous; but the veins, on the other hand, are so conspicuously darker than the fuliginous membrane that the resemblance is not so great as it otherwise would be; and, in other respects, the specimen conforms to the type of *T. pseudofasciata*. The two species agree in size.]

ANCONIA, (ἄγχω), *nov. gen.*

Head rather small, unusually smooth; space between the eyes above equal to twice the extreme width of the basal joint of the antennæ; fastigium nearly flat, a little transversely sulcate between the middle of the eyes, the lateral carinæ somewhat elevated but blunt, bent slightly inward and less elevated beyond the eyes, continuous with the outer margins of the flat frontal costa; the latter is as broad above as the basal joint of the antennæ, expands a little just above the ocellus, below which it contracts (in the ♀ to nearly half its greatest width), and remains of the same width nearly to the clypeus; in this contracted portion, it is sulcate; lateral foveolæ flat: lateral ocelli very large; eyes large, ovate, very prominent; antennæ moderately short.

inconspicuously enlarged on apical third. Pronotum expanding much, and almost uniformly toward the hinder extremity, the two lobes about equal, the anterior subcylindrical, the posterior nearly flat, with rather prominent but perfectly bluut lateral carinæ, the median carina scarcely visible on the posterior lobe, obsolete on the anterior lobe; anterior border minutely notched in the middle; posterior border obtusely angled, the angle rounded; deflected lobes almost longer than broad; mesosternal and metasternal lobes equidistant in both sexes. Tegmina very long and straight, with slight costal shoulder, the intercalary vein traversing nearly the middle of its area, the axillary uniting at its tip with the anal vein in the basal half of the wing. Hind femora very slender, but scarcely extending beyond the tip of the abdomen; none of the carinæ elevated.

This genus has the general aspect of *Trimerotropis*, but differs from it in nearly all its structural features. It has not a few points of resemblance to *Hadrotettix*, but can hardly be placed near that genus.

46. *Anconia integra*, nov. sp.—Apparently greenish yellow in life; apical half of antennæ faintly infuscated; head and most of anterior lobe of pronotum smooth, the posterior lobe profusely punctulate. Tegmina rather uniformly flecked with small, obscure, sometimes very obscure, fuscous spots; wings hyaline; the costal margin with a pallid stigma near the middle of the apical half of the wing, some of the veins near the apex infuscated. Hind femora pallid or hoary externally, sometimes with faint pre-median and post-median dusky bands, the edges of the geniculations testaceous ; hind tibiæ pale yellow, the apical half of the spines reddish changing to black. Length of body, ♂, 19ᵐᵐ; ♀, 32ᵐᵐ; of antennæ, ♂, 8ᵐᵐ; ♀, 10ᵐᵐ; of pronotum, ♂, 4ᵐᵐ; ♀, 6.5ᵐᵐ; of tegmina, ♂, 22ᵐᵐ; ♀, 32ᵐᵐ; of hind tibiæ, ♂, 10.5ᵐᵐ; ♀, 16ᵐᵐ.

1 ♂, 1 ♀, Mojave Desert, Southern California, No. 829, Dr. O. Loew.

47. *Acrolophitus hirtipes* (Say) Thom.—1 pupa, Southern Colorado, June 11–20, Lieut. W. L. Carpenter.

48. *Brachystola magna* (Gir.) Scudd.—7 ♂.6 ♀, Northern New Mexico, August to September, Lieut. W. L. Carpenter ; plains of Northern New Mexico, eastern slope, October 14–31, Lieut. W. L. Carpenter.

49. *Tettix acadicus* (*Tettigidea acadica* Scudd.).—1 ♂,1 ♀, 1 immature, Northern New Mexico, August to September, Lieut. W. L. Carpenter. This species has only been known hitherto by a single specimen from the Lake of the Woods, British America. By some oversight, it was placed by me in *Tettigidea*.

MANTIDES.

50. *Stagmomantis*, sp.—A single pupa of some species of this genus was brought from Colorado River, California ; it was taken July 20, by W. Summers, No. 863.

ENUMERATION AND INDEX OF SPECIES.

APPENDIX H 10.

NEW SPECIES OF COLEOPTERA, COLLECTED BY THE EXPEDITIONS FOR GEOGRAPHICAL SURVEYS WEST OF ONE HUNDREDTH MERIDIAN, IN CHARGE OF LIEUT. GEO. M. WHEELER, UNITED STATES ENGINEERS, BY JOHN L. LECONTE, M. D.

CREMASTOCHILUS, Knoch.

1. *E. Wheeleri.*—Brownish black, not shining; head feebly punctured, much dilated in front, broadly truncate; side angles rounded. Prothorax subquadrate, a little wider behind, sinuate on the sides; front angles acute, incurved, and densely hairy on the inner side; hind angles rounded, expanded posteriorly; middle third of disk depressed, covered with large, shallow punctures; lateral thirds separated by a shallow impression, more distinct near the base and tip, very sparsely punctured, rather shining and quite smooth behind the middle. Elytra with large, shallow, elongate punctures as usual; humeri shining; mesothoracic epimera not visible from above in consequence of the posterior expansion of the hind angles of the prothorax. Tibiæ compressed; front pair with two approximate teeth; middle pair with two distant teeth besides the apical one; hind pair with a small acute denticle about the middle; tarsi compressed, hind pair two-thirds as long as the tibiæ. Beneath feebly punctured, pubescent with brown hair; mentum deeply concave, subtriangular, bisinuate behind, with the side angles produced and rounded; hind margin feebly notched at the middle. Length, $10.6^{mm} = 0.42$ inch.
Northern New Mexico, Lieut. W. L. Carpenter. In the division of the disk of the prothorax into three parts, this species seems to be related to *C. saucius*. The dividing groove is, however, not well defined, and in other respects there is no resemblance. The form of the mentum is quite peculiar, and in a group where specific differences are of less magnitude would warrant the establishment of a separate genus.

PLECTRODES, Horn.

2. *P. Carpenteri.*—Brown, clothed above with dense, short, pale pubescence, with long hairs intermixed on the head and prothorax. Clypeus slightly broader in front; side angles broadly rounded; front margin subsinuate, narrowly reflexed. Prothorax with the sides much less rounded than in *P. pubescens.* Body beneath and legs clothed as in that species with very long hair; abdomen densely covered with very short, appressed pubescence. Length, $21^{mm} = 0.83$ inch. ♀ clypeus less prolonged, with the margin more widely reflexed; club of antennæ small; last joint of maxillary palpi smaller than in ♂, and tarsi a little shorter.
Los Angeles and Mohave Desert. Differs from *P. pubescens*, Horn (Trans. Am. Ent. Soc., I, 167), only by the characters mentioned above. It is a curious fact that all the specimens of that species collected at Visalia, Cal., were males. The females, as is the case with *Polyphylla variolosa*, probably remained on the ground or at the entrance of their subterranean dwellings.

ATHOUS, Esch.

3. *A. cribatus.*—Slender, dark brown, sparsely pubescent; front deeply, triangularly impressed; margin strongly reflexed. Prothorax one-third longer than wide; sides nearly straight; hind angles not prolonged, nor carinate, slightly divergent; surface coarsely and deeply punctured. Elytra finely striate; interspaces flat, punctured. Antennæ with the second joint one-third as long as the third, which is triangular and as long as the fourth. Tarsi long, slender, not lobed beneath. Length, $7.5-10^{mm} = 0.30-0.40$ inch.
Southern Colorado and Northern New Mexico; found also on Taos Peak. According to the synoptic table (Candèze, Elat., iii, 421), this species would be placed next to *A. reflexus*, from which it is abundantly distinct by the smaller size and coarsely punctured prothorax.

4. *A. simplex.*—Slender, brown, sparsely pubescent; head densely, coarsely punctured; front deeply impressed; margin narrowly reflexed. Prothorax longer than wide; sides rounded; hind angles not prolonged, reflexed, not carinate; disk shining, sparsely punctured. Elytra finely striate; interspaces flat, punctured. Antennæ with the third joint one-half longer than the second, narrower and shorter than the third. Tarsi long, slender, not lobed. Length, $7.5-11_{mm} = 0.30-0.43$ inch.
Colorado, Mr. B. D. Smith.

ASAPHES, Curley.

5. *A. soccifer.*—Slender, brown, sparsely pubescent; front deeply concave; head strongly punctured. Prothorax more than one-third longer than wide, less strongly punctured than the head; sides nearly straight; margin reflexed; hind angles acute.

prolonged, divergent, finely carinate near the side margin. Elytra finely striate; interspaces flat, punctulate; antennæ strongly serrate; second joint small; third triangular, equal to the fourth. Length, 14 ^{mm} = 0.55 inch.

Northern New Mexico. This species resembles in appearance *Corymbites pyrrhos*, but the hind angles of the prothorax are less prolonged, and the antennæ are more strongly serrate. It also resembles *Athous cucullatus*, but differs by the antennæ being more strongly serrate, and also by the front being not reflexed. It is an intermediate between these two species of different genera, though differing from both by the broader antennæ. The tarsal lobes are less developed than in the other species of *Asaphes*, and the anterior margin of the front is somewhat above the labrum, and rather well defined.

A careful examination of this species convinces me that the genera *Corymbites*, *Asaphes*, and *Athous*, as at present comprehended, should be united; an opinion which has been already expressed by my learned friend, Dr. Candèze (Elat., iii, 208), and which he would have announced more definitely had he not attached too great importance to the authority of Kirby, Germar, Lacordaire, and myself.

COLLOPS, Er.

6. *C. hirtellus.*—Above clothed with long, erect, black hair; head and prothorax greenish black, shining, feebly punctulate. Elytra blue, deeply but finely punctured. Epistoma, margin of labrum and first joint of antennæ testaceous: abdomen testaceous, with large, lateral, black spots; legs nearly black. Length, 4.5 , = 0.45 inch. Northern New Mexico, found on Taos Peak, 13,000 feet elevation.

7. *C. reflexus.*—Above clothed with long, erect, black hair; head and prothorax scarcely punctulate, black; side margin of the latter strongly reflexed, especially toward the base, brownish. Elytra very densely and finely punctured, without luster. Antennæ testaceous, in the ♀ spotted with brown. Legs black; abdomen pale testaceous, with the last segment black; epistomá and labrum testaceous. Length, 5 ^{mm} = 0.20 inch. Northern New Mexico.

MALACHIUS, Fabr.

8. *M. montanus.*—Elongate, greenish black, not polished, slightly pruinose, with very short pubescence; head with a deep impression; epistoma white. Prothorax wider than long, rounded on the sides, which are feebly reflexed toward the base, narrowly bordered on the sides with red. Elytra finely scabrous; apical margin pale beneath, and legs black. Length, 3.5 ^{mm} = 0.14 inch.

Northern New Mexico. The antennæ are pectinate in the ♂, and the elytra are not appendiculate. This species is related to *M. Ulkei*, Horn, but differs by the prothorax being only narrowly bordered with red, and by the apical margin of the elytra being pale.

PODABRUS, Fischer.

5. *P. lateralis.*—Elongate, slender, black, very finely pubescent; head feebly punctulate, opaque, alutaceous; in front of the eyes testaceous. Prothorax nearly as long as wide, subquadrate, rounded on the sides in front of the middle; front angles rounded; hind angles small, prominent; disk punctulate, longitudinally concave, with two large convexities, also broadly impressed transversely in front, with the apical margin reflexed; dorsal line finely impressed; sides narrowly margined and pale. Elytra finely scabrous, opaque. Beneath, legs and antennæ black. Length, 7 ^{mm} = 0.28 inch.

Colorado and Northern New Mexico; found on Mount Taos, at 13,000 feet elevation. The antennæ of the ♂ are rather longer and stouter, and the second joint comparatively smaller, than in the ♀.

Belongs with *P. lævicollis, puncticollis, &c.*, but is quite distinct by the characters given above. The claws are appendiculate.

HYDNOCERA, Newman.

9. *H. hamata.*—Black bronze, with a green reflexion on the head and prothorax; thinly clothed with erect white hairs. Elytra sparsely, not very strongly, punctured, with a large, common, pale, spot diverging from the suture, and broadly hooked behind the middle. Antennæ, palpi, and legs testaceous. Length 3.4^{mm} = 0.13 inch.

One specimen, Northern New Mexico. Very closely allied to *H. pallipennis*, but the head and prothorax are less opaque, less alutaceous, and more distinctly rugose; the elytral markings are also different; the arrangement of color might be equally well described by saying that the elytra are pale, with a side margin, and the apical fourth black; the black extends narrowly along the suture nearly to the middle, and from the side margin proceeds an oblique stripe ending behind the middle, midway between the lateral edge and the suture; the tips of the elytra are separately rounded and feebly serrate.

NOTOXUS, Fabr.

10. *N. digitatus.*—Elongate, brownish testaceous, clothed with fine pubescence and with many intermixed fine, long hairs. Head finely punctured, obliquely narrowed behind the eyes, truncate at base; hind angles rounded. Prothorax globose, finely punctulate; horn in front deeply concave, with but five large rounded teeth, one apical and two on each side; hind part of horn suddenly elevated; summit narrow, acutely margined, and with the edge not serrate. Elytra very finely punctulate, paler, with two irregular dusky bands connected by a longitudinal dusky line; tip subtruncate; sutural angle rounded. Length, 3^{mm} = 0.12 inch.

One specimen, Southern Colorado. This species is allied to *N. serratus*, but the horn is quite different by the small number of teeth, and the form of body is less elongate.

I have several other new species of this genus from the interior regions of the continent, and they would well repay the labor of preparing a revision and synoptic table.

MACROBASIS, Lec.

11. *M. murina*, Lec., Proc. Acad. Nat. Sc. Phil., 1853, p. 344 (*Cantharis*).—Several specimens of both sexes were found in Northern New Mexico; the males are quite similar to the two collected by me at Lake Superior, and which were considered by Dr. Horn as a variety of *M. unicolor*. The females, however, differ from that species by the second joint of the antennæ being but little shorter than the first, and nearly equal to the next two united. I am not prepared to say that this is a difference of specific value, for there are in several parts of the *Meloidæ* family indications of a flexibility of structure which we are not yet prepared to account for.

Catalogue of the Coleoptera collected by the explorations during 1875.

The collections were made in two parts of the country surveyed, which are so distant as to have but little zoological relation. I have therefore thought it more useful to prepare two separate lists; the first containing those species collected in California as far east as the Mohave Desert and as far north as Santa Barbara. Small collections from Santa Cruz Island are included, and do not exhibit anything peculiar or previously unknown. The second list contains species found in Southern Colorado and Northern New Mexico, mostly from the eastern foot-hills of the Rocky Mountains. Fourteen specimens were collected by Lieut. W. L. Carpenter on Taos Mountain, at an elevation of 13,000 feet. Three of them are new, but are found at lower elevations, and do not specially indicate arctic or subarctic affinities. These fourteen species are marked with an * in the following list.

I.—*Californian Coleoptera.*

B. Santa Barbara. Cr., Santa Cruz Island. M, Mohave Desert and Colorado River.

Omophron dentatum	Cr	Dermestes talpinus	Cr
Calosoma semiloeve	B	vulpinus	B
cancellatum	B	Helichus productus	M
Lebia cyanipennis	M	Tropisternus californicus	Cr, M
Calathus ruficollis	M	Hydrocharis glaucus	Cr
ruficollis, var	Cr, B	Philhydrus normatus	Tejon
Platynus brunneo-marginatus	Cr, B	perplexus	Tejon
maculicollis	M	Necropharus guttula	B
Pterostichus, n. sp.? (race of vicinus?)	B	Silpha ramosa	B, M
vicinus	B	lapponica	B
laetulus	Cr	Quedius explanatus	S. Cal
Amara californica	Cr	Thinopinus pictus	S. Cal
Chlaenius tricolor	M	Philonthus canecens	B
Anisodactylus consobrinus	Cr	Sinodendron rugosum	B
Bembidium Mannerheimii	Cr, B	Atænius stercorator	M
Hippodamia vittigera	B, Cr, M	Plectrodes Carpenteri	M
ambigua	M	Cyclocephala hirta	M
convergens		longula	M
Coccinella californica		Dichelonycha pusilla	B
Psyllobora taedata	M	Anorus piceus	M

I.—*Californian Coleoptera*—Continued.

Photinus (Ellychnia) facula	M	Noserus plicatus	B
(Pyropyga) californica	M	Cryptoglossa verrucosa	M
Telephorus tibialis	M	Coniontis viatica	Cr
Carpophilus pallipennis	Cr	subpubescens	Cr. B
Ditemnus obtusus	M	Eleodes armata	Angeles & M
Pristoscelis sordidus	M	acuticauda	B
Clerus quadrisignatus	M	dentipes	B
Polycaon Stoutii	Cr	producta	B
ovicollis	Cr	cordata	B
Amphicerus punctipennis	M	Cratidus osculans	Cr
Prionus californicus	B	Amphidora littoralis	M
Stenaspis solitaria	M	Eulabis pubescens	M
Xylotrechus insignis	B	obscura	B
Lema trilineata	M	Blapstinus pulverulentus	B
Chrysochus cebaltinus		Copidita quadrimaculata	B
Diabrotica trivittata		Cantharis vulnerata	S. Cal.
soror		Thricobaris mucorea	M
Phloeodes diabolicus	Cr, B	Scyphophorus yuccæ	M

II.—*Coleoptera of Southern Colorado and Northern New Mexico, collected in 1875, by Lieut. W. L. Carpenter, Ninth Infantry.*

Cicindela longilabris.
    pulchra.
    splendida (*race* amœna).
    purpurea (*race* Audubonii).
    Cimmarona (*var.* greenish bronze).
    12-guttata (*race* guttifera).
    repanda.
    cinctipennis.
    punctulata.
Notiophilus semistriatus.
Calosoma scrutator.
    calidum.
    obsoletum.
Carabus serratus.
    baccivorus (*race* Agassizii).
Cychrus elevatus.
Pasimachus elongatus.
    obsoletus.
    duplicatus *var.*
Dyschirius sphæricollis.
Loxopeza atriceps.
Lebia vividis.
Cymindis abstrusa *Lec.*
    brevipennis *Zimm.*
    cribricollis *Lec.*
Calathus dubius.
Platynus cupripennis.
    placidus.
    octocolus (Taos Mountain).
Evarthrus substriatus.
    constrictus.
Pterostichus protractus.
    Luczotii.
    (Poecilus) lucublandus.
Amara (Lirus) laticollis.
    (Bradytus) latior *Kirby.*
    hyperborea *Lec.*
    libera *Lec.*
    laevistriata *Putsg.*
    oregona *Lec.*
Amara polita.
    chalcea (Taos Mountain).
    interstitialis.
    terrestris.
    obesa.

Chlaenius laticollis.
    sericeus.
Agonoderus pallipes.
Harpalus amputatus.
    retractus.
    herbivagus.
    oblitus.
    basillaris.
Cratacanthus dubius.
Bembidium tetraglyptum.
Hydroporus striatellus.
Laccophilus decipiens.
Colymbetes binotatus.
Agabus obliteratus.
Gyrinus, not determined.
Tropisternus nimbatus.
Silpha lapponica.
    ramosa.
    truncata.
Creophilus villosus.
Dermestes marmoratus.
Trox scutellaris.
    Sonoræ.
Alhodius occidentalis.
    (Taos Mountain).
Hoplia laticollis.
Diplotaxis brevicollis.
Dermestes nubilus.
Erotylus Boisduvalii.
Carpophilus pallipennis.
Hippodamia quinquesignata.
    Lecontei.
    convergens (Taos Mountain).
    parenthesis (Taos Mountain).
    sinuata.
Coccinella trifasciata.
    9-notata.
    5-notata.
    prolongata.
    bipunctata (*var.* humeralis).
Exochomus marginipennis (*var.* æthiops).
Brachiacantha ursina.
Epilachna corrupta.
Hister abbreviatus.
Saprinus lugens.
    plenus.

II.—*Coleoptera of Southern Colorado and Northern New Mexico, &c*—Continued.

Dorcus mazana.
Onthophagus Hecate.
Phanæus carnifex.
Bolbocerus Lazarus.
Diplotaxis Haydeni.
Serica sericea.
Dichelongcha Backii.
    sulcata.
Tostegoptera lanceolata.
Lachnosterna fusca.
    and another species.
Listrochelus, probably new.
Polyphylla 10-lineata.
Euphoria Kernii.
Cremastochilus Wheeleri, *n. sp.*
Trichius affinis.
Aphonus pyriformis.
Chalcophora angulicollis.
Buprestis maculiventris, *race* rusticorum.
    Langii.
Melanophila miranda.
    longipes.
Melanophila Drummondi.
    gentilis.
Anthaxia inornata.
Chrysobothris triennaria.
Acmaeodera pulchella (*race* variegata).
Anelastes Druryi.
Alaus lusciosus = gorgops *Lec.*
Drasterius dorsalis.
Melanotus incertus.
Asaphes coracinus.
    soccifer, *n. sp.*
Corymbites, *n. sp.* (broken).
Athous cribratus, *n. sp.* (Taos Mountain).
Photinus (Ellychina) lacustris.
    (Pyropyga) thoracicus = flavicollis *Lec.*
    nigricans.
Chauliognathus basalis.
Podabrus lateralis, *n. sp.* (Taos Mountain).
Collops bipunctatus.
    reflexus, *n. sp.*
    hirtellus, *n. sp.* (Taos Mountain).
Trophimus æneipennis.
Malachius montanus.
Pristoscelis texanus.
Listrus analis.
Dolichosoma nigricornis.
Trichodes ornatus.
Clerus cordifer.
    moestus.
Hydnocera hamata, *n. sn.*
Amphicerus bicaudatus.
Prionus Californicus.
Homaesthesis emarginatus.
Batyle ignicollis.
    discoideus.
Leptura propinqua.
    convexa (Taos Mountain).
    canadensis (*race* cribripennis).
Pachyta liturata.
Acmaeops pratensis.
Typocerus brunnicornis.
Monilema annulatum.
Monohammus clamator.
Dectes spinosus.
Tetraopes canescens.

Bruchus prosopis.
    amicus (Arizona, in seeds of Circidium floridium).
Orsodacna childreni.
Coscinoptera vittigera.
Cryptocephalus confluens.
    notatus.
Chrysomela multiguttata.
    exclamationis.
    auripennis (Taos Mountain)
Gonioctena pallida (Taos Mountain).
Plagiodera lapponica.
    scripta.
Colaspis tristis.
Phyllobrotica decorata.
Adimonia externa.
Trirhabda convergens.
Monoxia debilis.
Disonycha triangularis.
    punctigera.
Disonycha glabrata.
Graptodera obliterata.
Orchestris albionica (Taos Mountain).
Odontota Walshii.
Epitragus canaliculatus.
Asida opaca.
    polita.
    convexicollis.
    rimata.
Coniontis ovalis.
Eleodes obscura.
    tricostata.
    obsoleta.
    extricata.
    hispilabris.
    nigrina.
    pimelioides.
Blapstinus pratensis.
Corphyra Lewisii.
Notoxus digitatus, *n. sp.*
Anaspis nigra (Taos Mountain).
Mordella scutellaris.
Meloe sublaevis.
Megetra vittata.
Cantharis vividana.
    sphaericollis.
Macrobasis immaculata.
    tricolor.
    murina.
Epicauta maculata.
    ferruginea.
Nemognatha apicalis.
Thecesternus humeralis.
Ophryastes vittatus.
    sulcirostris.
Pontaria rugicollis *Horn.*
Diaminus subaeneus *Horn.*
Epicaerus imbricatus.
Thacolepis inornata *Horn.*
Centrocleous angularis.
Cleonaspis lutulentus.
Stephanocleonus plumbeus, *Lec.*
Cleonus vittatus.
Dorytomus mucidus.
Laemosaccus plagiatus.
Balaninus nasicus.
Baris striatus.
Cossonus subareatus.

APPENDIX H 11.

REPORT ON THE ALPINE INSECT FAUNA OF COLORADO AND NEW MEXICO, SEASON OF 1875, BY LIEUT. W. L. CARPENTER, NINTH INFANTRY.

UNITED STATES ENGINEER OFFICE,
GEOGRAPHICAL SURVEYS WEST OF THE 100TH MERIDIAN,
Washington, D. C., May 1, 1876.

SIR: I have the honor to submit a report on the alpine insect fauna of Colorado and New Mexico. A separate entomological collection was made during the season of 1875, at high altitudes, which has proved of interest in its relation to geographical distribution and in the production of many rare and new species.

It is to be hoped that future collectors will not overlook this important field for zoological research, but will endeavor to increase our limited knowledge of the insects of this region.

Very respectfully, your obedient servant,

W. L. CARPENTER,
First Lieutenant Ninth Infantry.

Lieut. GEO. M. WHEELER,
Corps of Engineers, in Charge.

The mountain-ranges of the world produce a fauna of remarkable interest in its bearing upon the discussion of geographical distribution and the existence of varietal forms which have resulted from the great climatic changes through which the globe has passed.

In the elucidation of geological epochs, we find that even the lowest forms of life have a history which, could we but trace back through the dim ages which have intervened since the dawn of life, would throw a flood of light upon many subjects at present conjectural. When it is found that certain insects occur uniformly in the mountains of Asia, Europe, and America, at great elevations, and in British America at high latitudes, the geological significance of this fact becomes apparent; and the mind reverts to a period when the steady encroachment of a vast field of ice caused the extinction of delicate species and compelled the survivors to change their habitat for lofty mountains, which had become islands in a sea of ice. Here, amid new conditions of temperature and a modified flora, a few hardy species were perpetuated, which have preserved their alpine characteristics to the present time.

Thus we find certain insects on Mount Washington, New Hampshire, which are lost sight of as we journey westward through the great expanse of valley and plain to the foot of the Rocky Mountains; as we ascend them and approach the verge of the alpine flora, these same species re-appear with wonderful regularity, establishing the perfect identity of the insect-fauna of our mountains.

With the return of a genial climate, during the Champlain period, the ice receded to the north, releasing the fauna from its imprisoned state, and stimulating all life to again spread over the continent. The valleys and plains were once more filled with species which found the warmth of the new climate congenial to their tastes; these, spreading over the land and mingling with other forms which followed the retrogression of the ice from its southern limit, produced the fauna as it now exists. Other species, to which a boreal climate had become agreeable, finding the increasing temperature of the valleys distasteful, migrated to the northward in the path of the ice, which was slowly uncovering the country, until they found a suitable habitat in the arctic regions. The few species which remained upon the mountain-peaks lingered about their old haunts until the climate which we now have had become established and their retreat accordingly prevented. They are thus imprisoned in the mountains, contented with a modified climate so nearly resembling that voluntarily chosen by their relatives which have colonized the barren ground of the far north. Consequently we find these species dwelling in an extremely isolated range, although having a geographical distribution only dependent upon a requisite elevation above the level of the sea, or an equivalent high latitude and consequent congenial climate.

The alpine insect fauna of America should then be regarded as but a fragment of that which survived the geological changes which occurred at the close of the Tertiary and beginning of the Quaternary epochs. Although the mountain genera and species are nearly all represented in the arctic fauna by the same species or their analogues, yet the number inhabiting the latter region is greatly in excess of the former, as would naturally be supposed in view of the general migration of the alpine fauna, set free by the northward movement of an isothermal zone. With such similarity existing between them, the study of both faunæ becomes necessary in instituting any comparison between individuals of the same species for the purpose of determining varietal differences which are the natural outgrowth of the modifications of food and climate produced during ages of separation. The primary causes of differentiation, however, appear to be of a geological rather than meteorological nature; the climatic causes operating as a factor, which subsequently appeared as a necessary sequence of the termination of the Glacial period. This interesting fauna obviously possesses characteristics which at once attract attention and induce study in determining the great causes which have operated to produce such a remarkable distribution of life.

The observations of Dr. A. S. Packard, jr., on the geographical distribution of the moths of Colorado, have established the fact of the existence of a law of increase in the length of certain peripheral parts for western species. Professor S. F. Baird and Mr. J. A. Allen have shown that the same law of variation obtains in regard to the birds of North America. Dr. Packard states that the moths of the Pacific coast are generally larger than those of the Rocky Mountains, and almost invariably larger than the same species from New England and Labrador; and he considers the difference in growth to be due to the more genial climate and greater rain-fall of the western coast. While a difference of temperature and relative humidity exercise an influence over the growth and coloration of insects of sufficient importance to be accepted as a factor affecting their development, this is probably also due in some degree to the prevalence of high winds, which have operated to produce greater development of the wings as a natural result of greater habitual exertion in combating them; and perhaps also to the acquisition of nomadic habits rendered necessary in a comparatively barren region, where vegetation is not as exuberant as in more civilized localities which have been under cultivation for a long time. The swarming grasshopper, *Caloptenus spretus*, which by its periodic migrations, extending over 20° of latitude, proves so destructive to our agricultural interests, singularly confirms this law. This insect, having been accustomed to sustain long continued flights, has consequently developed more powerful wings than the same species which occurs in the Eastern States, in no respects differing from the western insect except in its habits of more local residence and shorter wings.

As regards variation of color, I believe that species from a cold climate, or where there is an unusual absence of sunshine, will be characterized by the predominance of somber-hued types, or will present a bleached appearance when compared with specimens from a warmer climate. In the arctic regions, the short summers seem poorly adapted to the development of vivid colors, and we accordingly find few of the beetles known to occur there which are remarkable for bright colors; the same fact is noticeable in those obtained from the western mountains above the timber-line. A notable exception to this rule, however, is *Carabus vietinghovii*, Adams, from Hudson Bay; a northern insect, which may well vie with tropical species in brilliancy of color. This insect may be regarded as an example of the effect of a short period of almost constant sunshine, such as prevails during the summer in this part of the circumpolar regions. More complete data regarding variation in color should be collated before the facts can be regarded as establishing any conclusive law of melanism affecting species inhabiting separate zoological zones.

Although I have not had an opportunity to institute extensive comparisons between collections from different mountain-regions, observations made in the West during several seasons have convinced me that, in general, the animal kingdom existing at great elevations is as dwarfish as the vegetable kingdom. This is especially apparent in the *Coleoptera, Hemiptera, Orthoptera, Neuroptera,* and *Arachnida*, the types of which are usually represented by the smaller species. But this does not appear to be the case with the *Lepidoptera, Hymenoptera,* and *Diptera*, which furnish large-sized forms, especially prominent among which are the *Papilionidæ, Argynnides, Apidæ,* and *Tabanidæ*.

The *Coleoptera* show a remarkably uniform distribution over Colorado, Oregon, and the circumpolar regions. So apparent is this, that a species obtained at a great elevation from Colorado may be looked for with almost certainty in these other localities. Further research will probably verify this distribution in other orders. The collection of spiders exhibits a striking resemblance to those found in the barren grounds of the far north, by Captain Back, R. N., whose collections contain four genera; all of which occur above timber-line, and have the same habits of living in the ground and rocky crevices, incumbent in regions of stunted and scant vegetation. Among the grasshoppers obtained by Captain Back, we find *Acridium sulphureum*, Pal. de Beauvois. (*Tomonotus sulphureus*, Fabr.), which is found in the mountains of Colorado; and although it has not yet been obtained above timber-line, it may be accepted as a true resident of that desolate region. From the *Fauna Boreali-Americana*, we have *Locusta leucostoma*, Kirby (*Caloptenus birittatus*, Uhler), found by Sir John Richardson, latitude 65° north, and which is also a western species. Among other mountain *Orthoptera*, the genus *Pezotettix* should be regarded as a typical alpine form; two species having been found at an elevation of 13,000 feet.

The *Hymenoptera* were among the insects earliest observed above timber-line. Frémont records the presence of bumblebees amidst snow and ice on the summit of the

Rocky Mountains during his overland journey. They are also known to be the first insects to appear in the early spring in the arctic regions.

The *Apidæ* are well represented in this collection, and have their prototypes reproduced in the arctic regions with remarkable fidelity.

Anthophora bomboides, Kirby, and *Bombus borealis*, Kirby, from latitude 65° north, occur also in the mountains of Colorado and New Mexico as alpine species. *Allantus basilaris*, Say, found above timber-line, is a species widely distributed and *à priori* ought to be found in the arctic regions by future collectors.

A few circumpolar species of butterflies which occur also in Colorado and New Mexico, appear to be among the very rarest of the western species; such are *Vanessa antiopa*, Linn; *Cynthia huntera*, Fabr. (*Pyrameis huntera*, Drury); and *Hipparchia nephele*, Kirby. Subsequent collections will probably disclose their existence in greater abundance along a zone following the Rocky Mountains northward, and bending eastward through the Lake Superior region to Hudson Bay and Canada, and branching westward to Oregon and Northwest British America.

In a region of sparse vegetation, the *Hemiptera* would necessarily be few in numbers; the insects of this order found in the arctic regions are represented in this collection only by the genus *Miris*, although *Lygæus recliratus* is a species to be looked for in British America, as it is one of those singular forms apparently endowed with a remarkable vitality, which has enabled it to survive great climatic changes, and consequently to acquire extensive geographical distribution.

Among the most interesting of the butterflies is *Chionobas semidea*, Say, which occurs in the Alps, Rocky Mountains, and on Mount Washington, N. H. Its presence in the mountains of Asia may be considered as extremely probable. The writer was informed by Dr. S. H. Scudder that in the Swiss Alps it appeared to be a sluggish insect, unable to sustain long flights, and consequently easily taken. In our western mountains, it has a strong, rapid flight, and is one of our most wary species, although quite common at an elevation of 14,000 feet.

Colias meadii, Edw., and *Argynnis freya*, Esper, are truly alpine butterflies; the former has its analogue, *C. hecla*, in the Arctic regions.

The southern Rocky Mountain chain appears to support a greater abundance of some orders of insects than the northern. A season's collecting above timber-line in Northern Colorado only produced five species of butterflies, while the mountains of New Mexico yielded sixteen species. The ratio of luxuriance, however, in the case of some other orders is largely in favor of the northern mountains, a result seemingly incongruous, but which undoubtedly bears directly upon the biology of each order of insects. Of all orders, the *Lepidoptera* is the one which most delights in a warm climate and bright sunshine, and any deviation from such a habitat should be regarded as an involuntary change, rendered obligatory by the slow substitution of a cold climate during ages of progress in their development. The temperature of the climate of the mountains of New Mexico is considerably warmer than that of the same elevations in Colorado; the alpine flora consequently extends into higher regions in New Mexico, making a suitable habitat for the *Lepidoptera*, *Hemiptera*, and *Diptera*, three of the most delicate orders of insects.

The *Coleoptera*, *Hymenoptera*, *Orthoptera*, and *Arachnida*, being more hardy, accordingly predominate in more northern regions.

A table is here presented, embodying the results observed.

	More abundant above timber-line in—	
Hymenoptera	Colorado.	
Lepidoptera		New Mexico.
Diptera		New Mexico.
Hemiptera		New Mexico.
Coleoptera	Colorado.	
Orthoptera	Colorado.	
Arachnida	Colorado.	
Myriapoda	Colorado.	

It is but quite recently that thorough alpine collections have been made, and our knowledge of this fauna is consequently not very extensive, being at present restricted to the mountains of Colorado and New Mexico. The northern Rocky Mountain chain offers a new field for the further investigation of this subject, which will undoubtedly produce many other species to be added to the catalogue, and establish still more conclusively the relationship existing between the alpine and the arctic faunæ.

This report is based upon collections made by the writer during the season of 1875, while employed with the Colorado section of this Survey. The results here presented should be regarded as but an imperfect outline of the entire fauna. Other collections ought to be made before sufficient material will have been secured to enable the naturalist to discriminate between actual residence and some few species which may owe their presence to storms or atmospheric currents, and thus would not properly belong to this *fauna*. But with each comparison of collections the constant recurrence of certain species will designate those which should be regarded as conterminous. It is a difficult matter to determine a vertical limit for alpine species, because a few forms occurring in the foot-hills sometimes encroach upon the verge of the alpine flora.

The altitude of 12,000 feet above the level of the sea, which is about 300 feet above timber-line, has been therefore selected as the elevation best calculated to yield a collection characteristically boreal; and all species below enumerated were accordingly taken at a considerable altitude above timber-line.

The following is a conspectus of the alpine insect fauna of Colorado and New Mexico :

HYMENOPTERA.

‖ Bombus borealis.
Bombus flavifrons.
Bombus rufocinctus.
Bombus ternarius, Say.
Allantus basilaris, Say.

Cryptus robustus, Cres.
Odynerus tigris, Sauss.
Lyda Carpenterii, Cress.
Ammophila robusta, Cress.

LEPIDOPTERA.

Papilio Asterias, Fabr.
Papilio Zolicaon, Boisd.
⸙ Parnassius Smintheus, Doub. 3 varieties.
Colias Keewaydin, Edw.
† Chionobas Semidea, Harris.
Chrysophanus Sivius, Edw.
Thanaos brizo, Boisd.
Thanaos persius, Scudd.
Hesperia Centamea, Rambur.
Colias meadii, Edw.
Argynnis Carpenterii, Edw., new species.
Argynnis helena, Edw.

Melitæa nubegina, Beh.
Phyciodes Carlotta, Reak.
Grapta Latyrus, Edw.
Euptychia Henshawi, Edw., new species.
Erebia Tyndarus, Edw.
Pieris occidentalis, Reak.
‖ Argynnis freya, Esper.
†‡ Arctia Quenselii, Paykull.
‡ Agrotis islandica, Staub.
Anarta melanopa, Thunb.
Plusia Hochenwarthi.

DIPTERA.

Syrphus obliquus, Say.
Musca erythrocephala, Meigs.
Bibio ———.
Tachinidæ ———.
Authomyidæ ———.

Tipula ———.
Pachyrrhina ———.
Melanostoma ———.
Sarcophagidæ ———.

COLEOPTERA.

* Carabus taedatus, Fabr.
*‡† Nebria Sahlbergii, Fisch.
*⸙ Platynus octocolus, Mann.
Amara chalcea.
* Amara terrestris. Lec.
†‡ Amara obtusa, Lec.
* Bembidium tetraglyptum, Mann.
Silpha ramosa, Say.
Hippodamia convergens.
Hippodamia parenthesis.
Aphodius coloradensis.
⸙ Coccinella transversogutta, Fald.
Athous cribratus, Lec., new species.
Podabrus lateralis, Lec., new species.
⸙†‡ Podabrus lævicollis, Kirby.

Collops cribosus, Lec.
Collops angustatus, Lec.
Collops hirtellus, Lec., new species.
*⸙†‖ Adoxus vitis, Linn.
*‖ Leptura convexa, Lec.
⸙ Gonioctena pallida, Linn.
Chrysomela auripennis, Say.
Chrysomela dissimilis, Say.
Trirhabda convergens, Lec.
Orchestris albionica, Lec.
⸙ Anaspis nigra, Hald.
* Alophus alternatus, Say.
Stereopalpus guttatus, Lec.
*⸙ Dendroctonus obesus, Mann.

HEMIPTERA.

Thyanta perditos, Fab., ♂ var.
Lygaeus reclivatus, Say.
Miris instabilis, Uhler.
Coriscus ferus, Linn.
Trapezonotus nebulosus, Fallen.

Agalliastes punctulatus, Uhler.
Geocorris bullatus, Say., var.
Orectoderus amocoens, Uhler.
Lygaens circumcinctus, Stal.
‖ Aradus americanus, Fab. (affinus).

* Found also in Alaska.
† Found also on Mount Washington, N. H.
‡ Found also in Labrador.

⸙ Found also in British America.
‖ Found also in Arctic regions.
** Found also in Oregon.

305

ORTHOPTERA.

Caloptenus spretus, Uhl.
Pezotettix stupefacta, Scudd., new species.
Anabrus coloradus, Thom.

Pezotettix Marshallii, Thos.
Platyphyma montana, Thos.
Gomphocerus Carpenterii, Thos.

ARANEINA.

Attoidæ, 1 species.
Drassoidæ, 5 species.
Theridoidæ, 1 species.

Thomisoidæ, 2 species.
Lycosoidæ, 14 species.

MOLLUSCA.

Isthania simplex, Gould.

Lonites nitidus, Müll.

MYRIOPODA.

Lithobiuus Americanus, Newport.

APPENDIX H. 12.

NOTES ON THE MAMMALS TAKEN AND OBSERVED IN CALIFORNIA, IN 1875 BY H. W. HENSHAW.

FELIDÆ.

1. *Felis concolor*, L.—Cougar; American Panther.

This formidable Cat appears to have disappeared almost entirely from the lower and more thickly settled districts, and is found now only in the heavily-timbered regions of the deep mountains. In winter only, and when the deep snows have rendered precarious its usual methods of obtaining food, does it descend from its mountain fastnesses and make known its presence by its depredations on the farmer and ranchman.

CANIDÆ.

2. *Canis latrans*, Say.—Coyote; Prairie Wolf.

The persistent efforts of sheep-raisers and farmers have resulted in the almost com plete extermination of this animal from many parts of California. Comparatively few were met with by our parties during the past season. As its name indicates, it is a rather exclusive inhabitant of the plains and lowlands. The mountains are not infre- quently visited by them in summer; they there finding, perhaps, the seclusion they court during the season of reproduction, as well as a greater abundance of food.

3. *Urocyon cinereo-argentatus*, (Screb.) Coues.—Gray Fox.

A fine female of this species was shot by Dr. Rothrock near Walker's Basin. This was the only specimen secured, though they were seen with sufficient frequency to war- rant the statement that they are numerous in Southern California. As with the Coyotes, strychnine administered by the sheep-herders has reduced their numbers very materially. They inhabit the wild mountainous districts, which are well timbered; woodland being more essential to the mode of life of this fox than to most others of the family.

No.	Sex.	Locality.	Date.	Collector.
719	♀ ad.	Walker's Basin, Cal	Aug. 27, 1875	Dr. J. T. Rothrock.

4. *Urocyon cinereo-argentatus*, (Screb.) Coues, var. *littoralis*, Baird.—Island Fox.

This, the smallest of our North American Foxes, was brought to the notice of naturalists through specimens obtained by Lieutenant Trowbridge on the island of San Miguel, off the coast of California. Though noticing the marked similarity in its external appearance when compared with the common Gray Fox (*Vulpes virginianus* of most authors), Professor Baird yet treated the present animal as a distinct species, though in his article in vol. VIII, Pacific Railroad Report, he expresses the doubt as to whether it is not merely a local race of that animal. An excellent opportunity was had during the past season of examining numerous specimens of this little Fox in the field, and, since returning, of making direct comparison of the skins of the two animals, as well as comparing the crania. The result shows an extremely close relationship, and, as remarked by Professor Baird, size alone appears to be the only external character of much importance in the discrimination between the two animals, the Island Fox being in round terms but about one-half the size of its congener. In color the discrepancies are quite unimportant, and are almost, if not quite, within the usual range of individual variation. Perhaps the Island Fox may be of a generally darker tone of coloration throughout. Specimens, however, taken in June, while they show a deeper tint of rufous, lack the decided glossy black mixture of the back and tail shown on a Gray Fox shot in August in the Sierra Nevadas, this being replaced by a blackish-brown color. This may be a purely seasonal difference.

A comparison of crania of the two animals shows no distinctive peculiarities of moment, those noted by Professor Baird appearing to be merely individual.

The islands in the Santa Barbara Channel, where alone, so far as is now known, this animal is found, have long been separated from the mainland, with which geological evidence shows they were formerly connected. Supposing them, at the time of their isolation from the mainland, to have been inhabited by the Gray Fox, which, as is well known, extends from the east quite to the Pacific coast, still retaining its typical form, we may readily assume that in the long interval that has since elapsed, pent up within a very circumscribed area and subject to greater or less changes in the conditions of life and climate, the animal has deteriorated in size to what we now find it, without having suffered other notable differentiation. Principally to a difference in food may, perhaps, be ascribed the diminution in size.

Taking into consideration the complete isolation of the two forms in question, and the fact that no perfect intergradation of size can be shown to connect them, we might perhaps be justified in according full specific rank to this Fox, and this while fully admitting the extreme probability of it having originally sprung from the allied species. I have thought best here, however, to consider it merely as a varietal form of the ordinary Gray Fox.

It was only upon the island of Santa Cruz that I had the opportunity of seeing this animal, though I was informed by good authority that all the group contained them in greater or less numbers. Upon all portions of Santa Cruz* they are abundant, and in certain parts they exist in almost incredible numbers. On the west, the high broken ridges descend in a somewhat gradual slope to the shore, and this portion is clothed with a scanty growth of cactus (prickly pear). This seemed to be a favorite resort of the Foxes, or else the large number seen here was due to the lack of undergrowth, and their consequent inability to find good hiding-places. In passing over the terrace-like plateaus, where the cactus-plants were tolerably thick, I had no difficulty in starting up one of these animals every few moments. No fewer than fifteen were seen in a two hours' walk. Of timidity they showed scarcely a trace, and fear of man had certainly never been inherited by the individuals I saw. The scanty shade afforded by the cacti was here their only protection from the hot sun, and snugly rolled up underneath these plants, I usually found them taking their noonday naps. Arousing themselves to the situation when my footsteps within a few feet of their retreat awoke them, they would quietly walk out a few steps. The most timid moved off at an easy trot, now and then throwing a glance backward, as if somewhat doubtful of our intentions. From the character of the droppings I concluded that their food, at least at this the summer season, must be largely insectivorous, and such proved to be the case. The beaches, too, doubtless supply more or less of their subsistence, though the tracks found upon these in the morning did not indicate that they resorted to them in very great numbers.

On one occasion I had an excellent opportunity of watching one of these little animals as he was busily engaged hunting his supper. His search, as long as I followed him, which he permitted me to do at a short distance, was limited to insects, especially grasshoppers, which he found on the open plain among the scanty herbage and under small stones. These, after a preliminary sniff told him that game was beneath, he readily turned over with his long snout.

Notwithstanding the fact that the island is crowded with sheep, I could not learn that any depredations had been traced to the Foxes. On the contrary, they appeared to be considered by the herders as perfectly harmless, although it seems most probable that the young lambs must suffer from their attacks. They prove a determined enemy to the Gulls, Cormorants, Guillemots, and other sea-birds that congregate on the islands for purposes of reproduction, so that these are forced from the main island, and compelled to deposit their eggs upon the little inaccessible islets contiguous. Quite a number of their nests which had been rifled of their contents came under my observation.

---

* This island is distant from the mainland about 16 miles, being the nearest. It is about 30 miles in length, with perhaps an average width of 5 miles.

In preparing the skins of some half dozen of these Foxes I noticed a curious fact, which may be worth mention. In each instance the interior surface of the hide was perforated by a fewer or greater number of the cactus spines. Of one, apparently an old patriarch, the hide was fairly coated with these spines, which, having penetrated, had worked through and had become disposed along the horizontal surface of the skin. So thick were they that a point of a knife-blade could hardly have been applied to the skin without touching one or more. Many had become soft and flexible with age, while those more recently introduced were hard and stiff,

They probably whelp quite early in the summer. I saw, however, but one litter of young. These, about six weeks old, I found June 3, close to the mouth of a subterranean burrow, and, coming suddenly upon them, I had no difficulty in catching one of them in my hands. The little fellow made no attempt to bite, but immediately began to howl most lustily, when the mother came trotting out from some retreat hard by, and, approaching to within a few feet, looked up into my face with a most earnest, pleading expression, which effectually deprived me of all desire to do injury either to her or her offspring. After a moment's silent pleading, she slowly walked off, keeping an eye on my every motion, and, having withdrawn a short distance, awaited till my departure gave her a chance to regain her progeny.

Measurements—fresh specimens.

	No. 3.	No. 4.	No. 5.
Tip of nose to canthus of eye	1.75	1.75	1.65
Tip of nose to ear	2.75	2.75	1.55
Tip of nose to occiput	2.00	1.86	2.00
Tip of nose to tail	14.25	13.65	13.25
Tail, from root to end of vertebra	10.50	9.00	10.50
Tail, from root to end of hairs	12.75	11.25	12.62
Arm, from elbow to end of claws	2.75	2.75	2.50
Leg, from knee-joint to end of claws	4.75	4.25	4.00

No.	Locality.	Date.	Collector.
1	Santa Cruz Island, Cal.	June 1	H. W. Henshaw.
2	......do.........	June 1	Do.
3	......do.........	June 1	Do.
4	......do.........	June 2	Do.
5	......do.........	June 2	Do.
720	......do.........	June 3	Do.
721	......do.........	June 3	Do.
994	......do.........	June 1	Do. (alcoholic.)

5. *Vulpes vulgaris pensylvanicus*, (Bodd.) Coues.—American Red Fox.

From sheep-herders I learned of the existence in the region about Mount Whitney of a Fox which, without hesitation, I refer to this species with the greater certainty, inasmuch as I had an opportunity of examining a young one which had been captured at the base of Mount Whitney. The animals were said to be quite numerous, and to live by preference along the borders of the extensive mountain-meadows.

MUSTELIDÆ.

6. *Mephitis mephitica*, Baird.—Common Skunk.

The Skunk appears to be as common in Southern California as anywhere in the East. It is most often found along the borders of the streams. One that I stumbled upon had arranged a rather neat nest in the tops of some dead rushes, a foot or so from the water, in which it was lying curled up fast asleep.

7. *Mephitis bicolor*, Gray.—Little-striped Skunk.

Though known to be a resident of California, none of this diminutive species were noted by our parties. A beautiful specimen was presented me by Captain Forney, of the Coast Survey, he having obtained it from the island of San Miguel, the outermost of the Santa Barbara group.

8. *Taxidea americana*, (Bodd.) Bd.—American Badger.

Throughout Southern California, the Badger is a very common animal, ranging upward to a height of at least 8,000 feet.

It was noticed that their burrows were frequently made with direct reference to the abundance of the "Gopher Squirrel" (*Spermophile harrisii*), and these no doubt constitute no small proportion of their food, a fact which might well be borne in mind by the farmers, to whom these squirrels are almost a deplorable scourge.

No.	Sex.	Locality.	Date.	Collector.
718	♀ ad.	Walker's Basin, Cal...................................	Aug. 28, 1875	Francis Klett.

URSIDÆ.

9. *Ursus horribilis*, Ord.—Grizzly Bear.

Perhaps few animals have suffered more from persistent and relentless warfare waged by man than this formidable Bear. To the sheep-owners especially, whose immense flocks under the care of one or two men are driven far into the heart of the mountain wilderness to pass the summer months, are these animals special objects of dread. Accordingly every means in their power are used for their extermination. A supply of strychnine is part of the outfit of every shepherd, and by means of this the number of Bears is each year diminished, till in many sections where formerly they were very abundant they have entirely disappeared. This is particularly the case with the Grizzly, whose nature seems to be far more savage, and who courts the secrecy of the deep wilderness far more assiduously than his congeners, the various varieties of the Black Bear.

Accordingly, in many thinly-settled regions, where the latter is still by no means uncommon, the Grizzly has entirely disappeared, having been killed out or forced to withdraw to more inaccessible sections.

In some sections, however, as in certain portions of the mountains near Fort Tejon, the Grizzlies are quite numerous, sufficiently so as to make deer-hunting on the mountain ridges, where the chaparral grows in almost impenetrable clumps, a matter of no little danger.

It is an indisputable fact that the temper of the Grizzly of the Sierra Nevadas and of the same species of the Rocky Mountains is very different. The latter seems to be an animal to be dreaded but little, if any, more than the Black, Brown, or Cinnamon Bears of the same region, and rarely has it been known to assume the initiative in a contest with man.

Very different is it in the Sierras, where stories of unprovoked attacks by Grizzlies are frequent, and not few are the lives lost in such encounters.

10. *Ursus americanus*, Pall.—Black Bear.

The various types of this Bear, known as the Black, Brown, and Cinnamon varieties are all found in California, and are more or less numerous.

The approach of winter sends down from the mountains many of these animals, which often congregate in some locality which proves favorable for a supply of food, and where earlier they do not appear at all. Thus, in the hills near Caliente their broad tracks were everywhere visible in the oak-groves, where they had descended in the night from their lurking places in the deep cañons and the dense chaparral thickets, to feast upon the rich harvest of acorns.

OVIDÆ.

11. *Ovis montana*, Cuv.—Rocky Mountain Sheep.

A small band of these animals was seen upon the summit of Mount Pinos, in the Coast range, and another near the top of Mount Whitney. Their tracks were frequently observed by the parties who ascended the lofty summits of the Sierras, though they are apparently less numerous in California than in many parts of the Rocky Mountain region.

12. *Antilocapra americana*, Ord.—Prong-horned Antelope.

The dry plains in various portions of Southern California are tolerably well stocked with Antelopes, bands of which may occasionally be seen close to the lesser-traveled roads. Their extreme wariness, as well as the desert nature of much of the country inhabited by them, serves to protect and prevent their extermination.

CERVIDÆ.

13. *Cervus canadensis*, Erx.—American Elk.

The Elk has almost entirely disappeared from Southern California, in some portions of which it existed in great numbers but a few years since. I was informed by relible authority that a few still remain, making their abode in the impenetrable tulle swamps that environ the Tulare and Kern Lakes.

14. *Cervus columbianus*, Rich.—Black-tailed Deer.

This is the prevailing Deer throughout the Coast range, as well as the Sierra Nevada, and is in many sections numerous, though we everywhere heard of their constantly-diminishing numbers.

The summer is passed among the high mountains, where the ridges clothed with chaparral form their retreat during the day, the bucks especially ascending to the bases of the highest peaks. At the beginning of cold weather, they gradually work down into the warm and sheltered valleys, there to pass the winter.*

Like the Bears they resort to the oak-groves to glean the crop of acorns, of which they are very fond.

TALPIDÆ.

15. *Scapanus townsendi*, Bach.—Oregon Mole.†

To this species is referred a mole taken at Santa Barbara.

Measurements.—Nose to root of tail, 5.00; tail to end of vertebra, 1.23; to end of hairs, 1.43; fore foot, 1.00; hind foot, 0.80; breadth of palm, 0.60.

No.	Locality.	Date.	Collector.
992	Santa Barbara, Cal	July —, 1875	H. W. Henshaw.

16. *Hesperomys* (*Vesperimus*) *americanus*, (Kerr) Coues.—White-footed Mouse.

This, the common species of the interior region, was seen in many localities throughout Southern California, and specimens secured.

It may be mentioned as of interest that these little animals appeared to figure largely as an item of fare of the large Brook-trout (*Salmo* ——?), which abounds in the north fork of the Kern River. Scarcely one of the trout was taken that one or more of these Mice was not found in its stomach, while from one fish of unusually large size no fewer than five were taken. The trout secure them at night as they run about the margin of the pools.

17. *Arvicola* (*Myonomes*) *riparius*, Ord.

A widely-distributed species, and one common in Southern California. Lives principally beneath rotten logs.

18. *Ochetidon longicauda*, (Bd.) Coues.

This little Mouse appears to be confined to California, no specimens having been taken, so far as I am aware, elsewhere. It there replaces the allied species *O. humilis*, the Harvest Mouse of the region east of the Rocky Mountains.

Measurements.—Specimen No. 207: Nose to occiput, 0.81; to ear, 0.67; to eye, 0.33; to tail, 2.03; tail vertebræ, 2.70; with hairs, 2.75; fore foot, 0.25; hind foot, 0.63; ear above notch, 0.50.

No.	Sex.	Locality.	Date.	Collector.
207	♂ ad.	Santa Barbara, Cal	July 6, 1875	H. W. Henshaw.
1300	♀ ad.	Mescal, Cal	May 1, 1875	Lieut.Eric Bergland.

19. *Dipodomys phillipsii ordi*, (Woodh.) Coues.—Kangaroo Rat.

A single specimen of this Rat was secured by Mr. Hasson near Los Angeles.

No.	Locality.	Date.	Collector.
993	Southern California..	July —, 1875	J. A. Hasson.

* Quite a number of Bats were secured during the season, but have not yet been identified.

† For the identification of this and others, especially the species of *Muridæ*, I am indebted to Dr. Coues.

AP. JJ—20

20. *Cricetodipus parvus*, Bd.—Least Kangaroo Mouse.

A single specimen from near Fort Tejon proves to be of this extremely rare and interesting species. This, the fourth specimen known, serves to substantiate more fully the specific distinction made between it and the single other member of the genus, *C. flavus*.

Measurements.—Nose to occiput, 0.90; to ear, 0.77; to eye, 1.47; to tail; 2.05; tail vertebræ, 2.50; with hairs, 2.75; fore foot, 0.25; hind foot, 0.65; ear above notch, 0.28.

GEOMYIDÆ.

21. *Thomomys (Talpoides(umbrinus*, (Rich.) Coues.—Black-faced Gopher.

This Gopher was found to be very abundant in many localities of Southern California, and its dispersion over the country would appear to be very general. Its habits are not very well known, chiefly from the fact that most of its existence is passed under ground. It moves about from place to place through long burrows made just beneath the surface of the ground; tunnels being driven outward every few feet to admit of foraging on the surface for all sorts of small seeds. The little heaps of earth with which it always closes the mouths of its burrows are noticeable in all directions where the animal is found, and sufficiently attest its activity during the night hours; for, I believe, its habits are chiefly nocturnal, though they may often be noticed at work in the early morning and late afternoon. In the fall, when perhaps the approach of winter hurries them in their efforts for laying up the needed supplies, their labor is continued during the sunny hours of the day.

Their sense of hearing is wonderfully keen, and the chief reliance to warn them of danger. When perfectly silent, a person may remain within a few feet of the mouth of their burrows and watch their motions at will.

No.	Sex.	Locality.	Date.	Collector.
188	♀	Santa Barbara, Cal	July 5, 1875	H. W. Henshaw.
339	—	Fort Tejon, Cal	Aug. 8, 1875	Do.
345	—	......do	Aug. 9, 1875	Do.
243	—	Santa Barbara, Cal	July —, 1875	Do.
995	—	......do	July —, 1875	Do.

SCIURIDÆ.—SQUIRRELS.

22. *Sciurus leporinus*, Aud. and Bach.—Californian Gray Squirrel.

This large and beautiful species was found to be extremely abundant in certain portions of the Coast range, and also in the Sierras. Its distribution over California appears to be very general. In summer I found them living almost exclusively among the pines, where they found an abundance of the seeds. The ground under the trees was often covered with the husks of these and with the cones half eaten. In the fall they resort very much to the oak-groves, whence they glean a rich harvest of acorns.

The chestnut patches at the base of the ears appear to be wanting in summer, as of quite a number of individuals taken in the worn summer pelage none possessed this conspicuous mark.

No.	Sex.	Locality.	Date.	Collector.
265	♀	Tejon Mountains, Cal	Aug. 3, 1875	H. W. Henshaw.
281	♂	.....do	Aug. 3, 1875	Do.
602	♂	Mountains near Kernville, Cal	Oct. 25, 1875	Do.

23. *Sciurus douglassii*, Bach.—Oregon Red Squirrel.

This Squirrel was found to be extremely numerous in the pineries of the Sierras, extending nearly, if not quite, up to the pine limit. It has many of the ways of the eastern Red Squirrel, with which its habits correspond most closely, but its notes are very different. These are so sweet and pleasing as to almost entitle them to be ranked as a song, and indeed their voices were mistaken for those of birds by more than one of our party.

No.	Sex.	Locality.	Date.	Collector.
446	♀	Near Mount Whitney, Cal	Sept. 6, 1875	H. W. Henshaw.
447	♀	......do	Sept. 6, 1875	Do.
458	♂	......do	Sept. 10, 1875	Do.
459	♂	......do	Sept. 10, 1875	Do.
742		......do	Sept. 10, 1875	Do.

24. *Tamias quadrivittatus*, (Say) Rich.—Four-striped Squirrel.

Distributed throughout California, where confined to the mountains. Abundant.

No.	Sex.	Locality.	Date.	Collector.
447	♂	Near Mount Whitney, Cal	Sept. 6, 1875	H. W. Henshaw.
448	♂	......do	Sept. 6, 1875	Do.
449	♀	......do	Sept. 10, 1875	Do.
449 A	♀	......do	Sept. 10, 1875	Do.
460	♂	......do	Sept. 10, 1875	Do.
721	♀	......do	Sept. 10, 1875	Do.

25. *Tamias lateralis*, (Say) Allen.—Rocky Mountain Chipmunk.

This is an other mountain species, which probably occurs throughout Southern California. It prefers rocky localities where underneath bowlders or fallen logs it excavates its habitations. Is rarely found other than in communities.

No.	Sex.	Locality.	Date.	Collector.
455	♂	Near Mount Whitney, Cal	Sept. 10, 1875	H. W. Henshaw.
456	♂	......do	Sept. 10, 1875	Do.
457		......do	Sept. 10, 1875	Do.

26. *Spermophilus harrisii*, Aud. and Bach.—Antelope Squirrel.

This little creature was noticed in several districts in Southern California. It lives in communities on the dry, sandy plains, where the herbage is of the scantiest kind. Its nature is shy and timid in the extreme, and so quick and agile are its motions that even when found some little distance from its hole it proves no easy task to shoot one. Usually their quick ears and roving eyes warn them of coming danger in time to secure an unimpeded retreat, and the glimpse of a white tail, which is held upright after the manner of an antelope, as its frightened owner disappears beneath the ground, is all one usually obtains.

No.	Sex.	Locality.	Date.	Collector.
632	♂	Valley of Kern River, Cal	Oct. 27, 1875	H. W. Henshaw.

27. *Spermophilus beecheyi*, Richardson.—California Ground Squirrel.

No animal is perhaps better known throughout California than this Squirrel, and insignificant as it may be thought from its small size, its numbers are so vast that it proves to be one of the greatest nuisances to the farmer, and in many sections renders successful agriculture all but impossible. They live together in large colonies, and in some portions of the State where their increase has been left unchecked, it seems, to one passing through, as if the country was one vast burrow, so rapidly do their colonies succeed one another. Once firmly established, the ireradication is a matter of extreme difficulty. It is rarely possible to completely destroy a large colony, and if a few pairs are left they soon multiply to such an extent as to make their ravages on corn-field or vegetable-patch seriously felt; in sections where their numbers are greatest, I was informed that not infrequently the loss by these little indefatigable thieves had reached one-half the corn-crop.

Their burrows are usually made in an uncultivated patch of ground, most often a pasture immediately adjoining cultivated ground, and their raids rarely extend more than 200 yards from home. Their ravages begin as soon as the first green leaf of cereal or vegetable appears above the ground, and cease only when the last kernel of grain has been carried away.

With regard to their rate of increase I was informed by a farmer, though seven or eight was the usual number produced at a birth, he had in one instance on opening a female found no less than thirteen embryos. As they are known to produce two or more litters during the year, their excessive multiplication will thus not appear surprising. Poisoning by strychnine appears to be the only sure means for their suppression. This is neither easy nor always safe, since in exposing poisoned grain there is always danger of killing horses and cattle; and as the farmer and ranch succeed each other by turns, the extermination of these pests fails from lack of the necessary cooperation.

No.	Sex.	Locality.	Date.	Collector.
174	♀ jun.	Santa Barbara, Cal.	July 1, 1875	H. W. Henshaw.
175	♀ jun.	do	July 1, 1875	Do.
283	♂ ad ..	do	July 7, 1875	Do.
239	♀	do	July 8, 1875	Do.
240	♀	do	July 8, 1875	Do.
512	♀ ad ..	Near Mount Whitney, Cal	Sept. 29, 1875	Do.

LEPORIDÆ.

28. *Lepus callotus*, Wagl.—Jackass Rabbit.

Numerous throughout Southern California.

29. *Lepus trowbridgei*, Bd.—Trowbridge's Hare.

A single specimen of this little Hare was obtained at Santa Barbara, where, however, it was not uncommon; it was the only species seen in that locality.

30. *Lepus sylvaticus*, Bach., var. *arizona*, Allen.—Arizona Gray Rabbit.

The locality where was taken one single example of this Rabbit was Kernville. They were very numerous here, living principally in the crevices of the rocks. I am inclined to believe that this form is spread over much of the southern half of the State. Many, presumably of this variety, were shot for food.

No.	Sex.	Locality.	Date.	Collector.
631	♂	Kernville, Cal.	Oct. 25, 1875	H. W. Henshaw.

Respectfully submitted.

H. W. HENSHAW.

Lieut. GEO. M. WHEELER,
Corps of Engineers, in Charge.

APPENDIX H 13.

REPORT ON THE OPERATIONS OF A SPECIAL PARTY FOR MAKING ETHNOLOGICAL RE-SEARCHES IN THE VICINITY OF SANTA BARBARA, CAL., WITH AN HISTORICAL ACCOUNT OF THE REGION EXPLORED, BY DR. H. C. YARROW, ACTING ASSISTANT SURGEON, UNITED STATES ARMY.

UNITED STATES ENGINEER OFFICE,
GEOGRAPHICAL SURVEYS WEST OF THE 100TH MERIDIAN,
Washington, D. C., December 18, 1875.

SIR: I have the honor to submit herewith a report of the operations of a special party under my charge detailed by you for the purpose of making ethnological researches in the vicinity of Santa Barbara, Cal. The report, as will be found, is prefaced by a short historical account of the region explored, as given by Cabrillo, a Portuguese, who visited the coast of California in 1542.

Very respectfully,

H. C. YARROW,
A. A. Surg. U. S. Army, Surg. and Zool. to Expedition.

Lieut. GEO. M. WHEELER,
Corps of Engineers, in Charge.

On the 27th day of June, 1542, Juan Rodriguez Cabrillo, a Portuguese navigator in the service of Spain, left the port of Navidad, New Spain, with two small vessels, the San Salvador and La Victoria, to explore the coast of California, which he sighted on the 2d of July. Proceeding along it, on the 7th of October he came in view of two islands some distance from the mainland, which he named after his vessels: these islands, lying in Santa Barbara Channel, southwest of San Pedro, and now known as San Clemente and Santa Catalina. On these islands, Cabrillo found many aborigines, who at first showed great fear of the Spaniards, but finally, becoming friendly, told him of numerous other Indians on the mainland. Resting here but two days, he set sail on the 9th. Shortly afterward, reaching a spacious bay and following its shore-line, he soon came upon a large village of Indians close to the sea-shore. Here his ships were visited by the savages in canoes, from the great number of which he called their town *Pueblo de las Canoas*. It would appear impossible to fix the exact site of this town, but circumstances point to the city of Santa Barbara as the locality. On the 13th, resuming his voyage, he passed near two large uninhabited islands, now known to be Santa Cruz and San Miguel, and anchored in front of an extremely fertile valley. Here he was visited by many natives coming to sell fish, who informed him that the whole coast was densely populated as far northward as Cabo de Galera, or Point Concepcion of the present day. Northwest from the Pueblo de las Canoas, he discovered two islands, which he named San Lucas, afterward known as San Bernardo, and which at the present day are supposed to be those of Santa Rosa and Santa Barbara. Point Concepcion was reached by this Portuguese navigator on November 1, after much suffering from cold, winds, and tempests. Anchoring near this place to obtain wood and water, he called the port *de las Sardinas*, from the abundance of fish thereabouts. Here were found many natives of most friendly disposition, one of whom, an old female, said to be the Queen of the Pueblos, came off to the captain's ship and remained two nights. Returning to the Island St. Lucas on account of bad weather, on the 3d of January Cabrillo died on the island called *la Posesion*, believed to be the present San Miguel. Of the manner of his death, and his notes in regard to the Indians he saw, we shall have occasion to speak hereafter. With this account of one of the earlier explorers of the region to be visited by ourselves, as a proper preliminary to a report of our own operations, we now proceed to give the latter in detail, first, however, briefly mentioning the circumstances which led to the exploration in question.

It is reported that some years ago the captain of one of the small schooners common to the Pacific coast returned from a visit to the island of San Nicholas, and stated having seen quantities of pots, stone implements, skulls, and divers sorts of ornaments on the surface of shell-heaps, which had been uncovered by storms, and exhibiting in proof of his assertions a number of these articles which he had brought with him, and which he distributed among his friends. It is reported that this captain again visited San Nicholas and its neighbor, Santa Catalina, and returned with a full schooner-load of relics, but this part of the tradition lacks confirmation.

Little attention was paid to this most valuable archæological discovery until 1872 and 1873, when Mr. W. G. W. Harford, of the United States Coast Survey, happened on the islands of San Miguel and Santa Rosa, lying to the northward and westward of the islands before mentioned. From these islands this gentleman procured a small but exceedingly valuable collection of interesting objects, which came into the hands of Mr. Wm. H. Dall, a most intelligent and enthusiastic collector, from which he deemed the locality of sufficient importance to visit it in person. This he did in the winter of 1873 and 1874. Mr. Dall visited San Miguel and Santa Catalina, but as his time was limited, no thorough examination was made of this mine of archæological wealth lying then temptingly open to view. He, however, procured many interesting specimens. During the same season, Mr. Paul Schumacher, well known for his investigations farther up the coast, discovered in the vicinity of San Luis Obispo and the Santa Maria River, deposits similar to those found on the islands. The results of these discoveries being communicated to the Smithsonian Institution, this establishment determined to make a thorough and exhaustive exploration of not only the mainland, but also of the islands; and, in the spring of 1875, Mr. Schumacher was named to conduct the work in behalf of the National Museum. By a fortunate coincidence, one of the parties of the Expedition for Explorations west of the One hundredth Meridian under the War Department, of which the writer was placed in charge, was about to visit the Pacific coast, and an arrangement was entered into whereby hearty co-operation and unity of effort were effected. Mr. Schumacher was to explore the islands, and the Exploring Expedition party the mainland along the coast from Santa Barbara north for a distance of 20 or 30 miles.

Leaving San Francisco June 4, after a pleasant sail of forty-eight hours, we arrived at Santa Barbara, the *Pueblo de las Canoas* of Cabrillo, and there found the other members of the party, consisting of Dr. J. T. Rothrock, botanist, and Mr. H. W. Henshaw, ornithologist, whom you had directed to assist in the enterprise. Arrangements were at once made to explore the neighborhood, and the day following that of our arrival we started, and under the guidance of the Rev. Stephen Bowers, whom we were informed had already made some excavations in the section about to be visited, for the ranch of T. Wallace More, near the little village called La Patera, some eight miles from Santa Barbara. Arrived at a spot where our guide informed us he had found a few bones and arrow-heads, the work, digging a trench in a north and south direction on a cliff overlooking the sea and probably 80 feet above it, was at once commenced. There were no indications that this locality had been used as a burial-place, but after digging a few feet, and beyond some loose bones that had been reinterred by Mr. Bowers on the occasion of his first visit, we came to an entire skeleton *in situ*. It was lying on the right side, facing the west, with the lower limbs drawn up toward the chin. No ornaments or utensils were found, but a quantity of marine shells were near the cranium. Continuing the excavation deeper, two other skeletons were discovered in a similar position to the first, and near them a few broken arrow-heads. These were removed and the excavation extended downward and backward from the sea-cliff, the labor being rewarded by the finding of seven other skeletons. These latter, however, were huddled together and gave no evidence that care had been taken in the burial of the bodies to place them in any particular position. Near by were a few shell-beads and other ornaments, and an abalone shell (*Haliotus splendens*) containing a red pigment. The bones were so friable as to crumble to pieces on exposure to the atmosphere, and on this account none could be secured. On excavating to a depth of 5 feet, a layer of marine shells was reached, under which was a firm stratum of yellow, sandy clay, beneath which, as our subsequent experience proved, burials were never made. After digging for several hours, and finding nothing further of special interest, the trench was refilled.

Moving around from place to place in the field, our attention was finally attracted to a depression in the center of it, some 200 yards from the sea-cliff, which on examination gave undoubted evidences of being a burial-place, ribs and vertebræ of whales being scattered about, and small inclosures found that had been made in the earth by setting up large flat stones on their sides. Digging into one of these inclosed areas, broken bones and some broken pestles and mortars were found, but nothing of special value. The excavation was continued to a depth of 3 feet only, which, as subsequently ascertained, was not sufficient. We left this locality for a time.

While engaged in the interesting search in question, Dr. Rothrock, who had strolled off some distance after botanical specimens, communicated to us that he had discovered, on the opposite side of a small *estero* to the northward, a locality which he believed to be a burial-place, founding his belief on the fact that he had seen a number of whales' ribs, placed so as to f rm arches over certain spots. As we well knew that the Santa Cruz Island burial-grounds were similarly marked, we anticipated a " good find," and, indeed, so richly were our anticipations rewarded that we named it the " *Big Bonanza*." The annexed diagram will give an idea of this place and the several other localities already mentioned.

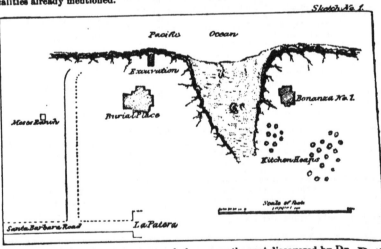

Sketch No. 1.

The next morning found us at an early hour near the spot discovered by Dr. Rothrock, and from the surface indications it could hardly be doubted that at some period it must have been a burial-place of note. The surface of the ground, instead of presenting the appearance of mounds, or hillocks, was rather depressed in a semicircular form, and in various spots ribs and vertebræ of whales had been partially buried in the ground, the ribs in some instances being placed together, as reported by Dr. Rothrock, in the form of arches. Selecting what appeared to be a favorable place, 20 feet from the edge of the cliff, fronting the *estero* shown in the map, a trench was commenced running due north and south. Two feet below the surface the first indications of burials were reached, quantities of broken bones being met with at every stroke of the spade, interspersed with pieces of whales' bones and decaying redwood. At a depth of 5 feet the first entire skeleton was found in position, and near it several others were subsequently uncovered; in all of them the head fronted northward, the face was downward, and the lower limbs were extended. Over the femur of one of the skeletons was a flat plate of steatite, a sort of soapstone, 12 or 14 inches square, with a hole in one end, which we called a "tortilla-stone," its probable use having been for cooking their cakes, or tortillas, the hole in the end serving to withdraw it from the fire when thoroughly heated. In rear of the skeleton, and to one side of the plate, was an olla, or jar, of steatite, broken, but containing some fine glass beads and human teeth, and behind this a stone pestle of symmetrical shape, about 3 feet in length, of a hard species of sandstone, and another plate of steatite, and two large ollas of over five gallons capacity, their mouths or apertures fronting north, and just above was a single cranium facing the cliff, face downward, and on top of it a single femur. Continuing the excavations toward the cliff, a small sandstone mortar was exhumed containing a mass of red paint, and in its immediate vicinity a large number of beads of glass and shell with ornaments made from the lamina of the abalone shell, which is common to this coast, being found in great abundance on the islands some 20 miles distant. Digging still farther, other skeletons were found in similar positions, but in many instances the lower limbs were flexed upon the body, while in a few cases the fingers of the right hand were in the mouth. One skeleton was that of a child, near which were found beads, ornaments, tortilla-stones, and two more ollas, one of which contained portions of the cranium of a child. This skeleton had apparently been wrapped in a kind of grass matting, as small portions were found attached to the bones and scattered near by. In the olla containing the head-bones of the child were a great number of small black seeds, smaller than mustard-seed, which were recognized by one of the laborers as a seed used by the present California Indians and natives in making demulcent drinks and eye-washes, the Spanish name being *chiya.*

A second trench, opened 40 feet from the first, yielded quite a number of excellent crania and other specimens, among which were fish-bones, crenated teeth, (of fossil shark possibly,) and a very large olla containing bones and covered on top with the epiphysis of a whale's vertebra. The following are the notes furnished by the gentleman in charge of the excavations at this point: First trench, 6 feet by 2, running north and south, trending to the westward. Indications of burials, whales' bones, and rocks set up vertically. Two and one-half feet below the surface found skeleton with face downward, head to the north. Three feet below surface reached a large flat stone, which being removed was found to cover ribs and shoulders of a female skeleton, head pointing north, body resting on left side. A small mortar was over the mouth, small sandstone mortar and pestle of fine workmanship near top of head. This locality proving rather unprolific, a second trench was commenced 40 feet below last, nearer cliff, and about of same size. Two feet below the surface to our great surprise a large steatite olla was discovered, which proved to contain the skeleton of an infant wrapped in matting. Unfortunately, upon exposure to the air, the bones crumbled away. Beneath the olla was a cranium, apex west, face north. Three feet below the surface were two skeletons in fair condition, with crania to the north. Our discoveries this day had developed so much of interest that it was not until darkness had overtaken us that we discontinued our work.

In order to give some idea of the amount of material recovered during the excavations, a record of each day's work follows:

June 10.—This morning began work shortly after sunrise at both trenches opened the day before, digging in a westerly direction in the first. In this, numbers of crania and bones were found in similar positions to the first met with, and also several fine ollas, tortilla-stones, mortars, and pestles. All these utensils were invariably in the immediate vicinity of the heads of the skeletons; in fact, in many instances the crania were covered by large mortars placed orifice down. In the second trench, the digging was in an easterly direction, and the first discovery that of a skeleton and a fragment of iron near the right hand, probably a knife or spear-head, which, archæologically speaking, was a source of great grief to us, our hope being that no remnants of Spanish civilization would be found in these graves. It could not be helped, however, although a great deal of pre-historic romance was at once destroyed. Near this skeleton was another, and by its side the first pipe met with, which was similar in appearance to a plain modern cigar-holder, and consisted of a tube of the stone called serpentine, 8 inches long, the diameter of the wider orifice being a little over an inch. At the smaller end was a mouth-piece formed from a piece of a bone of some large water-fowl, and cemented in place by asphaltum. How these pipes were used with any degree of comfort is impossible to surmise.

Continuing this excavation, the next discovery was a steatite olla containing a skull, differing in many respects from those found in the graves; if from one of the same tribe, it shows marked differentiation. Near the olla was a large sandstone mortar, over 2 feet in diameter, and behind it another olla containing more bones, and another pipe, 10½ inches in length, and near this latter article a smaller olla filled with red paint. It should have been mentioned that from this trench was procured a femur showing evidences of a fracture through the neck of the bone, which had become absorbed, the head uniting to the upper portion of the shaft between the greater and lesser trochanters. Further search revealed at the same depth a mortar, covered by the shoulder-blade of a whale, which also contained the skull of an infant covered with an abalone shell, while near by was paint, piece of iron, a nail, and various shell ornaments and beads. Near at hand, to the rear, were a broken mortar and pot underneath, which was a small olla, the whole covering the skull of a child; and a little deeper a skull resting upon a fine, large, pear-shaped steatite olla, the outside of reddish color. These remains appeared to have been inclosed in a sort of fence, as a plank and stakes of decayed redwood were near by. At the bottom of this trench, just above the firm clay, and under all the specimens just described, was a fine sandstone pestle 17½ inches in length.

June 11.—Continued in same trench, advancing in a northerly direction toward trench No. 1. At a depth of 4 feet were two skeletons, and near them was a square cake of red paint; alongside were two more skeletons, over one of which was a large mortar, mouth downward, and close by another similar utensil. Under this skeleton was an instrument of iron 14 inches in length, a long iron nail, and two pieces of redwood, much decayed. A little farther in was a small canoe carved from steatite. All the skeletons were face downward, heads to the north. In trench No. 1 the digging was continued in a southerly direction. The first object encountered was an enormous mortar, 27 inches in diameter, with its pestle near by. This article was on its side, the mouth toward the south; around it were no fewer than thirty crania, some in a fair state of preservation, and others very friable, broken, and worthless. Lying on top of this mortar, on further removal of the earth, was an almost entire skeleton, with fragments of long bones and of steatite pottery. As surmised by some of the party, the perfect skeleton was that of a chief, and the remains those of his slaves slain with him; which is at least a possible, if not a plausible, view of the case.

Experience by this time had taught us that nearly all the burial-places or spots had been carefully marked, since near the head of each skeleton were either bones of the whale or stakes of redwood.

Being obliged to leave for Los Angeles June 12, the work was continued by Mr. Bowers, who, up to June 25, secured the following articles from the two trenches in question, viz: 32 skulls, 24 large steatite ollas, 6 large mortars, 7 large pestles, 2 small serpentine cups, 7 tortilla-stones, 7 abalone shells, 3 iron knives, 4 stone arrow-heads, 1 iron ax of undoubted early Spanish manufacture, quantities of glass, shell beads, paint, shell ornaments, black seed of the character previously mentioned, 2 pipes, 2 soap-root brushes with asphaltum handles, and a copper pan 8 inches in diameter, which were found covering the top of a skull—the copper evidently having preserved a portion of the hair, which was quite black and silky, and not coarse, as is usually the case with Indians.

June 25.—The same excavation No. 2 was continued, and 3 crania were uncovered, also an olla containing the bones of a child, not far from which were 3 mortars and 2 ollas. Just above the stratum of clay the most interesting discovery was made of an entire skeleton, which had been buried in a redwood canoe, but which was so decayed that only a small portion could be preserved. Near the head of the canoe were a large olla and mortar, the mouths northward. On removing the skeleton, which was lying on its back, the bones fell to pieces. In the canoe, alongside of the skeleton, were 3 pestles, 2 pipes, an iron knife or dagger blade that had been wrapped in seal-skin or fur, and a stone implement of triangular form and about 6 inches in length, probably used as a file, or perhaps for boring out pipes.

June 26.—Trench No. 2 was abandoned and work resumed in No. 1, which yielded several crania in bad condition. Near a whalebone, standing on end, was an empty broken olla, and not far off a skeleton on its right side, legs drawn up, head facing west. On its right-hand side, near by, was a small highly-polished serpentine cup and a small mortar and pestle. After excavating awhile and finding nothing but broken bones, digging here was discontinued and an excavation commenced ten feet to the northward and near the edge of the cliff, but after going down 5 feet through kitchen refuse, ashes, bones, shells, it was filled up and work resumed in the same trench. Several hours' digging resulted in finding nothing, but finally the "lead" was once more struck. The first discovery was a skeleton, which, from the appearance of the pelvic bones, was that of a female, and near which were great quantities of beads, shell ornaments, and seeds. It was here we first encountered what at first sight appeared to be dried cloves, but which, on closer examination, proved to be ornaments of asphaltum, hollow in the center, and in some instances having at one end a small piece of dried grass or fiber, by means of which doubtless they were fashioned into necklaces. Some abalone shells were also found, in close proximity to which were the bones of a child. Another mortar was discovered, containing some bones in bad condition.

June 27.—Being Sunday, operations were suspended until the next day.

June 28.—Work was resumed at trench, No. 1 but for 6 or 8 feet nothing was met with save isolated bones. Digging to the southward, however, a skeleton was found with top of head to the northward, the position of which was nearly face downward. On its removal and beneath it was a large mortar, cavity down, slightly tipped, and facing west. In another direction, to the eastward, was a large sandstone mortar facing north, and beneath it a skull in good condition, while near by was a small olla containing ornaments of shell, beads, seeds, and paint. Deeper down, still another small olla was revealed, filled with the black seeds, and near it a small pestle. A number of crania and bones were also found, but all in bad condition. One of them, however, was particularly interesting from the fact of two arrow-points, one of a porphyritic stone, the other of obsidian, being imbedded in the outer table of the skull. From the position of the arrows it was inferred that the wounds were received by the person while lying down. Digging in a northerly direction in this trench, 8 or 10 more skeletons, all huddled together, were exhumed, also 2 small pestles, 2 mortars, and some abalone shells containing ornaments. In one of the larger of these shells were the head-bones of a young child, and near it two polished serpentine dishes, containing some of the clove-like asphaltum before alluded to. A broken dish had been neatly mended with asphaltum and probably sinew, as drilled holes were found in both pieces. Not far from these cups was found a leather(?) pouch curiously ornamented on the outside with circles of shell-discs.

On June 29, finding that our labor was not as richly repaid as formerly, further excavation in this locality was delegated to Mr. Shoemaker, who, having discovered only 6 crania, and these in poor condition, after six hours' faithful labor, the "Big Bonanza" was abandoned, and in the meanwhile the writer was prospecting.

Crossing the estero, and reaching the ranch of T. Wallace More, esq., we visited the asphaltum mine, from which it is probable the Indians whose resting-places we had been so ruthlessly disturbing, procured their supplies of this, to them, most precious material, since it must have been extensively used in fastening on their arrow-heads or spear-points, and in mending and filling up cracks and holes in their canoes. Not far from this mine, the spot was reached which has been mentioned as that where burials were indicated by whalebones and flat stones, and it was determined to explore it next. Near it was a depression, in which appeared to have been either a threshing-floor or dancing-place, oval-shaped and 60 feet long by 30 or 40 wide. It had been beaten or trodden down so firmly that no vegetation could flourish thereon. In the afternoon, not far from camp, one of the party discovered some fragments of human bones which had been thrown out of a squirrel-burrow, which circumstance led us to search for relics. Opening a trench 300 yards to the westward from camp, at a depth of 3 feet, some broken bones were found and one skull; near the latter were a quantity of beads and a matted mass of fur, apparently of either the seal or sea-lion. After some hours' fruitless labor, digging in this locality was discontinued. This was the only instance in our experience where the burial of but one individual had taken place.

On the following day, one of the laboring party, assured of finding something to repay further labor in the "Big Bonanza," urgently suggested the same, whereupon excavating was again entered upon at that place; and, curiously enough, after a little digging, a remarkably fine knife of obsidian was discovered, nearly 10 inches in length; a bone implement, similar in appearance to a sword-blade; and two pipes, one of them ornamented. This ornamented pipe was the first of the kind we had met with, and we congratulated ourselves upon having yielded to the suggestion of the workman.

July 1.—Resolved to excavate in the locality last discovered, and an early start was made. This trench in T. Wallace More's ranch was commenced 200 yards from the sea-cliff. At a depth of 2 feet broken bones were uncovered, and at 4 feet entire skeletons, which in many instances had been inclosed with flat stones, forming a kind of coffin. Some mortars and pestles were here also met with, as well as pipes, arrow-heads, and another fine spear of flint, and one of iron. After four days' hard work, with no other results than those mentioned, this trench was abandoned. It is doubtless probable that many more articles might have been found here, but the time that would be consumed in securing a few small articles was demanded where results would most likely prove richer and more interesting.

From Dr. Taylor, of La Patera, a gentleman who for years had studied the ethnology and archæology of the Pacific coast, we learned of the probable existence of burial-places at a spot some 12 or 15 miles up the coast, known as Dos Pueblos, Dr. Taylor having there seen the remains of numerous kitchen-heaps, inferred that a large population once lived in that locality, and that their dead would be found not far distant. Accordingly Dr. Rothrock and the writer started on a prospecting tour, and after a couple of hours' ride came in sight of the Dos Pueblos ranch, occupied by Mr. Welch and family. Making ourselves and object known to Mr. Welch, we received a welcome, and were invited to dig anywhere we might think proper. Mr. Welch showed us in his potato-patch numbers of broken bones that had been turned up by the plow; but being attracted by some whalebones partially imbedded in the earth of the sea-cliff near by, we immediately left the potato-patch, knowing from experience that where the whalebones are there also were graves. The position of these graves, as well as some others subsequently discovered, may be seen from the map. (See Sketch 2.)

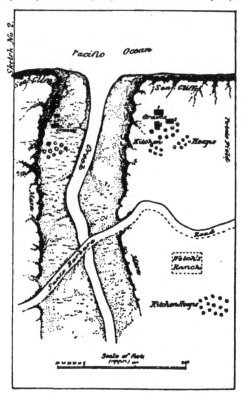

The next day it was determined to move the entire party to this locality and excavate, which was done, the first trench being made at the point marked 1, near the brow of the cliff, where were whalebones and large, flat stones. At a depth of 4½ feet, great quantities of bones were found huddled together, but no skeletons in a particular posture. In some instances, stone receptacles, similar to the one already described, were encountered, but from their infrequency this burial feature was apparently not common. All the bones were in a very bad state, much worse than those about La Patera, and but few were preserved. Throughout the graves, but not placed in particular position, were several large mortars, large and small ollas, pipes, beads, and ornaments, besides bone awls. In locality No. 2, the same class of articles was brought to light, but in larger number.

In the narrative of Cabrillo, by Bartolome Ferrel, this locality is called Dos Pueblos, from the fact of there being two towns on opposite sides of the creek, which runs down from the Santa Inez Mountains. These towns were densely populated with a mild, inoffensive people. We were informed by Mrs. Welch that she had heard from an aged Indian woman that two separate tribes, speaking different dialects, lived on opposite sides of the creek, which constituted the boundary-line between them, and that the tribes were not permitted to cross this creek without first obtaining each other's consent.

Continuing our excavations in No. 2, a long, straight pipe and a small mortar having a handle, (the first of its kind,) and containing red paint, were found, and near the latter a pipe only partially bored out. On the opposite side of the creek a trench was opened beneath a gigantic piece of whalebone, but several hours' work revealed nothing but broken bones, and it was abandoned and work resumed in Nos. 1, 2, and 3. During the 6th, 7th, 8th, 9th, and 10th, the excavating was continued, resulting in the discovery of mortars, ollas, pipes, &c., and curiously enough in No. 3 of no fewer than 30 skeletons which had been buried in sea-sand, and under which were 3 fine stone spear-heads and some fragments of iron. In No. 2 were several large ollas and mortars, and near the head of a skeleton, presumably that of a female, some china cups and saucers of very ancient shape. The time allotted to these explorations having now nearly expired, the remainder of our stay was devoted to filling up holes and packing the specimens. The specimens were roughly estimated as weighing from 10 to 15 tons.

Regarding the people of whom we have been speaking, and of whom no representative remains to tell of their history, but little could be learned; the crumbling bones and household gods we had so ruthlessly disturbed, were the only witnesses of the former existence of a once populous race; but beyond this they made no revelation, while careful examination of the entire literature of the Pacific coast proved fruitless in throwing light on these early generations. All the writers who speak of these aborigines, and it is but fair to state that few, if any, of them were possessed of original information on the subject, (having gathered their materials from Ferrel's narrative,) are of the opinion that they were friendly, peaceable, and inoffensive, which opinion is enforced by the absence in their graves of warlike implements to any extent. Cabrillo states that they were armed with bows, the arrows being pointed with flint heads, similar to those used by the Indians of New Spain; he also speaks of clubs, but mentions no other weapon. As to population, he states that on some of the islands there were no people, but that others were densely populated; the former we have not been able to identify. The Indians told him they had occasionally suffered from the attacks of warriors armed like the Spaniards, and from the fact that toward the middle of the eighteenth century the mission priests of Santa Barbara removed their savage parishioners from the islands to the mainland to escape the ravages of the Russians and their Kodiak allies, it is supposed that this warfare had been going on for a number of years. As to the extent of the population, we can form an idea only from the number of burials, at different points, and villages, the names of which have been handed down to us through Cabrillo. At a rough guess, our party must have exposed at their main trenches the remains of no fewer than 5,000 individuals, and, from what we have subsequently learned, there are hundreds of these burial-places along the coast.

With regard to the towns, the Indians informed Cabrillo that the whole coast was densely populated from the Pueblo de las Canoas to 12 leagues beyond the Cabo de Galera, (Point Concepcion,) and gave him the names of these towns. To the northward of their city was Xuco, Bis, Sopono, Alloc, Xabaagua, Xocotoc, Potoptuc, Nacbuc, Omlqueme, Misinagua, Misisopano, Elquis, Colve, Mugu, Xagna, Anacbuc, Partocac, Snauquiy, Omanmu, Gna, Asimu, Aguen, Casilic, Tucumu, Incpupu. These towns were in the immediate vicinity of the Pueblo de las Canoas. Near the Cabo de Galera, or Point Concepcion, as it is at present called, was the pueblo named by Cabrillo, "Pueblo de las Sardinas," in consequence of the great number of small fishes taken by the natives. In the neighborhood of this pueblo were the villages of Xixo, Cincacut, Cincut, Anacot, Maqumanoa, Paltated, Anacoat, Paltocac, Tocani, Opia, Opistopia, Nocos, Yutum, Iuiman, Micoma, and Garomisopona. These towns or villages were ruled over by the aged queen to whom reference has been made, the capital and seat of government said to have been Cincut. Cabrillo also gives us the names of some of the towns on the islands; for instance, on one of them, which he states is 15 leagues long, probably San Miguel, Niquipos, Maxul, Xugua, Nitre, Macano, Nimitapol. On other islands not intelligibly specified were the towns of Ciquimuymu, Nicalque, Limu, Zaco, Nimollolli, Niohochi, Coycoy, Estoloco, Niquesesquelua, Poele, Pisqueno, Pualnacatup, Patiquin, Mnoc, Patiqnilia, Nimumu, Piliaquay, and Lilibique. He also mentions that on an island south of Isle de la Posesion was one called Nicalque; on this were three towns, Nicoche, Coycoy, Coloco. From this extended list it may be inferred that a large population once lived in the region explored.

With regard to the time that these people disappeared we can only conjecture. From the mission records it appears that in 1823, the total number of Indians in the vicinity of Santa Barbara was upward of 900, but this census embraced all Indians, and not alone those from the islands and sea-coast. In 1875, the year in which we write, not a soul can be found to give any information as to the ancient inhabitants of this part of the coast. There is a tradition that many years ago a Mr. Neidifer, while on a trip to the island of Santa Cruz, discovered there, much to his surprise, an aged hag, and that he removed her to Santa Barbara, but no one could understand her language, and after a short time she died; also that she was a young girl at the time the Indians were removed to the mainland, and returning from the boats to seek her infant, in the hurry and confusion of the embarkation she was left behind; that when found she was clothed in furs ornamented with the feathers of birds. Doubtless this woman was the last survivor of the island tribes. As to the causes which led to the total extinction of this once populous race, there are no trustworthy data, and it would profit us but little to enter the wide field of speculation.

Of their manner of living little if anything is known. Cabrillo states that on most of the islands miserable huts existed, but on the mainland there were houses similar to those of the Indians of New Spain. On one of the islands, however, which he states was four leagues long, there were many good houses of wood. We are at a loss for further information on this point, but it is certain that the dwellings of these people were constructed of perishable materials and not of adobe bricks like the Pueblo Indians of New Mexico, since no trace can be discovered of such material, and it is hardly possible this would be the case in the short space of time since Cabrillo's visit. It is extremely probable, therefore, that they built their houses of timber, or else used the skins of animals slain in the chase. Referring to the matter of houses of wood up n the islands, some doubt might apparently be thrown upon this portion of Cabrillo's narrative, for at present no trees of a size sufficient for building purposes are found on the islands; but this author states that on the Isle de St. Augustin he saw trees 60 feet in height and of such girth that two men could not encircle them with their arms joined.

In their choice of localities for towns these ancient people showed the same degree of sagacity as that evinced by the American aborigines down to the present day. On the islands were myriads of water-birds and quantities of sea-lions and seals; the water fairly teemed with fishes and molluscous animals, affording a plentiful supply of food, and no doubt at the time they were occupied there was plenty of sweet water to be had, which, unfortunately, is not the case at present. On the mainland, at all the localities visited, the circumstances of environment must have been such as to render the struggle for existence extraordinarily easy. For instance, at Santa Barbara and up the coast, or what was called the Pueblos de las Canoas, the land is extremely fertile, and must have yielded good crops, for Cabrillo especially mentions that the Indians lived in a fertile valley, and had an abundance of corn and many cows. In addition to their pastoral pursuits, the Santa Inez Mountain afforded them game, and the waters, fishes, clams, mussels, &c. From the great quantities of shells found in the graves and kitchen-heaps, and the absence of mammalian bones in any quantity, it is fair to suppose that the tribes living near the seaside derived the greater portion of their sustenance from the waters. The favorite places for towns appear to have been not far from groves and near small mountain-streams. Anterior to 1542, these Indians must have been idolaters, but we have good reason for believing that after the advent of the mission priests many of them embraced the Roman Catholic religion, and faithfully followed its teachings. Cabrillo speaks of having seen on one of the islands (probably San Miguel) a temple of wood with paintings on its walls, and idols. San Miguel and some of the other islands have been carefully searched for this temple, but in vain.

It is hardly necessary to refer again to the different utensils found in the graves of these people, but it may be well to state that all the ollas, mortars, cups, pipes, and pestles met with were fashioned out of steatite, or magnesian mica, a sort of soapstone, consequently very soft, which alone was used for the ollas, sandstone of different degrees of hardness for the pestles and mortars, and serpentine for the cups and pipes. It is easy to understand that the ollas were readily carved from the soft soapstone-like material by means of stone knives, but how the gigantic and symmetrical mortars were hewn out with such rude tools is beyond our comprehension; yet they must have been easily procured, otherwise such lavish generosity in burying them with the dead would hardly have been possible. It is thought that the steatite articles were not made by the mainland Indians since no deposits of this mineral were at their disposal, but by the dwellers on the islands of Santa Catalina and Santa Rosa, where alone this mineral existed, and the supposition is that the islanders trafficked with those of the mainland for their commodities, giving in exchange utensils of steatite. The ollas were doubtless used for cooking, as many of them bear marks of fire, and the mortars for bruising grain, acorns, and grass-seeds, the smaller cups and basins for ordinary household purposes, and the pipes for smoking. Canoes are mentioned by Cabrillo, who states that some were small, holding only two or three persons, while others were of sufficient capacity for ten or twelve. These were probably hewn, not burned, from logs of redwood cast up by the waves. The one mentioned as discovered by our party containing a skeleton was, however, formed of three planks, which had been lashed together by sinew or cord, the joints being payed over with asphaltum. The ornaments and beads of domestic manufacture were made of the nacre of shells and of small shells, but the glass beads found were undoubtedly of European workmanship. There seems but little doubt that nets were used for trapping fishes, a small portion of what appeared to be mesh-work being found. Furs are spoken of as articles of clothing in Cabrillo's narrative, but beyond this nothing is known. In speaking of the employment of furs, mention is made of the long, fine, black, and beautiful hair of the natives; this statement is corroborated by the appearance of some hair found on the skull which we have spoken of as being found covered with a copper pan.

It was at first supposed that a certain design had been followed in the manner of interment, or rather of the posture in which the bodies were placed, but an examination of the notes already given will show that such was not the case, although most of the entire skeletons discovered at La Patera were in the same position, but those at Dos Pueblos were in all attitudes, consequently we infer that there was no regular mode of procedure. From the fact that so many loose and broken bones were found close to the surface of the earth, it is probable that the same spot had been used over and over again for burials, the remains of the previous occupants being shoveled out to make room for new-comers. Perhaps the utensils disinterred were also made to serve for more than one burial. A question in connection with the burials, which is yet to be satisfactorily answered, is, How were these people enabled to pass the heads of children, and even grown persons through the narrow openings in the ollas except in a mutilated condition. It is true that some savage tribes expose the bodies of their dead until the flesh is removed, but we know of no instance where savages are in the habit of cutting up their dead for burial purposes. It may be these people practiced the cutting method, or that finding bones in digging anew, these were thrown in the ollas simply as a ready means of their disposal.

In addition to the burial localities already mentioned, we are cognizant of others to the northward and southward of Santa Barbara and quite a number of them have already been explored, although doubtless others still remain *perdu* to excite further archæologic cupidity. Mr. Paul Schumacher has examined a number in the vicinity of San Luis Obispo and on the Santa Maria River, and Mr. Bowers quite a number in Santa Barbara and in the vicinity of Carpenteria, lying south of this city.

We have carefully consulted all available works which would tend to throw light on the history of these aborigines, but, with the exception of the narrative of Cabrillo, have found little pertaining to the subject. It may, perhaps, be interesting to give the full title of the rare and most entertaining manuscript from which we have so freely quoted, which was reproduced i, typography by the late Buckingham Smith, in his work entitled | Coleccion | de varios documentos | para la historia de la Florida | y tierras adyacentes :

Cabrillo's own title, or rather Ferrel's, is as follows : " Relacion, ó diario de la navegacion que hizo Juan Rodriguez Cabrillo con dos navios al descubrimiento del paso del mar del sur al norte, desde 27 de Junio de 1542 que salio del puerto de Navidad, hasta 14 de Abril del siguiente año que se restituyo á el, haviendo llegado hasta el altura de 44 grados, con la descripcion de la costa, puertos, ensenadoe é islas reconocio y sus distancias, en la estenacion de toda aquella costa." The death of Cabrillo, as already stated, occurred on the Isla de la Posesion, in the middle of January, 1543, and was caused by injuries received from a spar which fell from aloft and broke his arm near the shoulder. Before his death, he named as his successor Bartolome Ferrel, " Piloto mayor de los dichos navios," and to this successor we are indebted for all we know of the people under discussion.

In conclusion, it may with propriety be stated that we have here only endeavored to show the results of the exploratory work performed in the vicinity of Santa Barbara by the party sent out under the auspices of the expedition in your charge, and that no attempt has been designed toward solving questions appertaining thereto, more particularly in view of the fact that the entire subject will be fully and ably discussed by Professor F. W. Putnam, of the Peabody Museum of Archæology, Cambridge, to whom the entire collection has been submitted for examination and study, and who is perhaps better fitted for this most entertaining task than any other person in the country. In his hands we willingly leave the subject, confident that, with the rich materials gathered by us as a basis, he will elucidate many hitherto mysterious problems connected with the customs of this extinct race, and bring to light much of their now hidden history.

APPENDIX H 14.

NOTES UPON ETHNOLOGY OF SOUTHERN CALIFONIA AND ADJACENT REGIONS, BY DR. O. LOEW.

UNITED STATES ENGINEER OFFICE,
GEOGRAPHICAL SURVEYS WEST OF THE 100TH MERIDIAN,
Washington, D. C., February 19, 1876.

DEAR SIR : I have the honor to submit herewith a report upon the Indian tribes visited, their customs and relations, as well as old hieroglyphical writings upon rocks in Mono County, California. As ethnology is a matter of steadily increasing interest, I hope this contribution will not be without its value.

Very respectfully, your obedient servant,

OSCAR LOEW.

Lieut. GEO. M. WHEELER,
Corps of Engineers, in Charge.

During the field-season of 1875 you kindly gave me an opportunity to visit a number of localities in California that did not lie directly upon the route of Lieutenant Bergland's party, to which I was attached, and I therefore availed myself gladly of any occasion to collect facts upon ethnology whenever the regular duties permitted me to do so.

While on the expedition of Lieut. Bergland's party, the Payutes of Southern Nevada, the Hualapais of Northwestern Arizona, the Mohaves and Chemehuevis of the Colorado River Valley, the Kauvuyas and Takhtams of the vicinity of San Bernardino, were visited; and before the expedition started out, the Mission Indians of Santa Barbara and San Gabriel; and after the return from the field to Los Angeles, the Indians of San Juan Capistrano, San Diego, and of Mono and Inyo Counties, California. Vocabularies comprising from 200 up to nearly 400 words, and also sentences that may assist in establishing certain grammatical rules,* were collected of those languages, of which some have almost died out and now spoken by very few individuals, as is the case with the Kasuá of Santa Barbara and the Tobikhar of San Gabriel.

While of some of the languages (Mohave and Chemehuevis) a long list of words was collected by Lieut. A. W. Whipple during his exploration for a Pacific railroad route in 1853; of others, over two to three dozen words were barely known, as with the languages of San Diego, San Juan Capistrano, and San Gabriel. Of others again, no vocabularies had been published to this time, viz, Kasuá, (Santa Barbara,) Takhtam, (Serranos of San Bernardino,) and the Western Payutes, (Mono and Inyo Counties, California.) Hence I trust that the collection of vocabularies and sentences now made will prove of value and fill a gap in the philological knowledge of Indian idioms.

THE MISSION INDIANS.

The pious zeal of the Spanish priests drove them soon after the religious subjugation of Mexico into Southern California, but up to the end of the seventeenth century they had but little success; many were murdered, stoned to death, or cremated alive. It was mainly in the eighteenth century that they gained considerable headway. Some of the mission churches were built in the present century; that of San Juan Capistrano in 1806, destroyed by an earthquake in 1812, has remained in ruins ever since; and that of San Bernardino was built in 1822, whose ruins are now used for a sheepcorral. *Sic transit gloria mundi!* Nearly all the missions, hardly over forty in number, were in the coast counties. The most important of them were San Diego, San Juan Capistrano, San Luis Rey, Los Angeles, San Fernando, Santa Barbara, San Luis Obispo, and Monterey.

The Mohaves, Yumas, and Cocopahs resisted all attempts at conversion; they could not conceive the sometimes contradictory teachings of Loyola's followers, who preached different morals from what they practiced, and it appears that the Mohaves, like the Moquis, became the more averse to the Christian religion the greater the zeal and energy of the Jesuits in forcing their belief upon them.

The *Padres* generally tried to eradicate the original name of the tribes and substitute Spanish ones; the tribe of San Diego, for instance, is known under the name Diegueños, and their original name is forgotten, but, as the language indicates, the tribe forms a branch of the Yuma stock. Just north of this tribe lives another, that speaking a very different language, and without a uniform tribal name. They occupy about a dozen ranches† situated between the coast and the Coahuila Valley. The tribal name Netela, mentioned by Buschmann appears to be unknown there, at least all my questions were answered in the negative by the Indians of San Juan Capistrano and San Luis Rey. The former call themselves Akhátchma, the latter Gaitchim, (the Ketchis of Buschmann,) but the Spanish names have here also taken root, the names San Juaneños, San Luiseños, &c., being frequently used. Their language is closely related to that of the Kauvuyas, (Cowios, Coahuillas,) who live just east of theformer, and occupy a number of ranches in the San Jacinto Mountains and the adjacent Cabezon Valley. The Kauvuyas had also been converted by the Jesuits, and belonged, with the related tribe of the Takhtams,‡ to the mission of San Bernardino. One of the Kauvuyas told me that their forefathers used to burn their dead, but the padres abolished that practice, saying that "the Great Spirit would be displeased," (se enojaria Dios.) These Indians raise corn and watermelons, and serve as laborers with the whites.

Another tribe lives at San Gabriel, a town nine miles east of Los Angeles, but the full-blood, as well as the half-breeds, use more of the Spanish language than their own, which is known to some extent only by the two old chiefs living there. The name Kizh, given by Buschmann to this tribe, could not be verified with all my efforts; if this tribal name ever existed, it is now entirely forgotten. The old chief I visited called his tribe Tobikhar, or Spanish, Gabrileños. He was probably over ninety years

---

* The reader is referred to Gatschet's paper in this report. (App. H 16.)
† Among them the lately much talked of Temécula.
‡ Their Spanish name Serranos signifies "inhabitants of the Sierra."

of age, very weak, and suffering from a painful eye-disease. Among other statements he said that he had made a treaty with General Frémont in 1843. Many words I propounded to him he could not recall in his native language, and excused himself by saying, " we are now so far civilized that we have forgotten our own language," (somos tan civilisados que hemos olvidado nuestra lengua.) Still the collection of words obtained comprises about 200.

Another language nearly extinct is that of the Indians of Santa Barbara, on the coast. After much inquiry, an intelligent Indian * about three miles north of the town was found, who owned a large farm, and spoke, besides his own language, tolerably well Spanish and English. He called the original tribe Kasuá. The ruins of the old mission church are about three miles east of the town.

THE UNCONVERTED TRIBES.

One of the most numerous tribes in North America is that of the Payutes.† Indeed, this tribe, the main stock of the Shoshone family, has ramifications that reach very far. From the Mohave River ‡ in Southern California to Central Utah, from the Moqui towns to the northern boundary of Nevada, they are distributed in larger or smaller bands across valleys and mountains, and have many dialectical differences of language. While the party was encamped in the Colorado River Valley at Cottonwood Island, a great number of Payutes came daily into camp, and occasion was taken to collect over 350 words and many sentences; an easy matter if one meets an Indian speaking well Spanish or English. The vocabulary was again compared at El Dorado Cañon and Stone's Ferry.

The Chemehuevis live farther south, near the mouth of Bill Williams Fork, in the valley of the Colorado River. Their language is nearly identical with that of the Payutes of Southern Nevada. However, the language of the Payutes of Inyo and Mono Counties, California, shows very considerable differences; again, the dialect of Aurora, Nev., is differing considerably. The distinction is therefore made between the Southern Payutes, living in Southern Nevada and in the Colorado River Valley below the mouth of the Grand Cañon, and the Western Payutes, living in Mono and Inyo Counties, California. The Payutes are but little devoted to agriculture, some families raising watermelons, being exceptional cases. Their principal food consists of mesquite, beans, pine-nuts, lizards, vermin, grasshoppers, occasionally rats and rabbits, still rarer a deer or a mountain-sheep forms means of subsistence. Fish are not eaten because of a superstition. The Southern Payutes, who have, like the Mohaves, four blue vertical stripes on the chin, used to be a dangerous tribe. Camp Cady, on the Mohave River, was established on their account, and in a most desolate uninviting region; but the post was abandoned several years ago, the Indians having gone to the Colorado River. In 1864, over two hundred Payutes were surrounded at Owen's Lake by a party of whites and all drowned.

A tribe much superior to the Payutes is represented by the Mohaves, devoted to agriculture and but little to hunting. Lieut. A. W. Whipple, 1854, was the first who published details of the customs and language of this interesting tribe. This officer also selected the spot for the establishment of a military post in that region, and Fort Mohave was soon afterward built there. The Mohaves have seldom been troublesome to the whites, and the latter have in such cases been the cause of difficulty. In 1859, they killed some emigrants who had stolen corn and watermelons from their fields, which caused a fight between a company of soldiers under Captain Armistead and a band of Mohaves, whereby the former were repulsed and would have suffered heavy loss had not succor arrived from the fort at the critical moment. This tribe numbers about 3,000 souls, and is one of the tallest on the continent, surpassing in height the hunting tribes of the Payutes and Hualapais, the latter speaking a tongue closely related to Mohave and inhabiting the cool mountain regions of Northwestern Arizona. The Colorado River Valley, from Fort Mohave about 200 miles to the southward, with a very hot climate in summer-time, was the home of the Mohaves for many generations past. They live principally as vegetarians, using meat but very rarely. The Mongolian features are more marked with the Payutes than with them. The color of the skin is light-brown, their countenance is rather pleasant and even intelligent, and the physiognomies differ as much as among the white race. The front teeth are worn down to one-half the usual size, and flattened, showing that they are much used in masticating food. Bad teeth appear to be unknown there. In summer-time they live in open huts, in winter in holes dug in the ground and covered with branches. They have names for constellations, for some even the names of animals, a singular coincidence with the idea of the old oriental nations. Thus the Orion is called amó, (mountain-sheep;) Ursus Major, hatchá; Milky way, hatrhil-kuva-avunyú, (trail of heaven;) Venus, hamaú

---

* His Spanish name was Vincente Garcia.
† Spelled in various ways: Pa-utes, Pi-utes, Pai-utes, Pah-utas.
‡ They left the Mohave River but three years ago.

valtai, (the big star;) Jupiter or Saturnus, *hamood kavotanye*. According to the position of the "Great Bear" they judge the time at night, and know that its position is a different one at sunset at different times of the year.

The language is polysyllabic, melodious, and rich. There exist four words for "to eat," according to the food, and three words for "ant," according to the species: *Tohama thulye*, (little piss-ant;) *Hano-pó oka*, (large hairy ant;) *Horó-o*, (little black ant.) They have a separate word for "thinking," *ahieta*, and in expressing it put their fingers to the forehead, knowing well that brain-work and thinking are identical. Some of their words have eight syllables, for instance, *melago-pénya-hanólye*, the throat-bone, thyroid cartilage. Although they have no law against polygamy, most of them have but one wife. The women are well treated, and by no means like slaves, a moral feeling in the families generally being observed. Of course there are exceptions to the rule, exceptions that become conspicuous with those Indians that live just around the white settlements.

The Mohaves have a myth of a great flood, during which their forefathers lived upon the neighboring mountains. They are very superstitious. Dreams are ascribed to the influence of deceased friends. If one dies upon a trail, his spirit will hover there to harm those passing by at night. To avoid this, another trail is made, leading far around the bewitched spot. After the death of a man, the whole family bathe for four days, with little interruption, in the river, and a horse is killed in order to enable the spirit to ride to heaven. The heaven, *okidmbova*, is situated in a hot and dry valley west of the Mohave range; while the hell, *avikvomé*, is on the top of a big mountain where it is cold and rainy, (Dead Mountain, forty miles north of Fort Mohave.) They believe in a good and bad spirit. The custom of cremation is very old with them. Upon inquiry why the dead are not interred, as among white people, they laughed, and said, "It stinks bad." During the ceremony, all the clothes of the deceased and of his relatives are burned. If a medicine-man predicts three times falsely he is invariably strangulated. Several years ago such a medicine-man was only saved by the interference of the military authorities of the post. Another provided himself with a pistol, having resolved not to submit to the punishment for his unfortunate diagnosis.

COMPARISON OF LANGUAGES.—HIEROGLYPHICAL WRITINGS.

As to the origin of the Indians, many theories have been offered, of which the most probable is that of Asiatic descent; especially the marked Mongolian features of some tribes are favorable to this conception, so ably treated by Mr. H. Howe Bancroft in his "Native Races of the Pacific States," and by Oscar Peschel in his "Voelkerkunde," to which works the reader is referred. What I desire to call attention to, however, are several points not heretofore treated with the desirable minuteness, chiefly on account of want of proper knowledge of some of the Californian languages; I mean affinities between some of the idioms with the Japanese and Chinese. It is true some attempts have been made to prove a relationship between a Mexican language (Otomé) and the Chinese; however, those efforts become ridiculous to the eye of the critical examiner.[*]

When collecting vocabularies in Southern California, I was struck with the Kauvuya word *tam-yat*, for sun, resembling much the Chinese *yat-tau*, for sun. This led me to compare all the vocabularies I collected on your expeditions with Japanese and Chinese, with which languages I am acquainted to some small extent However, it was with the apprehension of touching a field outside of my sphere that I commenced this work. I therefore declare expressly that I leave it to the professed philologist to decide whether the similarities of words contained in the following table are to be ascribed to accidental coincidence. I further declare that among the eighteen languages of California, New Mexico, Nevada, and Arizona, compared with Japanese, only the Paynte offered some striking similarities of words with this idiom. About one dozen more words could be added than contained in the following table, but as the similarities between these words are confined to one single syllable, they were discarded.

---

[*] See Bancroft, Native Races, vol. iii.
[†] About infusion of Japanese blood into the California Indians, and drifting of Japanese vessels from the Asiatic to the American coast, see Bancroft, Native Races, vol. v, p. 52.

Table showing Indian words similar to Chinese or Japanese.

English.	Chinese.	Japanese.	Southern Payute and Western Payute, and Móqui.	Mohave.	Takhtam and Kauvuya, (San Bernardino, Cal.)	(Jaitchim, (San Juan Capistrano, Cal.)	Tobikhar and Kasuá, (San Gabriel and Santa Barbara, Cal.)
woman	nu		na-intai, (S. P.)			nu-itmol	nurt, (To.)
child	tau	ko	tatsiv, (S. P.)				ko-ar, (To.)
head	mo			mo-gora			
hair	to	omote	muta-gav, (S. P.)	i-to			san-tugh, (Kas.)
face	sum					mo-shun	
belly	uk, kan°		gap, (S. P.)		he-son, (Kau.)		
heart	shau-tchif		malatchi, (Móq.)				
house		ya					ya, (Kas.)
finger		ama			hu-yal, (Kau.)		
arrow				amaya			
heaven	yat, or yat-tau		tan-vabita, (S. P.)		tam-yat, (Kau and Ta.)		tamet, (To.)
sun	fung	foei	potaiv, (S. P.)	hamoóé	ho, (Ta)	‡‡	
star							
wind	tso	kava	voa-kave, (W. P.)		akav, (Ta.,) (Cal.)		tan-egto, (Kas.)
bark							
grass	tso	mual	mubita, (S. P.)				
insects {beetles, crickets}	tai	hata-hata	hué-tata, (W. P.)	vat-tai	hué-hata, (Jemes)		
grasshopper	siu	taito	tu-u-tai-e, (W. P.)	tahana			
large		hana-hada					
small							
very	tso	tau-kurl	tsa-rai, (S. P.)	tchego-varum	havun, (Kau.)		
to make		varai					
to laugh		haya					
quick	fai						

* Used as numerative or classifier of houses. † The sons of hand.

AP. JJ—21

In preparing the foregoing list, 350 words of Payute and Mohave served for comparison, and from 200 to 300 of the others. The similarities of words for *sun*, *star*, and *heaven* are certainly remarkable.

In comparing a number of otherwise totally different languages of the West, we are struck by the fact that a few words, especially for *water*, *hand*, and *bird*, seem to have the same roots in most of them; moreover, roots we find again in other countries.

In comparing the word for *hand*, we find—

Name of tribe.	Country.	Word for "hand."
Tehua	New Mexico	*ma.*
Taos	do	*ma-nena.*
Querez	do	*shka-mastsi.*
Isleta	do	*man.*
Jemez	do	*ma-tash.*
Moqnis	Arizona	*ma-khde.*
Aztec	Mexico	*ma-itl.*
Ute	Utah	*mti.*
Payute	Nevada	*mo-om.*
Tobikhar, (San Gabriel)	California	*a-man.*
Kauvuya	do	*ne-ma-to-e.*
Gaitchim	do	*ne-ma-vuitchaig.*
Digger	do	*ac-mut.*
Wihinasht*	do	*i-mai.*
Nocehenam*	do	*ma.*
Meidoo*	do	*ma-ma.*
Shoshone	Nevada	*ma-shitu.*
Comanche	Texas	*ma-shpa.*
Cahita	Mexico	*ma-ma.*
Cora	do	*moa-ma-ti.*
Lombok†	Java	*li-ma.*
††	Hawai Islands	*lima.*
††	Timor Islands	*lima.*
††	Celebes Islands	*lima.*
Tonga†	Friendship Islands	*lima.*
††	Formosa Islands	*rima.*
††	New Zealand	*balima.*
††	New Guinea	*limangh.*
Satahuan†	West Polynesia	*galoi-ma.*
Saparua†	Molucca Islands	*ri-mani.*
Caffre†	Africa	*mandha.*
Romans	Italy	*manus.*

* Taken from Bancroft, "Native Races," vol. iii.   † Taken from Balbi, Tableaux des Langues.

While the syllable *ma* for *hand* can be traced in America as far south as Brazil, it is not found with Indians east of the Rocky Mountains, nor with the Esquimaux. Of the eighty African languages I compared, only the Caffre language contains it again.* Neither is this syllable found in the Semitic, Finnish, Bas Basque, Caucasian, Central and North Asiatic languages, nor in the Sanscrit, (*hasta*,) but again in the Latin word *manus*. As regards Chinese (*hand = shau*) and Japanese, (*te*,) the analogy is wanting; but there is a word in the latter language, *mat*, meaning a handle.

Another word common to a great many Indian languages is that for *water*.

Name of tribe.	Country.	Word for water.
Payute	Nevada	pa.
Kauvuya	California	pal.
Gaitchim	do	pal.
Ute	Utah	pa.
Tobikhar	California	par.
Isleta	New Mexico	pa.
Jemez	do	pa.
Tehua	do	po.
Weithspek	Oregon	pa-ha.
Mohave	Arizona	aha.
Sidney	Australia	bado.
Abal	Philippine Islands	bahi.
	Java	pa-niu.
Zend	Persia	apo.
Sanscrit	India	apa.
Guzerate	do	pani.
Roumains	Wallachia	apa.
Cholo	Central America	pay-to.

* These remarkable resemblances appear to give support to A. Murray's geological hypothesis of a now submerged continent in the Indian Ocean, the supposed cradle of mankind.

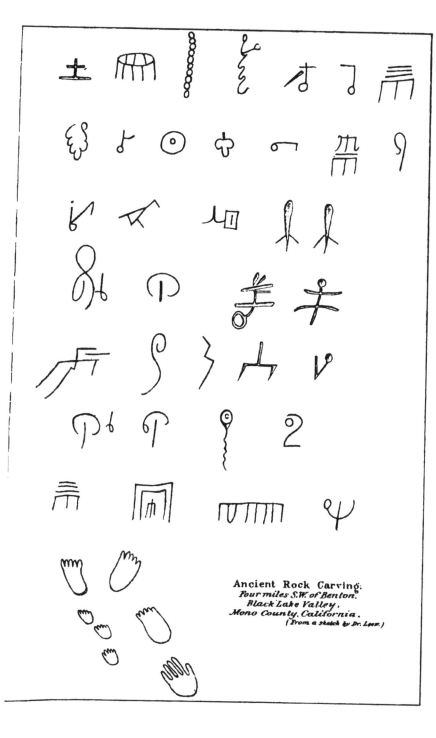

Ancient Rock Carving.
Four miles S.W. of Benton.
Black Lake Valley,
Mono County, California.
(From a sketch by Dr. Lear.)

Like the syllable *ma* for *hand*, so is the syllable *pa* for *water* wanting in the Indian languages EAST of the Rocky Mountains. This syllable is *not* found with the Semitic, Central Asiatic, nor African languages.

In the Chinese *shui* and the Japanese *mitsu* we miss the resemblance; but in both languages occur words with a certain relationship: in Japanese *pai* means a drinking-cup, the Chinese *pui*.

Furthermore, the syllable *tsi* for "bird" occurs in almost every Indian language west of the Rocky Mountains; again, the same syllable *tsi* means "male birds" in Chinese, as well as Japanese.

The formation of the plural in Payute is nearly the same as in Japanese; both languages use the syllables *bara* and *gara* to express the plural form:

Payute: gan (house.) Plural, ganigara.
        kanab.   kanabara.
Japanese: fito (man.)   fitogara.
          matsu (fir.)   matsubara.

Again, the *prepositions* become *postpositions* in both languages.

Payute: nuni gan upa ne.
        I   house in am.

Japanese: watakusi iye-no utsi-ni ari.
          I   house   in   am.

If the Indians are of Asiatic origin, it is but natural that, if an affinity to Asiatic nations is traceable at all, the most success would be expected in examining those tribes that are still in closest proximity to the Asiatic coast; that is, the tribes of the Pacific States.

Historical facts are unfortunately not known proving these migrations, but the possibility cannot be questioned. That, however, in comparatively recent time, that is, in the beginning of our era, Japanese and Chinese visited the American coast is beyond a reasonble doubt. An old Chinese work relating to the discovery of a country 20,000 li to the eastward, by the Buddhist priest Hoel-Shin, has been translated, and given rise to much discussion.[*]

In connection with this, it may be interesting to reproduce here a singular inscription containing several distinct Chinese characters on basaltic rocks in Black Lake Valley, about four miles southwest of the town of Benton, Mono County, California. Mr. Richard Decker, of the town, called my attention to this inscription. Certainly, I never saw a similar one in New Mexico or Arizona. Should the striking resemblance of some of the charcters to Chinese symbols be a mere accident, or give proof of early Chinese explorers? The inscription is scratched in the basalt surface with some sharp instrument, and is evidently of great age.

The Indians say that it was a mystery with their fathers who made these inscriptions. The rocks upon which they are seen lie regularly upon each other; and the true order of sequence of the symbols cannot be ascertained. The latter are from 3 to 5 inches high.

The Chinese-resembling symbols in the inscription are the following:

No. 1.—⊥ = to,   the earth.
No. 2.—+ = shau, the hand, (contracted form.)
No. 3.—⟁ = shi, an omen.
No. 4.—⌐ = min, a cave, a house.
No. 5.—⊓ = kwang, a desert, and ⇑, kan, the clothing.

The direction of the feet at the bottom of the page giving the description is exactly toward the northwest.

About 20 miles farther south from this locality exist two other similar inscriptions, according to Mr. Richard Decker, aforementioned.

---

[*] The reader is referred to the recently published work "Fusang, or the discovery of America by a Chinese Buddhist priest in the fifth century," by Charles G. Leland. The work is very ably written. It contains but one remark from the author with which I cannot agree, when he speaks, page 136, of Coronado and the "Quivira River" in connection with *Northern California.* We know that Quivira was a town in *Eastern New Mexico* visited by Coronado in 1542. Coronado never was in California.

APPENDIX H 15.

ON THE PHYSIOLOGICAL EFFECTS OF A VERY HOT CLIMATE, BY DR. O. LOEW.

UNITED STATES ENGINEER OFFICE,
GEOGRAPHICAL SURVEYS WEST OF ONE HUNDREDTH MERIDIAN,
Washington, D. C., February 9, 1876.

DEAR SIR: I have the honor to submit herewith a report upon the physiological effects of a hot and dry climate, a subject deserving attention in connection with the exploration or occupation of the Colorado Valley, and one not heretofore treated upon to any great extent.

Very respectfully, your obedient servant,

OSCAR LOEW.

Lieut. GEO. M. WHEELER,
Corps of Engineers, in Charge.

It is in but comparatively few regions of the earth that the temperature of the air rises above blood-heat for weeks and months in succession; hence, our knowledge of the physiological changes produced by it are quite meager.

When it is considered that under ordinary circumstances the whole tendency of the human system is directed toward keeping its temperature above that of the surrounding air, the task is suddenly reversed in the hot deserts, where the thermometer rises for considerable periods daily up to 110° to 116° F., in some cases up to 120°, while the normal temperature of the blood is 98°.5 F. That this task is at first not solved with precision is because the system only gradually accommodates itself entirely to the new conditions, and apparently after losing twelve to fifteen per cent. of its bodily weight. What a change in the conception of hot and cool is undergone in such a climate, when it is found *agreeably cool* in the evening, when the thermometer has descended from 110° to 94° F.! Observations on pulsation and respiration were frequently made, and it was found that the former was generally much increased, while the latter showed but a slight, sometimes no increase.

The following table shows some of these observations:

	July 30.		July 31.	
	Pulsation.	Respiration.	Pulsation.	Respiration.
A	75	23	80	12
B	78	16	80	25
C	73	19	80	21
D	80	14	65	22
E	76	18	72	22
F	56	12	60	17
G	72	18		
H	96	18		

The temperature of the air was 106° F. in the afternoon when the observations were made. G and H were two Payute Indians, who came into camp, one of about twenty-five, the other about sixty years of age. On July 31, a hot wind was blowing all the afternoon, and the observations were made after a march of 15 miles to El Dorado Cañon, on the Colorado River. A, B, and C, were at work upon the mesa and exposed to the hot sun and wind, while D, E, and F had taken a bath in the river and remained in the cool shade on the shore. F was a very phlegmatic person, and his pulsation and respiration were always lower than the others. The normal number of respirations during the day is considered to be 18 per minute; that of pulsation 60 per minute.

The temperature of the body was taken on various occasions and found, to be increased in very hot afternoons about 1° F., but at one time, after a march of 20 miles in a scorching heat, it was found with B increased fully 2°, (taken under the tongue.)

If it is considered that the temperature of the body is but 98.°5 F., it must be a matter of surprise that in a heat of 110° to 116° F. it is not increased more than one or two degrees above the normal temperature, although inner heat is being produced by breathing and oxidation that is continually going on in the blood. The system has, therefore, to contest against two sources of heat, the interior and the exterior.

That a decrease in the assimilation of food forms one condition is doubtless true; but he principal factor is an enormous evaporation from the body, which to determine quantitatively appeared a matter of great interest. From a series of observations made during very hot days, (108° to 114° F.,) measuring the amount of water drank and the volume of urine secreted, the conclusion was arrived at that two liters of water leave the body in the gaseous state during the twelve day hours. If, however, engaged in heavy work, such as packing mules, climbing mountains, &c., the amount is nearly doubled. The volume of urine was found to average only one-twelfth to one-fourteenth of water drank. The latter had generally 70° F.; but that the lower temperature of the drinking water has but very little to do with the cooling of the body may be shown by the following calculation:

Let it be supposed that a man of 70 kilograms weight drinks 2 liters of water of 70° F., and that the specific heat of flesh is the same as that of water, which is very nearly correct, as the body consists of about 75 per cent. of water, we have the following equation:

$$\frac{70 \times 98.5 + 2 \times 70}{72} = 97.7$$

The blood temperature would therefore be decreased to 97°.7 F., or not more than 0.8°. That this is not sufficient cooling effect for twelve hours in such a hot climate is evident. But how different the result if we calculate the amount of heat becoming latent by conversion of two liters of water into vapor! According to Regnault, the latent heat of one gram water converted into vapor of a certain temperature is expressed by the following formula:

$$W = 606.5 + 0.305\ t.$$

t is here the temperature of blood-heat $= 36°.8$ C., (98°.5 F.) Hence, the number of calories $= W$, becoming latent by evaporation of two liters of water, is equal to 1,235,000; which, should the evaporation suddenly take place, would suffice to depress the temperature of the body (70 kilograms weight) for 17°.6 C., or 31°.7 F.! But this cooling effect does not take place at once; the two liters evaporate gradually during twelve hours; hence, the cooling effect per hour is $= 2°.6$ F. This, therefore, would be the amount which the temperature of the body would be raised if no more water were available for evaporation.

How soon a person must succumb from thirst in such a hot climate becomes evident. The first symptom is delirium, and when arrived at that state the efforts to save are rarely crowned with success. A number of cases were related to me where helpless sufferers have been picked up, but died after one or two days, in spite of all care bestowed upon them. Even old pioneers, miners in the mountains, who were well acquainted with the dangers of crossing the desert in a new direction, deviating from the old trails leading to water, perished, overcome with thirst and fatigue; not by sunstroke, which is almost unknown there. It appears that this latter calamity only takes place when the hot atmosphere is at the same time charged with humidity, interfering with the free evaporation from the body.

It is evident that by drinking large quantities of water, the blood must acquire a high degree of dilution; hence all the juices, those of digestion included, the gastric and pancreatic, must be more diluted than usual, and the power of digestion weakened. Therefore, but a limited amount of food is assimilated, no matter how much is eaten, and a great deal leaves the body undigested, with the feces, which generally are of a thin consistency.

As another consequence we observe the decrease of muscular power; every exertion requires an increase of combustion, whose result, the heat, has to be converted into mechanical force. But everything tends here to keep the combustion at a low state. Hence it is preferable to be vegetarian in this climate, as a consumption of meat produces an increase in the number of blood-corpuscles, the absorbers and carriers of the oxygen, the oxidizing surface. The Mohave Indians inhabiting the hot Colorado Valley below Fort Mohave are exclusively vegetarians.

It is worth mentioning that after consuming fatty matter a considerable portion is exuded unchanged by the skin. It was also observed that while meat increases the thirst immensely, fat suppresses it considerably. Repeatedly I was assured, that in crossing the desert, alcoholic drinks and tobacco have a very injurious effect, as a person using them succumbs much sooner. A resident of Saint Thomas, a little oasis on the Muddy Creek in Southern Nevada, stated that two or three days previous to undertaking a desert trip he abstained from any stimulating material.

It is a question of interest in what proportion stand the quantities of water evaporated by the lungs and by the skin.

The average volume of air inhaled during twelve hours is, according to Carpenter's Physiology, 5,000 liters. This volume leaves the lungs perfectly saturated with moisture* at blood temperature; hence the maximum am unt of water removable by exha-

---

* At least in the hot desert climate, where the inhaled air is hotter than the body. The repeated introduction of the wet bulb of the psychrometer into the nostrils substantiated this fact.

lation is 210 grams. We have seen above that two liters $= 2{,}000$ grams evaporate from the body in twelve hours. Hence the remaining 1,790 grams must take their way through the pores of the skin, and the quantities stand in the proportion as $1:8.5$, while under ordinary circumstances the relation is but $1:0.66$ (!).

It appears, therefore, that the main evaporation, hence the main cooling effect, takes place by the skin, and that evaporation by the lungs exercises but little influence on the temperature of the body.

APPENDIX H 16.

ANALYTICAL REPORT ON ELEVEN IDIOMS SPOKEN IN SOUTHERN CALIFORNIA, NEVADA, AND ON THE LOWER COLORADO RIVER, THEIR PHONETIC ELEMENTS, GRAMMATICAL STRUCTURE, AND MUTUAL AFFINITIES, BY ALB. S. GATSCHET.

NEW YORK CITY, *April* 3, 1876.

SIR: I have the honor to submit herewith a linguistic report on the subject of Indian languages, of which vocabularies and sentences have been collected by members of your survey during the summer months of 1875. These idioms are enumerated in the order in which they were commented upon: the Kasuá, Kauvuya, Takhtam or Serrano, Gaitchin, Kizh, Southern Payute, Chemehuevi, Western Payute, Mohave, Hualapai, and Diegeño. Four of them, the Takhtam, Chemehuevi, Western Payute, and Hualapai, were, up to this day, only known *by name* to the scientific world.

In my report I took care to dwell mainly on such points as seemed most important from a linguistic point of view, and would give the best idea of the characteristics and peculiarities of each idiom. In two instances, where the affinity of the idioms were unknown or doubtful, I have tried to establish their genealogical connection by etymological comparisons with neighboring idioms.

A fact not mentioned in express words in my report is, that the commonly admitted affinity between the Yuma and the Pima dialects does not exist at all. Except a few similarly sounding terms, I have been unable to find any traces pointing in the direction of this theory, which was started only on account of the vicinity of both language-families; and, in fact, the Yuma stock of aborigines is thoroughly independent for itself, and disconnected from others, as well in race as in its form of speech.

I remain, sir, most respectfully, your obedient servant,

ALB. S. GATSCHET.

Lieut. GEO. M. WHEELER,
Corps of Engineers, in Charge.

The territory visited in 1875 by that section of your expedition of which Dr. Oscar Loew was a member is inclosed by the Pacific Ocean on the west, the Colorado River on the east, and by the thirty-fifth parallel on the north. This wide-stretching, mountainous, or rugged section of Southern California is the abode of a number of Indian tribes, once numerous and powerful, whose dialects Dr. Loew has undertaken to study during the short stays made at each of the Indian settlements.

On examining carefully his notations, vocabularies, and collections of sentences, which extend over eleven idioms, I have arrived at the following classification:

The dialects studied by Loew belong to only *three distinct families* of aboriginal languages; to the Santa Barbara, the Shoshonee, and the Yuma family.

The *family of Santa Barbara* languages seems to extend only over a small portion of the coast and the interior, and a dialect of it was also spoken on Santa Cruz Island. Dr. Oscar Loew has studied the dialect of the *Kasuá*, also called Cashwah or Cieneguita Indians, near Santa Barbara.

The *Shoshonee family* extends over an enormous inland area, from the Columbia River, Montana, and the British Possessions, through the great interior basin, down to the southwestern corner of the United States. It comprises the idioms of the Bannocks or Pa-vasht, of the real Shoshonees or Snake Indians, the Utahs, the Pa-Utes or Payutes, the Kauvuyas or Cahuillos, the Comanches. Other languages, as that of the Kiowas and Moquis, have borrowed so extensively from the Shoshonee stock of words, that they appear to be dialects of that family.

The Shoshonee dialects studied by Dr. Loew in 1875 belong to two branches. Of the Kauvuya branch in California, he studied the Kauvuya, the Takhtam or Serrano, the Gaitchin or Kechi, and the Tobikhar; of the Payute dialects, he has transmitted notations from the Southern Payute and the Chemehuevi, spoken on Colorado River and west of it, and from the Western Payute, spoken in Mono and Inyo Counties, California.

The *Yuma family* of dialects has received its denomination from the most populous of all the tribes using these idioms as their means of intercommunication. Those studied by Dr. Loew are the Mohave, the Hualapai and the Diegeño, and in 1873 he took a vocabulary of the Tonto or Gohun. The other Yuma dialects are the Cocopa, Maricopa, the Cosino, the Yabapai, the Comoyei, and that of the name-giving tribe, the Yuma, who are called Cuchans by their neighbors. The Yuma tribes live on the Gila and Lower Colorado Rivers, and in the Colorado Desert.

THE SANTA BARBARA STOCK.

KASUÁ.

The full extent of the territory in which the idioms of the Santa Barbara family is spoken is unknown to us at present. On the north it is bordered by the idiom spoken at the mission of San Luis Obispo, by that of San Miguel, and probably also by the dialects of the Tatche or Telame Indians, whose intricate grammar and difficult pronunciation has been transmitted to us by the careful notation of Padre B. Sitjar (Vocabulario de la Lengua de los naturales de la Mision de San Antonio, Alta California, in Shea's Linguistics, New York, 1861). To the east and southeast it borders on various Kauvuya idioms to be described below, perhaps also on some Payute dialects ; and when following on the map the location of these neighboring idioms, we must conclude that the space allotted to the Santa Barbara family is comparatively narrow.

A vocabulary taken at Santa Barbara by Horatio Hale, of the United States Exploring Expedition, and reprinted in Transact. of Am. Ethnolog. Society, vol. II, page 129, (1848,) does not, though very short and imperfect, differ essentially from Loew's; and we may, therefore, conclude that the Indians seen by Hale virtually spoke a dialect almost identical with that now prevailing among the Kasuá at Cieneguita, 3 miles from Santa Barbara mission. But the dialect observed on Santa Cruz Island by Padre Antonio F. Jimeno, and carefully noted by him on November 4, 1856, shows much difference from that of the mainland; still the great number of roots in which both coincide prove them to be offshoots of substantially the same linguistic family. By a few examples submitted, every reader may be enabled to judge for himself of the differences exhibited by the three vocabularies.

	Kasud.	Santa Barbara.	Santa Cruz Island.
	O. Loew.	Hale.	Rev. Jimeno.
my forehead	pi khsi		pi gatshe
beard	sats-ûs		tchatses
arrow	yá	yah	yhush
sun	álish	alishákhua	tannum
moon	ávueigh	aguai	o-uei
night	salkukh	sulkuhu	aughemei
leaf	skáp		hulncappa
water	ó	oh	mihie
meat	sáman		shomun
cold	sakh-tatakh	sokhton	aktaw

The cardinal numerals agree in all three vocabularies, the figures 1 and 5 excepted

A close examination of Loew's *Kasuá* vocabulary, and of the sentences transmitted by him, shows the following phonetic components:

Vowels : u, û, o, a, ä, e, i.—û is a surd vowel, equal to u in English *lump, thumb.*

Diphthongs : au ; ui, oi, ei.

Consonants : k, t, p; g, b, (b very scarce ;) kh, gh; s, sh; h, y, (the German jod,) v; n, m ; l.

The sounds ts, kh, gh, sh frequently occur, especially at the end of words. d, f, r do not occur at all in the Kasuá dialect, whose words terminate as often in vowels as in consonants, and show a marked tendency to monosyllabism.

On Santa Cruz Island, plurals are mostly formed by reduplication of the first sylla ble, as in twopau, *bow,* plural two-two-pau, *bows.* In Kasuá we have a few faint indi cations of a plural being formed by the addition of a syllable: sgut-*et, female breasts* gaikhus-e, *nails,* skam, *wings,* compared with skab, *feathers,* but we do not discover at present any plurals or duals formed by reduplication. But still this sort of grammatical synthesis, which occupies such a prominent place in the languages of the Pacific coast, is observed in some Kasuá appellatives which possess a collective meaning: túptá-ûp, *forest ;* shik-shép-shu, *ice;* and in a few verbs, evidently endowed in former times with an iterative signification: pekhpetch, *to sing ;* ptiptá-ulgh, *to speak ;* ksak-alálan, *to cry ;* perhaps also, palpat, *to run.* Verbs frequently commence with a p, the transitives as well as the intransitives.

One of the most frequent endings forming substantives is -*sh* ; it occurs in knosh, *head,* nokhsh, *nose,* ñash, *tobacco-pipe,* and is found in the shape of -*tch* in Santa Cruz Island. Other terminal forms are -*p,* -*gh,* etc., and the two following:

game, *brother*   gamute, *sister*
koko, *father*   khone, *mother*

Most topographical and meteorological terms commence with *s*.

Numerals follow the quinary counting system, and ordinals are formed in the following manner : ishgomo *two*, kumusk *second ;* masgh *three*, kamaskh *third*.

Possessive pronouns precede their substantives : uk pu, *my hand ;* u pu, *thy* and *his* hand.

In the conjugation of verbs, the negative particle ke, ke-á, precedes the verb; ke tchámon, *I do not know ;* the particle of the future tense is shá, and is inserted between the personal pronoun and the verb. The particle of the preterit is double; moe . . . . uash; moe pa shu-un uash, *thou hast eaten.* This *uash* also occurs in the terms for *old ;* pagó-uvash; for *morning :* vash-nákhi-et, which also means *to-morrow.*

Our knowledge of the Santa Barbara family of languages has been until now so restricted that the solution of the problem, what linguistic relations it bears to other American languages, could not be attempted with any hope of arriving at the truth. The painstaking labors of O. Loew have now enabled us to investigate this curious idiom more exhaustively, and, as nothing has yet been published concerning it, I intend to expatiate more fully on its affinities. and to draw all the conclusions that can be drawn safely from the material presently available.

The purpose of linguistic comparisons of roots, word-stems, and words belonging to different languages, and showing some similarity in sound and signification, is to find out whether the objects compared are borrowed, or whether they are cognate or not cognate. To do this with safety, the phonetic rules of these languages must have been reduced to a system, and where such systems are yet wanting, as here and in all the Californian languages, only empirical rules can be followed.

The Tatché language of the mission of San Antonio corresponds in the following terms :

	San Antonio.	Kasuá.	Santa Cruz Island.
father	ecco	koko	
chest, breast	tch^cúuo	ko'-ugh	
blood	akáta	akhóles	aughyoulish
sea	sh'kem	shkámin	
		(Hale : skahanúhui)	
hare	kól	kú'n	
large, great	kátcha	khá-akh	
small	skitano	tstáne-ugh	
bone	ekhakó		ikukuie
dog	ótcho		wutchu
to drink	kátcheme		tchakmil

The idiom of San Luis Obispo would, if we had a more comprehensive vocabulary of it, show many more affinities than the ones we subjoin here :

	San Luis Obispo.	Kasuá.	Santa Cruz Island.
ear	p'ta	'tu	thú
salt	tepu	tip	
hand	pu	pu	pu, (*plur.* púpu)
man	h'lmono		alamu-un
two	eshin		ishum
three	misha		maseghe

The Mutsun language, spoken in a large tract of territory around San Juan Bautista, does not show any similarities beyond the following :

	Mutsun.	Kasuá.	Santa Cruz Island.
two	utagin	ishgómo	
nose	us		ishtono

Further to the north, the idiom of the root-digging Wintoons, who live on the upper Sacramento River, corresponds in the following terms :

	Wintoon.	Kasuá.	Santa Cruz Island.
teeth	si	sú	(tcha-) sa
ears	tumut	'tu	

and the Klamath-Modoc in the negative particle *ka*-i, not; Kasuá ke, ke-á ; perhaps also in ó-ush, *lake ;* Kasuá, ó-ukeke.

The distant Pima language, spoken on the Gila River and south of it, shows striking analogies in two terms :

mukat, *far off, distant ;* Kasuá, mú-úkbk.

ni kuna, n'-kuna, *my husband ;* Kasuá, kunivu-e.

It may be reasonably expected that the wide-stretching Shoshonee family, which has even sent a few offshoots down to the barren coast of the southern part of the State of

California, has exercised a powerful influence on the Santa Barbara stock of words. A few may be traced, indeed, to the Kauvuya branch; others seem related to Kiowa and the Pueblos, inasmuch as these two idioms are themselves largely impregnated with Shoshonee words.

Takhtam, tóvuat; Kauvuya, továt, *pine-tree;* Kasuá, tómolgh.

Payute and Chemehuevi, kaiv, *mountain;* Kasuá, khûp, *stone.*

Western Payutes, kauvó, *hair;* Kasuá, okvó-'n.

Kiowa, kóh'-, *mother;* Kasuá, khóne.

Moqui, tsi-i; Tehua, tchi-i, *bird;* Kasuá, tchnivu-e.

Moqui, shuki, *nails;* Kasuá, gsikhua-e.

Moqui, pehue, *to sleep;* Kasuá, pué.

Affinities observed between Kasuá and the neighboring Kizh or Tobikhar will be given below.

Santa Barbara has borrowed from Yuma the term for *chief:* kvátai in Diegeño, (vatéga in Hualapai,) *large, great,* occurs in Santa Cruz Island as ghotah, in San Antonio as kvátai, *chief.*

gámutum, *girls,* in Hualapai, turns up in Kasuá as gamute, *sister.*

A few word roots occur almost in all, or at least in a large number of western languages, with equal or similar signification:

Kasuá, –'tu, *ear;* Wintoon, tumut, (plural :) Kiowa, tá-ati.

Kasuá, ke, ke-a, *not, no ;* Kizh, khai, *not ;* Paynte, gatch ; Chemeh. katch.

Kasuá, tip, *salt;* S. L. O., tepe ; Maya, táab; occurs in the signification of *rock, stone,* as tipl, timpi, tamp, tu-ump, tub-'e in the Shoshonee dialects.

Kasuá, nó ; 'o, J, pron. pers., occurs in western languages as no, noma, nû-û, nù-ni, na, and in many other similar forms.

From Spanish, Kasuá has borrowed the words plata, *silver;* kavay, *horse;* and the use of the article *el,* which is changed into *il.*

From all these word-resemblances and real affinities, no linguist will feel justified to pronounce the Santa Barbara family *cognate* to any of the surrounding idioms, as they are not conclusive enough to prove this. We are sadly in want of the most important criterion for such researches, viz., of reliable grammars and texts; and, while these are wanting, all we can admit is, that the languages in question have simply *borrowed* from each other to a certain extent. There seems to exist, however, a pretty close relation between Kasuá and the neighboring idioms of San Luis Obispo and San Antonio, which deserves to be followed up.

The mission of Santa Barbara was founded on December 4, 1786, and the Indians settled around it were called Silpaleels or Saughpileels, Aswalthatans, &c., all of them using dialects slightly varying from each other. The Indians living around Santa Inez Mission also spoke a dialect of the Santa Barbara family, and their tribes were called Alahulapas, Akachumas, Jonatas, Cascellis, &c. Spanish priests have left us a few liturgic texts of the Santa Barbara as well as of the Santa Inez idioms, and the Lord's prayer is given in Duflot de Mofra's Explorations, vol. II, page 388.

THE SHOSHONEE STOCK.

KAUVUYA.

This is, according to O. Loew, the correct form of the name of the tribe inhabiting the Cabezon or Coahuila Valley, which lies between the San Bernardino range and the San Jacinto Mountains. They are variously called Cawéos, Cavios, Kavayos, and by Mexicans Coahuila, Cahuillos. Their language, combined with that of the neighboring Takhtam, Serranos or Mountaineers, and the dialects of a few coast tribes, forms the Kauvuya branch of the Shoshonee family of languages.

Vo*ɯ*els: u, o, a, ä, e, i (pronounced as in Italian.)

Diphthongs: au, iu ; ui, ai; vowels are *not* nasalized.

Consonants: k, t, p ; g, b, (b scarce,) kh ; s, sh, h, y, (the German jod,) v ; n, m ; l. The sounds d, f, r are wanting entirely ; kh is the *rough* guttural sound of k in the Spanish *ojo' dejar.*

Combinations of consonants like bs, tch, khk, ksh, are frequently observed.

Kauvuya syllables are generally built up of the combinations: consonant + vowel, or consonant + vowel + consonant.

Syllables made up of one vowel only are not frequent, though the Kauvuyas, as most other Indian tribes, like to drawl out simple vowels by doubling, repeating, or varying them. Thus pem (*these*) becomes pe-em ; kil (*not*) ki-il, &c.

Case-inflection is formed here, as elsewhere, by adding to nouns postpositions as suffixes : pal, *water;* pá-aga, *in the water;* tumuet, in Serrano, *mountain;* tamikan, in Kauvuya, *on the mountain.* A possessive case does not appear from the sentences given. The objective or accusative case does not differ from the nominative, but is generally placed after the verb, except in interrogative sentences.

The almost universal termination for the plural of nouns is um, which, in a few instances, diverges into –*em,* –*im,* (and –*on?*) The ending -*sh* seems to form collective appelatives. When assuming the termination of the plural, many nouns insert a new vowel or alter the vowel of the last syllable into a diphthong or another vowel, thus producing a change similar to the *Umlaut* in German, and to the irregular English plurals in goose, *geese ;* louse, *lice ;* man, *men ;* cow, *kine.* We subjoin some instances of Kauvuya plurals:

turtle	ayil	ayilum
fly	a-avat	ava-atum
bird	vigitmol	vigitmoilum
many	mete-uet	mete-etchim
hare	ta'vut	tavutim
boy	tiat	tigitum
fish	ki-ul	kiulǒm

Adjectives assume the plural form as well as substantives do, even when used as predicates or connected with a noun.

Derivatives are formed from roots or stems by the addition of the following terminations:

–at, –it, –ot: sogat, *deer ;* alvat, *crow ;*
panyit, *egg ;* vuyit, *grasshopper ;*
huminot, *meat.*
–uet (in Gaitchin, –ut:) pokauet, *snake, lizard ;* isuet, *wolf.*
–il: auvil, *blood ;* ingil, *salt ;* nietchil, *woman ;* manyil, *moon.*
–mol: nauishmol, *girl ;* tapa-amol, *cup ;* nakhánmol, *old.*
–ish: kauvish, *rock, stone.*
–liu: ne gi, *my house ;* ne giliu, *my friend.*

In Kauvuya, the numerals strictly follow the quinary counting-system, which they do not in the cognate idioms of the Serranos and Gaitchins.

The terms for parts of the human frame and for consanguinity always prefix the possessive "mine," whose form is determined by the quality of the initial syllable of the following noun, thus appearing under the variable shapes of n'-, na, ne, ni, no, nu.

The interrogative pronoun and particle is *mi–*, as appears from the subjoined list of pronouns and adverbs, to which *mi–* is prefixed:

mi, *what ?* bakhe, *who ?* mi keats, *how many ?* mi pá-akh, *when ?* mi vákh, *where ?* mi ikhone, *why ?* mi vakha, *wherefrom ? whence ?* mi vikin, *whereto ?* mi yákhon, *how ?*

From Loew's Kauvuya sentences, I add a few scraps, to the purpose of showing the mode of conjugating verbs :

te, *to see ;* men téokve, *I see ;* pin téokval, *I have seen you ;* pe téokval, *you have seen ;* té-e, *look here !* gopka, *to sleep ;* hen gopka, *I shall sleep ;* hen gopkale, ne gopkalet, *I have slept ;* kilia hen gopkale, *I have not slept.*

–al forms verbal adjectives nearly equivalent to our participles in –*ing :* pin ni aukal mukha-a, *I have rheumatism ;* literally: "this I having sickness".

pe, pen, pin is prefixed to all transitive or active verbs, and seems to point out a relation of the subject to the outside world ; *hen* is prefixed to all intransitive and reflective verbs, and shows a relation of the verb to its subject only, as we observe also in the Greek medium and many Latin deponentia ; *hen* may therefore properly be interpreted by *himself, herself, oneself.*

Verbs also assume the plural endings of the nouns: nitchika, *I go ;* nitchi-im, *we go.* Of the Kauvuya dialect, Mr. Loew has transmitted a considerable amount of words and sentences. In taking his notes, he closely followed, in this dialect, as well as in all the others, the graphic method recommended by Turner and Hale, who by their scientific studies were prompted to adopt the Italian pronunciation for most of the letters representing the sounds of their phonetic systems.

TAKHTAM.

This is the general name by which the Indians inhabiting the hills around San Bernardino, Cal., call themselves, and it may be properly used to designate their dialect also. Takhtam simply means *men,* being the plural form of takht, *man.* This word occurs in many Shoshonee languages, and sometimes not only signifies *man,* but also *young man.* The Spanish-speaking population calls the Takhtam *Serranos* or *Mountaineers,* a term frequently used in Mexico to distinguish also dialects of the hill regions from cognate ones of the adjacent plains, and derived from *sierra,* mountain-range.

The Takhtam dialect seems to differ from Kauvuya more in the dictionary than in the grammatical forms. It has the same vowels and does not nasalize them, but as for consonants it differs from it in the following peculiarities:

R occurs in Takhtam as well as in Gaitchin, but less frequently ; *f* only in vú-ungaiftch, *rain,* which could be rendered just as well by vú-ungaivtch. I find *d* only in hamd, *grass,* as a terminal sound, and *h* is only found when commencing words. Their *sh* is pronounced down in the throat ; the deep guttural *kh* also occurs here. We find combinations of consonants like *kv, mk, ts, tch, tohk,* occurring more or less frequently. The accent scarcely ever rests on the terminal syllable of a word.

The endings of the plural form of nouns are *–im, –am,* as in Kauvuya, but *–um* does not occur here. Cases are formed by suffixing post-positions to substantives: kitch, *house;* katchúka, *in the house.* Many adjectives are composed with the prefix: akup-, akopo-, kopin-, meaning plurality or abundance, (*much, many.*) Adjectives of colors are formed by means of the suffix -anka, -inka, -inkum, &c. In derivative nouns, the following terminals are most frequently observed; *-tch* being the most common of all, and in fact a substitute for the definite article *the:*

    -tch : kitch, *house;* in Kauvuya, gish.
        tokuvtch, *sky;* in Kauvuya, tokovas.
        á-aetch, *good;* in Kauvuya, átsa-e.
    -at : tamyat, *sun;* kotchat, *wood.*
    -et : túmuet, *rock;* mó-umet, *sea.* .
    -it : shuvuit, *wind.*
    -ut : honùt, *bear;* in Kauvuya, hunuet.

In most verbs, we observe the ending *-kin, -kain,* which corresponds to the *-ka,* and probably also to the *-kal* in Kauvuya verbs, *-al* alternating with *-ain.*

GAITCHIN.

This dialect of the Kauvuya branch of Shoshonee languages is spoken on the coast of the Pacific Ocean at San Juan Capistrano and at San Luis Rey, and, according to Loew's statements, at some distance from the coast at Pala, Temecula and, environs. We possess two old vocabularies taken at San Juan Capistrano from Indians who called themselves *Akátchma,* said to mean "pyramid hill," or "ant hill," and gave to their dialect the appellation of Netela, evidently né täle, no täle, *my speech, my language.*

We have also a few words collected from the San Luiseños Indians, or aborigines settled around San Luis Rey de Francia Mission, which slightly differ from Loew's Gaitchin words, and were said to belong to the *Kechi* language. Gaitchin, Kechi is derived from gitch, kitch *house,* or *settlement,* and consequently identical with "Kízh."

O. Loew obtained his words and sentences from an Indian living near San Juan Capistrano Mission, but hailing from San Luis Rey.

Vowels and diphthongs are the same as in Kauvuya and Takhtam. Of consonants, *d* and *f* do not occur at all, *r* is not found very often and is alternating with *l*; *b* is found only in bi-it, *younger sister.* Words generally show consonantal endings, those in *k, l, t, tch,* being the most common of all.

The accent generally rests on the penultima, though it is often laid on the last syllable of the word-stem, as in magát, *large, great,* vué, *two,* vosá, *four.*

In substantives and adjectives, the plural ends in *-um,* (in San J. Cap. in *-um, –öm, -om, -am,*) and the verb also assumes a plural form, *-otum, -von.*

Adjectives do not drop their plural endings when joined to a noun in the plural.

Nouns are inflected by postpositions in the same manner as in the cognate dialects of the Kauvuya branch: kauitch, *mountain;* kauvi-nga, *in the mountain;* kauvi-ik, *on the mountain;* mout. *sea;* mŏm-nga, *in the sea;* pushún-nga, *inside;* pesá-onga, *out doors.*

Further case-inflections appear in the endings *-am* and *-ov* of the following sentences: Gitcham gùmùk, *on the other side of house;* na-á-atch auvólov húikhnunga, *the horse is larger than the dog;* gitch meaning *house,* and aual *dog.*

Terminals for derivative nouns are as follows: -itch (the most frequent): yumi-itch, *forest;* vunú-itch, *river,* &c., and in some adjectives designating colors, in nangvitch, *deaf,* &c.

    -al : hungal, *wind;* ókhal, *earth.*
    -at : tomat, *lightning.*
    -ut : shovó-ut, *winter;* vokhá-ut, *frog.*
    -mol, -mul : amayomol, *young;* kavá-amal, *cup;* olú-umul, *small;* titchmol, *butterfly.*
    -ant, -ont : vuymkhant, *heary;* tchórokhont, *round.*
    -ev, -ov : emengev, *ripe;* poló-ov, *costly.*

A gradation of the adjective is effected by adding the terms *more* and *very, most;* magát, *great;* magat huikhnunga, *greater;* vam huikhnunga magàt, *greatest,* and in addition to this the gradation is made more apparent by a circumscriptive sentence.

The numerals exhibit elements of the quaternary counting-system, ($2 \times 2 = 4$; $4 \times 2 = 8$,) the other figures resting on the quinary method of numeration.

The interrogative particle is mi, me.

The subject-pronoun is prefixed to the stem of the verb; the negative particle *kai* is inserted before the verb or stands at the head of the sentence. The particles of the *preterit* tense are : omn . . . gat, or amn . . . gat; those of the *future* : ivi . . . lot, or ati-i . . . let.

more, *to kill :* non amo moregat, *I have killed.*
non ivi morelot, *I shall kill.*
non kai moregat, *I have not killed.*
telévna, *to see :* telévnak, *to see something* (*k* sign of object).
telévtchok, *to talk to somebody*, (viz. to see somebody.)

Father Boscana has left an interesting sketch of the Capistrano Indians, their history, customs, manners, and mythology, in his Chinigchinich, or "World-Maker." Robinson translated it from the Spanish, and published it as an appendix to his "Life in California," 12mo, New York, 1846. The only text of the Gaitchin language given by him is an Indian popular song of five lines, which has been republished in the elaborate treatise of Professor Buschmann, "Traces of the Aztec Language," on page 546. The Lord's Prayer was transmitted by the explorer Duflot de Mofras in 1842 with that of the Kizh.

KIZH.

Of this dialect we possess three vocabularies: that of Dr. Coulter, (1841;) of the Exploring Expedition, collected by H. Hale and published in 1846; and that of Osc. Loew, (1875.) All three were taken at the mission of San Gabriel; but the Lord's Prayer, taken by Mofras, II, 393–4, at San Fernando, proves that various sub-dialects of Kizh are spoken through the whole vicinity of Los Angeles. Neither the term *Kizh* nor *Netela* are known on the spot to designate any particular language or tribe, kizh meaning simply *houses.* The remnants of the once populous tribes or bands settled around San Gabriel Mission call themselves Tobikhars, (meaning *settlers*, from tóba, *to sit*, tobakharó, *to stand* in Kizh) and speak almost universally Spanish. Having adopted the name Gaitchin for the Southern coast dialect, we may just as well use Kizh, which has the same signification of "houses" as a name for the northern twin-idiom.

At first sight, Kizh seems to differ considerably from Gaitchin, Takhtam, and Kauvuya; but a careful comparison of all the vocabularies now available shows that a real affinity exists between the four. The following terms are rendered by the same radical in all the four idioms: father, mother, ear, nose, teeth, arm and hand, heart, arrow, house, heaven, sun, moon, star, water, mountain, bear, fish; I, thou, to drink; one. two, three, four. Kizh agrees at least with two of these dialects in the following important terms: mouth, breast, sea, salt, stone, deer, wolf, fox, rattlesnake, to eat, to kill; and in many of them a close coincidence is observed between Kizh and the Northern Shoshonee dialects on Columbia River and in Montana, the Utah, Payute, Moqui, Comanche, and even the Kiowa. Some words not found in the southern branches occur only in Kizh and the Northern Shoshonee.

It might be with propriety objected to the statement that Kizh is a Shoshonee idiom, the circumstance that the Kizh grammar differs widely from that of the Shoshonee languages; that these latter do not employ reduplication of the first syllable as a means of grammatical synthesis; that they lack the sound r, or employ it very rarely; that their possessive pronoun *mine* is *na, ni, nu,* and not *a,* as in Kizh, and that they do not generally place it before the parts of the human body or the degrees of consanguinity. To these objections we reply as follows: The *a* in Kizh is nothing else but the *na* with apheresis of the initial *n,* and this pronoun sounds *ni* in Kizh before the terms of consanguinity. The northern Shoshonees really *do* prefix the *mine* to the terms of the human limbs and to *father, mother,* &c. The scarcity of the r in other idioms proves nothing, since they employ other sounds in its stead, and Kizh lacks *l* almost entirely. Reduplication also occurs in Shoshonee dialects, though not generally to render the idea of plurality as in Kizh. We quote the following instances of reduplication from the Kauvuya branch:

In Kauvuya: yuyuma, *cold ;* sasaymol, *duck ;* vévonkon, *rain.*
In Takhtam: votchevnetch, *old.*
In Southern Payute: mobits, *fool;* momobits, *fools.*
In Gaitchin: magát, *great,* plural mámt, probably contracted from mamagat.

It is true that the reduplicative plural is a peculiar feature of the languages spoken along the Pacific coast of North America, and it occurs in this quality in Selish, Klamath, Island of Santa Cruz, and probably in many other Californian idioms; also in Pima, Aztec, Tarahumara, and in Tepeguana.

In the elements of verbal inflection, numerals, and in the degrees of consanguinity, Kizh agrees closely with Gaitchin, to which it bears the closest resemblance of all the Kauvuya dialects. But what languages have furnished to Kizh its words not traceable in the other Shoshonee dialects ?

Many of them must be, nevertheless, of Shoshonee origin, for we are yet very far from being acquainted with all the Shoshonee words, word-stems, and radicals.

For the others, it may be safely asserted that Kizh did not borrow to any extent from the Yuma family. From the neighboring family of Santa Barbara it probably borrowed the *extensive* use of the reduplicative plural, a grammatical figure not inherent to the Kauvuya family, and an affinity is traceable only in the following words:

	Kizh.	Kasuá.
blood	khain	akhóles
fox	khaúr	khus

San Antonio coincides with Kizh in:

Kizh, voshó, *dog;* S. A., ótcho, ótch; yait, *alive;* S. A., (kakhoo-) yota.

Kizh agrees with other Western idioms in:

uiti, *boy;* Wintoon, uéta, *man;* uéta-ela, *boy.*
tchábo, *fire;* on Sacramento R., ça, sa; Maya, káak.
tam. *teeth;* Pima, tatami.
(pa-) vahe, *six;* Maya, uác, (*seven*, uuc; *eight*, uaxab.)
tota, *stone;* Pima, hotië, hota; Heve, tet, Azt., tetl.
yu-uit, *great;* Taos, ya-á.

The affinities of Kizh and Gaitchin to Aztec, and to four languages spoken in the northern Mexican provinces, have been pointed out by Prof. J. C. E. Buschmann in a very erudite paper, entitled "Die Sprachen Kizh und Netela." We refer to the words demonstrated by him to be cognate with Aztec, and only present the subsequent ones

Kizh.	Aztec.
otsó-o, *cold*	ytztic, (Shosh., utshuin)
mahar, *fire*	macuilli
mukánakh, *to kill*	macmiqui
pukitcha, *to steal*	itchtequi
(pau-) enatch, *to cry*	(t-) enotza

Buschmann seems willing to admit that the noun-endings -t, -ta, -te, -ti, -ts, -tch, in Kizh, replace or closely correspond to the Aztec terminals -tl, -tli, and shows four ways of forming plurals in Kizh:

(1) by reduplicating the initial syllable, as in haikh, *mountain,* pl. hahaikh; tchinuit *small,* pl. tchitchinui; (2) by syncope; (3) by affixing -nôt, -rôt; and (4) by affixing the Gaitchin terminal -om, -öm.

Ordinals differ somewhat from cardinals.

The particle of the future tense -on is suffixed to the verb; that of the preterit, yamo-, prefixed to it.

The language of this tribe does not sound unharmoniously to the ear, and shows a vigorous, energetic constitution in its words and sentences.

PAYUTE BRANCH.

Passing from the Kauvuya branch to the Payute branch of Shoshonee languages, we are struck, when first glancing over Loew's very complete vocabularies, with the preponderance of deep-sounding vowels, as o, u, and a, over the high-pitched e, i; and o, u, often assume a darker shade by being pronounced surd, (ù, o,) or by being nasalized, (ñ, ö, ü, ū.) This pronunciation of the three vowels is also peculiar to the Utah, and occurs in many of the Pueblo idioms of New Mexico. In addition to this, we perceive in the Payute dialects a frequent occurrence of a vocalic r, marked ṛ, and in the dialect of Mono and Inyo Counties, Cal., a buzzing s, marked ṣ.

The three dialects studied by Loew almost entirely lack the sounds of d and f; b and r occur frequently in word-terminals, and there seem to be interchangeable.

Payute is evidently a sister language of Utah, and bears close relationship to it. It extends over the whole of Nevada and parts of the adjacent States and Territories.

O. Loew has taken words and sentences of the Southern Payutes on the Colorado River, of the Chemehuevis settled on a reserve on the western shore of Colorado River, and of the Western Payutes roaming in Mono and Inyo Counties, California.

Although these three do not differ widely among themselves, greater discrepancies will be probably observed between these Southern and the *Northern* dialects of Nevada, when we will be in possession of linguistic materials from these parts.

In order to exhibit more plainly the dialectic differences between the Southern and Western Payute, the Chemehuevi, and the Uintah-Utah, I subjoin a comparative table of words.

	Southern Payute.	Chemehuevi.	Western Payute.	Uintah-Utah.
body	nó-uav	nó-uan	nu-um	ningovh
teeth	távuamb	tauvamb	tava	taua
hand	mo-om	mu-um	vu-èla	mû
bone	a-óv	ó-oan	oho	
bow	atch	atch	óde	á-ats
snow	ne-ovav	novab	nevave	nevavai
fire	kun	ku-un, kun	kosh	k'-un
rock, stone,	tûmp	tu-ump, tump	túbé'e	timb
fly	mubitch	mobitch	m-úivi	múpu
who ?	hangi	hangá	hagó	hang
yes !	ē-ē	û-û	hû-ū	ū-vay
no !	gatch	katch	karú-u	kats
to eat	tokai	tokara	tûgate	teke

For want of space, I have to refrain from extending this table over all the other Shoshonee languages and dialects. To do this would certaiuly be very instructive and also furnish materials from which to derive phonetic laws for the whole Shoshonee family.

SOUTHERN PAYUTES.

The words and sentences given by O. Loew were gathered from Indians living at the little mining town of Yvanpah, west of Colorado River, Nevada, compared in Cottonwood Island and at Stone's Ferry, both settlements being located on Colorado River. Some more words were added on the last-mentioned place.

Vowels: u, ū, o, a, e, i.

Nasalized vowels: ā, ē, ō, ū.

Diphthongs: au; i, ei, ui.

Consonants: k, t, p; g, d, (occurs only in pa-ubd, *blood*,) b; kh; s, sh; h, y, (the German j,) v; ng, n, m; r, ŗ. l and f do not occur.

In words having no derivative ending, the accent mostly rests on the peualtima; and, in words provided with such a termination, it commonly rests on the syllable pre ceding it.

In this idiom, as in Kizh, we notice several modes of forming the plural of nouns, and singularly enough even cardinal numbers show a singular and a plural form.

This curious circumstance might be explained through the law of analogy; but probably the plural of the numeral has here a distributive meaning, like *quini*, *deni*, in Latin.

Plurals in –atum: avan, *many*, *much*, avá –atum.

in –im, –am: pa-átsiv, *louse*, pa-átaivim; hun, *rat*, hunam.

in –vun, –um: tukibun, *friend*; tukibuvun, pay-ay, *three*, pa-ayum.

in –uts: narávungg, *sheep*, narávunguts.

in –aŗa: bivinump, *cup*, hivinumpaŗa; sovib, *cottonwood-tree*, sovíbaŗa.

All these various endings can be easily reduced to three original forms: –atum, (or –itum); –uts (or –its); –aŗa.

The first of them changes into –itum, –otum, etc., the penultima being always short and indistinctly uttered; or it collapses, by dropping the –at, –it, into –am, –im, –om, –un, etc. The second terminal, –uts, probably corresponds to the collective –tch in Kizh and Gaitchin; the third; –aŗa, evidently is the adjective avan, *many*, *much*, having altered its pronunciation into ava, aua, aŗa.

When adjectives and numerals are joined to substantives expressing inanimate objects, they are liable to drop their plural endings. No separate form exists for ordinal numbers.

The most frequent derivative termination in nouns is: –ab, probably equivalent to –ob and –ub; pa-uyab, *mud*; kanab, *large willow*; movitob, *narines*; angúai –urub, *leather strap*. Other endings are:

–ib, –iv: anókuib, *a kind of squash*; pigiv, *bread*.

–av: haiko-ótsav, *bottle*, and in many parts of the human frame.

–at: móbuat, *fool*.

–au: puŗuan, *skin*; vuytsan, *calf of leg*.

–ash, –ats, –atch: shuyush, *one*; tauats, *man*; na-ûbitch, *wet*.

–ump: aŗump, *tongue*; po-onump, *lead-pencil*.

In nouns, a case-inflection is observed as in the Kauvuya dialects: p'-a, *water*: pa-upa, *in the water*; kaiv, *mountain*; kaiv-umbay, *on the mountain*.

The subject-pronoun prefixed to the verb is frequently omitted when there is no doubt of the meaning of the sentence.

Negative sentences begin with the negative particle, and positive (not interrogative) sentences generally with the predicate, and when the subject is not expressed, with the object: pa-ai avan hiviga, *I have drunk much water*.

Tenses are formed after the following model: nuni tokay, *I eat*; nuni tokayan, *I have eaten*; katchun tokayan, *I have not eaten*; nuni tekavan, *I shall eat*; katchun teka-vau-va, *I shall not eat*.

This Paynte sub-dialect does not differ half as much from the Southern Paynte than Spanish does from Portuguese, and many of the differences observed in Loew's vocabularies between.the two seem to depend only on the individual pronunciation of the Indians from whom he obtained his information. Chemehuevi has frequently *p* and *tch*, where S. P. has *b* and *ts*. Like the Southern Payntes, the Chemehuevis do not prefix the possessive *mine* to the degrees of consanguinity and the parts of the human body as the Western Payntes do, who abbreviate the *ni* into *i*.

The terms for numerals, colors, man's limbs, and in fact the great majority of all the terms noted by Loew, radically agree in both dialects, and from this we can infer that their grammatical structure may be of the same type also, though no sentences of the Chemehuevi are at present submitted for examination.

The dialect spoken in the extensive mountain-tracts of Mono and Inyo Counties, California, and some adjacent parts of Nevada, diverges considerably from the Southern Paynte, and seems to have retained many terms in common with the neighboring idiom of the Western Shoshonees or Snake Indians. The personal appearance of the Western Payntes, especially their features, vividly recall to our mind the Mongolian type of mankind. Their deportment does not offend our ideas of propriety, and their faces bear a friendly, often intelligent, expression. Some of the aborigines are earning wages from American settlers, but the majority lead a wretched life by feeding on pine-nuts, roots, worms, and lizards.

Mr. Loew collected the main part of his linguistic material in Benton, Mono County. The sentences and a few terms were taken in Aurora, a little mining town of Inyo County, on the borders of California and Nevada. A few dialectic variations can be traced between the idioms of both places.

Vowels : u, ŭ, o, a, e, i.
Nasalized vowels : ā, ē, ū.
Diphthongs : au ; ai, oi, ui.
Consonants : k, t, p ; g, b ; s, ṭ, (or ss,) sh ; h, y, (the German j,) v ; n, m ; r, ṛ.

Western Paynte, therefore, lacks the consonantal sounds of *f*, *th*, (which occurs in Mohave,) *kh*, *l*; and *d* may be said to be wanting also, for it occurs only in 6de, *bow.*

Syllables generally begin with consonants, but terminate as often in vowels as in consonants. *v* seems to alternate with *b* and *p* and *ts*, *tch* of the southern dialects often turns up as *r* in Western Paynte.

Of derivative endings of nouns, the most frequent is *-ve*, as in Zuñi : toyáve, *mountain* ; ováve, *salt* ; vóve, *wood.* Other terminations are :
–ut : nugut, *goose* ; tunś-agut, *great spirit.*
–ib : tuvib, *sand* ; toshumib, *midnight.*
–sh : agúsh, *feathers* ; agish, *grasshopper.*

Western Paynte must have dropped long ago the plural ending observed in almost all the Shoshonee languages, (–*um* or –*im*, –*dm* ;) pagve, *fish* ; vahai pagve, *two fish.* In a few words, however, we notice that plural forms have been retained, as in num, *man* ; plural, nś-ana ; and the ending –*im*, –*itim* re-appears in the plural forms of verbs, as in koinú-itim, *to hunt*—said of many persons hunting, or of many animals hunted.

The names for the colors end in –nagite, except that of *yellow*, which exhibits the contracted form eahanite.

The interrogative pronouns and particles are as follows :
hayó-o, *what?*
hino-oy, hino-oytn : *how many?*
banágue, *whence! wherefrom?* hinó-ue, *when?* o-u hŭ-ut, *whereto?*

Tenses and negative sentences are formed in this manner :
To drink, hivít : *I shall drink*, hiví nŭ.
      I have drunk, hivívai nŭ.
      I have not drunk, garo-o nŭ hiví.
To sleep, ŭvnit : *I have slept*, (already,) nŭ vi tushu hapíṣu.
      I shall sleep, mi-asha haví.

Many transitive and intransitive verbs end in –at (or –it, –ut): yaróhat, *to speak, talk*, in the Aurora subdialect : yarú-a ; navágiat, *to swim* ; kvatohat, *to fall* ; voagit, *to work* ; hŭvi- -erut, *to sing* ; in the Benton subdialect the majority of all verbs seems to have this termination, which in the plural form is increased by –*im.* From the lengthy trisyllabic or quadrisyllabic forms of most verbs we may readily infer that they are compounds of the root, with some pronominal affixes, nouns or fragments of nouns.

THE YUMA STOCK.

Owing to the patient labors of Dr. Loew, the Yuma group in its totality of dialects will become .one of the best known of all the language-families of Western North America when the collections of words and sentences made by him will be made public. Loew has studied four of its dialects, while before him only the Mohave and the Yuma proper, (or Yuma-Cuchan, as I call it,) were known to a certain extent, and a few vocables only had been published of the Diegeño (Comoyei) and Maricopa. (See Reports on Pacific Railroad, vol. III.)

The dialects which constitute the Yuma family of languages are spoken east and west of the Lower Colorado and on Gila River. The Yuma family has kept itself pretty independent from extraneous influence, for it did adopt only a very few terms, if any, from the neighboring Sauta Barbara, Kauvuya, Payute, Pueblo, Apache, Pima, from Opata, and other Sonora dialects.

Owing to the prevalence of the vocalic element, Yuma is sonorous and not unpleasant to ears unaccustomed to aboriginal speech. Though words often end in consonants, vocalic terminations prevail in initial syllables and in syllables of the middle part of the word. The elements of which Yuma syllables are mainly made up are a consonant followed by a vowel. The counting system is the quinary one, and the numbers from *six* to *ten* disagree considerably in the different dialects.

The words of the six dialects of which we have the vocabularies illustrate and explain each other mutually, and many forms can be truly understood only by referring to a parallel from another dialect. To show their phonetic differences, the best means will be to quote some terms coinciding in their radicals.

	Mohave.	Hualapai.	Diegeño.	Cuchan.	Tonto.	Maricopa.
nose	ihu	yaiya	khu	ibós	hu	yehe-utche
beard	yavume	yavenime-e	alemé	yabo-íne	yanimi	yebomits
hand		sal	i-salgh	i-sáltche	shála	
arrow	ipá	apá-a	bal	n'yepá	apa	
knife	akhkvue	kva-a	akhgoś		akvá	
sun	anyá	inyá-a	inyá	n'yatch	nyá	n'yats
fire	ú-ana	tuga	á-ua	aa-wó	ho-o	áhutch
water	akha	ahá-a	akhá	ahá	aha	
earth	amata	nat	mat	omút	mata	
stone	aví	uvi	ú-uil	oví	vui	
black	vanilgh	niágh	nilgh	n'yulk	nya	
large	vatá-im	vatéga	kvatai	otaike	vete	betátchi
I	inie-pa	anyá-a	inyau	n'yat	nya-a	inyáts
two	havik	hovak	óak	havik	uake	
to drink	akhathim	akhathiga	kisi	hasúe	hasi	

We now turn our attention to the Mohave dialect, of which about 120 sentences and over 400 words were transmitted by Dr. O. Loew.

MOHAVE.

The individuals using this dialect are at present located upon two reservations. About 1,540 Mohaves, 600 Hualapais, 540 Chemehuevis, 180 Cocopas, and as many Kauvuyas are tilling the ground in the Colorado River agency on the eastern shore of the river; and about 400 Mojaves were removed in 1875, with 678 Tontos and 500 Cuchans from Camp Verde to the White Mountain reserve on the Gila River. They are a peaceably disposed, laborious set of Indians, who seem to have forgotten the fierce wars formerly waged by them against their aggressive neighbors. They tattoo the whole of their body in various colors. Their name is also written Mahhaos, Mo-óav, in Spanish Mojaves.

They do not nasalize or alter their *vowels*, which are to the number of five, u, o, a, e, i, and five diphthongs: au; ai, ei, ui, oi. They possess all our *consonants* except f, and though they have a very complete series of them, they rarely double them. The series is as follows:

	Not aspirated.	Aspirated.	Spirants.	Nasals.	R and l sounds.
Gutturals:	k, g	kh, gh	h	ng	
Palatals:	tch		ç, y		
Linguals:			sh		r, ŗ, l
Dentals:	t, d	th	s	n	
Labials:	p, b		v	m	

Heterogeneous vowels often meet, and produce hiatus: á-uva, *tobacco*; kahu-eilk, etc r and d seldom occur. No other consonants can end a word but the following: -g, -gh -k, -l, -m, -n, -p.

We find in this dialect the following combinations of consonants: bk, lk, tk, tht. mk, rk, vk, lg, shg, thp, gv, nqv, ngb, mb, all of which are of easy pronunciation.

The accent generally rests on the final syllable of the word-stem; inflective terminations usually are not accentuated. In many terms, the accentuation is dubious, but generally rests on the same syllable through all the dialects.

Of derivative endings, the most frequent for substantives is -a, preceded by a consonant, (-ta, -ya etc.;) for instance, vu-úga, *thunder*; asha, *bird*; vayaniya, *wine*; avuyá, *door*; amata, *earth*; huksara, *wolf*. -k is also very frequent.

-ik, -lk very frequently terminate adjectives; as, tauvanik, *low*; hibilk, *hot*; nakvimulk, *rich*. Another termination is -um, which occurs very often: ara-árum, *deep*; akú-utchum, *ripe*.

Substantives do not assume any sign of the plural, and it is doubtful whether adjectives and pronouns do; as, inyep ido namasávum, *my teeth are white*; makatitum, *who, which* (pl.); ataik, *much*, (sing.); ataim, *many* (plur.) shows a form probably contracted from ataikum. Adjectives and numerals are placed after the noun which they qualify.

A gradation of the adjective is effected by a circumscriptive sentence or by the particles *táhana, nimka-amk, more*; the superlative repeats the *táhana* or *tahán* twice or three times.

In regard to case-inflection, no distinct mark exists for the possessive and dative case except the position of the words. The accusative is rendered by prefixing -entch to the direct object of the sentence, and by placing this object between the subject of the sentence and the transitive verb. Relations expressed by our prepositions are rendered by postpositions: ava liuvá-aga, *in the house*; avá matareigh, *outside of the house*.

The pronoun entch, intch, abbreviated itchi, tchi,-tch, is a demonstrative, and in compound nouns and verbs means *somebody* or *somewhat, something*. Substantives composed with it are itchi-halyúluve, *stove, oven*, (viz. *something-smoking*;) itch-auyo-orahaga, *inkstand*, (viz. somewhat-writing-liquid;) Verbs: tcha-koark, *to speak*, verbally "something-say;" tchi-kiauk, *to bite*, verbally "something bite." This element is one of the most frequently occurring parts of Mohave speech, and also serves to form accusatives, as mentioned above, and in this quality means *him, her, it, them*. As the definite *article* it is frequently suffixed to nouns, as in ipá, *man*; ipátch, *certain man just spoken of*; gutch, *what?* contracted from ka-entch, literally, what-thing, or what-it?

There are three other demonstrative pronouns which are used in similar combinations: *ti, inya*, and *pa*.

Personal object-pronouns are suffixed to the verb; subject-pronouns are frequently omitted when there is no doubt about the meaning of the sentence.

The elements of verbal inflection are as follows:

Iyéma, *I go*; match'm iyema, *thou goest*; hovatch iyema, *he goes*; inyetch iyema, *we go*; match'm iyema má-ama, *you go*; tcha-am't iyema, *they go*; iyema, *I will go*; iyema tétchuma, *I have gone*; iye-em potchuma, *I did go*; iyemota, *I do not go*; iyemotum tétchuma (or: iye-em mo-ot e-ep tétchuma) or iyemotum pótchuma, *I have not gone*.

The negative particle *mot* is incorporated into the verb, and also serves as privative particle in the derivation of adjectives: ithperum, *strong*; hithpermutum, *weak*; tétchuma and pótchuma are composed of three pronominal roots: ti, entch, ma; pa, entch, ma, and are intended to mark a past tense more or less remote.

Concerning the modes in which verbs are composed in Mohave, we frequently find a syllable *hi-*, prefixed to the stem of reflective and intransitive verbs, as hilgivak, *to ride*; hitchibisk, *to fall*, etc. This particle seems to form verbs equivalent to the medio-passive verbs in Greek.

Of the verbal terminations, -um is the most frequent, and occurs in tapuyum, *to kill*; tchegovárum, *to laugh*; kotá-akum, *to open* (a door;) besides this we find a large number of verbs ending in -k, or more explicitly in -ák, (akhoák, *to smoke*,)-ók (hiók, *to romit*,)-isk -ilk, -eilk, etc., which often have the accent on the last syllable.

A large number of verbs is formed directly from nouns, for instance:

mata, *earth*; matahúilk, *to dig a hole*.

oyá *air, breath*; tchoho-ik, *to whistle, blow*.

agóaga, *deer*; gógo, *fox*; ha-ilguág, *to hunt*.

To show more clearly the mode of word-composition in Mohave, I add a few groups of words centering around one root and arranged etymologically.

AHAT, HATA, ANIMAL, BEAST: ahát, ahat-o-ólove, *horse*; hata-ghlal, *saddle*; hati; ánik, *bit of horse*; ahat-kagham, *spur*; hatchóra, akhatchóra, *dog*; makho-háta, *bear-amo-njo-hat, *domestic, tame sheep*; maguá-kuiniu-hata, *hog*; in Hualapai akhániga, *alive*.

MATA, AMATA, EARTH, GROUND: amata-tchikvara (in Diegeño), *meadow, prairie*; maták (Mohave), *north*; matagó-opa, *hole*; matahúilk, *to dig a hole*; mathé, *mud*; matara, *outside*; matana, *inside*; matuma, *inwardly*; matmaguilya, *skin* (as the enclosing substance).

IL, THREAD, in Diegeño, *wood*; ivu-il, *grass*, in Hualapai, vila in Tonto; ilvi, *green, light green*; avo-ilpo, *pole, stick*; si-vilya, *feathers*; ilya, final syllable in tree names.

AKHA, WATER: aha-tchopa, *well, water, pump*; akhathim, athim, *to drink, to drink water*; akh-mata, *squash, pumpkin*; akh ké-el, opposite (viz. beyond the water); ahávam, *wet*; kható, *island*; nu-há-vuk, *cloud*.

HUALAPAI

This dialect is closely related to Mohave, since the tribe of the Wallpais, Wallapais, or, according to Spanish orthography, Hualapais, have constantly lived in close contiguity and intercourse with the Mohaves. In the spring of 1874, 580 Hualapais came to the Colorado River reservation, where they live together with about 1,540 Mohaves and many other Indians.

The lexicon of this dialect shows many terms in which it differs from Mohave and the other dialects. But the prefixes, suffixes, derivative endings, &c., are substantially the same, showing many dialectic variations, however. So we observe that the Hualapai terminal -aga is in Mohave -aga; -ega becomes -é or -um; -oga turns up as -ank, u-uga as -ug; koark, *to speak*, appears as koank in Hualapai; harabk as hatabok, *bee*; mailbó as malú-n, *tobacco-pipe*. In a good number of terms, H. coincides entirely with Tonto, or more closely than with Mohave.

Being in want of the material requisite to construct a complete grammar of this Yuma dialect, hitherto almost unknown, I subjoin the few sentences given by Dr. Loew illustrating the inflection of the verb, in which the auxiliary verb *I go*, *miama*, is used to designate the future tense.

kvimago, *I eat.*
miama kvimago, *I shall eat*, (viz, "I go eat.")
kvimago vam, *I have eaten just now.*
kvimago kuré, *I have eaten some time ago.*
kvimago ta ópaka, *I will not eat.*
kutchu kanaba, *What do you want?*

vam in Hual. means *now, to-day*, and kuré occurs in Diegeño as okur: *distant, far off.* The negative particle ta is found also in tuya, *nothing.*

DIEGEÑO.

The Indians of the Yuma stock belonging to this warlike race were called so from the vicinity of the seaport San Diego, in Southern California, which will be the terminus of the Southern Pacific Railroad. The correct form for this name would be Diegueños, or San Diegunos, but Diegeños is now generally adopted. Some travelers have asserted that the Diegeños were identical with the Comoyei, or Comoyas, inhabiting some desert plains between that port and the mouth of the Colorado River, but the words taken from both prove that in their language, at least, some difference exists.

Diegeño and Cuchan exhibit many radical discrepancies in the vocabulary, and it is not improbable that the languages of the Californian peninsula have in former times influenced their stock of words; and a few expressions are traceable to Sonora sources.

Diegeño words more frequently end in consonants than those of Tonto, Mohave, and Hualapai, but the consonantal combinations and the grouping of the sounds are substantially the same as in Mohave. The gutturals *gh* and *kh* occur very frequently, but *th* of the Mohave, which is pronounced just like the English *th*, is not found. Among two hundred terms I find *r* occurring only in three, viz, sepir, *strong;* kitchur, *cold, winter;* okur, *distant.*

The accent not unfrequently rests on the final syllable of nouns as well as of verbs.

The parts of the human body assume the prefixed pronoun -*i*,(" mine"), but nothing of the kind is observed in the degrees of consanguinity.

Of compound nouns we notice: akhá-kvan, *river*, viz. " *large-water;*" uma-teté, *mountain*, viz. "rock- above ;" amata-tchikvara, *meadow, prairie*, viz. " ground-which-large ;" khá-ailgh, *sea*, viz. " water-salt."

Numerals from six to nine are composed with nio-, niu-, and Loew's numbers differ largely from those given by Whipple in the reports. These latter were probably taken from a Comoyei Indian.

No sentences or conjugations are at present available from which to construct paradigms or syntactic rules for the Diegeño dialect, and from all what may be inferred from the vocabularies, it must differ in this respect considerably from the Mohave and from Yuma-Cuchan, of which Lieutenant Whipple has given us some phraseology.

Undoubtedly the several Yuma dialects have borrowed a few words from nations speaking various other languages, as it is observed all over America, but in general this family kept itself more free from such importations than many other Indian races. A faint relationship, not heretofore mentioned by any investigator, exists between Yuma and the dialects of the peninsula of California. This connection deserves to be followed up as closely as the scanty material which we possess of the peninsular idioms will allow, and in this way an ancient immigration of some Yuma tribes into this deserted and barren stretch of land may be traced out and proved by linguistic research. I will here only point out the following similarities:

Cochimí: amat, amet, ammet, *earth;* Mohave: amat; Cuchan, omút.
Cochimí: ama, amma, ambayujúp; Waikuru: datembá, *heaven, sky.* Mohave: amaya; Cuchan, ammai.
Cochimí: maba, *upon, above.* Mohave: amail, *above.*
Laymonió: litsi, *to drink;* Diegeño, kisi; Cuchan: asi.

I conclude this brief notice on the eleven idioms studied in 1875 by Dr. Oscar Loew with the remark that, when his collections of words and sentences shall have appeared in print, careful comparative studies of their contents will undoubtedly throw more light on the origin and peculiarities of these languages than I have been able to give within the short space allotted to me in these pages. From other travelers or from residents on the Colorado River and its tributaries we may soon expect further contributions to the linguistic information gathered up to this day among the interesting tribes settled there. Then a new era will dawn upon the elucidation of the linguistic treasures still hidden near the lofty cañons of that majestic western stream.

INDEX.

INDEX TO GEOGRAPHICAL NAMES.

INDEX TO TECHNICAL NAMES.

MISCELLANEOUS.

O